T0317492

VERTICAL 3D MEMORY TECHNOLOGIES

VERTICAL 3D MEMORY TECHNOLOGIES

Betty Prince
CEO, Memory Strategies International, Texas, USA

This edition first published 2014

Registered office
John Wiley & Sons Ltd, The Atrium, Southern Gate, Chichester, West Sussex, PO19 8SQ, United Kingdom

For details of our global editorial offices, for customer services and for information about how to apply for permission to reuse the copyright material in this book please see our website at www.wiley.com.

Library of Congress Cataloging-in-Publication Data

Prince, Betty.
Vertical 3D memory technologies / Betty Prince.
pages cm
Includes bibliographical references and index.
ISBN 978-1-118-76045-1 (cloth)
1. Three-dimensional integrated circuits. 2. Semiconductor storage devices.
I. Title.
TK7874.893.P75 2014
621.39′732–dc23

2014016397

A catalogue record for this book is available from the British Library.

ISBN: 978-1-118-76051-2

Set in 10/12 pt TimesLTStd-Roman by Thomson Digital, Noida, India

1 2014

Contents

Acknowledgments

I would like to thank all of those who contributed information and offered suggestions for this book. In particular I would like to thank my husband, Joe Hartigan, who put up with the long hours spent writing it and who used his experience as a memory engineer to check the facts in Chapter 1. I would also like to thank the many engineers and researchers throughout the memory industry who gave me permission to use their material in this book.

I would like to thank David Prince, my colleague at Memory Strategies International and my son, who spent time helping gather information for this book, discussing its content with me, and reading it carefully. David's background in physics as well as his years spent teaching physics helped in checking the accuracy and readability of the entire manuscript.

I would like to thank Dr. Chih-Yuan Lu of Macronix for his encouragement during the writing of the book, for modifying his figure for use on the cover of the book, and for his permission to use many of the figures in Chapter 4. I would also like to thank Dr. Andrew Walker of Schiltron for reading and commenting on Chapter 4, for suggesting additional technical references to enhance the book, and for permission to use material from his published critical analysis of vertical memories in Chapter 3.

Dr. Stephan Menzel, of the Forschungszentrum Jülich, deserves special mention for his very careful reading of the material in Chapter 5, for his many helpful suggestions, detailed comments, and excellent suggested references, many of which were included in the final version of the chapter. I would also like to thank Dr. H.S. Philip Wong of Stanford University for his suggestions and encouragement as well as for his many technical publications on cross-point arrays, which helped me master the material in Chapter 5, and for permission to use his material.

I would like to thank Dr. Subramanian Iyer of IBM for his detailed reading of Chapter 6, his many suggestions, helpful references, and for permission to use IBM materials in this and other chapters of the book. I would also like to thank the four anonymous readers that my editor used to read Chapter 6 in the early stages of preparation of this manuscript. Their thoughtful comments and our discussions contributed to the preparation and direction of the entire book.

Ultimately, the book in its entirety along with any errors or omissions is my responsibility.

1

Basic Memory Device Trends Toward the Vertical

1.1 Overview of 3D Vertical Memory Book

This book explores the current trend toward building electronic system chips in three dimensions (3D) and focuses on the memory part of these systems. This move to 3D is part of a long trend toward performance improvement and cost reduction of memories and memory system chips.

Thirty years ago it was thought that if the chips could just be scaled and more transistors added every few years, the cost would continue to drop and the performance and capacity of the chips would continue to increase. The industry then struggled with the effect of scaling to small dimensions on the functionality and reliability of the memory technology. Along the way dynamic RAMs (DRAMs) replaced static RAMs (SRAMs) as the high-volume memory component. Twenty years ago the memory wall became the challenge. This gap in performance between DRAM memory technology and fast processor technology was solved by the clocked synchronous DRAM. Nonvolatile memories were developed. The quest for fast, high-density, nonvolatile memories became more urgent, so the NAND flash was invented, made synchronous, and became the mainstream memory component. Meanwhile the ability to integrate millions of transistors on scaled chips led to an increased effort to merge the memories and processors on the same chip. The many advantages of embedded memory on chip were explored and systems-on-chip became prevalent. Now systems-on-chip exhibit some of the same circuit issues that printed circuit boards with mounted chips in packages used to have. Redesigning these large, integrated chips into the third dimension should permit buses to be shortened and functions moved closer together to increase performance. System form factor can be reduced, and lower power consumption can permit smaller, lighter-weight batteries to be used in the handheld systems required today.

This first chapter reviews these trends that have brought us to the point of moving into the third dimension. Chapter 2 focuses on vertical fin-shape field-effect transistors (FinFETs) used as flash memories both with silicon-on-insulator (SOI) and bulk substrates and on making stacked memories on multiple layers of single-crystal silicon. Chapter 3 discusses the advantages of gate-all-around nanowire nonvolatile memories, both with single-crystalline

Vertical 3D Memory Technologies, First Edition. Betty Prince.

substrate and with polysilicon core. Chapter 4 explores the vertical channel NAND flash with both charge trapping and floating gate cells as well as stacked vertical gate NAND flash. These technologies promise high levels of nonvolatile memory integration in a small cube of silicon. Chapter 5 discusses the use of minimal-dimension memory cells in stacked, cross-point arrays using the new resistive memory technologies. Chapter 6 focuses on the trend of stacked packaging technology for DRAM systems using through-silicon-vias and microbumps to migrate into a chip process technology resulting in high-density cubes of DRAM system chips.

1.2 Moore's Law and Scaling

In the past 40 years electronics for data storage has moved from vacuum tubes to discrete devices to integrated circuits. It has moved from bipolar technology to complementary metal–oxide–silicon (CMOS), from standalone memories to embedded memories to embedded systems on chip. It is now poised to move into the third dimension. This move brings with it opportunities and challenges. It opens a new and complex dimension in process technology and 3D design that only the computers, which have been a product of our journey through the development of electronics, can deal with along with their human handlers.

Much of the trend in the electronics industry has been driven by the concept of Moore's law [1], which says that the number of transistors on an integrated circuit chip doubles approximately every two years. This is illustrated in Figure 1.1, which shows the Intel CPU transistor count trend during the era of traditional metal–oxide–silicon field-effect transistor (MOSFET) scaling [2]. Because the individual silicon wafer is the unit of measurement of production in the semiconductor industry, this law normally ends up meaning that the number of bits on a wafer must increase over time. This can occur by the wafer getting larger, the size of the chip shrinking, or the bit capacity increasing. Technology scaling and wafer-size increases result from engineering improvements in the technology. Chip capacity and performance

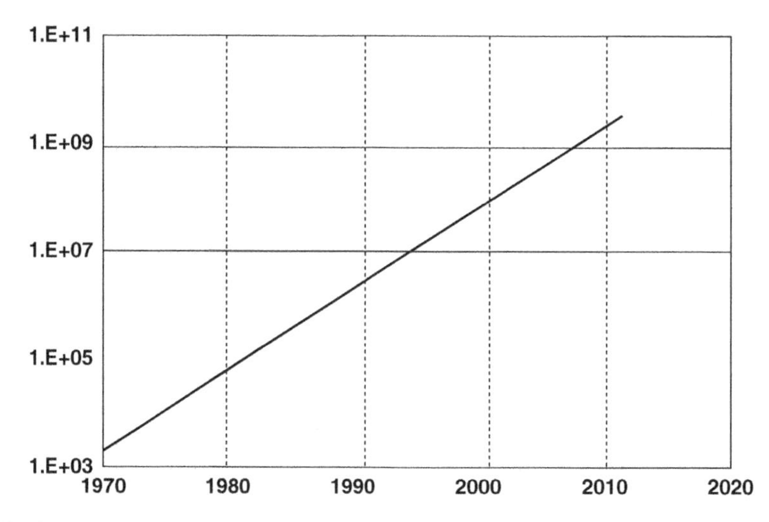

Figure 1.1 Illustration of Moore's law showing transistor count trend In Intel CPUs (Based on M. Bohr, IEDM, 2011 [2].)

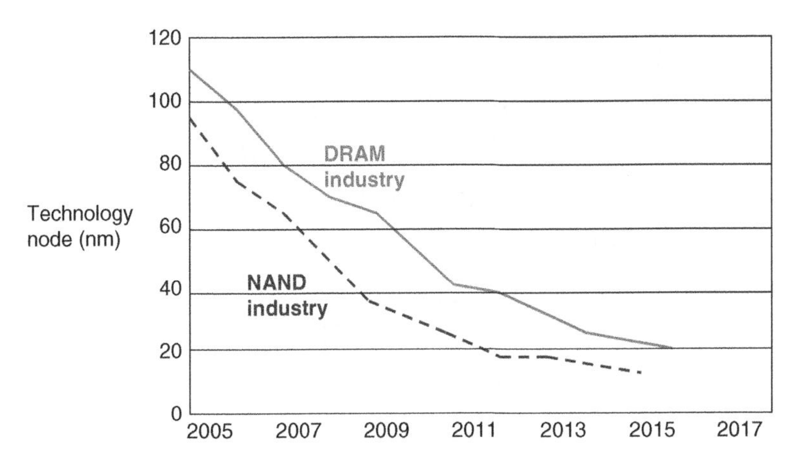

Figure 1.2 Scaling trends of DRAM and NAND flash 2005–2015. (Based on N. Chandrasekaran, (Micron), IEDM, December 2013 [3].)

increases are driven by the demands of the application. These application demands are driving the move to 3D vertical memories.

Scaling the dimensions of the circuitry on the chip is the method that has been used to shrink the size of the chip over the past 30 years or so. Scaling the dimensions has become increasingly expensive such that the required cost reduction is harder to obtain. Some memory cell technologies permit multiple bits to be stored in a unit cell area, which increases the capacity. Figure 1.2 illustrates the trend in scaling of the mainstream DRAM and NAND flash memories over the past 10 years [3].

Memory storage devices tend to be useful test chips as process drivers for the technology because memories are repetitive devices that require thousands or even millions of tiny, identical circuits to each work as designed. This permits low-level faults to be analyzed statistically with great accuracy. The trend to 3D has started with memory.

There are a finite number of types of memory devices that have been with us for the last 30–40 years and are still the mainstream memories today. These are the static RAM, dynamic RAM, and nonvolatile memories. While innovative, emerging memories have always been around, none has as yet replaced these three as the mainstays of semiconductor data storage.

1.3 Early RAM 3D Memory

1.3.1 SRAM as the First 3D Memory

The static RAMs were the first integrated circuit (IC) memory device. Historically, their chief attributes have been their fast access time as well as their stability, low power consumption in standby mode, and compatibility with CMOS logic, as they are composed of six logic transistors. Their historical stability and low power consumption has been due to their configuration from CMOS latches. The six-transistor cell CMOS static RAM is made of two cross-coupled CMOS inverters with access transistors added that permit the data stored in the latch to be read and permit new data to be written. A six-transistor SRAM with NMOS storage transistors, NMOS access transistors, and PMOS load transistors is shown in Figure 1.3.

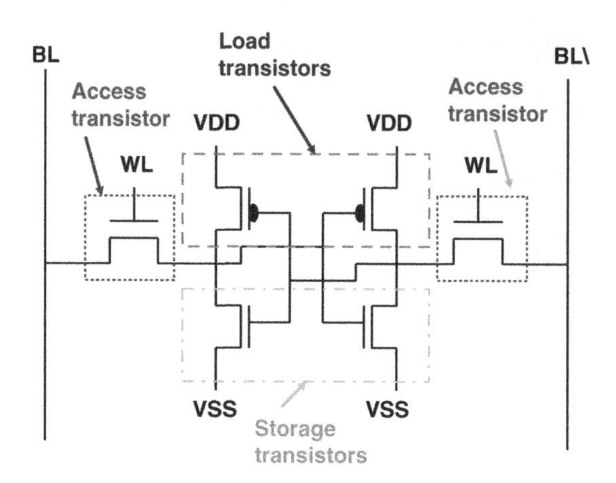

Figure 1.3 Six-transistor SRAM with access, load, and storage transistors noted.

The data is read from an SRAM starting with bit-line and $\overline{\text{bit-line}}$ high. The desired word-line is selected to open the access transistors, and the cell then pulls one of the bit-lines low. The differential signal is detected on bit-line and $\overline{\text{bit-line}}$, amplified by the sense amplifier, and read out through the output buffer. To write into the cell, data is placed on the bit-line and $\overline{\text{data}}$ on the $\overline{\text{bit-line}}$, and then the word-line is activated. The cell is forced to flip into the state represented on the bit-lines, and the new state is stored in the flipflop.

One of the transistors in the CMOS inverters of an SRAM is always off, which has historically limited the static leakage path through the SRAM and given it both its stability and its very low standby power dissipation and retention capability at low voltage. The trend toward lowering the power supply voltage in scaled SRAMs has reduced cell stability, usually measured as static noise margin (SNM). It has also increased the subthreshold leakage and, as a result, increased the static power dissipation. Thinner-scaled gate oxide increased the junction leakage, while shorter channel length caused reduced gate control, resulting in short-channel effects. Process variability made it more difficult for the matched transistors in the SRAM to be identical so that the latch is turned off. An eight-transistor cell has been developed to improve read stability, but it increases the cell size [4].

The development of double polysilicon technology in the late 1970s led to using one layer of polysilicon for load resistors to replace the PMOS load transistors in the six-transistor SRAM. These load resistors were stacked over the four NMOS transistors in the substrate [5]. This memory was fast, but it was difficult to tune the resistivity of the load resistors. In the late 1980s several companies used the new thin-film transistor (TFT) polysilicon technology to make stacked PMOS load transistors in the second layer of polysilicon [5]. These TFT PMOS transistors were stacked over the four NMOS transistors. These were the first 3D SRAMs. A schematic cross-section of one of these polysilicon load transistor SRAMs is shown in Figure 1.4 [6]. This six-transistor SRAM cell used a bottom-gated polysilicon transistor stacked over NMOS transistors in the silicon substrate.

More recent efforts have been made to stack both the two PMOS load transistors and the two NMOS pass transistors over the two pull-down NMOS transistors that remain in the silicon substrate. This allows the SRAM cell to occupy the space of two transistors on the chip rather

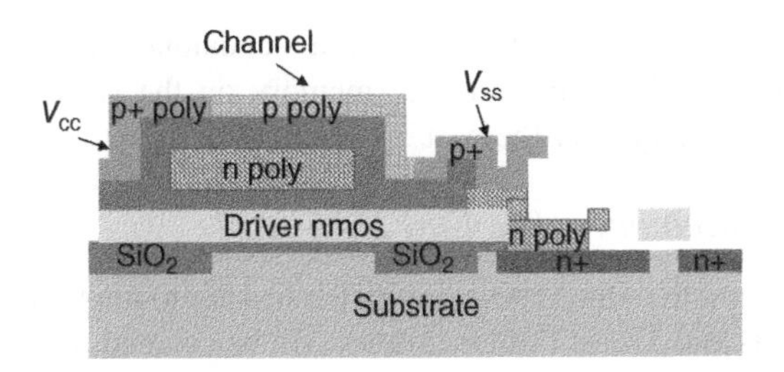

Figure 1.4 Schematic cross-section of inverted polysilicon PMOS load transistor for a 3D SRAM. (Based on S. Ikeda *et al.*, (Hitachi), IEDM, December 1988 [6].)

than six. Even more important, the relaxation of the scaling node means that the two transistors can be more perfectly matched and some of the original benefit of the SRAM regained. In addition, other latches and circuits in the logic part of the chip, initially in the periphery of the SRAM, can also be redesigned in 3D and stacked. This two-transistor SRAM with four stacked transistors is discussed in Chapter 2.

Because the SRAM is made of logic transistors, it requires less additional processing to integrate onto the logic chip. As the number of transistors possible on a chip has increased, performance has been improved, active power decreased, and system footprint reduced by moving more of the SRAM onto the processor chip. An illustration using the eight-core 32 nm "Godson-3B1500" processor chip from Loongson Technology in Figure 1.5 shows the various SRAM caches on a high-performance processor chip [7]. Last-level cache has 8MB, and a 128kB cache in each core totals 9MB of SRAM cache on the chip. Of the 1.14 billion transistors in the 182.5 mm^2 chip, about half are in the various SRAM caches.

As a standalone memory, the six-transistor CMOS SRAM was not able to compete with the much more cost-effective one-transistor, one-capacitor (1T1C) DRAM in the standalone memory market where cost is the main driver of volume. The chip size of an embedded memory is not as important to process cost as its ease of processing, so the silicon consumed by

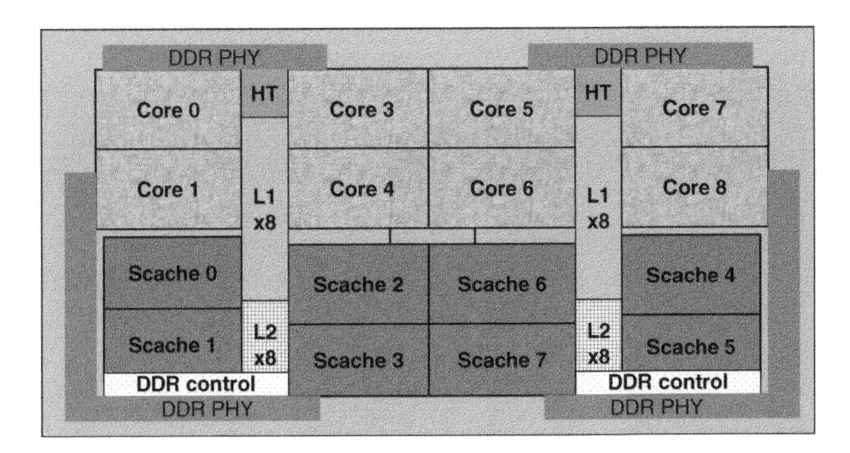

Figure 1.5 Illustration of 8-core 32 nm processor with 9MB of on-chip SRAM cache. (Based on W. Hu *et al.*, (CAS, Loongson Technology), ISSCC, February 2013 [7].)

the six transistors becomes less important than in a standalone memory. Because there are performance benefits to having processor and memory on the same chip, the SRAM has become an embedded memory over the past 10 years.

1.3.2 An Early 3D Memory—The FinFET SRAM

The important scaling benefit of the vertical FinFET transistor to improve the characteristics of both embedded SRAMs and also of flash memories will be covered in Chapter 2. The FinFET was first discussed in December of 1999 by Chenming Hu and his team at the University of California, Berkeley [8]. This first FinFET device was a PMOS FET on SOI substrate. It was a self-aligned vertical double-gate MOSFET and was intended to suppress the short-channel effect. The gate length was 45 nm. It evolved from a folded channel MOSFET.

The first memory device that benefitted from the development of the vertical 3D FinFET transistor was the SRAM because it is made of logic transistors. The channel of a FinFET transistor is a vertical fin etched from the silicon substrate, doped for the source and drain, with thermal gate oxide and gate polysilicon defined on the center of the fin.

A FinFET transistor used in an early vertical SRAM was discussed by TI, Philips, and IMEC in June of 2005 and is shown in Figure 1.6 [9]. The FinFET transistor solved the short-channel effect problem by changing the gate length (L_g) from a lateral lithographic issue to a fin length issue and by making the gate width (W_g) a 3D fin vertical issue, thereby providing sufficient on-current, which improved the static noise margin (SNM). A high dielectric constant (Hi-κ) Ta_2N–SiON gate oxide increased the capacitance, resulting in a higher threshold voltage (V_{th}), which improved cell stability. The cell size was reduced from 0.314 to 0.274 μm^2 in the same technology. The six FinFET transistors could be matched in an SRAM to solve many of the scaling issues.

1.3.3 Early Progress in 3D DRAM Trench and Stack Capacitors

Another memory device that developed 3D process capability was the DRAM. The DRAM cell is just a low-leakage access transistor in series with a large capacitor. The data is stored on the

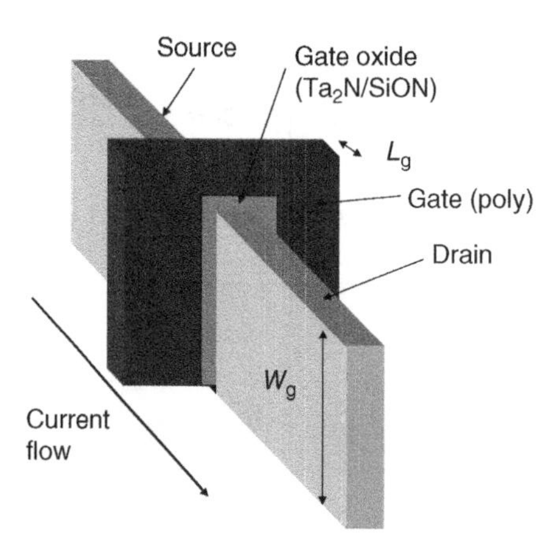

Figure 1.6 Vertical FinFET SRAM transistor with TaN gate stack. (Based on L. Witters *et al.*, (Texas Instruments, Philips, IMEC), VLSI Technology Symposium, June 2005 [9].)

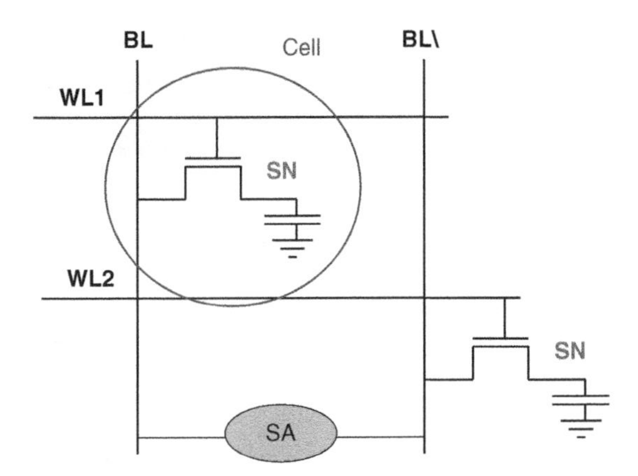

Figure 1.7 Basic circuit configuration of a 1T1C DRAM cell and array.

storage node between the capacitor and the access transistor, as shown in Figure 1.7, which illustrates the basic circuit configuration of a 1T1C DRAM cell and array. The capacitor initially was formed on the surface of the MOS substrate.

Internally the DRAM has not changed over the 40 years of its existence. It stores data in the storage node of a 1T1C cell. This data is accessed by raising the word-line of the selected cell, which causes the charge stored on the capacitor to feed out onto the bit-line, and from there to a sense amplifier normally connected to an adjacent bit-line for reference. Before closing the word-line, the data must be restored to the storage node of the cell or the cell be written with new data. The bit-lines must then be precharged to prepare for the next operation. A read and a refresh are essentially the same operation. The DRAM has five basic operations: read, restore, precharge, write, and refresh.

Sufficient charge must be stored in the capacitor to be sensed relative to the capacitance of the bit-line. As the capacitor was scaled to smaller dimensions, however, its capacitance fell ($C = \kappa \times A/d$), where κ is the dielectric constant of the material between the plates, A is the area of the capacitor plate, and d is the distance between the plates. The capacitance could be increased either by using a higher dielectric constant material or by increasing the area of the capacitor plate. The solution taken for increasing the area of the capacitor was either to drop the capacitor into a trench or to stack it over the surface of the wafer as shown in Figure 1.8.

The 3D processes required to make these vertical capacitors gave us the trench processes used to create the TSV described in Chapter 6 and to make the vertical channel NAND flash memories described in Chapter 4.

The DRAMs advantage was its small cell size. Its disadvantage was its slow bit access time. While an entire word-line of data was accessed on every cycle, initially only one bit at a time came out on the output bus. This was solved at first by making the output wider, which involved dividing up the array and accessing multiple open word-lines at one time. This made the area overhead, and hence the size of the chip, larger and more expensive. Wide input/output (I/O) DRAMs were not area efficient, and they still accessed only a fraction of the data available on the open word-line.

This issue of a data bottleneck with DRAMs was called "the memory wall" and indicated that the DRAM was not providing data fast enough for the processor. A two-step solution was

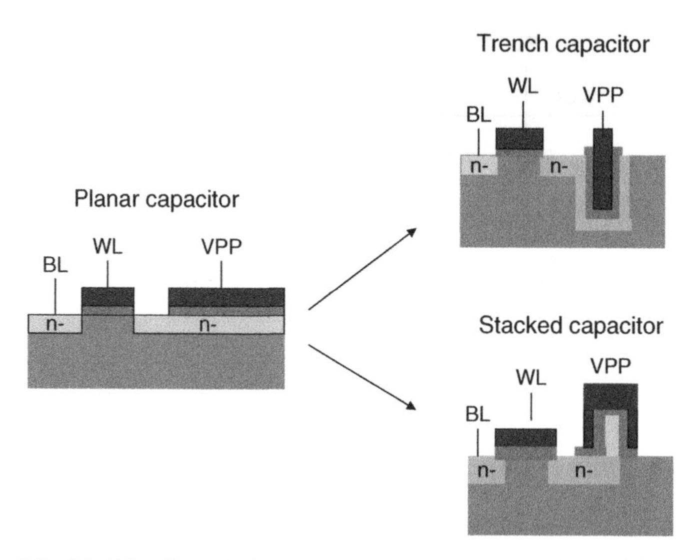

Figure 1.8 DRAM cell trends from planar capacitor to trench and stacked capacitor.

developed. First, the DRAM was made synchronous, or clocked so the data could be accessed on the system clock. This made the DRAM work better in the system that was already clocked. Second, the DRAM was divided up into separate wide I/O DRAMs, called banks, integrated on a single chip. This permitted multiple "banks," which were separate DRAMs to be accessed simultaneously. Their clocked output data was transmitted to the output of the DRAM on wide internal busses where it could be interleaved and clocked out rapidly. The interleaved, clocked data was called double data rate or DDR.

The data on the DRAM could then be accessed at a rate more compatible with the requirements of the system. In the process, the DRAM itself had become a memory system chip with multiple DRAMs, registers, and other control logic all integrated on the chip. This did increase the chip size but resulted in significantly improved performance.

A schematic block diagram of a double data rate synchronous DRAM (DDR SDRAM) is shown in Figure 1.9 [10]. This figure shows the DDR SDRAM interface, the SDRAM command interface, and the underlying DRAM array, which has four independent DRAM banks all integrated on a single chip. It illustrates the extent to which the SDRAM had become an integrated DRAM with logic chip.

As technology scaling continued, the 3D DRAM capacitor was stacked higher and trenched deeper, while in some cases high-κ material was used for the cell dielectric to help increase the capacitance without increasing the lateral area of the DRAM cell.

In June of 2011, Hitachi described a $4F^2$ cell area stacked capacitor DRAM in 40 nm technology that had a 10 fF cell capacitance [11]. A schematic cross-section of the $4F^2$ vertical channel transistor cell with the bit-line buried in the substrate is compared to the conventional $6F^2$ stacked capacitor cell in Figure 1.10 [11]. The $4F^2$ cell is 33% smaller than the conventional stacked $6F^2$ cell, which reduced the area of the memory array. The $6F^2$ cell capacitance was 16 fF, and the $4F^2$ cell capacitance was 10 fF. Conventional wisdom was that the capacitance needed to be around 20 fF for sufficient read-signal voltage for stable sense operation. The stacked capacitor DRAM has been primarily used for standalone DRAM.

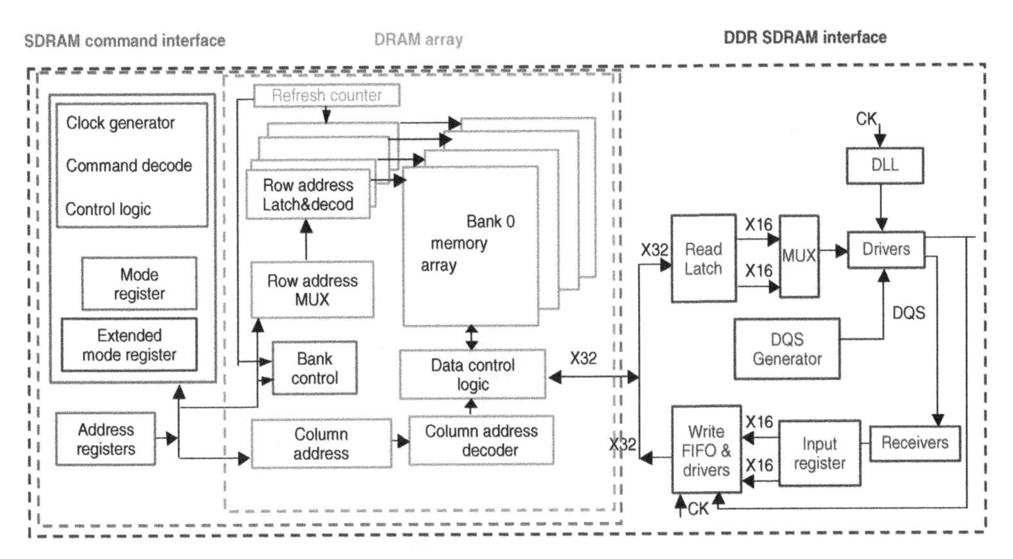

Figure 1.9 Schematic block diagram of basic double data rate (DDR) SDRAM. (Based on B. Prince, *High Performance Memories*, John Wiley & Sons, Ltd, 1999 [10].)

The trench capacitor DRAM continued to be developed for use in embedded memory. A schematic cross-section and illustration of the 40:1 aspect ratio of the SOI 3D deep trench DRAM cell used by IBM as Level 3 cache in its Power7™ Microprocessor was illustrated in June of 2010 by IBM and is shown in Figure 1.11 [12].

In February of 2010, IBM described further the SOI deep trench capacitor 1Mb eDRAM macro on this microprocessor [13]. A schematic block diagram of the Power7™ microprocessor in Figure 1.12 shows the SRAM L2 cache and DRAM L3 cache along with the eight cores and the memory controllers. The eDRAM cell size was 0.0672 μm^2. The eDRAM macro

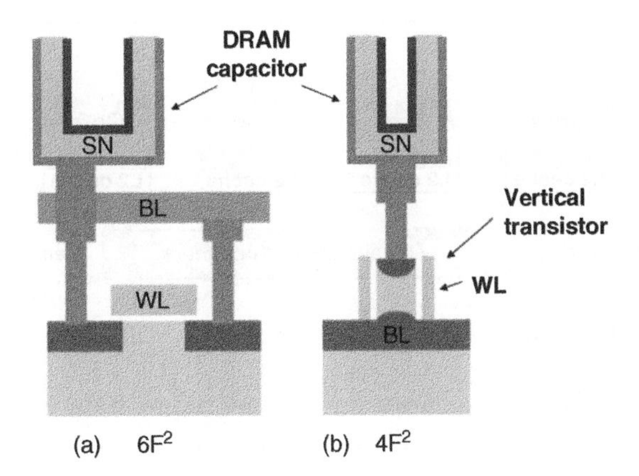

Figure 1.10 Stacked 3D DRAM cells in 40 nm technology (a) $6F^2$ 16 fF conventional eDRAM cell; and (b) $4F^2$ 10 fF vertical channel transistor pillar cell with buried bit-line. (Based on Y. Yanagawa *et al.*, (Hitachi), VLSI Circuits, June 2011 [11].)

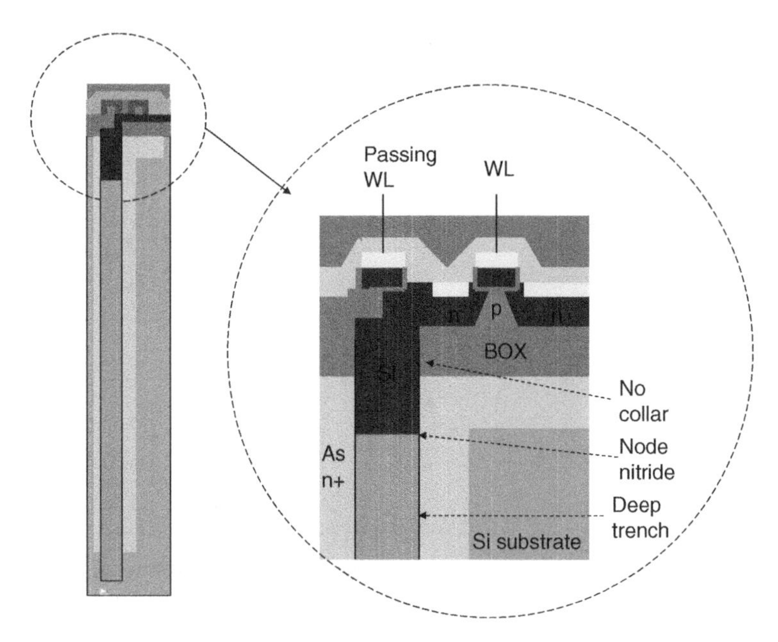

Figure 1.11 Deep trench 3D capacitor DRAM used in microprocessor L3 cache. (Based on K. Agarwal *et al.*, (IBM), VLSI Circuits Symposium, June 2010 [12].)

was made in 45 nm fully depleted SOI technology. Thirty-two macros were used per core supporting eight cores for a 32MB L3 on-chip cache in the 567 mm^2 microprocessor die. The deep trench had 25 times more capacitance than planar DRAM capacitor structures had, and it reduced on-chip voltage island supply noise. The 1Mb macro was made of four 292 K subarrays that were organized 264 word-lines $\times$ 1200 bit-lines. There was a consolidated

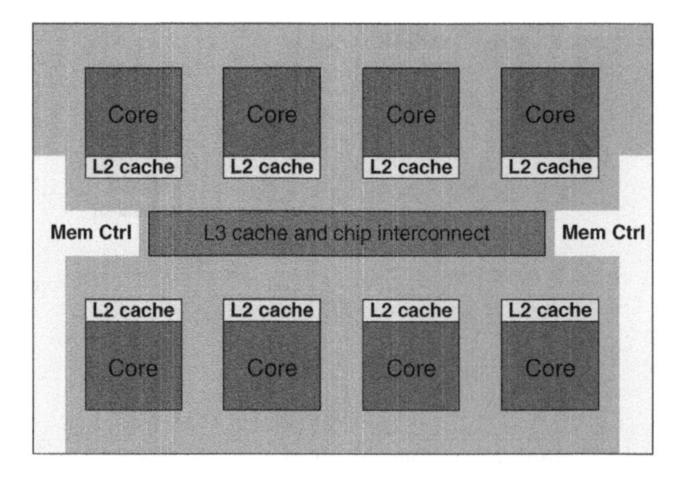

Figure 1.12 Schematic block diagram of a microprocessor with embedded DRAM L3 cache. (Based on J. Barth *et al.*, (2011) (IBM), *IEEE Journal of Solid-State Circuits*, 46(1), 64 [14].)

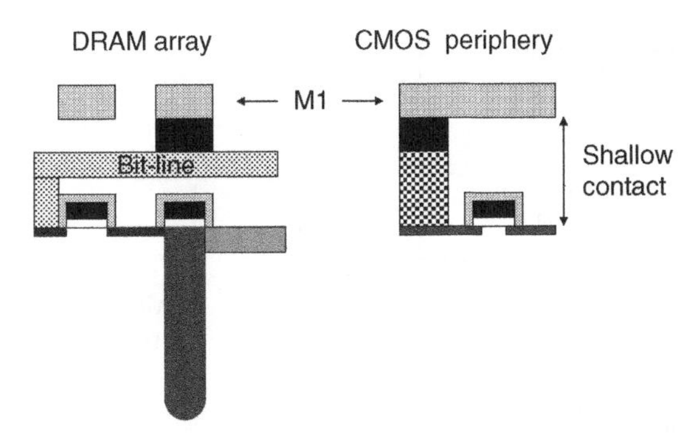

Figure 1.13 Trench eDRAM as starting substrate for CMOS logic process. (Based on S.S. Iyer, *et al*, (2005) IBM, *Journal of Research and Development*, 49(2.3), 333 [15].)

control logic and 146 I/Os where the inputs and outputs were pipelined. There were two row address paths to permit concurrent refresh of a second subarray. Late selection was offered to support set associative cache designs. In order to have a high transfer ratio, an 18 fF deep trench cell was used together with a 3.5 fF single-ended local bit-line. The DRAM macro used a 1.05 V power supply and had a 1.7 nm cycle time and a 1.35 nm access time.

In Chapter 6, a 3D two-chip TSV stacked system is explored, which includes a 45 nm eDRAM and logic blocks from this processor's L3 cache [14].

One of the aspects of the on-chip DRAM with trench was the potential for processing the trench first and using the substrate with trench as the starting substrate for the logic, which included the logic circuits in the periphery of the DRAM. This eliminated any effect the processing of the trench might have on the characteristics and performance of the logic transistors. It also leveled the surface of the chip so that the access transistor for the DRAM cell was in the same plane as the other logic transistors on the chip. Figure 1.13 illustrates using the trench eDRAM as the starting substrate for CMOS logic [15]. The wafer with the DRAM trench became the starting wafer for the conventional logic process. The DRAM capacitor still has capacitance greater than 20 fF.

Chapter 4 describes a 3D vertical gate stacked flash memory that used this old DRAM technique of dropping the array into a trench for anneal before processing the more sensitive parts of the stacked array.

A microprocessor chip could then be run very fast because the processor cores and memory could be integrated closely with high-speed buses on the same chip. The SRAM L1 cache could be integrated with the processor using the on-chip advantage of the wide I/O. The L2 and L3 caches could be large blocks of synchronous SRAM or DRAM, collecting data from the fast DDR SDRAM main memory and sending it to the processor or L1 cache SRAM. A significant part of the memory hierarchy was now integrated onto the chip, which improved both the performance and power dissipation of the system.

1.3.4 3D as the Next Step for Embedded RAM

Before leaving the topic of embedded memories, let's recall why embedded memories were heralded a few years ago as such a good idea for solving system issues. The first system problem solved by embedded RAM was the ability to reduce system form factor. Merging the SRAM and DRAM with the processor reduced package count and board size, which was

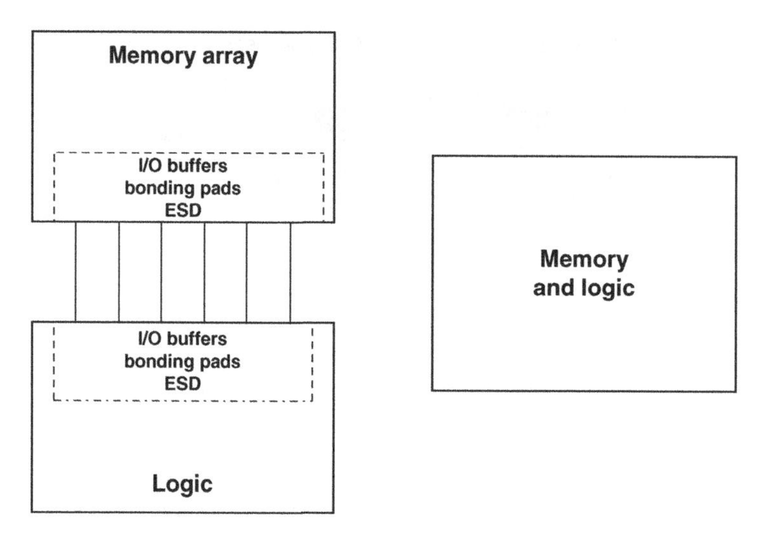

Figure 1.14 System form factor for (a) separate memory and logic chips; and (b) embedded memory in logic chips.

critical in a world moving to portable handheld systems. I/O circuitry in the memory and logic chips, such as I/O buffers, bonding pads, and ESD circuitry, could be eliminated. Figure 1.14 illustrates the reduction in system form factor made possible by embedding memory in logic.

Integrating the RAM with the processor also reduced active power consumption by permitting wider on-chip buses, which could have the same bandwidth as off-chip buses but with reduced speed because bandwidth equals bus speed times bus width. A lower power consumption meant the weight of the battery could be reduced and the life of the battery extended. It also meant that the cost of cooling the high-speed processor could be reduced.

The integration of wide internal buses between RAM and processor on a single chip meant that there were fewer external I/Os and wires, which reduced system electromagnetic interference (EMI). Additional I/O circuitry duplicated on separate chips in separate packages was avoided, and ground bounce was reduced as was the need for custom bus and port configurations.

The ability to configure exactly the memory that is required on chip also eliminated silicon wasted on standard memory chip sizes. In addition, many logic chips were I/O limited. Because of the wide I/Os on the exterior of chips containing only a small amount of logic, the silicon was not used efficiently and the system footprint was increased by the large numbers of chips on the printed circuit board. At the same time, the transistors were getting smaller and faster and more of them could fit on each chip, so system chips became feasible. As a result, the system-on-chip (SoC) with processor and embedded memory increased in size and functionality and developed many of the bus routing and interference issues that the system previously had. Resistive and capacitive issues began to occur for long, thin on-chip busses. Some of the same issues that drove the integration of the SoC were now occurring on the system chip.

The next level of gaining back the advantages of integration of systems chips can come by moving the circuits into 3D. Smaller-footprint system chips can be made, moving us back onto the curve for Moore's law. High-speed, wide, resistive-capacitive buses between processor and memory can, in 3D, again be shortened to reduce interference. Some of the advantages of embedded memories can be regained at the current tighter geometries by using 3D effectively.

Chapter 6 explores the initial gains of through-silicon-vias (TSVs), which permit wide memory buses to be connected locally in 3D with the appropriate logic circuit. The advantage is higher-bandwidth buses and smaller footprints. The challenges are redesigning the circuits to take full advantage of the benefits of the move to the third dimension. Initially in 2.5D technology, which uses interposers to redistribute the interconnects between standard chips, these vias are isolated on separate parts of the chip, where the large copper TSVs can't interfere too much with the sensitive logic and memory circuitry. As we learn more about using these vias and see the gains of redesign for 3D, the interconnects could be more direct so the advantages will multiply.

1.4 Early Nonvolatile Memories Evolve to 3D

1.4.1 NOR Flash Memory—Both Standalone and Embedded

There is a significant advantage to be gained by having programmable nonvolatile memory in the system as well as on the system chip. Early work on a field-programmable ROM was done by Dov Frohman-Bentchkowsky in 1971 at Intel, resulting in the development of the erasable programmable read-only memory (EPROM) [16]. This device could be programmed in the package but not electrically erased and reprogrammed. The floating gate flash erase memory was first presented by Fujio Matsuoka of Toshiba in December of 1984 [17]. The term *flash* was used to indicate that a block of cells in the device could be erased at one time rather than having individual bit erase capability. Intel developed and produced the first single-transistor-cell electrically flash erasable memory. Previous electrically erasable memories had been made with large two-transistor-cell chips called electrically erasable programmable read-only memories (EEPROMs), which had a large cell size, so they were low capacity and not cost effective.

The single-transistor-cell flash memory chips were bit programmable and bulk (flash) erasable in the system. These chips had a single-stack control gate and floating gate, as shown in Figure 1.15(a), which resulted in small cell size. The devices could be programmed by channel hot electron injection (CHEI) from the substrate to the floating gate at the drain side of the junction, which resulted in a high-current program, and could be erased in the system by Fowler-Nordheim tunneling of the electron from the floating gate to the source. This technique avoided stressing the same side of the tunnel oxide for both operations and improved reliability.

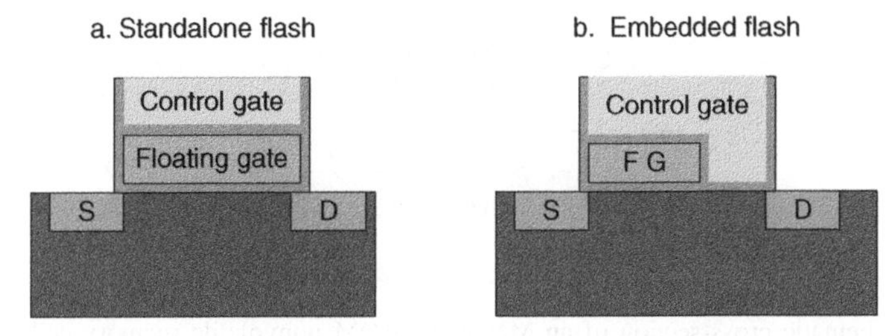

Figure 1.15 Flash nonvolatile memories: (a) stacked gate standalone flash; and (b) split gate embedded flash.

Because the floating gate could be over-erased, leaving the channel in depletion mode so that it leaked when the gate was intended to be off, this device used an iterative erase procedure to carefully define the voltage level of the erased state.

While the flash memory chips were initially made in volume production as standalone memory chips, there were advantages to integrating them onto the processor chip. For embedded flash memory arrays, a split control gate cell was often used as shown in Figure 1.15(b). This split control gate cell had a simplified erase because the control gate could turn the channel off, but the cell size was increased. This device used source-side CHEI from the substrate for programming, which used less current than the drain-side CHEI did in the standalone flash device. Erase was by poly-to-poly Fowler-Nordheim tunneling to the control gate, which was a thick oxide process and therefore lower in cost to make and control.

A recent potential alternative for the NOR flash memory is the phase-change memory (PCM). This part is in low-volume production today. It works by heating a calcogenide material, causing a transition between a high-resistance state and a low-resistance state. Its main advantage over the NOR flash is in a faster transition between states. The PCM consumes significant energy per bit and has issues with bit density [18]. It is unclear if this technology will transition into a volume production memory or become another of the many alternative memory technologies that have appeared over the past 30 years but failed to replace the high-volume mainstream memory technologies.

1.4.2 The Charge-Trapping EEPROM

Nonvolatile MOS memories have also been around as long as SRAMs and DRAMs. The first in-system, reprogrammable nonvolatile memories were called electrically alterable ROMs (EAROMs) or metal–nitride–oxide–silicon (MNOS) ROMs. MNOS reprogrammable ROMS were reported as early as 1969 by Dov Frohman-Bentchkowsky when he was at Fairchild Semiconductor [19].

P-channel MNOS EAROMs used silicon nitride (Si_3N_4) charge-trapping data storage, which was programmed and deprogrammed by Fowler-Nordheim tunneling through the tunneling oxide (SiO_2) between the substrate and the Si_3N_4, where the charges were trapped and stored. These low-capacity devices were used primarily in industrial and consumer circuits to store small amounts of data. They had a tunneling oxide and a nitride charge-trapping layer with an aluminum gate. A schematic cross-section of an early MNOS cell is shown in Figure 1.16 [20].

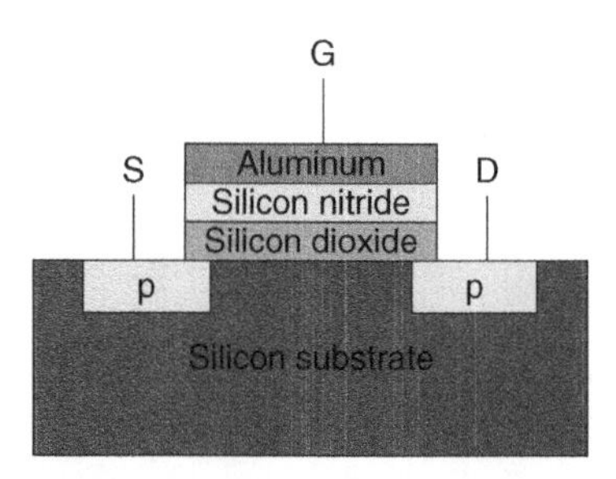

Figure 1.16 Schematic cross-section of an MNOS EAROM nonvolatile memory cell. (Based on B. Prince and G. Due-Gunderson, *Semiconductor Memories*, 1983, Figure 7.13, John Wiley & Sons, Ltd [20].)

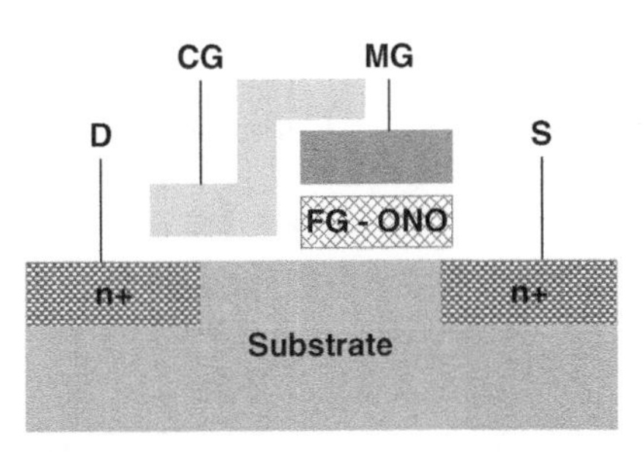

Figure 1.17 MONOS split gate flash memory cell. (Based on T. Tanaka (Hitachi), VLSI Technology Symposium, June 2003 [23].)

By 1989, MNOS charge-trapping EEPROMs with polysilicon gates, 28 nm Si_2N_4, and 1.6 nm SiO_2 were in volume production at Hitachi with yields equivalent to those of its SRAM lines [21]. In 1983, the Electrotechnical Laboratory in Japan suggested adding a blocking oxide to improve the reliability of the device, making the first MONOS electrically erasable programmable read-only memory (EEPROM) [22].

In 2003, Hitachi discussed a 512kB MONOS split gate flash memory intended for embedding in a microcontroller made in 180 nm CMOS technology [23]. The split gate memory cell is illustrated in Figure 1.17 [23]. This cell permitted the read path of the module to be made of low-voltage transistors similar to those used in the CPU core. Random access read for the module was 34 MHz. Program time for a 64 kB block was less than 4 ms, and erase time was less than 11 nm. The area of the module was 5.4 mm^2.

The MONOS cell was programmed using source-side hot electron injection, where the charge is stored near the ONO film edge on the side of the control gate. Erase was done by thick oxide electron tunneling to the memory gate (MG). This cell does not require high voltage for either the drain/bit-line or the control gate (CG) in the read and retention operation, making it compatible with the CMOS logic process. Higher voltages for program and erase can be applied to the source and the memory gate.

In January of 2014, Renesas described its 40 nm split gate MONOS n-channel memory, which is similar in function to the Hitachi split gate memory [24]. A schematic cross-section of the Renesas 40 nm MONOS flash memory is shown in Figure 1.18 [24]. This memory also permits high voltage to be applied for program and erase on the source side and the memory gate, but it uses logic voltage levels for read on the word-line/control gate. This device is intended for embedded flash memory in automotive microcontrollers. It was estimated by Renesas that MCUs with embedded flash memories accounted for about 70% of the 14 billion MCUs shipped in 2011.

Another charge-trapping memory also in production is a two-bit-per-cell NOR flash from Spansion [25]. In Chapters 2 to 4, nitride charge-trapping flash memory technology is shown as used in some of the vertical 3D NAND flash memories currently being developed.

1.4.3 Thin-Film Transistor Takes Nonvolatile Memory into 3D

The next development required for the charge-trapping technology to be used in the 3D vertical stacked flash technology was the development of the TFT, which could be made entirely by deposition of materials on a substrate. The TFT was developed originally by consumer

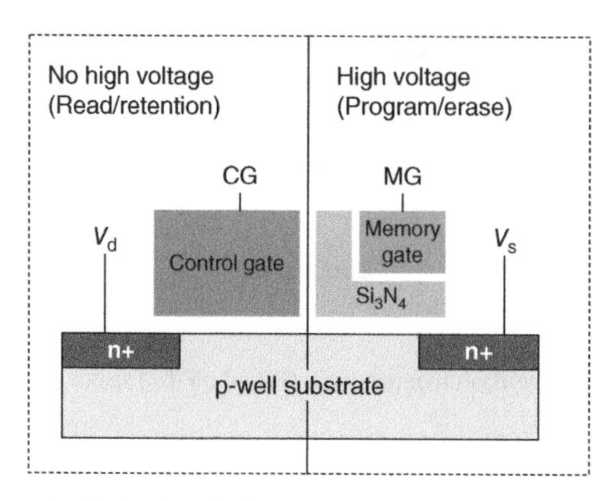

Figure 1.18 Split gate embedded MONOS flash memory cell. (Based on T. Kono, (2014) (Renesas) *IEEE Journal of Solid-State Circuits*, 49(1), 54 [24].)

companies for use in the circuitry made on glass around the outside of flat panel displays. Development of cost-effective 3D stacked layers of silicon circuits depended on this advance.

TFT transistors require a good-quality silicon substrate to make a good-quality memory cell. A deposited TFT nonvolatile memory cell can be made with a charge-trapping technology by depositing nitride over thermal oxide on a single-crystal silicon substrate. The formation of a 3D stacked single-crystal silicon substrate is described in Chapter 2. The NAND flash is ideal for stacking nonvolatile memory because there is a single bit-line access for an entire string of memory cells. Several layers can be stacked with a charge-trapping NAND transistor string in each layer with a single bit-line contact to substrate. This concept was described by Samsung in December of 2006 [26]. A schematic cross-section of a vertical stacked NAND string using 3D single-crystal silicon stacked substrates is shown in Figure 1.19. The bit-line and common source line (CSL) are formed through the second active layer. Well bias is simultaneously applied by the CSL.

Technology development of TFT charge-trapping memory using a gate-all-around (GAA) nanowire substrate is tracked in Chapter 3. The GAA memory uses the radial properties of the scaled cylindrical nanowire and their effect on the electric field. The asymmetry between the

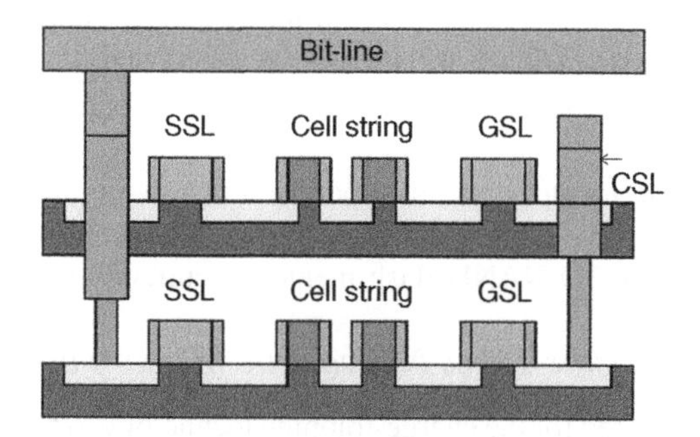

Figure 1.19 Stacked NAND string using 3D single-crystal silicon stacked substrates. (Based on S-M Jung (Samsung), IEDM, December 2006 [26].)

electric field of the tunneling oxide and that of the blocking oxide due to the radial design of the GAA meant that a fair-quality charge trapping transistor could be made with a polysilicon substrate. It was also discovered that shrinking the size of the GAA memory meant that the size of the polysilicon grains was similar to the size of the channel; the channel was effectively single-crystal silicon without grain boundaries to be crossed. These GAA memory devices are explored in Chapters 3 and 4.

1.4.4 3D Microcontroller Stacks with Embedded SRAM and EEPROM

Microcontrollers are ubiquitous today. By 2015, 20 billion U.S. dollars worth of MCU are expected to be shipped worldwide. All of them have RAM memory, and many have nonvolatile memory, EEPROM, or flash memory, which is named for its property of having a bulk, or flash, erase. These devices have multiple buses between MCU and memory cores including flash memory and SRAM.

In Chapter 6, examples of stacking using TSV and microbumps will illustrate how stacking can be used to create 3D chips with very short buses connecting various memories with their core processor. This type of stacking can save cost by reducing the size of the footprint of a high-performance MCU chip. It can also result in increased performance without needing to integrate the memory onto the MCU chip by shortening the interconnects between various circuits. An example of the block diagram for a high-performance 2D flash MCU for an automobile that used the Renesas RH850 core is shown in Figure 1.20 [27] along with a potential redesign for 3D, showing how the footprint might be reduced and the length of the buses between memory and logic be shortened [27].

1.4.5 NAND Flash Memory as an Ideal 3D Memory

An innovation that improved the capability of flash memory to scale was the development of the NAND flash by Toshiba in 1987 [28]. The NAND flash memory storage cell has only the

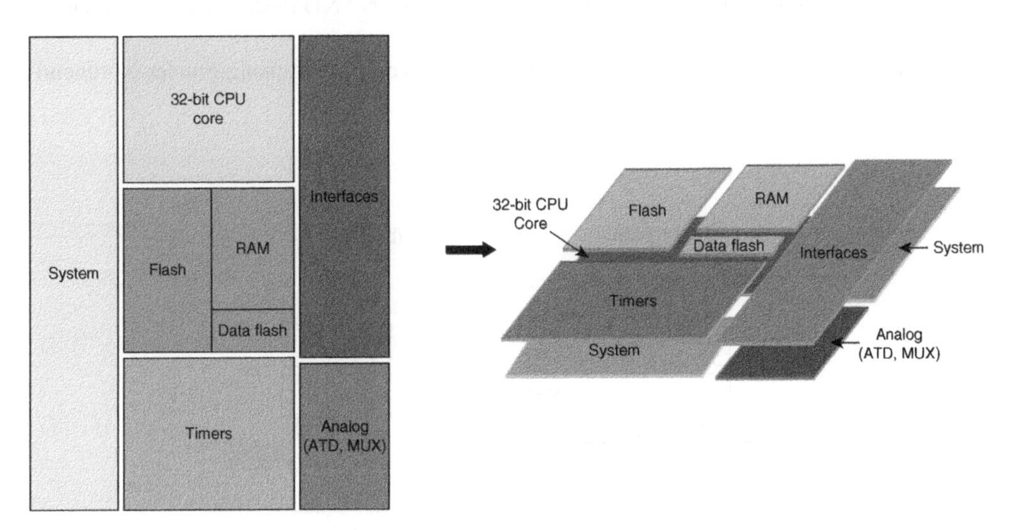

Figure 1.20 Block illustration of an automotive flash MCU showing 2D and possible 3D configuration with smaller footprint and shorter interconnects.

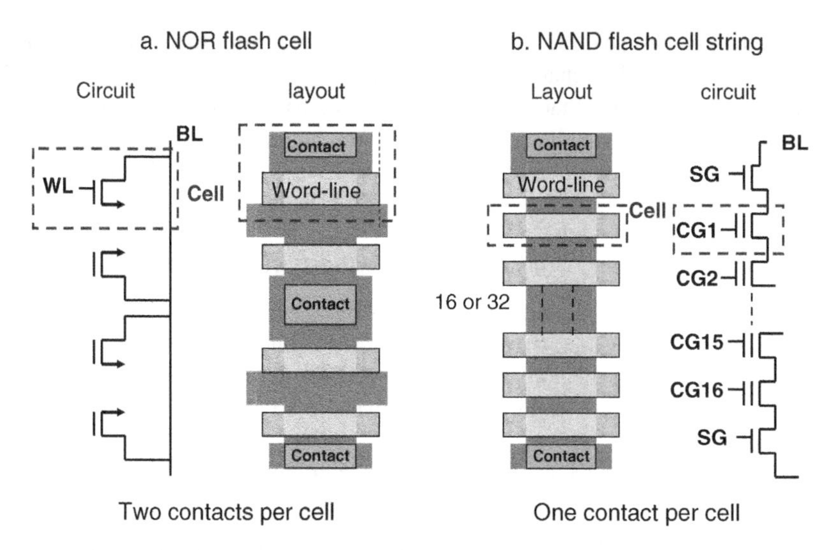

Figure 1.21 Schematic circuit diagram and schematic cross-section for (a) NOR flash showing two contacts per cell; and (b) NAND flash showing one contact per cell. (Based on F. Masuoka, (Toshiba) IEDM, December 1987 [28].)

word-line contact. A common bit-line runs through an entire string of cells so that an individual bit-line contact for each cell is unnecessary. This means that the storage cells are much smaller than for the NOR flash because two space-consuming contacts are not required for each cell. It also means that a single bit-line contact could access an entire string of cells. The number of cells on the NAND string has extended, as the technology has permitted, from 8 to 128 bits. Figure 1.21 illustrates schematic cross-sections and a top-down view layout of (a) the NOR flash cell with two contacts per cell and (b) the much smaller NAND flash cell with one contact per cell [28].

Program and erase for the NAND flash memory is accomplished using Fowler-Nordheim tunneling from and to the substrate, as shown in Figure 1.22.

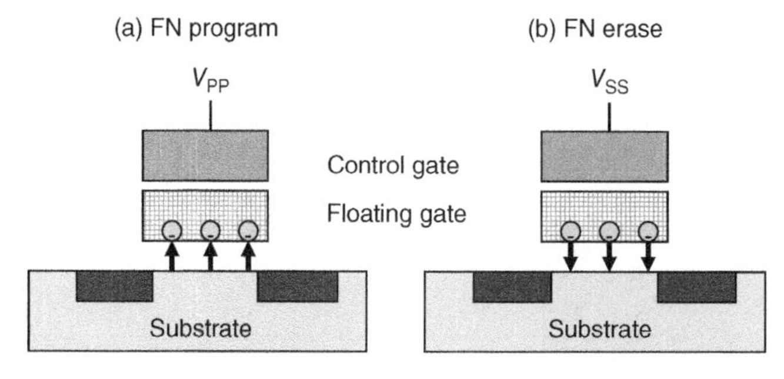

Figure 1.22 Illustration of Fowler-Nordheim tunneling used for (a) program; and (b) erase of a NAND flash memory cell.

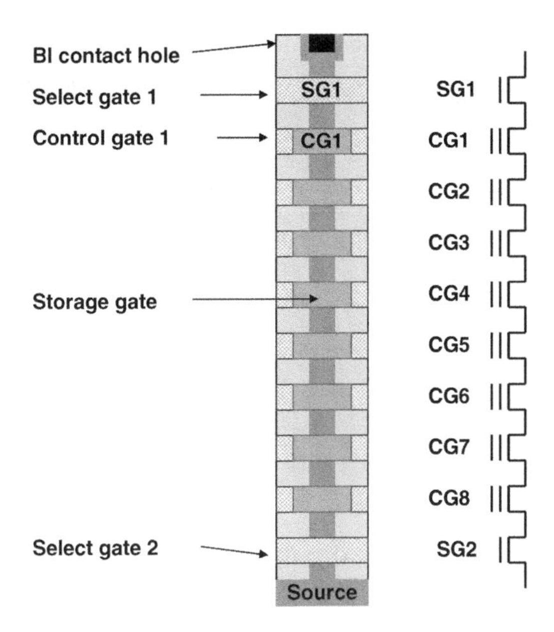

Figure 1.23 Layout and equivalent circuit diagram of an 8-bit NAND flash string.

A schematic of the layout and equivalent circuit of a NAND cell are shown in Figure 1.23. In this NAND string, eight cells are shown formed in series between two select gates. The first select gate (SG1) handles selectivity. The second select gate (SG2) prevents cell current through the string during a program operation. The storage transistors can be either floating gate or charge trapping. The control gates of the storage transistors are word-lines.

To bulk erase the NAND cell, all control gates are grounded and a high voltage is applied to the substrate and p-well, with the source and bit-lines floating. Erase can be performed on the whole chip or on selected blocks. A page is all the cells on one word-line and for one I/O width. A block is all the pages for a NAND string depth. An example of a block in an $8\,\mathrm{M} \times 8$ bit NAND flash is shown in Figure 1.24. In this NAND flash, a string is 16 bits long, a word-line is 512 bits long, and the I/O is 8 bits. A page is $512\ \text{bits} \times 8\ \text{bits}$, and a block is $512\ \text{bits} \times 8\ \text{bits} \times 16\ \text{bits}$. In the 64Mb device, there are 1024 blocks or 66Mb.

In the program operation, the p-well is grounded, a high voltage is applied to the selected control gate/word-line, and a lower voltage is applied to the unselected control gate to reduce program disturb.

It was reported by SanDisk that the NAND flash memory industry has seen a 100-fold increase in density in the last decade and a 50 000-fold cumulative cost reduction over 20 years. This has fueled growth in the mass storage market and is expected to influence the emerging SSD market [29].

The development of the charge-trapping flash memory eliminated many of the issues of the double-polysilicon floating gate for a 3D stacked NAND flash. Chapter 2 describes a charge trapping NAND flash that stacked two layers of single-crystal silicon made with an epitaxial-like process with a NAND string in each layer and local 3D connections. Chapter 3 covers the development of the gate-all-around memory with a polysilicon substrate nanowire that can be used vertically to handle an entire 64-bit NAND string in the lateral space of one transistor.

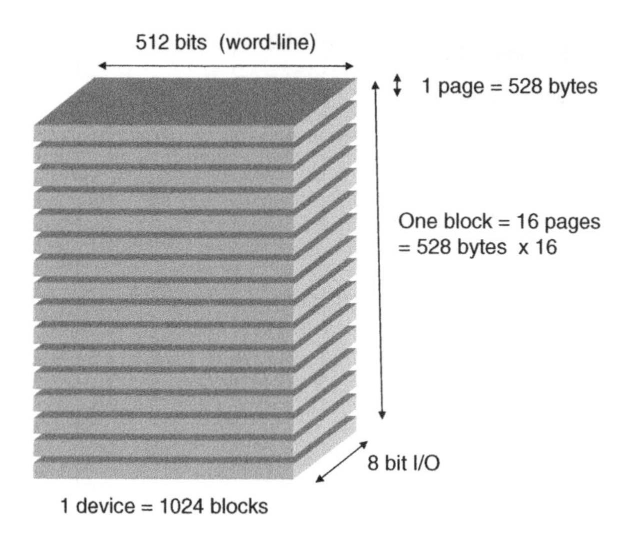

Figure 1.24 8 M × 8 NAND flash memory array organization.

In Chapter 4, both the vertical channel and the vertical gate NAND flash string are explored for large vertical arrays that are expected to produce terrabit NAND flash memory.

1.5 3D Cross-Point Arrays with Resistance RAM

Cross-point arrays, also called crossbar arrays, have seen significant research effort in the past few years. These 3D arrays have the attribute of being inherently stackable and composed of very compact 4–6F^2 memory cells. The final vision of the large-scale 3D memory is a stackable, addressable array of 4F^2 memory cells with minimal peripheral area for memory management. Any two-terminal resistance change memory can be used in a cross-point array. These memories are normally referred to as resistance RAMs (ReRAM). Several ReRAM types are discussed in the context of forming cross-point memory arrays in Chapter 5. A schematic top view of a cross-point array indicating the 4F^2 cell is shown in Figure 1.25.

Metal–oxide ReRAMs began to be widely reported in about 2004. They show promise configured in a 3D cross-point array architecture. These ReRAMs have been studied in various materials that will be discussed in Chapter 5 along with the cross-point array architecture. A ReRAM can be configured as either a bipolar or a unipolar operating memory device. If it switches by reversing the polarity of the applied voltage, then it is considered bipolar, and if it switches using only different voltage pulses applied in a single direction, then it is considered unipolar. The filamentary ReRAM has a low power per bit. The cell is stochastic and difficult to control because only a few atoms are involved in the distribution [3].

The PCM has also been described in a cross-point array configuration. This resistance-type memory device works by joule heating of a chalcogenide material followed by different cooling rates that determine whether the material ends up in a crystalline, low-resistance state or an amorphous, high-resistance state.

Chapter 5 is devoted to the 3D cross-point array memory technology and the ReRAMs and selector devices used in them. Characteristics of two-terminal memory devices that could use the cross-point array configuration are nonlinearity and bipolar or unipolar switching.

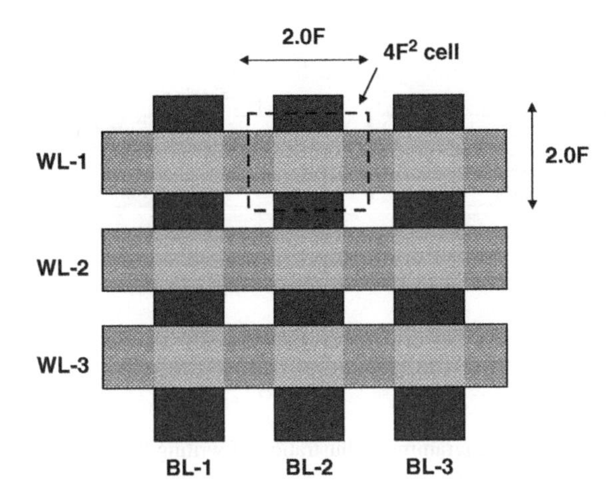

Figure 1.25 Schematic top view of a $4F^2$ cross-point memory array.

1.6 STT-MTJ Resistance Switches in 3D

A spin-RAM that is programmed by interaction of a spin-transfer torque (STT) current and the magnetic moment of the magnetic layers in a magnetic tunnel junction (MTJ) was first described by Sony in December of 2005 [30]. An external magnetic field was not necessary to switch the state of the device, differentiating it from the earlier field-programmable MRAMs. A 4Kb spin-RAM was made on a four-level, metal 180 nm CMOS process. Write speed was 2 ns, and write current was 200 μA. The spin-torque programming reduced the write current to about 5% of the power required to write a field-programmable MTJ MRAM device. Working prototypes were expected in a few years. The theory of the STT-RAM was that the direction of the free magnetic layer in an MTJ device could be reversed by direct injection of spin-polarized electrons. Figure 1.26 illustrates STT switching in an MTJ [30].

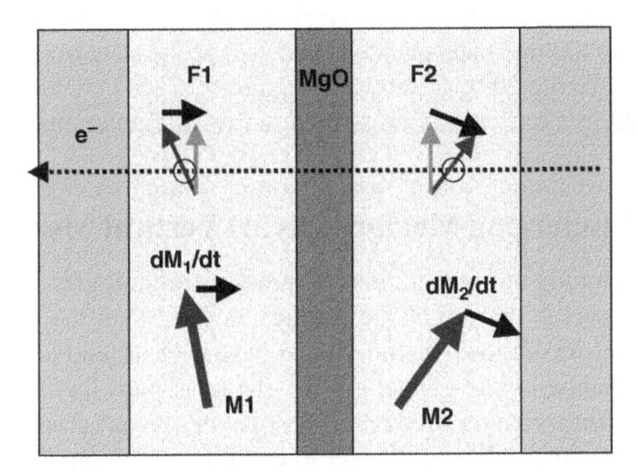

Figure 1.26 Illustration of spin-transfer torque switching in an MTJ. (Based on M. Hosomi *et al.*, (Sony), IEDM, December 2005 [30].)

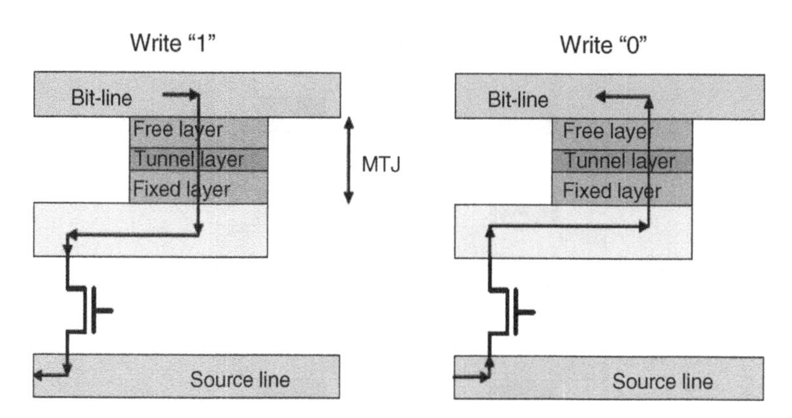

Figure 1.27 STT MTJ MRAM programming illustration of writing "1" and writing "0." (Based on F. Tabrizi, EE Times, April 2007 [31].)

The MTJ is made of a pinned layer (F_1), an MgO tunnel barrier layer, and a free magnetic layer (F_2). The mechanism used to switch the states is the spin of an electron, which has two states: "up" or "down." As a result, this memory is often referred to as a spintronics device. When a spin-polarized electron flows from F_2 to F_1, the spin direction is rotated by the direction of the magnetic moments of the two magnetic layers. The rotating of the spin direction of the electrons creates a spin torque, dM_1/dt and dM_2/dt, which acts on the magnetic moment of M_1 and M_2. When the torque is sufficiently large, the magnetization of the free layer F_2 (M_2) is reversed. This changes the magnetization of F_1 and F_2 from parallel to antiparallel, which reduces the amount of current that can flow through the MTJ. If the electron current flows in the other direction through the MTJ, the spin of the electrons can reverse the state from antiparallel to parallel so that more current flows through the MTJ. The STT-MTJ memory is, therefore, a resistance change memory that uses the direction of current flow through the device to switch the memory state between a high-resistance state (HRS) and a low-resistance state (LRS).

A current that flows from the free magnetic layer to the fixed magnetic layer can change the state of the MTJ element from antiparallel to parallel (write "1"). A current that flows from the fixed magnetic layer to the free magnetic layer can change the state of the MTJ element from parallel to antiparallel (write "0"), as shown in Figure 1.27 [31]. A few STT-MTJ research devices are discussed in Chapter 5 in cross-point array configurations and in Chapter 6 in stacked TSV chips.

1.7 The Role of Emerging Memories in 3D Vertical Memories

In the process of developing new generations of memory technologies, a number of attempts have been made to develop more scalable memory devices than the SRAM, DRAM, and flash memories. These conventional memory types have been around for more than 30 years, essentially without change in basic function. For the most part, the "emerging memories" have not yet made it into volume production despite significant amounts of development effort. It appears to date that it has been safer to continue developing and scaling the three historical memories than to invest huge amounts on a gamble on new and untried technologies. In this category are the ferroelectric RAM, the field-programmable magnetic

RAM, various single-transistor memories, the negative resistance RAM, the nanotube RAM, and so on [18]. Some of these emerging memories have moved into small-scale production, primarily at startup companies, while others have effectively vanished. The ovonic memory, later called the phase-change memory, has entered low-volume production at several large companies and has also been discussed in a few development papers for cross-point arrays.

The ReRAM, even though a relatively new device, appears to have potential for use in the cross-point arrays and is discussed in this 3D variation in Chapter 5. These ReRAM devices began to be described in about 2003 and are still being studied in a variety of different material technologies. A description of various ReRAMs in cross-point arrays is given in Chapter 5. These cross-point array resistive memory devices are in early development and may take significant effort to develop into a mainstream memory technology.

The STT-MTJ memory is also a fairly recent memory device that has been described in various 3D configurations such as a folded gate memory in Chapter 2, in cross-point arrays in Chapter 5, and in stacked TSV chips in Chapter 6. It is still in early development and is expected to take significant effort to develop into a mainstream memory technology.

There is little indication at this time that any of the emerging memory devices will replace the historical SRAM, DRAM, and flash in volume production in 3D memories in the near future.

References

1. Moore, G. (1965) Cramming more components onto integrated circuits. *Electronics*, **38**(8), 114.
2. Bohr, M. (December 2011) The evolution of scaling from the homogeneous era to the heterogeneous era. IEDM.
3. Chandrasekaran, N. (December 2013) Challenges in 3D memory manufacturing and process integration (Micron). IEDM.
4. Chang, L. *et al.* (June 2005) Stable SRAM cell design for the 32 nm node and beyond, (IBM). VLSI Technology Symposium.
5. Prince, B. (1995) *Semiconductor Memories: A Handbook of Design, Manufacture, and Application*, 2nd edn, John Wiley & Sons, Ltd.
6. Ikeda, S. *et al.* (December 1988) A polysilicon transistor technology for large capacity SRAMs, (Hitachi). IEDM.
7. Hu, W. *et al.* (February 2013) Godson-3B1500: A 32 nm 1.35GHz 40W 172.8 GFLOPS 8-core processor, (Chinese Academy of Science, Loongson Technology). ISSCC.
8. Huang, X. *et al.* (December 1999) Sub 50-nm FinFET: PMOS, (University of California, Berkeley, Lawrence Berkeley National Laboratory). IEDM, p. 67.
9. Witters, L. *et al.* (June 2005) Integration of tall triple-gate devices with inserted-TaxNy gate in a 0.274 μm^2 6T-SRAM cell and advanced CMOS logic circuits, (Texas Instruments, Philips, IMEC). VLSI Technology Symposium.
10. Prince, B. (1999) *High Performance Memories*, John Wiley & Sons, Ltd, Figure 6.45, p. 186.
11. Yanagawa, Y., Sekiguchi, T., Kotabe, A. *et al.* (June 2011) In-substrate-bitline sense amplifier with array-noise-gating scheme for low-noise 4F2 DRAM array operable at 10 fF cell capacitance, (Hitachi). VLSI Circuits Symposium.
12. Agarwal, K., Hayes, J., Berth, J. *et al.* (June 2010) In-situ measurement of variability in 45-nm SOI embedded DRAM arrays, (IBM). VLSI Circuits Symposium.
13. Barth, J. *et al.* (February 2010) A 45 nm SOI embedded DRAM macro for POWER7^{TM} 32MB on-chip L3 cache, (IBM). ISSCC.
14. Barth, J. *et al.* (2011) A 45 nm SOI embedded DRAM macro for the POWER7TM processor 32 MByte on-chip L3 cache, (IBM). *IEEE Journal of Solid-State Circuits*, **46**(1), 64.
15. Iyer, S.S. *et al.* (2005) Embedded DRAM: Technology platform for the Blue Gene/L chip. *IBM Journal of Research and Development*, **49**(2.3), 333.
16. Frohman-Bentchkowsky, D. (1971) Memory behavior in a floating-gate avalanche -injection MOS (FAMOS) structure, (Intel). *Applied Physics Letters*, **18**(8), 332.
17. Masuoka, F., Asano, M., Iwahashi, H. *et al.* (December 1984) A new flash E2PROM cell using triple polysilicon technology, (Toshiba). IEDM.

18. Prince, B. (2002) *Emerging Memories Technologies and Trends*, Kluver Academic Publishers.
19. Frohman-Bentchkowsky, D. (1969) An integrated metal-nitride-oxide-silicon (MNOS) memory, (Fairchild Semiconductor). *Proceedings of the IEEE*, **57**(6), 1190.
20. Prince, B. and Due-Gunderson, G. (1983) *Semiconductor Memories*, John Wiley & Sons, Ltd.
21. Kamigaki, Y. *et al.* (1989) Yield and reliability of MNOS EEPROM products, (Hitachi). *IEEE Journal of Solid-State Circuits*, **24**(6), 1714.
22. Suzuki, E., Hiraishi, H., Ishii, Kenichi, and Hayashi, Y. (1983) A low-voltage alterable EEPROM with metal-oxide-nitride-oxide-semiconductor (MONOS) structures, (Electrotechnical Laboratory Ibaraki Japan, Citizen Watch Co.). *IEEE Transactions on Education*, **ED-30**(2), 122.
23. Tanaka, T., Tanikawa, H., Yamaki, T., and Umeboto, Y. (June 2003) A 512kB MONOS type flash memory module embedded in a microcontroller, (Hitachi). VLSI Technology Symposium.
24. Kono, T. *et al.* (2014) 40-nm embedded split-gate MONOS (SG-MONOS) flash macros for automotive with 160-MHz random access for code and endurance over 10 M cycles for data at the junction temperature of 170 °C, (Renesas). *IEEE Journal of Solid-State Circuits*, **49**(1), 154.
25. Spansion (2009) Spansion Launches MirrorBit(R) SPI Multi-I/O Flash Memory with Up to 40 MB/s Read Performance. Press release, May 20.
26. Jung, S.-M. (December 2006) Three dimensionally stacked NAND flash memory technology using stacking single crystal Si layers on ILD and TANOS structure for beyond 30 nm node, (Samsung). IEDM.
27. Renesas (2014) *RH850 Family (Automotive Only)*, http://www.renesas.com/products/mpumcu/rh850 (accessed 14 May 2014).
28. Masuoka, F., Momodomi, M., Iwata, Y., and Shirota, R. (December 1987) New ultra high density EPROM and flash EEPORM with NAND structure cell, (Toshiba). IEDM.
29. Quader, K.N. (20 May 2012) Flash memory at a cross-road: Challenges & opportunities, (SanDisk). IMW.
30. Hosomi, M. *et al.* (December 2005) A novel nonvolatile memory with spin torque transfer magnetization switching: Spin RAM, (Sony). IEDM.
31. Tabrizi, F. (2007) Taking Hold of Embedded Memory Management, (Grandis). EE Times (April 23).

2

3D Memory Using Double-Gate, Folded, TFT, and Stacked Crystal Silicon

2.1 Introduction

An early step in moving from a planar memory technology to a vertical technology has been moving from a planar gate to a vertical gate device such as the fin-shaped field-effect transistor (FinFET) flash memory or other folded vertical memory structure. In the case of the FinFET, the short-channel effect, which is a limitation on the channel length created by scaling, is eliminated. The vertical FinFET also offers the potential for a double or triple gate, which enhances transistor and memory 3D characteristics.

An illustration of a FinFET transistor structure is shown in Figure 2.1. The vertical silicon fin forms the channel in which the current flows laterally. The transistor gate oxide and the polysilicon gate are patterned on the fin. The channel length is determined by the thickness of the gate, and the channel width is determined by the height of the fin. This structure permits increased channel length, which improves gate control, eliminating short-channel effects, and increases ON current by permitting increased channel width. It also provides a double gate with one gate on each side of the fin, which enhances gate control.

Thin-film transistors (TFTs) are also discussed in this chapter. TFTs can be used to form 3D vertical devices such as stacked double-gate NAND flash strings. To be able to stack memory devices with either planar or vertical gates, it is necessary to improve the quality of the polysilicon channels in the stacked substrate.

The development of single-crystalline stacked substrates makes it possible to have high-quality memory devices on stacked substrates. This development is discussed along with memory architectures that use single-crystalline stacked substrates.

Vertical 3D Memory Technologies, First Edition. Betty Prince.

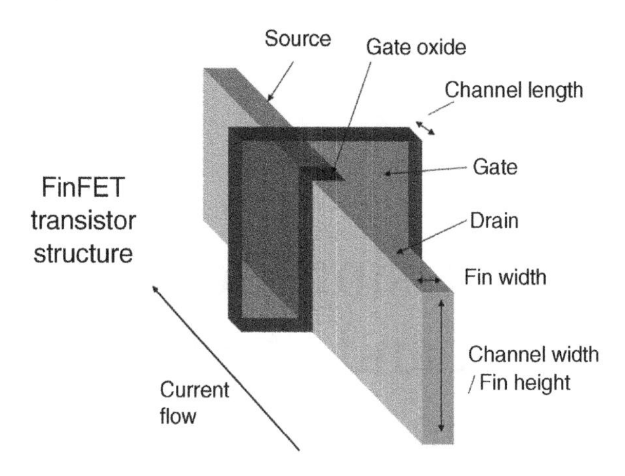

Figure 2.1 Schematic of a 3D FinFET transistor structure.

2.2 FinFET—Early Vertical Memories

2.2.1 Early FD-SOI FinFET Charge-Trapping Flash Memory

A thin fin of single-crystal silicon can be etched from bulk silicon or silicon-on-insulator (SOI) silicon. Schematic cross-sections of (a) an SOI FinFET transistor and (b) a bulk FinFET transistor are shown in Figure 2.2. The current flows laterally in the fin. The FinFET transistor has a higher cell current and reduced short-channel effects compared with a planar transistor at the same technology node.

An early, fully depleted SOI (FD-SOI) FinFET charge trapping (CT) flash memory was reported in December of 2003 by the University of California, Berkeley, and the Lawrence Berkeley National Lab [1]. This device was found scalable to a gate length (L_g) of 40 nm. The SOI FinFET memory device had no body contact but had good program/erase characteristics and good endurance along with long data retention time and low read disturb. Devices were made on both (100) and (110) silicon surfaces. A double-gate FinFET was proposed to suppress short-channel effects in a scaled sub–50 nm CMOS flash technology. The thin fin

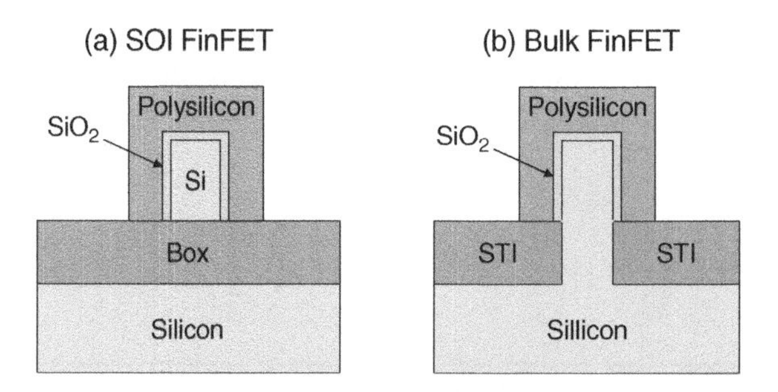

Figure 2.2 Schematic cross-section of a FinFET transistor in (a) SOI; and (b) bulk silicon.

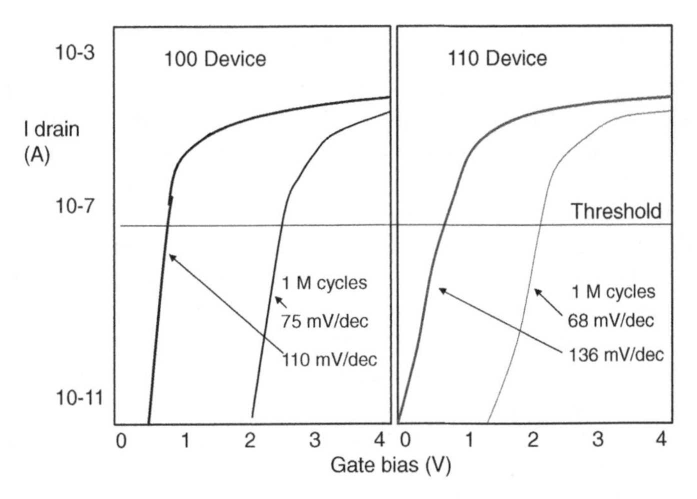

Figure 2.3 Illustration of subthreshold swing of (110) and (100) FinFET CT devices at 1 M cycles showing degradation of (110) devices. (Based on P. Xuan *et al.* (University of California Berkeley, Lawrence Berkeley National Lab), IEDM, December 2003 [1].)

body eliminated subsurface leakage paths and reduced drain-induced barrier lowering (DIBL), which is a reduction of the threshold voltage (V_{th}) of the transistor at higher drain voltages. An SiO_2–Si_3N_4–SiO_2 (ONO) stack was deposited on the silicon fin to form the memory device. Charges trapped in the silicon nitride layer were used to change the threshold voltage of the memory transistor. Performance characteristics were comparable to those of bulk silicon charge-trapping memory devices. The primary channels were along the vertical fin sidewalls. The crystal orientation of the sidewalls could be changed between (110) and (100) by altering the fin orientation with respect to the major wafer flat.

The fabricated FinFET devices had a large V_{th} window, from 0.9 to 2.9 V. This was achieved with a 10 ms/−11 V erase pulse and a 5 ms/+10 V program pulse. Erase characteristics were comparable to those of bulk silicon charge-trapping memories. After 1 million program/erase (P/E) cycles, the (100) devices showed little mobility degradation. Both (100) and (110) orientation FinFET memory devices showed good endurance up to 1 million program/erase cycles without degradation of the V_{th} window. The I_d–V_g characteristics showed an increase in subthreshold swing after 1 million P/E cycles, as shown in Figure 2.3 [1]. This increase was from 75 to 110 mV/dec for the (100) device and from 68 to 136 mV/dec for the (110) memory. The (100) device showed less degradation after 1 million cycles. The conclusion was reached that there were fewer interface traps generated after 1 million P/E cycles in the (100) device than in the (110) device.

Endurance was up to 1 million P/E cycles without degradation, and retention characteristics showed a greater than 1.4 V_{th} window after 10 years at 85 °C. The ratio of read current to leakage current was greater than 10^6, which improved the detection of cell current during a read operation. The silicon crystal orientation had an effect on the memory characteristics. Due to thicker tunnel oxide, the devices with (110) fin sidewall surfaces had slower P/E speeds but better retention than devices with (100) surfaces. The (100) devices were more resistant to P/E stress. Simulations showed the potential for scalability to a 40 nm gate length or less [1].

2.2.2 *FinFET Charge-Trapping Memory on Bulk Silicon*

A FinFET charge-trapping memory on bulk silicon made using a damascene gate process was discussed in December of 2004 by Samsung [2]. The FinFET had the advantages over scaled CMOS of having good immunity to short-channel effects and compatibility with a conventional CMOS process. Issues with the conventional FinFET fabrication included poor controllability of threshold voltage and increased gate etch. These issues were solved by using a damascene gate process with metal gate, which eliminated the gate-etch step. Damascene is a metal inlay process that was developed for use with copper interconnects because copper cannot be patterned by photoresist masking and plasma etch. In the damascene process, the dielectric is patterned with open trenches followed by deposition of a thick layer of metal in the trenches. Any extra metal is removed by chemical mechanical planarization (CMP).

A drawback of the planar structure for a charge-trapping memory is the difficulty in finding an optimum tunneling oxide thickness for both fast P/E speed and good data retention. In the FinFET structure, however, the transverse electric field to the gate oxide provides an opportunity to scale the tunnel oxide thickness while retaining good data retention. Damascene gate FinFET charge-trapping memories can have scaled tunnel oxide for high speed and low power. Devices with $W_{fin} = 30$ nm and $L_g = 50$ nm showed threshold shifts of about 4 V at 1 µs/12 V for program and at 50 µs/−12 V for erase. Data retention showed a more than 2.4 V threshold window after 10 years of reading as shown in Figure 2.4. Endurance testing showed that V_{th} windows maintained up to 10^4 P/E cycles [2].

An issue with body-tied bulk silicon FinFET flash memory is that the coupling ratio is lower than that of a planar flash memory, as illustrated in Figure 2.5 [3]. The coupling ratio is the ratio of the capacitive coupling of the gate to the charge storage layer divided by the capacitive coupling of the charge storage layer to the silicon channel. Fast programming requires a high coupling ratio or a high gate voltage.

A possible solution to increase the coupling ratio was discussed by Samsung in June of 2005 [3]. A 256Mb body-tied FinFET flash memory using hafnium (Hf) silicate interpoly dielectric (IPD) was made and compared with a FinFET flash memory using ONO.

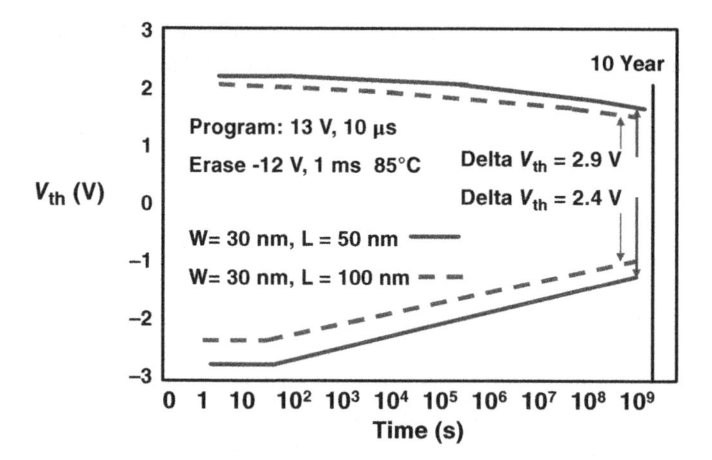

Figure 2.4 Data retention characteristics showing >2.4 V V_{th} window after 10 years reading. (Based on C.W. Oh *et al.*, (Samsung), IEDM, December 2004 [2].)

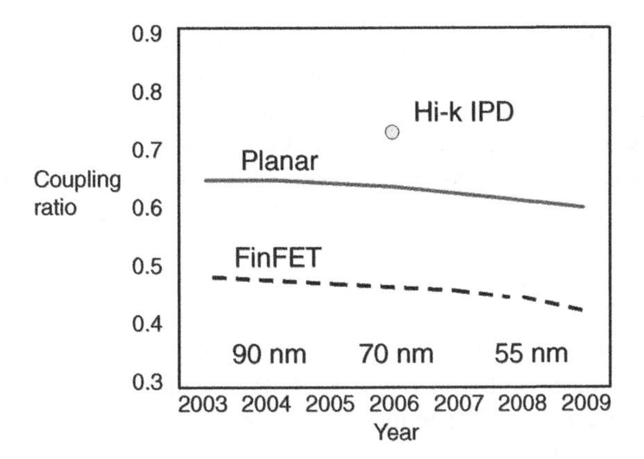

Figure 2.5 Coupling ratio by year for planar and FinFET charge-trapping memory. (Based on B.S. Cho *et al.*, (Samsung), VLSI Technology Symposium, June 2005 [3].)

There are two methods to increase the coupling ratio of the FinFET flash memory. One is reducing the tunnel oxide capacitance, and the other is increasing the IPD capacitance. By using Hf silicate (HfO_6Si_2) for the IPD, the higher dielectric constant increases the capacitance, which improves the coupling ratio, as shown in Figure 2.5 for the high-κ IPD. This, in turn, enhances the channel hot electron injection (CHEI) programming speed and reduces the operating voltage. Hot hole injection (HHI) erase gives a higher erase speed than Fowler-Nordheim tunneling does, without degrading the endurance characteristics. Key parameters of the body-tied silicon FinFET flash technology with Hf silicate IPD include the following: cell gate length of 160 nm, fin width of 60 nm, and fin height of 100 nm. Tunnel oxide thickness was 7.5 nm.

The formation and characterization of a 20 nm gate length FinFET SONOS flash on bulk substrate was described in December of 2005 by TSMC and National Chaio Tung University [4]. A 3D schematic of the formation process for the bulk FinFET charge-trapping (CT) flash device structure is shown Figure 2.6 [4].

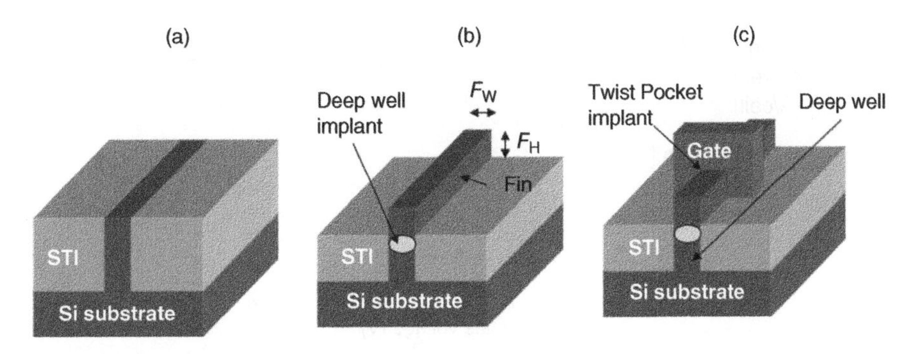

Figure 2.6 Bulk FinFET CT flash memory cell structure. (Based on J.R. Hwang *et al.*, (TSMC, NCTU), IEDM, December 2005 [4).

The process included (a) fin patterning using a planar stacked trench isolation (STI) process followed by (b) an STI oxide recess to define the fin channel, and (c) a deep well implant to isolate the fin from the substrate. Fin height and width were 40 nm. Thermal oxide was grown, followed by LP-CVD nitride deposition, and TEOS top oxide to form the CT ONO layer. Polysilicon deposition was followed by gate patterning with 193 nm lithography to form the gate. LDD and twist-pocket implants were done followed by source and drain extensions intended to avoid punch-through during program and erase and for a faster P/E speed and larger V_{th} window. Contacts followed with a copper and low-κ back-end process, completing the steps.

FinFET CT flash cells with vertical gates and gate length down to 20 nm were made on bulk silicon substrates and successfully operated. A 2 V P/E window was achieved with high P/E speed of $T_p = 10\,\mu s$ and $T_e = 1$ ms. The memory was programmed with CHEI, and erase was done using band-to-band HHI.

Multilevel storage was confirmed for the threshold voltage window $\Delta V_{th} > 4$ V with program and erase times of 1 ms. Experimental multilevel transfer characteristics of the 20 nm bulk FinFET SONOS at a reverse read bias of $V_d = 1.5$ V are shown in Figure 2.7 [4]. The transistor was programmed with $V_g = 7.0$ V, $V_d = 4.0$ V, and program times (T_p) of 0.4 μs, 30 μs, and 1 ms. Erase was performed with $V_g = -4.0$ V, $V_d = 4.3$ V, and $T_e = 1$ ms.

A 1.5 V window remained after 10 years at room temperature. Operating voltage was <7 V. Gate disturb issues were reduced by applying an appropriate bias on unselected bit-lines. In a NOR flash array, program disturb can occur in cells neighboring the selected cell on both the bit-line and the word-line. Read disturb can disturb the neighboring cell on the word-line.

In this 20 nm gate CT bulk FinFET flash cell, program was done with an inhibit voltage of 3 V on the unselected bit-lines and 0 V on the unselected word-lines. Good immunity from program disturb was shown up to 10 s. Figure 2.8(a) illustrates the inhibit voltage used on neighboring cells to the selected cell on the bit-line and word-line during program. For a read, 0 V on unselected word-lines and bit-lines during a 4.5 V disturb resulted in minimal read

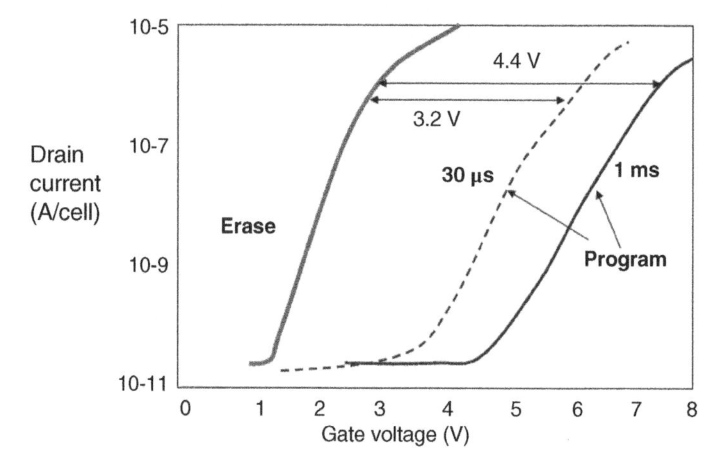

Figure 2.7 Bulk 20 nm CT FinFET flash multilevel transfer characteristics. (Based on J.R. Hwang *et al.*, (TSMC, NCTU), IEDM, December 2005 [4].)

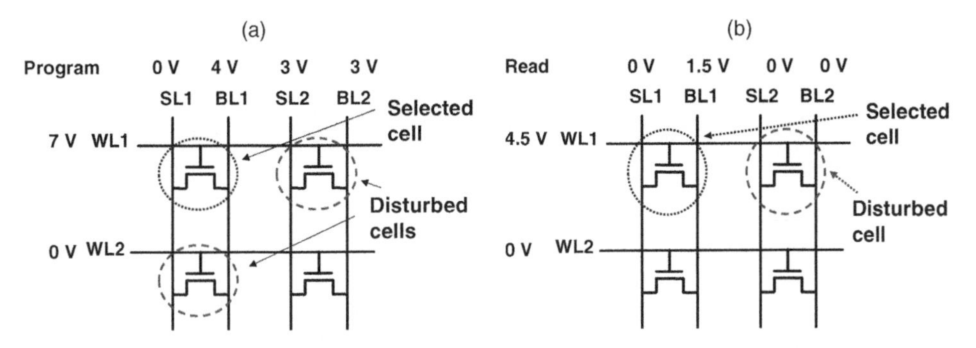

Figure 2.8 Circuit configuration for a 20 nm gate bulk FinFET SONOS NOR flash array for inhibit of (a) program disturb; and (b) read disturb. (Based on J.R. Hwang *et al.*, (TSMC, NCTU), IEDM, December 2005 [4].)

disturb up to 10 s. Figure 2.8(b) illustrates inhibit voltage used on unselected cells on the bit-line during read of the selected cell [4].

Data retention for the FinFET CT flash memory on bulk silicon was discussed in March of 2006 by Samsung [5]. Charge-loss mechanisms were studied and a comparison made between the thermal-accelerated and field-accelerated lifetimes. Thermal acceleration was already widely used for lifetime prediction for CT devices; however, the charge-loss mechanism is also dependant on the electric field. It was believed that Schottky emission and direct tunneling were dominant at high temperature and field-dependent mechanisms such as trap-to-band, trap-assist, or Poole-Frenkel emission were dominant at room temperature. These four charge-loss mechanisms are illustrated in the band diagram in Figure 2.9 [5].

The vertical channel CT FinFET on bulk silicon had a final fin width of 2.5 nm and a tunneling oxide thickness of about 4 nm. Nitride and blocking oxide thicknesses were about 6 nm each. The long-term data retention lifetime of the vertical gate CT memory device was predicted and compared to the result due only to temperature acceleration. The temperature retention lifetime of the ONO stacked device without gate voltage was about 35 years using only a thermal acceleration factor. In the case that both temperature and electric field were used for the acceleration, the retention lifetime was about 10 years.

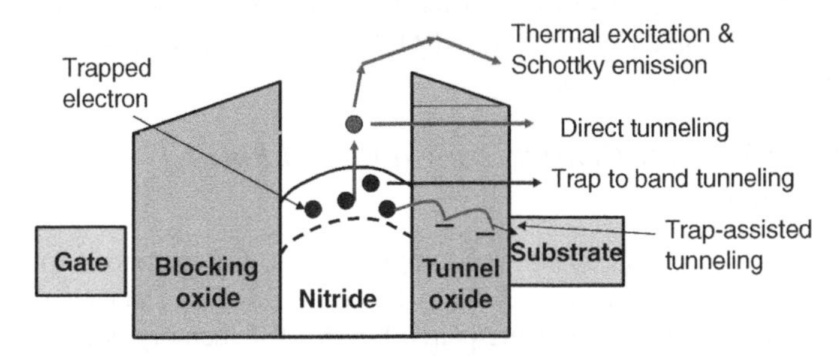

Figure 2.9 Schematic band diagram of CT FinFET memory showing major charge loss mechanisms. (Based on J.J. Lee *et al.*, (Samsung), IRPS, March 2006 [5].)

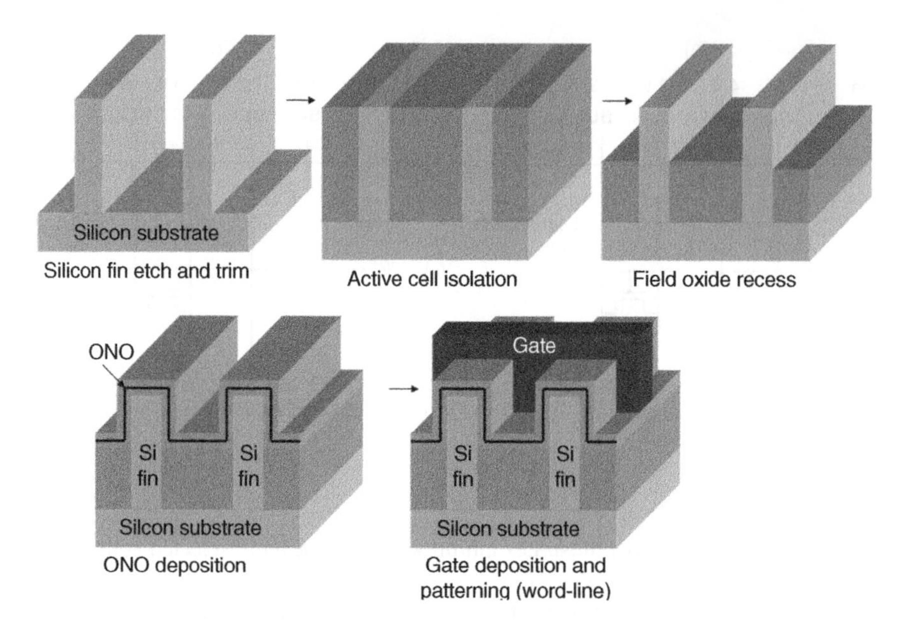

Figure 2.10 Process steps for CT body-tied FinFET structure. (Based on S.K. Sung *et al.* (2006) (Samsung), *IEEE Transactions on Nanotechnology*, 5(3),74 [6].)

A fully integrated NOR flash memory array using a body-tied FinFET structure on bulk silicon was described by Samsung in May of 2006 [6]. The body-tied FinFET structure process is schematically illustrated in Figure 2.10 [6].

The same authors in June of 2006 [7] discussed using a P+ polysilicon gate and an Al_xO_y high-κ blocking dielectric for a NAND FinFET CT memory array. A GSL/SSL transistor was added for the NAND string to obtain a multigigabit NAND flash memory. Issues in 50 nm planar NAND flash technology had been the ON–OFF current ratio of the cell transistors, the punch-through characteristics of the SSL transistors, and the floating gate coupling interference between cells. The CT technology on the FinFET device resolved the floating gate coupling disturb issue, and the increased channel width improved the ON current of the memory cell. The FinFET also improved the punch-through characteristics of the SSL transistors. A P+ gate was used for threshold voltage control of the GSL/SSL NAND string transistors. The P+ gate was also used on the FinFET memory cell to enhance the erase characteristics by adding to the effect of the Al_xO_y blocking dielectric in suppressing electron back-tunneling from the control gate.

2.2.3 Doubling Memory Density Using a Paired FinFET Bit-Line Structure

Dividing the fin of a finFET in 50 nm technology into two CT memories with an insulator between them was discussed in June of 2006 by Samsung [8]. Two channels were formed on the outer surface of the silicon, producing paired FinFET CT memory devices. The configuration is shown in the schematic cross-section comparison of the (a) normal FinFET CT memory and (b) paired FinFET memory device structure in Figure 2.11 [8].

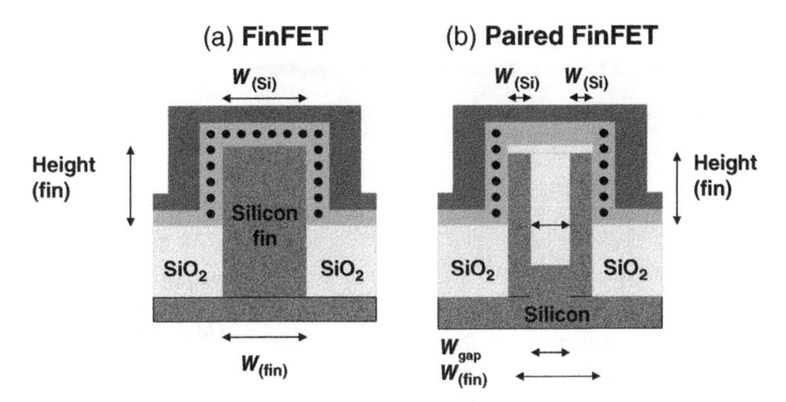

Figure 2.11 Schematic cross-section comparison of (a) normal FinFET; and (b) paired FinFET memory cell. (Based on S. Kim *et al.*, (Samsung), VLSI Technology Symposium, June 2006 [8].)

A set of paired FinFET memory devices were fabricated and showed good P/E characteristics. Independent programming on each storage node was shown. A top-down view comparison of the bit density per unit area of a FinFET and a paired FinFET is shown in Figure 2.12 [8].

The active silicon is split by an insulating layer and then a storage layer, after which the gate is formed on it so that each side of the structure forms an independent cell with a common gate. The average bit-line pitch is assumed to be 20 nm. The on-cell current is lower than for the FinFET. This is because the paired device uses only one side of the fin but can be higher than a planar device, as the channel width depends on the height of the fin, which can be increased. A CT FinFET device was integrated for both the memory cell array and for the GSL/SSL transistor of the NAND string. A P+ gate was used for threshold voltage control of the GSL/SSL transistors, and Al_xO_y blocking dielectric was used to suppress electron back-tunneling from the control gate.

A high-density multibit vertical gate NAND flash array using a 32-paired FinFET NAND flash cell string in 63 nm CMOS technology was discussed by Samsung in June of 2008 [9]. The CT NAND FinFET flash array used two paired sidewall channels on a single bit-line.

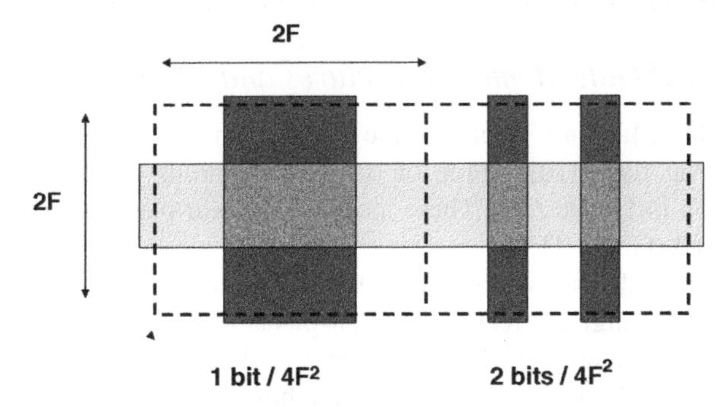

Figure 2.12 Top-down comparison of bit density per area of FinFET and paired FinFET. (Based on S. Kim *et al.*, (Samsung), VLSI Technology Symposium, June 2006 [8].)

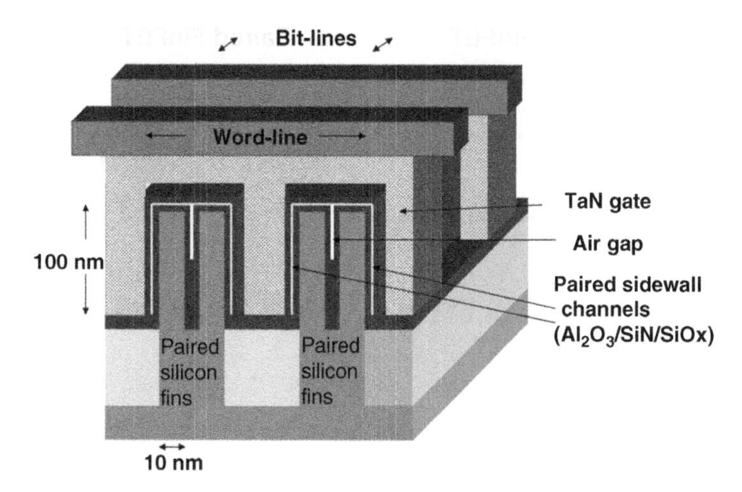

Figure 2.13 Schematic diagram of paired FinFET vertical NAND flash array structure. (Based on J.M. Koo *et al.*, (Samsung), VLSI Technology Symposium, June 2008) [9].)

The field oxide was recessed about 100 nm. SiN and a small air gap were used to improve isolation between the two sidewall channels. A TANOS (TaN–Al_2O_3–SiN–SiO_x–Si) CT cell structure was used for the memory stack. The high dielectric constant blocking oxide and high work-function metal gate helped provide acceptable P/E properties and Vth distribution characteristics for multilevel operation and also helped to reduce charge-coupling interference in the array. A schematic diagram of the vertical NAND flash array structure with paired FinFET sidewall channels is shown in Figure 2.13 [9].

P/E characteristics of the integrated array cell were determined. The program voltage and time for reaching $V_{th} = 3$ V was 18.5 V for 100 μs. The erase voltage and time for $V_{th} = -3$ V were 19 V and 10 ms, respectively. This was considered a sufficient V_{th} window for multilevel cell operation. Programmed cells could be distinguished from unprogrammed cells. The string cells were stressed by 1.2×10^3 P/E cycles then baked at 200 °C for 2 hours. The resulting Vth window charge loss was 0.5 V. The 32-cell string array was successfully integrated on a sub–10 nm paired fin body with a 100 nm sidewall channel using a 63 nm NAND flash technology. No program disturb from adjacent cells was seen during a checkerboard test pattern.

2.2.4 Other Folded Gate Memory Structures and Characteristics

In December of 2009, Macronix discussed the effect of the STI edge fringing field on the scaling of CT NAND flash [10]. Three of the configurations of NAND flash that were considered are shown in Figure 2.14. These included (a) near-planar, (b) body-tied FinFET, and (c)gate-all-around (GAA) TFT.

Edge fringing field analysis was done by 3D TCAD simulation during Fowler-Nordheim operation. A very small fringing effect for the near-planar device occurred at the STI corners. For the body-tied FinFET, a large field effect occurred at the rounded tip of the fin, resulting in a much higher P/E speed. In the bottom neck region, however, the bottom oxide electric field was degraded, which caused a parasitic leakage path when the device was programmed. For the GAA TFT, the fringing effect occurred at the corners of the rectangular nanowire. Because the

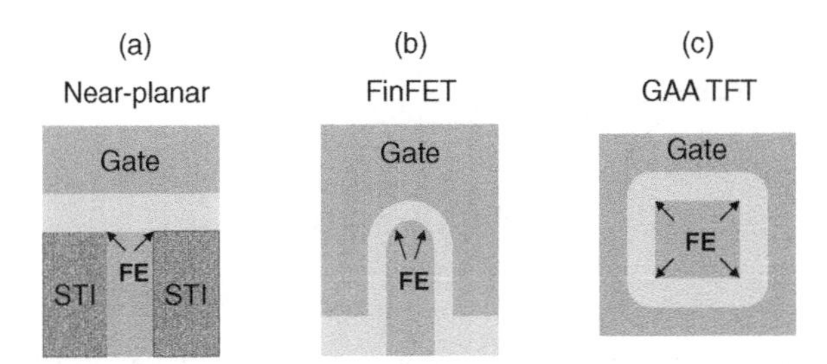

Figure 2.14 Configuration and fringing field location for (a) near-planar; (b) FinFET; and (c) GAA TFT CT structures. (Based on H.T. Lue *et al.*, (Macronix), IEDM, December 2009 [10].)

fringing for FinFET and GAA TFT occurs in the corners, it leads to a higher P/E speed and large memory window.

A CT NAND flash memory, which used an arch-shaped silicon fin and extended word-lines to improve virtual source/drain performance, was discussed in December of 2010 by Seoul National University [11]. The arch-shaped fin is shown in Figure 2.15 [11].

The arch shape concentrated the electric field, which resulted in higher electron concentration in the virtual source/drain region and a higher ON-state cell current. The extended WL process improved the short-channel effect immunity and also improved the I–V characteristics. An arch NAND flash array was made and characterized. The device showed a small short-channel effect and high ON-state cell current [11].

A nonplanar folded triangle-shaped gate flash memory was described in October of 2012 by the University of California, Riverside, and Nanjing University [12]. The floating storage gate was a high density of $1.5 \times 10^{12}\,\mathrm{cm}^{-3}$ NiSi nanocrystals (NCs). The technology used a triangle-shaped silicon nanowire array as the memory transistor channel. This permitted a denser NC array than would be possible with a lateral planar NC surface. A schematic diagram of the triangular nanowire array is shown in Figure 2.16 [12]. The nanocrystals were embedded in the control oxide, which was covered in turn by the gate (which is not shown). The cross-section of the silicon nanowire is a quasi-isosceles triangle.

The advantage of the cell structure is in the number of NCs in the projected lateral plane, which is nearly double that possible in a planar structure. This can increase the integration density of memory chips without reducing the number of NCs per cell. The memory showed good program, erase, and retention characteristics, which suggested that vertical nonplanar

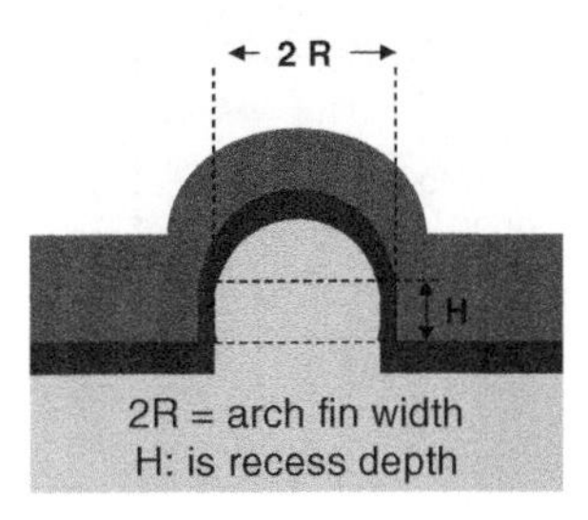

Figure 2.15 Arch-shaped silicon Fin in CT NAND flash cell. (Based on W. Kim *et al.* (2011) (Seoul National University), *IEEE Electron Device Letters*, 31(12) [11].)

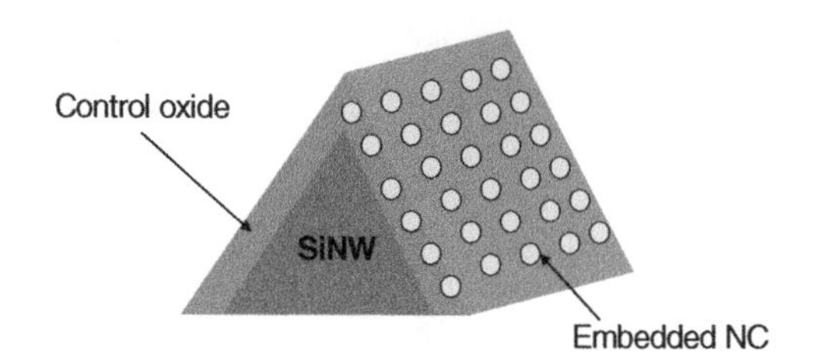

Figure 2.16 Schematic view of triangular nanowire array. (Based on J.J. Ren *et al.* (2012) (University of California Riverside, Nanjing University), *IEEE Electron Device Letters*, 33(10), 1390 [12].)

devices can extend the NC memory scaling limits without compromising the device performance.

Magnetic tunnel junction (MTJ) memory devices can also benefit from 3D folding of the cell. A folded 3D spin-transfer torque (STT) MTJ-RAM cell structure, which can be used with in-plane magnetic anisotropy materials, was described in June of 2012 by Samsung [13]. A schematic cross-section of a 3D folded cell is shown in Figure 2.17, where the free layer is folded over the MgO tunneling layer to improve memory characteristics. The folding can be done on two sides or four sides.

Conventional planar STT-MTJ cells with in-plane magnetic anisotropy can scale to about 40 nm before the free-layer volume falls to the point that the thermal stability is too low (defined as thermal stability <60). Perpendicular magnetic anisotropy (PMA) cells maintain sufficient thermal stability but have issues with low tunnel magnetoresistance (TMR), high switching current, and degradations from etch damage. In this study, the free layer was folded so the 3D MTJ cell structure was able to retain a large free-layer volume without an increase of cell footprint. This permitted scaling the cells laterally even though in-plane magnetic anisotropy materials were used. The required thermal stability of >60 was confirmed in a 3D STT-MTJ cell with $15 \times 30\,\text{nm}^2$ area. A comparison of switching current between a conventional STT-MTJ device and the 3D STT-MTJ device showed that the thermal stability was adequate and switching current for the four-times-folded version was the smallest.

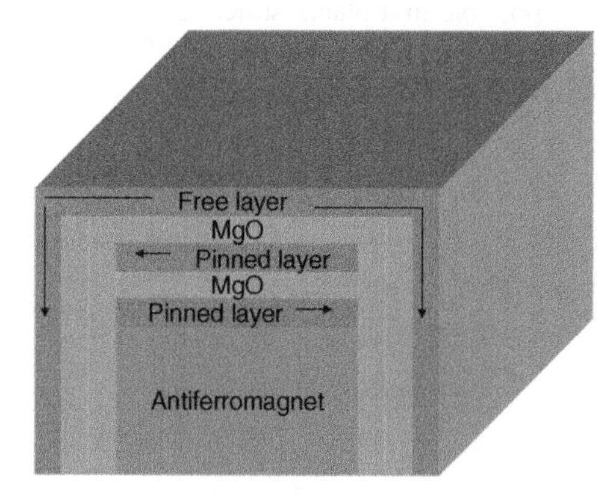

Figure 2.17 Schematic cross-section of 3D folded free-layer cell with two-sided folding. (Based on S. Lee *et al.*, (Samsung), VLSI Technology Symposium, June 2012 [13].)

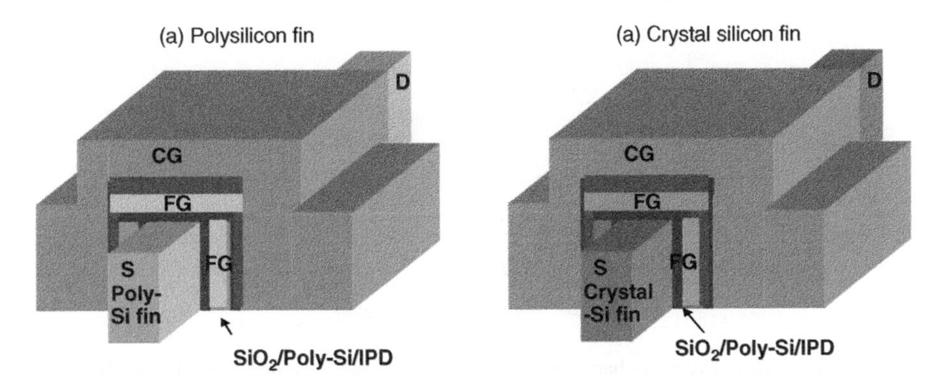

Figure 2.18 Polysilicon and single-crystal silicon fin channel tri-gate flash memories. (Based on Y.X. Liu *et al.*, (AIST), ESSDERC, 12 September 2011 [14].)

The thermal stability of the 3D MTJ cell increased with the increase of side folding length. For thermal stability $E_b/k_BT > 60$, a linear length of 26 nm was found sufficient.

A variability comparison was done for the threshold voltage of scaled floating gate tri-gate FinFET flash memories with polysilicon channels and those with single-crystal silicon channels. This study was done by AIST in September of 2011 [14]. The polysilicon grain size in this study was 20 to 70 nm, which was comparable to the channel area. As a result, there is some probability of polysilicon single-grain channels without grain boundaries. The gate length of the channel was about 76 nm.

Some superior I_d–V_g characteristics were found in the scaled polysilicon channel tri-gate cell with Lg down to 54 nm or less. The standard deviation of the V_{th} of the polysilicon channel FinFET cell was three times higher than that of crystal channel FinFETs at the same gate oxide thickness. The V_{th} of the polysilicon channel flash became comparable to crystal channel devices after one P/E cycle. It was found that punch-through voltage of the polysilicon channel tri-gate flash was as high as 4.6 V for L_g down to 76 nm. A schematic illustration of the two memory types is shown in Figure 2.18 [14].

A further study of floating gate, double-gate, and tri-gate FinFET flash memory using SOI-based fins was described in October of 2012 by AIST and Meiji University [15]. Gate structure dependence of V_{th} variability and SCE immunity was studied. The starting material used was (110) oriented SOI wafers. The silicon fin channels were 17–19 nm thick, and the gate length was 76 nm. The floating gate material was a layer of 30 nm thick n+ polysilicon. It was confirmed that smaller V_{th} variations, better SCE immunity, and a larger memory window were obtained in the tri-gate SOI FinFET flash than in the double gate.

2.3 Double-Gate and Tri-Gate Flash

2.3.1 Vertical Channel Double Floating Gate Flash Memory

An early double floating gate flash cell where the current flows vertically was discussed in December of 2007 by Intel and Stanford University [16]. It had a sub-50 nm body thickness and a vertical channel between double floating gates. This cell used conventional materials, was potentially scalable to an $8F^2$ cell size, and was expected to scale beyond the 32 nm technology node. It was made on a bulk silicon substrate but was not a conventional FinFET

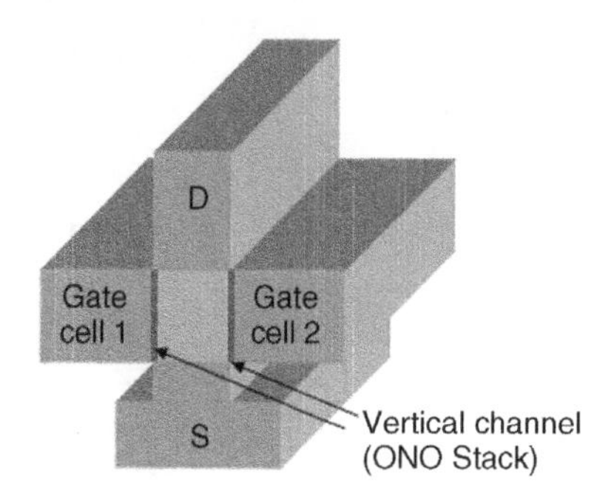

Figure 2.19 Schematic of vertical channel double floating gate sub–50 nm NOR flash cell. (Based on H. Cho *et al.*, (Stanford University, Intel), IEDM, December 2007 [16].)

because the drain and source were respectively above and below the vertical pillar channel, with the current flowing vertically, as shown in the schematic illustration in Figure 2.19 [16].

The cell that was characterized had L_g of 200 nm, silicon fin/channel thickness of 80 nm, and oxide thickness (T_{ox}) of 10 nm. Endurance characteristics were similar to those of a planar floating gate flash cell. The thin-body vertical channel with symmetric double gates was capable of multibit operation due to a charge interference effect between the two cells sharing a single thin body. This effect could be used to store up to three different voltage levels in the same cell. For single-bit storage per cell, this effect could be minimized by controlling the back gate bias during programming. The device was scalable because the vertical gate length dimension was eliminated from the critical path. The architecture was staggered so that two adjacent cells were not controlled by the same word-lines and bit-lines. A 3D schematic of the staggered cell array is shown in Figure 2.20 [16].

2.3.2 *Early Double- and Tri-Gate FinFET Charge-Trapping Flash Memories*

The advantages of the FinFET architecture in making 10 nm feature size double-gate and tri-gate CT flash memory devices were discussed in December of 2007 by CNR-IMM,

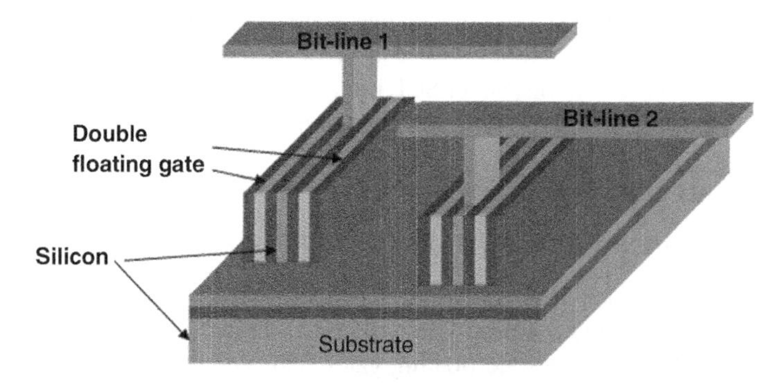

Figure 2.20 3D double-gate vertical channel staggered architecture sub–50 nm flash memory. (Based on H. Cho *et al.*, (Stanford University, Intel), IEDM, December 2007 [16].)

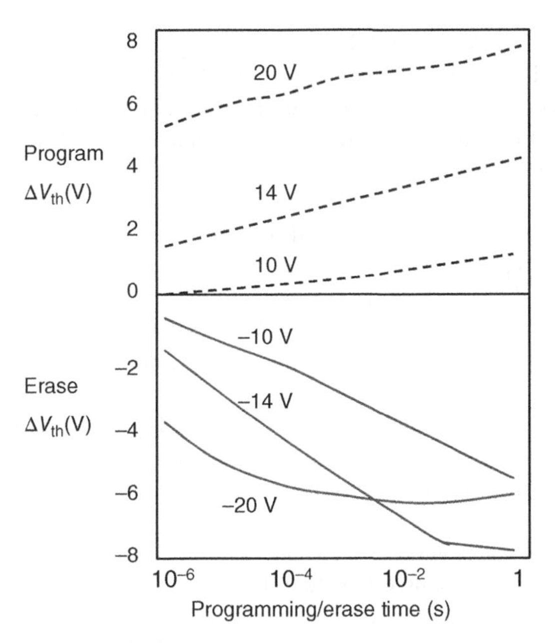

Figure 2.21 FinFET CT flash cell (a) program; and (b) erase curves using FN tunneling. (S. Lombardo *et al.*, (CNR-IMM, STMicroelectronics, CEA-LETI), IEDM, December 2007 [17].)

STMicroelectronics, and CEA-LETI [17]. Performance advantages in P/E characteristics and reliability were demonstrated for the FinFET type of architecture. Issues for scaled floating gate flash, such as stress-induced leakage current (SILC) defects and parasitic capacitive coupling, were suppressed in the CT FinFET architecture. The double-gate and tri-gate structures showed significant improvement in V_{th} window due to the distribution of the electric field in the gate dielectric stack. This permitted a thicker tunnel oxide, which resulted in better data retention.

These FinFET flash devices were made on SOI wafers using deep ultraviolet (DUV) lithography except for the shortest devices, which were made using e-beam lithography. Both 5–8 nm thick silicon nitride stacks and 10–15 nm thick ONO stacks with embedded silicon nanocrystals were used as CT layers. Devices with channel lengths of 30 nm and channel widths of 10 nm were fabricated and showed high curvature in the dielectrics at the fin corners. The ON current at $V_g - V_{th} = 1.5$ V and $V_{ds} = 1$ V was greater than 20 μA for the double-gate devices and did not depend on fin height because it was primarily affected by the fin corners.

Program and erase curves by Fowler-Nordheim tunneling in the CT FinFET flash cell with an ONO of 4.5 nm (bottom oxide)/3.2 nm(N)/8 nm(top oxide) are shown in Figure 2.21 [17]. Erase saturation occurred only at very high voltages, which indicates effective suppression of electron back-tunneling from the gate. Endurance testing showed a Vth window of about 3 V remained after 10 years at 250 °C.

2.3.3 Double-Gate Dopant-Segregated Schottky Barrier CT FinFET Flash

A NAND flash memory with a dopant-segregated Schottky barrier (DSSB) FinFET CT cell was discussed in January of 2009 by KAIST, ETRI, and the National Nanofab Center Korea [18]. The device had a nickel-silicided source and drain and was integrated in a FinFET with 30–50 nm fin width. While the CT memories normally show programming time in the range of microseconds, this DSSB flash device had a programming time on the order of nanoseconds with low programming voltage. To program the device, hot electrons were triggered by sharp band bending at the dopant-segregated source/drain region. A V_{th} shift of 4.5 V was achieved in 100 ns.

The same authors reported further in March of 2009 on the DSSB FinFET flash memory for NOR flash applications [19]. The sharp DSSB contact on the source side could generate hot electrons and be used to provide high injection efficiency at low voltage without constraint on the choice of gate and drain voltage. The DSSB FinFET devices could be made on the same SOI substrate as a conventional FinFET SONOS device; however, the process flow for the DSSB device included formation of gate spacers and the dopant-segregated silicided source and drain.

Hot electrons had different injection points for the DSSB flash memory and for the conventional flash memory device for CHEI programming bias of $V_{gs} > 0$ and $V_{ds} > 0$. For the conventional flash memory the hot electrons were generated near the drain side, while for the DSSB flash there was an abrupt band bending near the source-side region, which provided a high lateral field to generate source-side hot electrons even at low voltage and a high vertical field due to the large gate-to-source potential difference. This is illustrated in Figure 2.22, which shows the generation and injection point of hot electrons in each case [19].

In June of 2009, the same authors discussed a DSSB CT NOR flash memory in the form of a vertical double gate [20]. They found that source-side injection caused by sharp energy band bending in this device resulted in fast programming, giving a V_{th} shift of 4.2 V in 320 ns with bias $V_{gs}/V_{ds} = 7\,V/3\,V$. With a narrow fin width, a faster program was achieved for the DSSB device than for a conventional device due to low parasitic resistance and better gate control. With a fin width of 30 nm, the parasitic resistance of the DSSB part was three times lower than that of a conventional device. The programmed cell was shown to be free of drain disturbance characteristics. A schematic of the DSSB CT NOR flash cell with vertical double gate is shown in Figure 2.23 [20].

The double gates used DSSB source-side injection into the two ONO layers for programming. This is illustrated from a top-down perspective in Figure 2.24 [20]. This schematic of the NOR flash cell shows the high-efficiency DSSB source-side injection programming to both

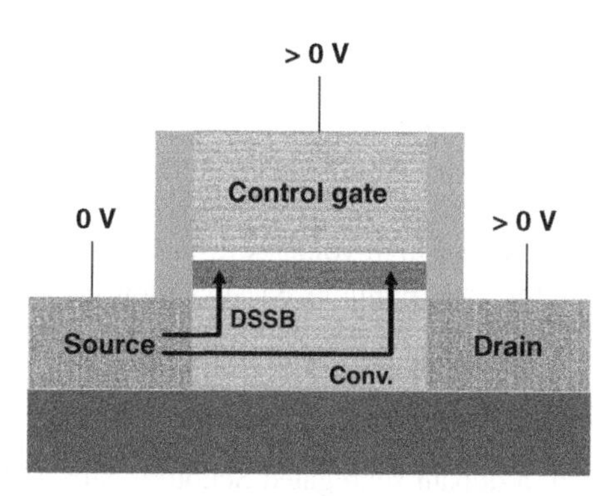

Figure 2.22 Generation and injection point of hot electrons for DSSB and conventional CHEI programming. (Based on S.J. Choi *et al.* (2009) *IEEE Electron Device Letters*, 30(3), 265 [19].)

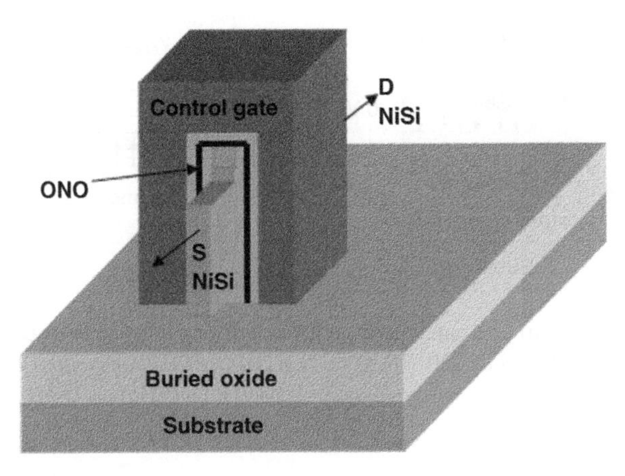

Figure 2.23 Schematic diagram of DSSB CT flash device with double vertical gate. (Based on S.J. Choi *et al.*, (KAIST), VLSI Technology Symposium, June 2009 [20].)

gates compared with the low-efficiency drain-side injection of the conventional CHEI process [20].

In August of 2010, the same authors discussed a p-channel DSSB NAND flash memory based on an SOI FinFET structure that had the potential for bit-by-bit operation by using a symmetric P/E operation [21]. This mechanism was based on programming using injected holes into the nitride charge storage layer due to Fowler-Nordheim tunneling enhanced by sharpened energy band bending at the DSSB source and drain junctions. Tunneled electrons from the silicon channel to the nitride storage layer were used as an erase method. A 4 V threshold voltage window and post 10^3 cycle data retention at 85 °C of about 3.5 V were found in a P/E time of 3.2 μs with 12 V erase and −14 V program. The P/E operations are illustrated in the schematic diagrams in Figure 2.25 [21].

The programming speed of this p-channel device was degraded compared to the n-channel DSSB FinFET CT device. However, it was comparable to that of the n-channel conventional CT device, and the erase speed was enhanced by the FN tunneling from channel to CT site.

A possible NAND-type flash architecture to implement the p-channel DSSB FinFET CT array is shown in Figure 2.26 [21]. For inhibited program, the source/drain voltage ($V_{s/d}$) of −6 V is used to reduce the vertical electric field. For inhibited erase, $V_{s/d} = 6$ V is induced.

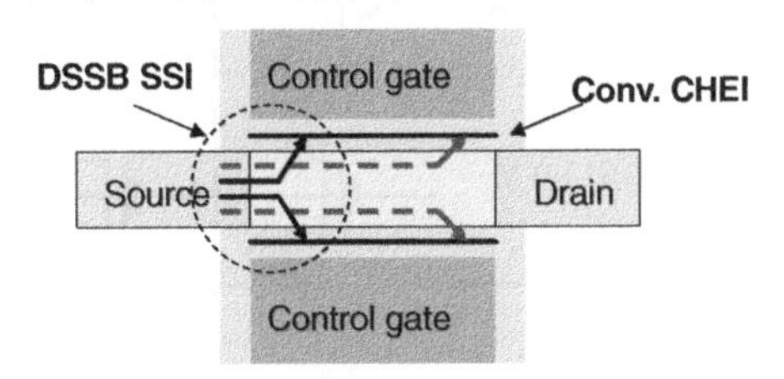

Figure 2.24 Top-down view schematic of NOR flash cell showing DSSB source-side injection. (Based on S.J. Choi *et al.*, (KAIST), VLSI Technology Symposium, June 2009, [20].)

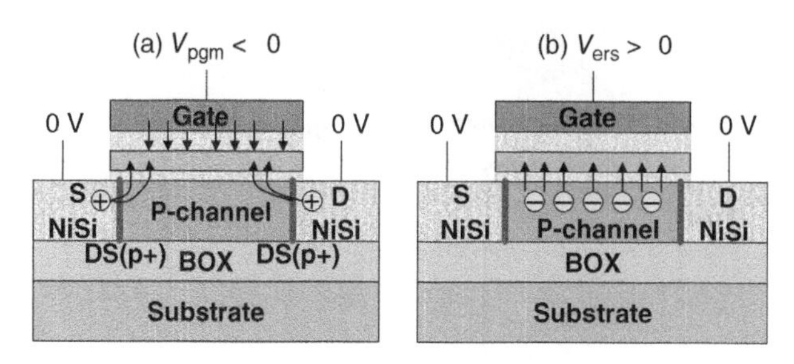

Figure 2.25 Illustration of (a) program; and (b) erase operations for p-channel DSSB FinFET NAND flash cell. (Based on S.J. Choi *et al.* (2010) (KAIST, ETRI), *IEEE Transactions on Electron Devices*, 57(8), 137 [21].)

Program voltage is −14 V, and erase voltage is 12 V. Using this configuration, the cells can be efficiently programmed or erased bit by bit.

2.3.4 Independent Double-Gate FinFET CT Flash Memory

A FinFET CT flash memory with independent double gates that stored four physical bits was discussed in August of 2009 by Samsung and Seoul National University [22]. The device had sufficient V_{th} window to also store multilevel bits. It had two sidewall gates that shared one

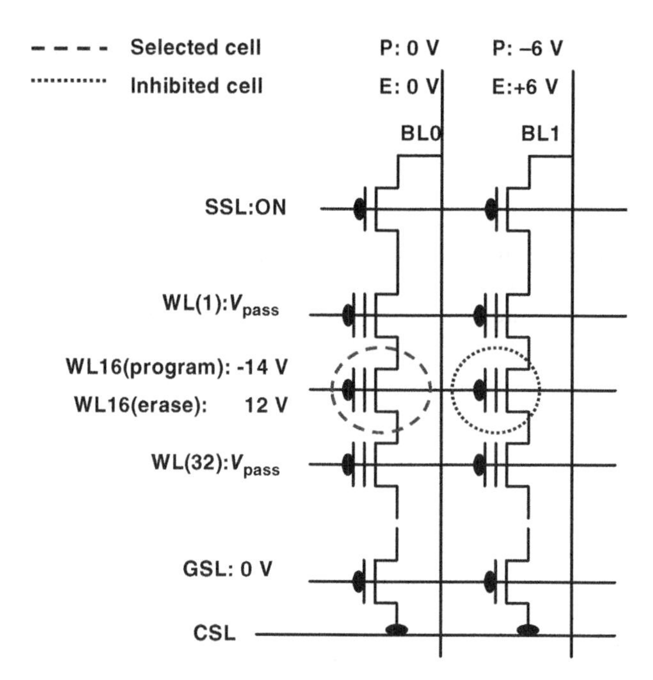

Figure 2.26 Possible configuration for program/erase in a p-channel DSSB FinFET NAND flash array. (Based on S.J. Choi *et al.* (2010) (KAIST, ETRI), *IEEE Transactions on Electron Devices*, 57(8), 137 [21].)

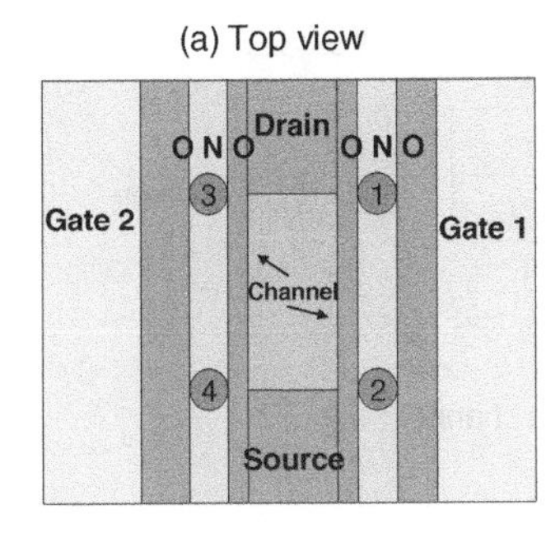

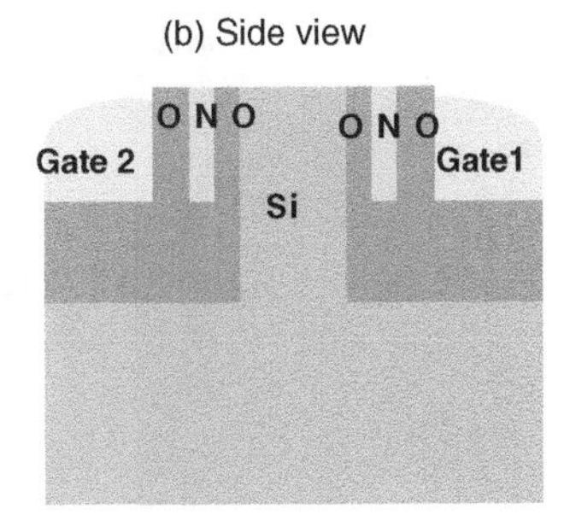

Figure 2.27 Independent double-gate Fin CT flash showing four stored bits. (Based on J.G. Yun *et al.* (2009) (Samsung, Seoul National University), *IEEE Transactions on Electron Devices*, 56(8) [22].)

silicon fin channel. The device was free from floating body effects, as the channel was directly connected to the body in the 3D fin structure. Sidewall spacer patterning was used to make the narrow silicon fin width. The two gates were on each side of the fin and controlled the two side channels. The silicon body was doped with boron. The process was CMOS compatible. Program was by CHEI, and erase was by BB HHI. For write, $V_{g1} = 7$ V and $V_d = 8$ V were used for 10 µs. For erase, $V_{g1} = -7$ V and $V_d = 8$ V were used for 10 µs. Reverse read was used to program the multiple bits with $V_d = 15$ V (or $V_s = 1.5$ V). Figure 2.27 shows (a) a top view and (b) a side view of the cell [22].

The cell suffered from paired-cell interference (PCI), as the fin was narrowed in the 3D devices. PCI is defined as the injected charges on one side of the channel influencing the stored bit on the opposite side of the channel. This interference from the electrons stored in the bit across the channel indicates that the bits facing across the silicon fin are not independent of each other but are electrically paired or coupled. This effect becomes more pronounced as the fin becomes thinner. The PCI can be reduced by making the fin thicker or possibly by inserting low-κ dielectrics or an air gap in the middle of the fin.

2.4 Thin-Film Transistor (TFT) Nonvolatile Memory with Polysilicon Channels

2.4.1 Independent Double-Gate Memory with TFT and Polysilicon Channels

A sub–50 nm vertically stacked independent double-gate (IDG) TFT CT nonvolatile memory with up to 64 cells in series strings was discussed in December of 2008 by Schiltron Corporation [23]. This dual-gate MOS transistor cell string was intended to replace the conventional NAND flash transistor string in TFT CT memory technology. A schematic cross-section of the conventional NAND flash string is shown in Figure 2.28(a). A challenge in the NAND flash string is to be sure that all pass cells are conducting when a single cell is being read. Because the individual states of the cells are not known, a read-pass voltage must be higher than the highest possible programmed V_{th}. Another challenge is to be sure a sufficiently large worst-case string current is used when all cells are in the highest programmed state and the selected cell being read is erased. This also requires a high read-pass voltage. A similar analysis can be used for program-pass. Another challenge is to be sure that programming never results in a V_{th} above the read-pass voltage.

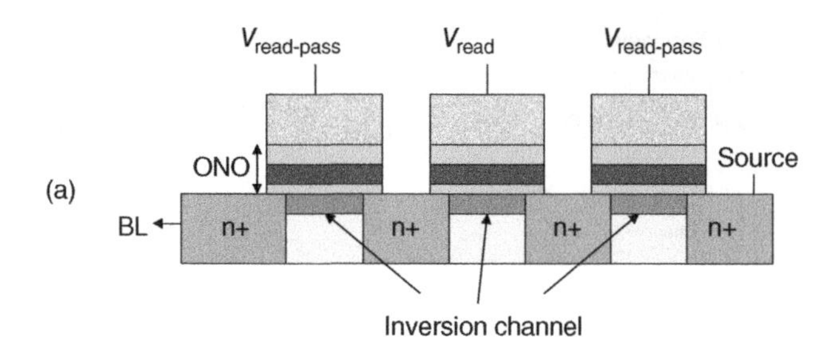

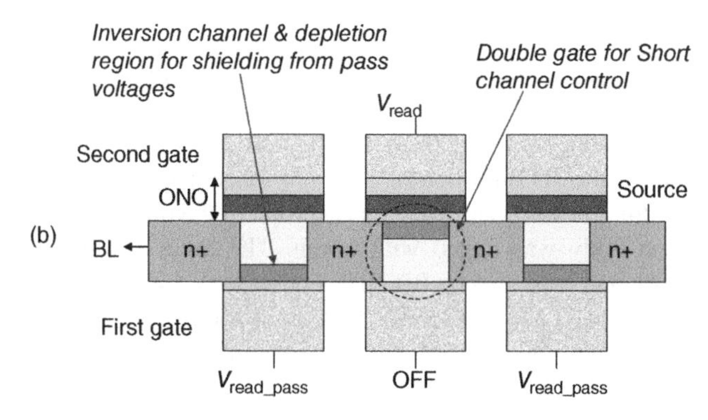

Figure 2.28 Illustration of read in NAND flash string: (a) schematic of conventional NAND flash cell string in TFT CT technology; and (b) schematic of double-gate TFT NAND flash CT cell. (Based on A.J. Walker, (Schiltron), IEDM, December 2008 [23].)

A double-gate TFT CT flash memory string intended to solve these issues was discussed. The basic concept of the double-gate TFT CT flash memory is shown in Figure 2.28(b). Read- and program-pass disturbs were effectively shielded, and good endurance and data retention were shown. The inversion channel and depletion region were shown to provide shielding of memory charge from pass voltages. The double-gate structure permitted electrostatic interaction between top and bottom devices, which resulted in good short-channel control. An ONO stack was used for the tunneling oxide for the second gate and a single tunnel oxide for the first gate.

The TFTs were combined in a series of strings of up to 64 cells. These devices had $L_g = 48$ nm, $W_g = 45$ nm, and 35 nm channel thickness. Target applications were in sub–50 nm storage class memory intended to replace NAND flash.

The read characteristics of a vertically stacked IDG polysilicon nanowire flash memory were discussed further in November of 2011 by NCTU [24]. The device was characterized with tunneling oxide on one gate and with an ONO stack as the charge-storage medium in the other gate. Because of the IDG configuration, the shift in the transfer characteristics due to a change in the amount of stored charge could be sensed using two different modes. In these modes, a sweeping bias was applied to one gate and a fixed bias to the other gate. It was shown that a larger memory window is obtained when the gate on the ONO side is used as the driving gate.

The memory window when the ONO side gate is used as the driving gate was found to be independent of the bias applied to the control gate. Due to this result, a new flash structure using

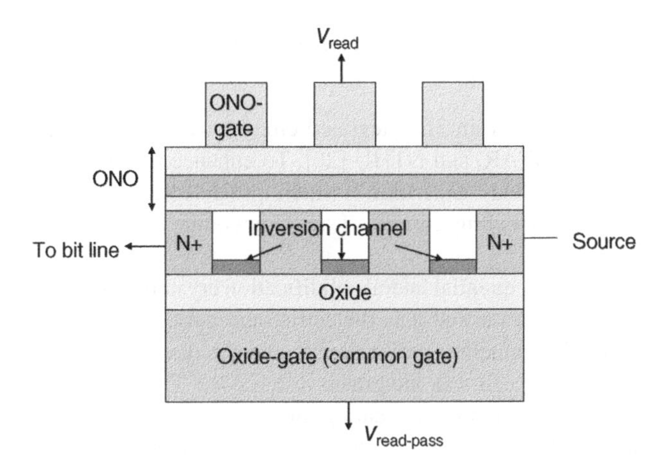

Figure 2.29 Read operation of new flash structure using IDG cells with common control gate. (Based on H.C. Lin *et al.* (2011) (National Chiao Tung University), *IEEE Transactions on Electron Devices*, 58(11) [24].)

IDG cells with a common control gate on the oxide side was proposed. This structure is shown in Figure 2.29 [24]. The common control gate was shown to simplify the device fabrication, which reduced the cost of manufacturing.

A CT flash device using both the nanowire concept and IDGs was discussed in June of 2010 by NCTU and the National Nano Device Labs of Taiwan [25]. A 3D diagram of the device is shown in Figure 2.30 [25].

The double-gate output current performance enhancement was investigated. The back-gate bias effect was used to probe the impact on programming efficiency of the separate gates. Reduced nanowire thickness was found to lead to stronger back-gate effects. The IDG nanowire device showed operational advantages over single-gated devices. For a very thin channel where strong gate coupling existed, an improved programming speed was achieved by applying an appropriate control gate bias [25].

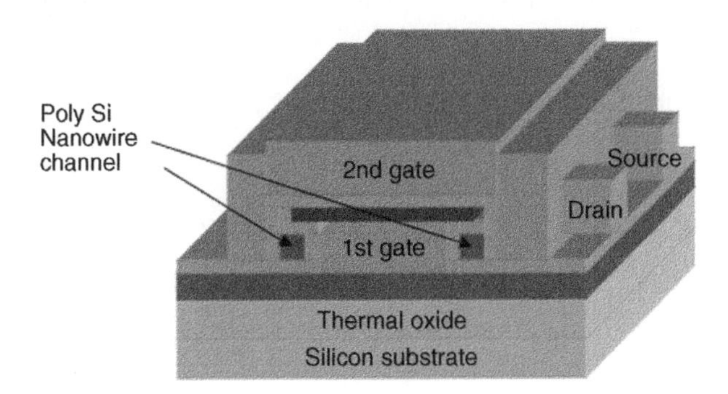

Figure 2.30 3D diagram of CT flash memory using nanowire and independent double gates. (Based on W.C. Chen, (National Chiao Tung University, National Nano Device Laboratories), VLSI Technology Symposium, June 2010, [25].)

2.4.2 TFT Polysilicon Channel NV Memory Using Silicon Protrusions to Enhance Performance

A TFT NAND flash cell for use in 3D integrated circuits was discussed in June of 2007 by researchers from ITRI, A*STAR, and NTHU [26]. To enhance performance, the device used silicon protrusions, also called "asperities," which projected into the thick tunnel oxide and provided a path for tunneling carriers. Low-temperature polycrystalline silicon TFTs were used.

The TFT CT device used sequential lateral solidification crystallization with a fully depleted polysilicon channel, an ONO stacked gate dielectric, and a metal gate. The cell used silicon protrusions to enhance the tunneling performance, and the device showed good nonvolatile memory characteristics. This cell was integrated into a 3×3 TFT p-type NAND array using sequential lateral solidification laser annealing of a 50 nm thick amorphous silicon film deposited by plasma-enhanced CVD and was furnace-annealed on a glass substrate. The height of the silicon protrusions were about 80 nm at the main grain boundaries. Spacing and height of the silicon protrusions could be controlled by varying the thickness of the amorphous silicon. The ONO gate dielectric used 15 nm thick tunneling oxide and 25 nm thick nitride. Blocking oxide was 30 nm. Dimensions of the TFT devices were $3 \times 3\,\mu m^2$. Good cycling endurance and data retention was reported with negligible program and read disturbance. A schematic cross-section of the TFT nonvolatile cell that highlights a silicon protrusion is shown in Figure 2.31 [26].

If $V_{read} = -3$ V and $V_{pass} = -6$ V, then the read current of the TFT cells at a $V_g = -0.5$ V for a string of three is greater than 1 μA. Unselected cells have a small gate disturb effect due to the isolated body and severe drain disturbance due to the strong electric field between the word-line and bit-line, which reduces the cell V_{th}. The reliability was found sufficient for the application.

2.4.3 An Improved Polysilicon Channel TFT for Vertical Transistor NAND Flash

Making an improved polysilicon channel in a vertical transistor NAND flash memory was discussed in April of 2011 by Samsung [27]. TFTs were used with very thin 7.7 to 18.5 nm polysilicon to investigate the properties of the polysilicon material. The TFT charge-transfer

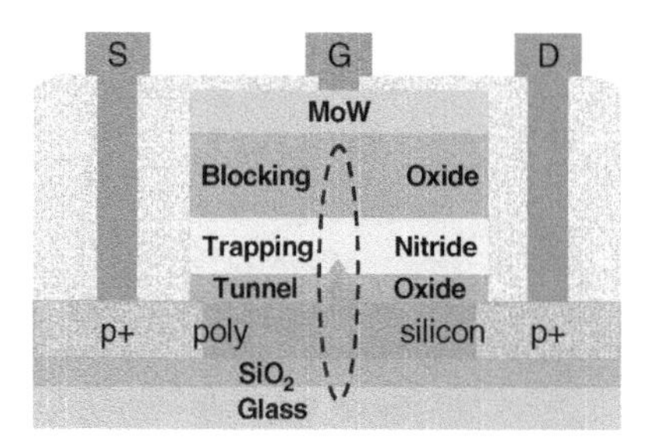

Figure 2.31 Schematic cross-section of TFT nonvolatile cell highlighting a silicon protrusion. (Based on H.T. Chen (2007) (ITRI, STAR-NTHU), *IEEE Electron Device Letters*, 28(6) [26].)

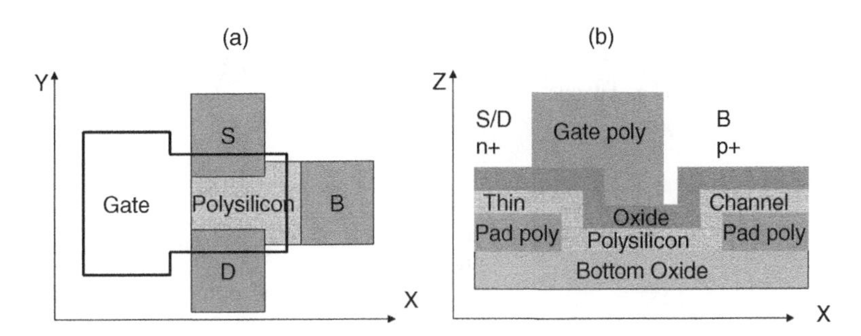

Figure 2.32 Test TFT with thin polysilicon architecture: (a) top-down view (x–y); and (b) schematic cross-section (x–z). (Based on B. Kim *et al.*, (Samsung), IRPS, April 2011 [27].)

characteristics, including ON current and effective mobility, were found to be dominated not by the thickness of the polysilicon but by the grain size of the polysilicon channel. As the channel thickness was decreased while maintaining the grain size, the subthreshold TFT characteristics were improved without degradation of ON current and reliability properties.

The vertical transistor NAND used a thinner polysilicon channel to improve its controllability. The electrical properties of a <20.0 nm TFT for vertical NAND flash were discussed. A top-down view and schematic cross-section of the experimental TFT with the thin polysilicon channel are shown in Figure 2.32 [27]. The gate and pad polysilicon are patterned to minimize the external resistance of the polysilicon channel.

A 50 nm thick amorphous silicon (a-Si) was deposited by LP-CVD on 100 nm wet oxide. Undoped a-Si was deposited by LP-CVD for a channel layer followed by solid-phase crystallization. The active regions were defined, and an 8 nm SiO_2 layer was deposited as gate dielectric. Another a-Si layer was then deposited and patterned for the gate dielectric. Ion implants were added to source, drain, and body. Grain structure of the polysilicon channel was analyzed using electron back-scattered diffraction. The grain size was found to decrease with the decrease in polysilicon thickness.

The electrical properties of the thin polysilicon TFT were investigated. Measurements showed that the electrical property is predominantly affected by grain size not by polysilicon thickness. This means that increasing grain size is important for large ON current and improved reliability in polysilicon TFT.

2.4.4 Polysilicon TFT CT Memory with Vacuum Tunneling and Al_2O_3 Blocking Oxide

A polysilicon TFT-based TiN–alumina-nitride–vacuum-silicon (TANVAS) CT memory with high-κ blocking oxide and vacuum tunneling layer with a field-enhanced nanowire (FEN) was discussed in August of 2011 by National Chiao Tung University [28]. The vacuum layer had a very low dielectric constant (κ) and was intended to replace the tunnel oxide layer. Due to the large κ-value difference between the blocking and tunneling layers, this material stack had significant field enhancement across the tunneling layer, which improved program and erase efficiency. Because defect creation was suppressed in the tunneling layer, it also showed superior retention. This TFT CT memory is thought

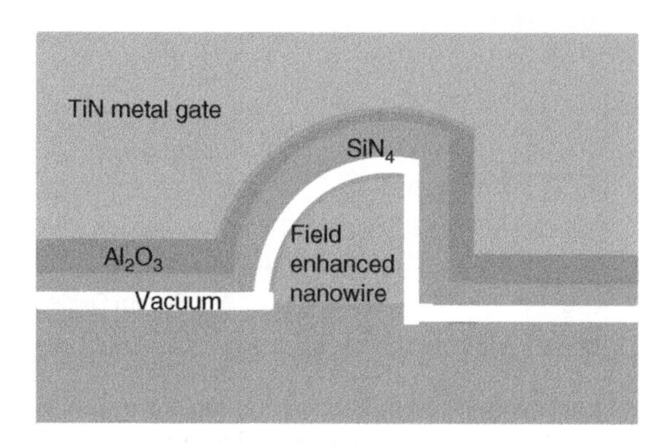

Figure 2.33 Cross-section of FEN with vacuum tunneling layer. (Based on C.Y. Wu *et al.* (2011) (National Chiao Tung University), *IEEE Electron Device Letters*, 32(8), 1095 [28].)

promising for 3D flash applications. A cross-section of the FEN with the vacuum tunneling layer is shown in Figure 2.33 [28].

The FEN structure was made by growing thermal SiO_2 on a single-crystal silicon wafer. An etch-stop layer of Si_3N_4 and a sacrificial layer of TEOS SiO_2 were sequentially deposited using LPCVD. The sacrificial SiO_2 layer was etched using RIE, and a layer of thick amorphous silicon film was deposited. The FEN structure was formed by etching the TEOS strips using diluted HF. A 3 nm TEOS SiO_2 film and an 11 nm Si_3N_4 film were deposited as tunneling and CT layers. An 11 nm Al_2O_3 film was deposited as the blocking layer. Finally, a TiN metal gate was deposited by PVD. Following the gate patterning, the TiN, Al_2O_3, and Si_3N_4 layers were etched by RIE, leaving the 3 nm TEOS SiO_2 film on the polysilicon nanowires. After implantation, a layer of Si_3N_4 was deposited and then etched back by RIE to form spacers that wrap around the Al_2O_3 blocking layer. The tunneling oxide was then side-etched off with diluted buffered oxide etch and a passivation oxide deposited to form a vacuum tunneling layer.

Due to the vacuum tunneling, the TANVAS device showed a larger V_{th} shift of 3.78 V in 10 ms compared to 2.62 V for TANOS, both at 13 V. The improvement was due to the local field enhancement of the tunneling layer as the κ-ratio was increased for the blocking/tunneling layers. The improved retention was thought to be due to the vacuum tunneling layer, which formed no defects under cycling.

2.4.5 Graphene Channel NV Memory with Al_2O_3–HfO_x–Al_2O_3 Storage Layer

In January of 2012, Seoul University discussed the functionality of graphene-based NV memory devices using a single-layer graphene channel and an Al_2O_3–HfO_x–Al_2O_3 charge storage layer. The impact of the gate material work function (WF) on the memory characteristics was studied using different types of metals including Ti with WF 4.3 eV and Ni with WF 5.2 eV. The ambipolar carrier conduction of graphene caused an enlargement of the memory window, which was 4.5 V for a Ti gate and 9.1 V for a Ni gate. The increase in memory window is attributed to the change in the flat-band condition and to suppression of electron back-injection in the gate stack [29].

2.5 Double-Gate Vertical Channel Flash Memory with Engineered Tunnel Layer

2.5.1 Double-Gate Vertical Single-Crystal Silicon Channel with Engineered Tunnel Layer

In June of 2010, Seoul National University discussed a high-density double-gate flash memory that used a single-crystal silicon channel and a bandgap-engineered (BE) ONONO tunneling layer. For high density, two NAND cells in the string are put together with a common vertical channel as shown in Figure 2.34 [30]. This provides a half-level reduction in footprint.

With the vertical channel, the channel length was controlled by anisotropic dry etch, which suppressed short-channel effects and increased the sensing margin. A BE layer of ONO replaced the tunnel oxide. The fin structures for the bit-line and word-line direction were formed by sidewall spacer patterning. After the fin pillar was formed, the ONONO multilayer stack was deposited.

Read operation was performed using mirror symmetrical biasing. Due to the very thin channel, there was a potential for leakage current to flow between n+ junctions at the bottom of the silicon pillar by punch through, but a deep channel implant suppressed the leakage. The array is organized as a cross-point array. A top-down view of a conventional single-gate cross-point array memory with $4F^2$ area is shown in Figure 2.35(a), and a top-down view of the double half-gate BE flash with $2F^2$ area cell is shown in Figure 2.35(b) [30].

A charge trap folded NAND flash memory with body-tied single-crystal silicon channel and BE storage layer was discussed further in January of 2011 by Seoul National University and Stanford University [31]. In this case, two memory cells are put together with a common vertical pillar silicon channel. This enabled implementation in a 30 nm technology. The array was made by folding the conventional 2D flash memory and is designated a folded NAND (FNAND). The stack structure used ONO to replace the tunnel oxide. The fin structures and the word-line and bit-line were formed by sidewall spacer patterning instead of photolithography. A schematic illustration of the charge trap folded NAND flash memory is shown in Figure 2.36 [31].

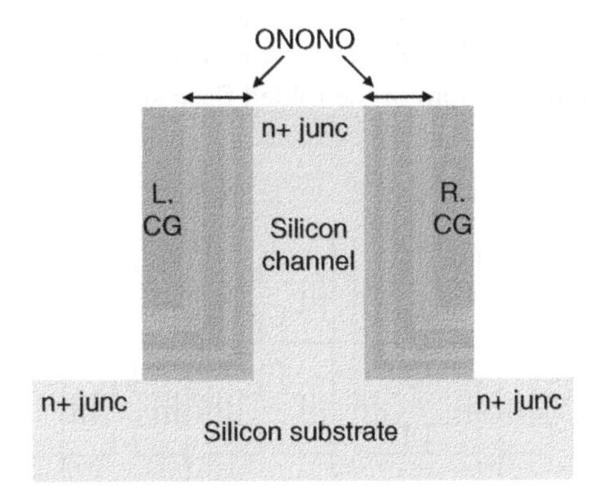

Figure 2.34 Vertical channel BE CT NAND flash. (Based on S. Cho *et al.*, (Seoul National University), Device Research Conference, June 2010 [30].)

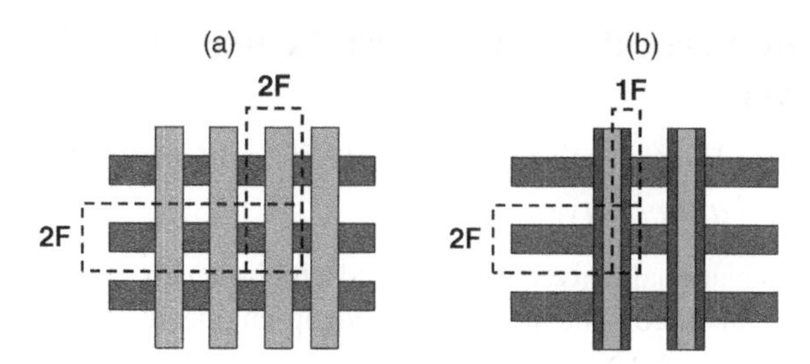

Figure 2.35 Top view of a (a) single-gate; and (b) double-gate BE flash laid out as $4F^2$ cross-point array. (Based on S. Cho *et al.*, (Seoul National University), Device Research Conference, June 2010 [30].)

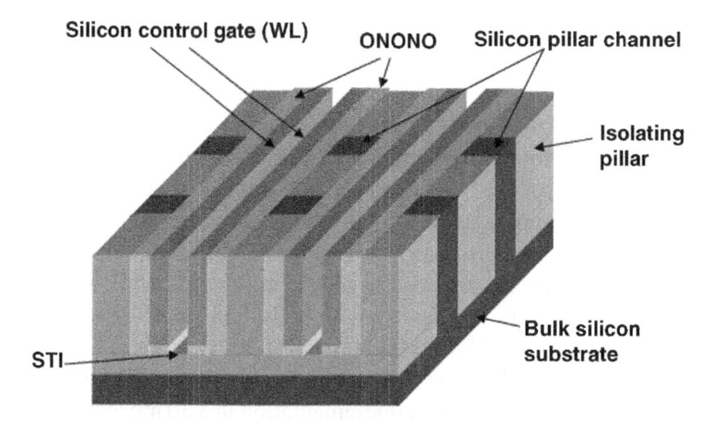

Figure 2.36 Charge trap folded NAND flash memory. (Based on S. Cho *et al.*, (Seoul National University, Stanford University), *IEEE Electron Device Letters*, 58(2), 288 [31].)

Advantages of the folded NAND flash are as follows: sidewall control gates improve the array density, short channel effects can be suppressed, and there is a larger sensing margin because device channel length is determined by the fin height along the word-line direction. In addition, the device is stackable because the channel can be formed by consecutive deposition of silicon channels. Two cells were folded to have a common pillar channel. Fine-line patterning used sidewall spacers. The T_{fin}/T_{pillar} was 80 nm/80 nm. The device showed 2-bit operation. A schematic circuit diagram of the folded flash string is shown in Figure 2.37 [31].

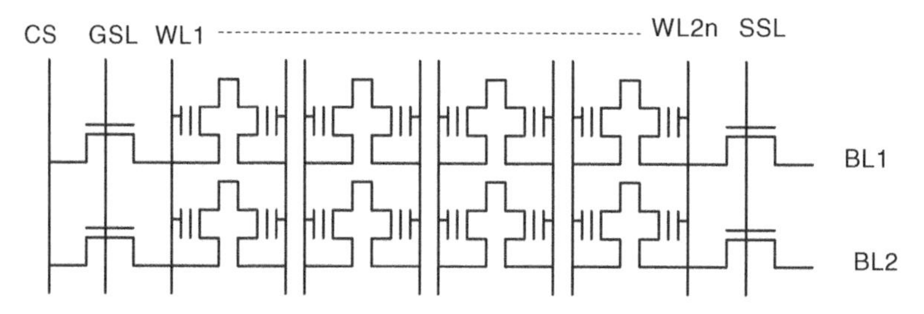

Figure 2.37 Schematic circuit diagram of folded flash string. (Based on S. Cho *et al.* (2011) (Seoul National University, Stanford University), *IEEE Transactions on Electron Devices*, 58(2), 288 [31].)

A body-tied FinFET BE NAND flash device was shown by Macronix in May of 2007 [32]. The FinFET structure used a 30 nm single-crystal fin width. The FinFET CT technology with a BE tunnel layer had better retention and erase speed than a floating gate FinFET memory technology that suffered from a low gate-coupling ratio and severe floating gate disturb between adjacent memory cells. The CT FinFET memory was immune to the floating gate coupling issues and therefore scaled well. It could also operate with as few as 500 electrons without erratic behavior and was easier to fabricate on a FinFET transistor than a floating gate technology. A BE FinFET CT memory can use a very thin tunnel barrier consisting of a thin ONO layer along with a thicker nitride storage layer. This thin ONO tunnel barrier permitted fast hot hole erase and good data retention. The BE FinFET CT memory had fast P/E speeds and good data retention as well as scaling to below the 30 nm node with good interference immunity [32].

2.5.2 Polysilicon Substrate TFT CT NAND with Engineered Tunnel Layer

A polysilicon TFT CT flash memory with an engineered tunnel barrier was discussed in February of 2012 by Kwangwoon University [33]. The device was made for system-on-panel applications. The device was processed on a glass substrate using low-temperature processes. An amorphous silicon film on the glass substrate was crystallized using excimer laser annealing. The engineered tunnel barrier was ONO with a high-κ HfO_2 CT layer and an Al_2O_3 blocking layer as shown in the schematic of the gate stack in Figure 2.38 [33].

The polysilicon TFT CT flash with an engineered tunnel barrier showed good memory characteristics such as an initial memory window of 9.5 V and an extrapolated window after 10 years of 3.96 V. Program was at 23 V and 100 ms. Erase was at −26 V and 100 ms. Endurance after 10^3 cycles showed the window reduced from 7.2 to 4.4 V.

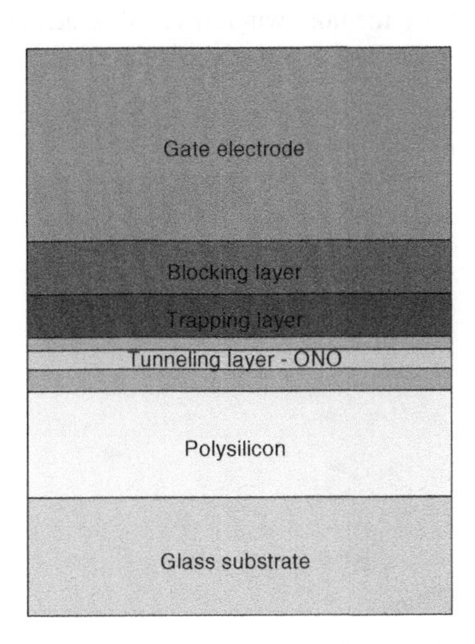

Figure 2.38 Stack of polysilicon TFT CT flash with engineered tunnel barrier. (Based on H.W. You and W.J. Cho (2012) (Kwangwoon University), *IEEE Electron Device Letters*, 33(2), 170 [33].)

2.5.3 *Polysilicon Channel Double-Layer Stacked TFT NAND Bandgap-Engineered Flash*

A double-layer TFT NAND flash memory that used a fully depleted polycrystalline channel and a BE tunneling layer was discussed in December of 2006 by Macronix [34]. A tri-gate P+ polysilicon gate was integrated into the NAND array. The array had a bottom and a top layer and was intended for 3D multilayer stacking. A cross-sectional view of the double-layer stacking in the channel length direction of the double-layer TFT NAND array is shown in Figure 2.39 [34]. The bottom layer showed no sign of reliability degradation with respect to the top layer. This showed the potential for stacking of further layers. A many-layer stacked vertical channel array using this technology is discussed further in Chapter 4 [34].

The TFT device has a 60 nm thick polysilicon channel that was deposited as an amorphous silicon layer by low-pressure chemical vapor deposition (LP-CVD) followed by a low-temperature thermal anneal at 600 °C for crystallization. A multilayer ONO tunneling dielectric was used for easy hole tunneling during erase while eliminating direct tunneling leakage during retention. Following deposition of the SiN trapping layer and top blocking oxide, a heavily doped P+ poly gate was used to suppress gate injection during Fowler-Nordheim erase. P/E cycling on this early device showed degradation after 1 K cycles. The read current was enhanced by increasing V_{pass}. For $V_{pass} = 7$ V and $V_{read} = 1$ V, the read current for a 16-cell NAND string array was about 1 μA.

The BE tunnel layer was introduced by Macronix in December of 2005 for use as a nontrapping tunneling dielectric [35]. They designated it "BE-SONOS." A very thin ONO tunnel layer provided a modulated tunnel barrier that suppressed direct tunneling at low electric field during data retention but permitted efficient hole tunneling erase in a high electric field. The <2 nm nitride had negligible CT, which made it suitable for use as a tunnel barrier. This technology has hole tunneling erase but is immune to the retention issues of conventional SONOS. By using a P+ poly gate, a depletion mode memory device with $V_{th} < 0$, such as that used in a NAND string, and >6 V memory window could be achieved. A NOR flash could be made by using an N+ poly gate. The Fowler-Nordheim tunneling program attains a >7 V

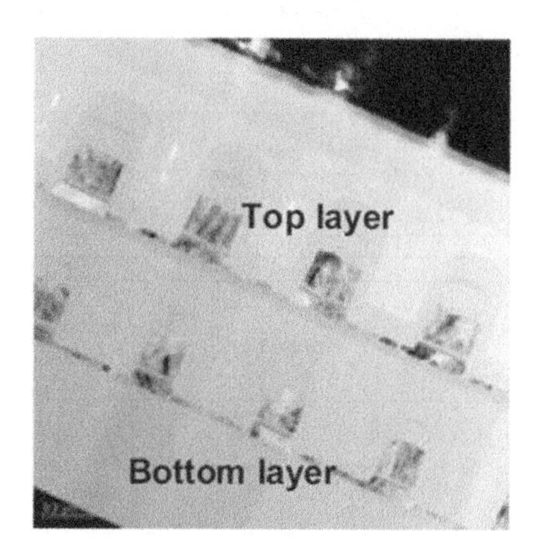

Figure 2.39 Cross-section of TFT NAND double-layer stacking in the channel length direction. (E.K. Lai *et al.*, (Macronix), IEDM, December 2006, with permission of IEEE [34].)

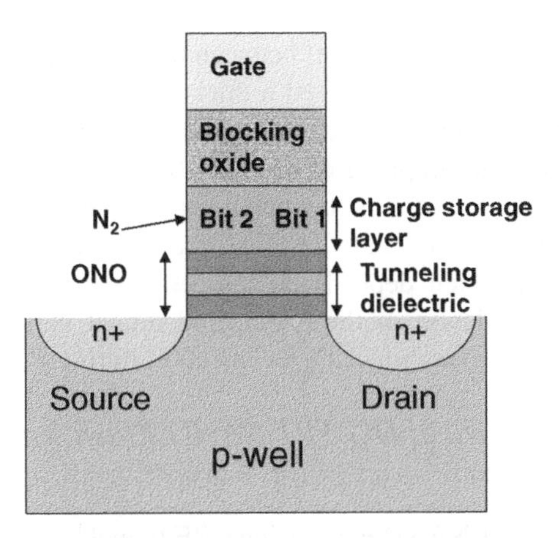

Figure 2.40 Schematic cross-section of the BE tunnel NAND cell. (Based on H.T. Lue *et al.*, (Macronix), IEDM, December 2005 [35].)

memory window in <10 ms. A schematic cross-section of a BE tunnel layer device is shown in Figure 2.40 [35]. The <2 nm ONO tunnel dielectric provides an efficient hole tunneling barrier for erase without requiring a high-κ gate dielectric to avoid back-tunneling. The thicker CT layer was 7 nm, and a 9 nm SiO_2 layer was the blocking oxide.

The concept for BE tunnel technology is shown in the following energy band diagrams. At high electric field during erase, the band offset reduces the hole tunneling barrier so it is just that of the O_1 layer as shown in Figure 2.41(a). During data retention, the full O_1–N_1–O_2 barrier stack prohibits both electron detrapping and hole back-tunneling as shown in Figure 2.41(b).

The reliability of the BE tunnel technology was discussed further in April of 2007 by Macronix [36]. The $O_1N_1O_2$ tunnel layer acts as a BE tunneling barrier that provides efficient hole tunneling for erase but eliminates direct tunneling leakage. It overcomes the early

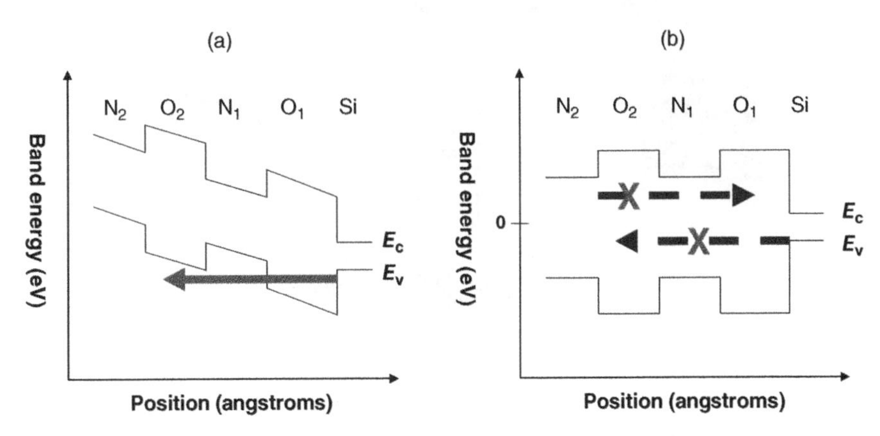

Figure 2.41 Bandgap engineering concept for SONOS: (a) band offset occurs at high electric field; and (b) no band offset occurs in data retention mode. (Based on H.T. Lue *et al.*, (Macronix), IEDM, December 2005 [35].)

limitation of SONOS, which was the trade-off between fast erase speed due to a thin tunneling oxide and good data retention, which required a thicker tunneling oxide.

The nontrapping behavior of the ONO tunnel barrier was verified under constant current stress and showed no significant trapping even after 1000 s of continuous stress. The I_g–V_g curves of the ONO barrier at various temperatures showed the current insensitive to temperature. This supports the fact that the measured current is tunneling current rather than Frenkel-Poole current, which depends on temperature. A 150 °C bake retention test showed no appreciable degradation of the engineered tunnel barrier after 10^3 hours. The heavily doped P+ polysilicon gate reduced gate injection during Fowler-Nordheim erase.

2.5.4 Eight-Layer 3D Vertical DG TFT NAND Flash with Junctionless Buried Channel

An eight-layer vertical gate TFT NAND Flash using a BE tunnel layer and a junctionless buried channel was discussed in June of 2010 by Macronix [37]. The array was made on a 75 nm half-pitch. A buried channel n-type well was used to improve the read current of the TFT NAND and to permit junction-free structure. In this layout the conventional word-lines and bit-lines were grouped into horizontal stacked planes. The bit-line contact was replaced by the SSL. The intercept of the three planes defined by the word-line, bit-line, and SSL determines the memory cell. Figure 2.42 shows a scanning electron microscope (SEM) image of two of the eight stacked flash memory devices [37]. Each device is a double-gate TFT with BE tunnel layer. The channels are all n-type polysilicon without additional junction implantation. The poly-silicon channel thickness is 18 nm, and the buried oxide that isolates the channel in the Z direction is 40 nm. The SSL and GSL devices have longer channel lengths [37].

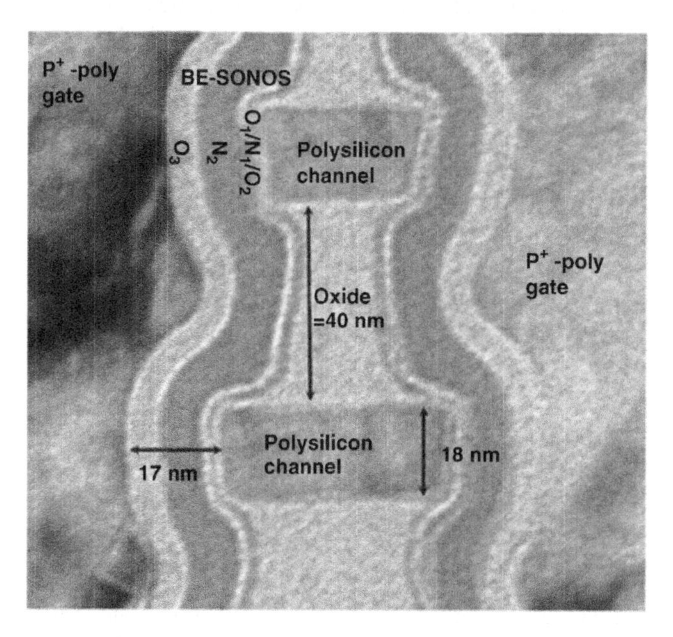

Figure 2.42 Illustration of two of eight stacked double-gate flash memory devices. (H.T. Lue *et al.*, (Macronix), VLSI Technology Symposium, June 2010 [37], with permission of IEEE.)

Interference between cells in the vertical Z direction was studied. This Z interference is a result of the fringing field induced by the potential change, because the body is floating in the 3D vertical gate NAND array. These stacked vertical gate arrays are discussed in more detail in Chapter 4.

2.5.5 *Variability in Polysilicon TFT for 3D Vertical Gate NAND Flash*

Variability in the polysilicon TFT used for 3D vertical gate NAND flash was discussed by Macronix in December of 2012 [38]. A 37.5 nm half-pitch vertical gate NAND flash was fabricated, and the random grain boundary effect was modeled. This effect is due to the grain boundary creating interface states that result in large local band bending. This band bending creates a surface potential barrier. Two effects were defined: the gate-induced grain barrier lowering (GIGBL) effect and the drain-induced grain barrier lowering (DIGBL) effect.

The DIGBL effect is a physical mechanism that affects subthreshold behavior. It was found that narrower bit-lines and larger word-line critical dimensions are critical parameters for providing a tight V_{th} distribution and large memory window. An asymmetry of reverse read and forward read was found for the TFT device. The physical mechanism for this asymmetry was explained by the DIGBL effect. It was found that this asymmetry could be used to determine lateral grain barrier trap location and the interface trap density [38].

2.6 Stacked Gated Twin-Bit (SGTB) CT Flash

A stacked gated twin-bit (SGTB) CT flash memory made in a trench was discussed in March of 2012 by Seoul National University, Stanford, and Samsung [39]. A gated twin-bit memory with a cut-off gate and two memory nodes for a single word-line was described.

Polysilicon gates were then stacked vertically resulting in high density with a $4F^2$ cell size. $2N$ memory nodes are integrated where N is the number of stacked gates. The fabrication and electrical characteristics were discussed. A schematic cross-section of the gated twin-bit memory array is shown in Figure 2.43 [39]. The CT layer is wrapped around the control gate so that there are two physical bits for each cell.

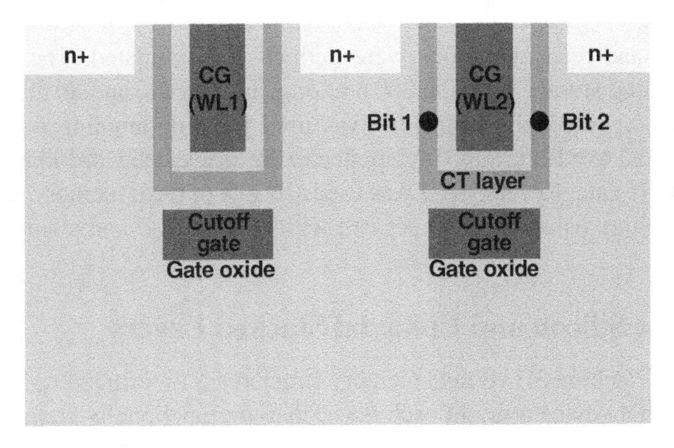

Figure 2.43 Schematic cross-section of gate twin-bit memory array in a trench. (Based on W.B. Shim *et al.*, (Seoul National University, Stanford University, Samsung), *IEEE Transactions on Nanotechnology*, March 2012 [39].)

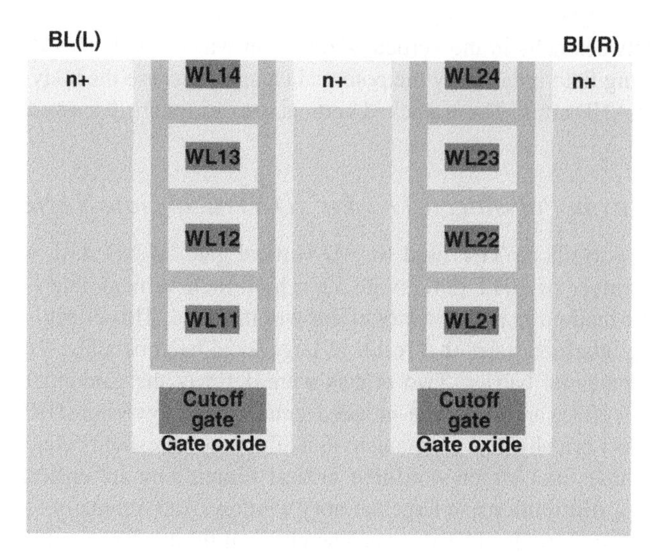

Figure 2.44 Schematic of stacked-gate twin-bit SONOS. (Based on W.B. Shim *et al.* (2012) (Seoul National University, Stanford University, Samsung), *IEEE Transactions on Nanotechnology*, 11(2), 307 [39].)

A cut-off gate is under each word-line. During programming, a low bias is applied to the cut-off gate so the channel potential on one side is not transferred to the other side. The channel potential can be 0 V on one side while the other side floats so that both sides can be programmed independently. A bidirectional read is used to read each cell [39].

To make a more highly integrated device, word-lines can be stacked in the vertical direction. Deep trenches can be formed to obtain sufficient height for multiple stacked gates. Dielectric is deposited and polysilicon gap filling and etch back are used to form a stacked twin-bit device.

Each word-line has two memory nodes operated independently by voltage-biasing the cut-off gate, which is similar to the approach used with the single-layer structure. This is illustrated in Figure 2.44 [39].

Program, erase, and read operations were defined and simulated. The simulated SGTB device had 40 nm long storage nodes, a 200 nm long cut-off gate, and 80 nm deep source and drain regions. Fowler-Nordheim tunneling was used for programming with a $V_{pgm} = 19$ V applied to the selected word-line and 10 V applied to the unselected word-lines. There were 0 V applied to the cut-off gate. Reverse read was used. 2*N* bits of flash memory are integrated in a $4F^2$ stacked technology using two bits per cell with reverse read and a single-crystal silicon channel [39].

2.7 Crystalline Silicon and Epitaxial Stacked Layers

Another method of stacking 3D vertical memory structures is to form multiple layers of single-crystal silicon, with a device array in each layer, that are functionally connected through the layers. This section explores this single-crystal silicon stacking technology, which was originally developed for stacking four of the six transistors of a static RAM then later used to stack two layers of conventional NAND flash.

2.7.1 *Stacked Crystalline Silicon Layer TFT for Six-Transistor SRAM Cell Technology*

A stacked static RAM (SRAM) technology that allowed the load PMOS and pass NMOS transistors to be stacked over the planar pull-down NMOS transistors in order to reduce the cell size was reported by Samsung in December of 2004 [40]. The technology was called the stacked single-crystal silicon thin-film transistor (S^3 TFT) technology. Using the S^3 TFT cell technology, it is possible to shrink the six-transistor (6T) SRAM memory cell to one-third of the size of the conventional planar 6T SRAM cell; however, the benefit is diminished if the periphery of the SRAM is not also reduced in size.

The S^3 technology was also used to stack the transistors of the peripheral CMOS logic, a task which required some 3D logic design organization. The peripheral logic transistors were stacked on the interlayer dielectric to save layout area, maximize array efficiency, and minimize the chip size. The resulting SRAM cell was 0.16 μm^2 in area in 80 nm design rules. A circuit schematic of a 6T SRAM cell showing the two common nodes is illustrated in Figure 2.45 [40].

In the double-stacked SRAM cell, every critical pattern layer consists of two transistors and is simple and nearly repeated on each layer. The word-lines, which are the gates of the pass transistors, are connected throughout the whole row of the cell array [40].

The node contacts are vertically connected so that local interconnect layers for cross-coupling of nodes and gates are not needed, because all nodes and gates are connected through just one node contact hole that can be aligned vertically through all three layers. This is illustrated in Figure 2.46, which shows the three transistors on one side of the SRAM and the single contact node that connects them to each other and to the gates of the load and pull-down transistors on the other side of the SRAM cell [40].

In this double-stacked cell, every critical pattern layer is simple and overlays well, resulting in lithographic regularity, which maximizes the resolution. Lithography was done using 193 nm ArF. The top-down view SEM image of the four layers in Figure 2.47 shows the regularity [40]. The SEM image of the active region for pull down is shown in Figure 2.47(a). The channel silicon and gate for the load PMOS is shown in Figure 2.47(b). The top-level pass transistors of the double-stacked cell are shown in Figure 2.47(c) along with the word-lines,

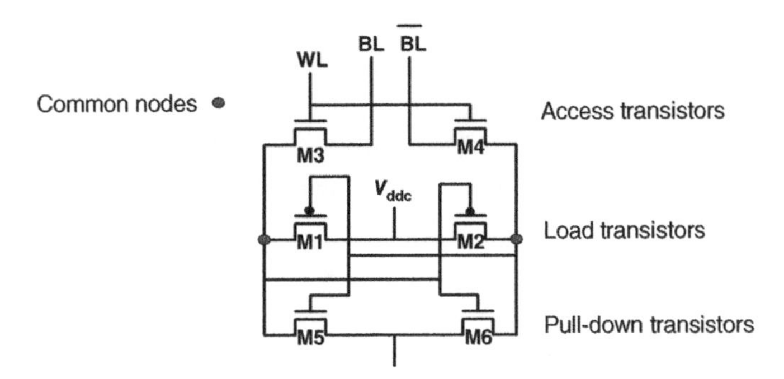

Figure 2.45 Circuit schematic of a six-transistor SRAM showing common nodes. (Based on S.M. Jung *et al.*, (Samsung), VLSI Technology Symposium, June 2004 [40].)

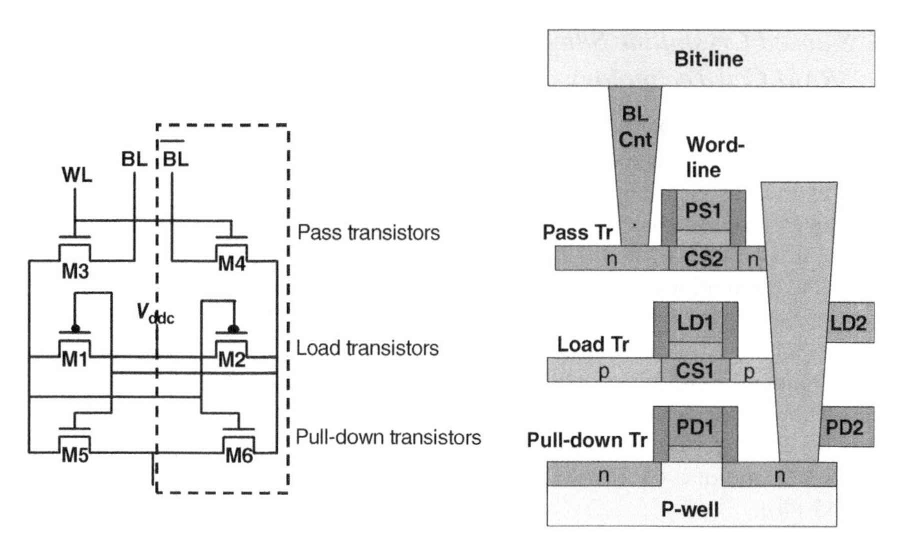

Figure 2.46 Illustration of vertical connections in two-tiered SRAM (a) schematic circuit diagram of SRAM cell; and (b) schematic cross-section of one-half of SRAM cell. (Based on S.M. Jung *et al.*, (Samsung), VLSI Technology Symposium, June 2004 [40].)

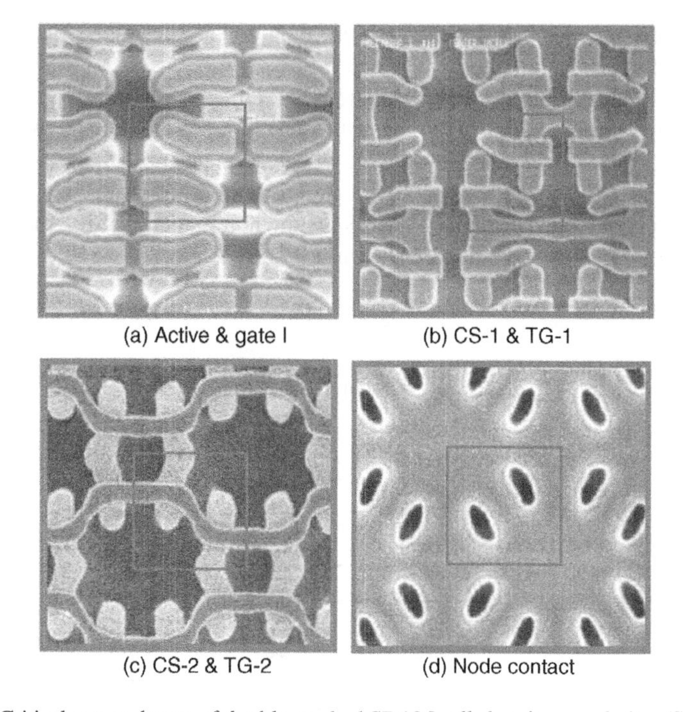

Figure 2.47 Critical pattern layers of double-stacked SRAM cell showing regularity. (S.M. Jung *et al.*, (Samsung), VLSI Technology Symposium, June 2004, with permission of IEEE [40].)

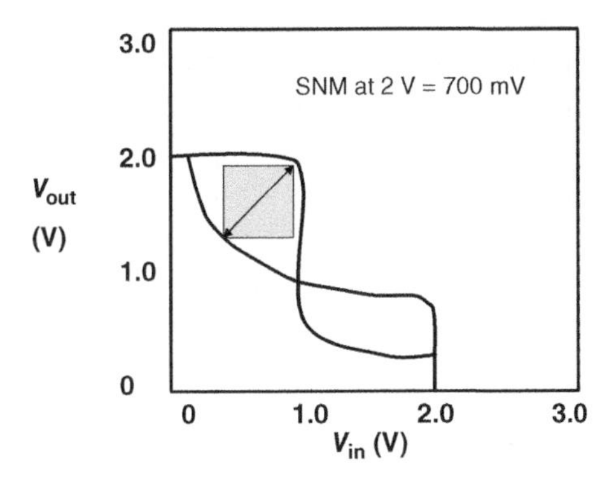

Figure 2.48 Static noise margin (butterfly) curves of double-stacked S^3 SRAM cell. (Based on S.M. Jung *et al.*, (Samsung), VLSI Technology Symposium, June 2004 [40].)

which are gates of the pass transistors, and are waved and connected through the entire row of the cell. Figure 2.47(d) shows the vertical node contacts.

The V_{th} of the pass transistor is 0.6 V, which is lower than that of a bulk transistor with the same I_{off} due to the fully depleted channel. DIBL was 100 mV/V and subthreshold swing was comparable to that of a bulk transistor. The V_{th} and off leakage of the pass transistor was well controlled. The static noise margin curve of the double-stacked S^3 SRAM cell is shown in the butterfly curves in Figure 2.48 [40].

A 65 nm stacked S^3 high-speed SRAM made with double-stacked cell transistors (SSTFT) with 0.16 µm^2 (25 F^2) cell size was discussed by Samsung in June of 2005. [41] This cell also stacked the load PMOS and the pass transistor NMOS of the SRAM cell over the planar pull-down transistor. A sketch of the SEM of a cross-sectional image of the stacked SRAM cell array and periphery is shown in Figure 2.49 [41].

When originally reported, this cell was intended for low-power applications, and the technology did not use salicide and high-performance transistors. In this study high-performance CMOS transistors with 65 nm gate length and 1.6 nm gate oxide were used along with

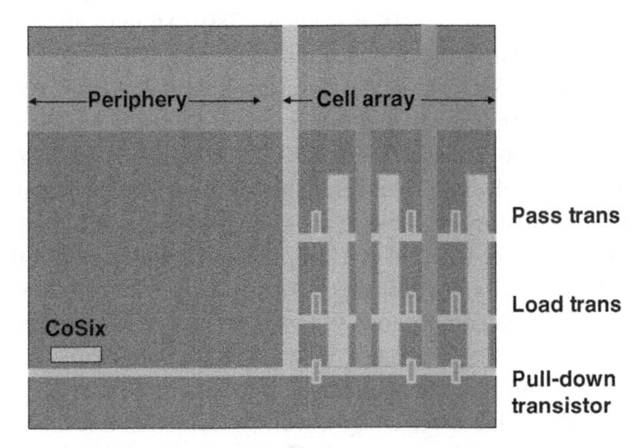

Figure 2.49 Sketch of SEM image of six-transistor SRAM cell with the pass and load transistors stacked over the pull-down transistors. (Based on S.M. Jung *et al.*, VLSI Technology Symposium, June 2005 [41].)

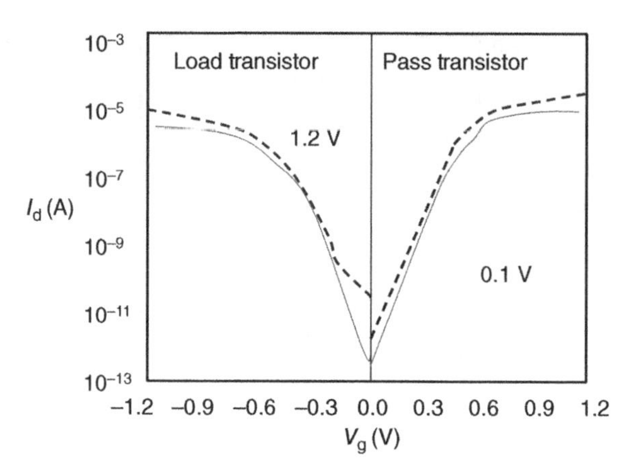

Figure 2.50 I_d-V_g characteristics of stacked pass NMOS and PMOS transistors. (Based on S.M. Jung *et al.*, VLSI Technology Symposium, June 2005 [41].)

low-resistance CVD Co for the contact holes. The cell uses 80 nm patterning with 193 nm ArF lithography. It was targeted at high-density embedded SRAM applications.

To form the cell transistor stack, three repeats of the transistor formation process were required. This included thermal oxidation, silicon film formation, and activation of dopants. The low thermal silicon film was annealed with a spike rapid thermal anneal (RTA) to form single-crystal silicon. To reduce the thermal budget of this process, low thermal gate oxide layers 1.6 nm thick were grown by plasma oxidation at 400 °C. Other stacked processes were also below 650 °C. The contacts of the cell were connected by etching vertically through the active silicon film of the stacked pass and load transistors. A novel CVD Co deposition technique was used to reduce the contact resistance and fill the deep, small contact holes. The I_d–V_g characteristics of the stacked pass NMOS and stacked load PMOS transistors are shown in Figure 2.50. V_{th} of the pass NMOS is 0.38 V, and V_{th} of the load PMOS is 0.35 V [41].

Electrical characteristics are similar to those of the planar bulk transistor because the channel silicon is single-crystal film. The static noise margin is 282 mA at $V_{cc} = 1.2$ V. Beta is >2.0. Drive currents, at $V_{dd} = 1.2$ V and $I_{off} = 100$ nA/μm, are 780 μA/μm for the NMOS and 340 μA/μm for the PMOS. Operational frequency is >500MHz [41].

A 500 MHz DDR 72Mb 3D SRAM made with laser-induced epitaxial crystal silicon growth on amorphous dielectric silicon dioxide was discussed by Samsung in February of 2010 [42]. The cell size was 36 F^2 or 0.36 μm^2 in 100 nm CMOS technology. The SRAM cell was three layers of TFT transistors. The electrical characteristics of the stacked n-channel pass transistors and the p-channel load transistor are close to those of the planar bulk transistors.

A critical process was the laser-induced crystallization technique used for the stacked silicon layers, which was called lateral epitaxial silicon growth (LEG) or selective epitaxial growth (SEG). This was done by melting amorphous silicon films and solidifying and crystallizing the molten silicon film from the single-crystal silicon seeds grown from the silicon substrate through contact holes by selective epitaxial silicon deposition. This produced perfect single-crystal silicon films on the oxide layer using a well-proven silicon technology.

The laser crystallization technique uses SEG from single-crystal silicon through the seeding process. This is illustrated in Figure 2.51 [42]. First, seed contact holes are made

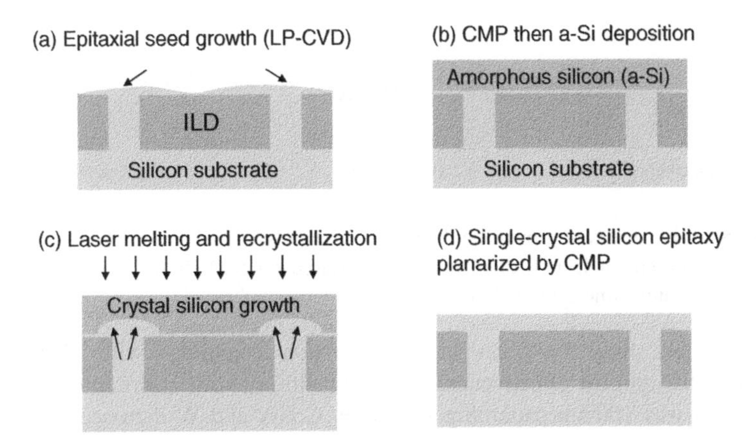

Figure 2.51 Laser crystallization by selective epitaxial growth from seed silicon. (Based on S.M. Jung *et al.* (2010) (Samsung), *IEEE Transactions on Electron Devices*, 57(2), 474 [42].)

through the ILD layer using lithographic patterning of holes at regular distances as shown in Figure 2.51(a). Epitaxial crystal silicon is vertically grown in the seeding contact holes with LP-CVD. When the silicon reaches the top of the contact holes, it grows laterally along the oxide surface in all directions. This film has many irregularities, so it is polished smooth with a chemical-mechanical-planarization (CMP) process. An LP-CVD amorphous silicon film is then deposited on the seed silicon, and the ILD oxide layers as shown in Figure 2.51 (b). This surface is irradiated by an Nd:YAG laser with a wavelength of 532 nm and pulse duration of 150 ns to melt the a-Si layer, as shown in Figure 2.51(c). The heat is conducted through the seed contact holes, and the crystallization process spreads from the seed silicon through the amorphous silicon, producing a single-crystal silicon film epitaxially grown from single-crystal seed silicon by melting and recrystallization by laser irradiation, as shown in Figure 2.51(d).

Using this technology, a double-stacked 3D SRAM cell was made by stacking the p-channel TFT load transistors and the NMOS TFT access transistors over the NMOS pull-down transistors, which were in the base silicon. The 80 to 100 F^2 6T planar SRAM cell area was reduced to 36 F^2 in 100 nm design rules, resulting in a 0.36 μm^2 cell size. The critical pattern layers were simple and nearly repeated in each layer. The node contacts were vertically contacted. Cross-coupled layers were not needed because the nodes and gate were connected though a single node contact hole that was vertically aligned through the entire structure. Because the silicon film and transistor formation process had to be repeated twice, it was critical to minimize the heat budget to avoid degrading the SRAM cell transistor characteristics. Low-temperature plasma gate oxidation, low-temperature thin-film deposition, and spike rapid thermal anneal were used for this reason.

2.7.2 *Stacked Silicon Layer S^3 Process for Production SRAM*

The double-stacked SRAM technology made with a production 100 nm process and a 0.36 μm^2 cell size was discussed in June of 2007 by Samsung [43]. Previously this development process was targeted at high- density 288Mb embedded SRAMs made with ArF lithography at the

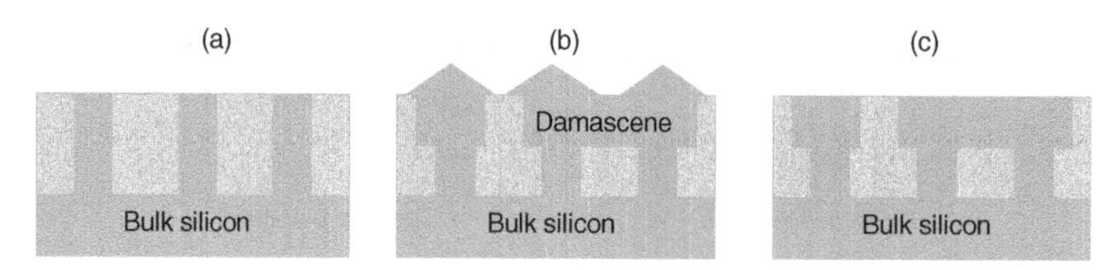

Figure 2.52 Selective epitaxial growth on damascene channels: (a) seed contact growth; (b) epitaxial growth in damascene channel; and (c) planarization. (Based on S.M. Jung *et al.*, (Samsung), VLSI Technology Symposium, June 2007 [43].)

65 nm node. In this study the target was a production technology with 72Mb embedded SRAMs with focus on cost effectiveness and compatibility with conventional CMOS. The lithography used was 100 nm patterning with KrF. CoSix and W damascene interconnects were used.

Selective epitaxial growth (SEG) was used for stacking the silicon layers on the interlayer dielectric (ILD). Seed contacts formed on the bulk silicon were followed by damascene patterning to be filled with the silicon epitaxial layers grown from bulk silicon in the seed contacts. The epitaxial layer grows initially in the seed holes then spreads laterally at the top of the seed contact. It is then planarized with CMP as shown in Figure 2.52 [43].

Low thermal budget was maintained by using low-temperature plasma gate oxidation, low thermal thin-film deposition, and spike rapid thermal anneal. All process temperatures were below 650 °C. Co salicide was used in the periphery only, and a tungsten shunt word-line scheme was used. Tungsten (W) shunt layers were formed on the gates of the cell pass transistors using damascene patterning, and W CMP was used to reduce the resistance of the word-line instead of silicide gate structures. A vertical SEM image cross-section of the two silicon layers grown laterally from seed contacts in the S^3 SRAM cell and the tungsten shunt word-line is shown in Figure 2.53 [43].

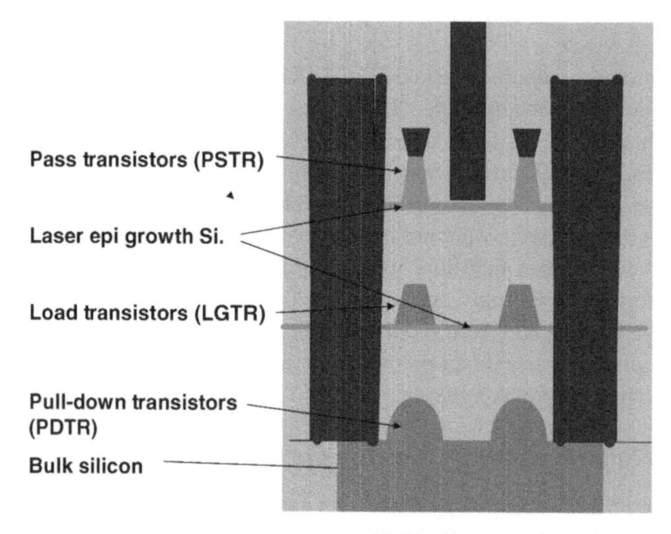

Figure 2.53 Sketch of vertical SEM image of double S^3 SRAM with tungsten shunt word-line. (based on S.M. Jung *et al.*, (Samsung), VLSI Technology Symposium, June 2007 [43].)

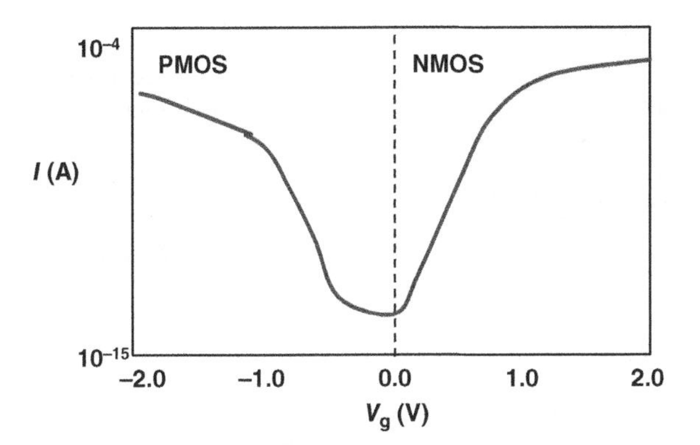

Figure 2.54 Illustration of I–V curve of PFET and NFET with $V_d = 2.0$ V showing SOI-type quality. (Based on Y.H. Son *et al.*, (Samsung), VLSI Technology Symposium, June 2007 [44].)

Drive currents at 1.2 V and $I_{off} = 30$ nA were 720 μA/μm for NMOS and 320 μA/μm for PMOS.

Electrical characteristics of the stacked pass NMOS and load PMOS transistors are close to those of planar bulk transistors due to the single-crystal channel films. Noise margin at $V_{cc} = 1.2$ V is 282 mV.

A 6T stacked SRAM cell with cell size of 0.16 μm^2 using 193 nm lithography made on 100 nm CMOS substrates was discussed by Samsung in June of 2007 [44]. LEG is used to form a single-crystal silicon layer over oxide so that the pass gate NMOS and load PMOS transistors could be stacked over oxide.

With the LEG process, the energy density of the laser beam and the seed formation are key factors in determining the crystal quality of the silicon layer formed on oxide. High-density LEG SRAMs with stacked transistors worked with standby current of less than 0.3 μA/Mb. The stacked SRAMs are made with LEG silicon films.

The I–V characteristics of the PFET with LEG thickness of 20 nm and NFETs with LEG thickness of 40 nm measured at 85 °C and $V_d = 1.0$ V and $V_d = 2.0$ V were evaluated, and the LEG film was determined to be of SOI-compatible quality. An illustration is shown in Figure 2.54 [44].

A production type 72Mb double-stacked 500MHz separate input/output (I/O) SRAM with an automatic cell bias scheme and adaptive block redundancy was discussed by Samsung in February of 2008 [45]. To attain the density required for mobile applications, this SRAM is made in a double-stacked S^3 technology. The design used conventional 100 nm mass production technology, but the cell size was increased from 25 F^2 to 36 F^2. CoSix layers and a Tungsten shunt word-line are used for the high-speed operation. The design included an automatic cell bias for managing the SRAM cell current by controlling cell bias and used word-line pulse width regulation for adjusting word-line pulse-width according to cycle time. The 72Mb chip was organized as 2 M × 36, 4 M × 18, or 8 M × 9. Cycle time was 500MHz, and data rate was 72Gb/s DDR with separate I/O. Latency was 5 ns. Power supply was internal 1.2 V and external 1.8 V. The cell size was 0.36 mm^2, and chip size was 70.5 mm^2.

The design included an adaptive block redundancy to handle defects due to the new cell technology and to improve yield. The ratio of lump-type defects could have been as high as 30% because the cell had a complex vertical structure. An adaptive block redundancy was

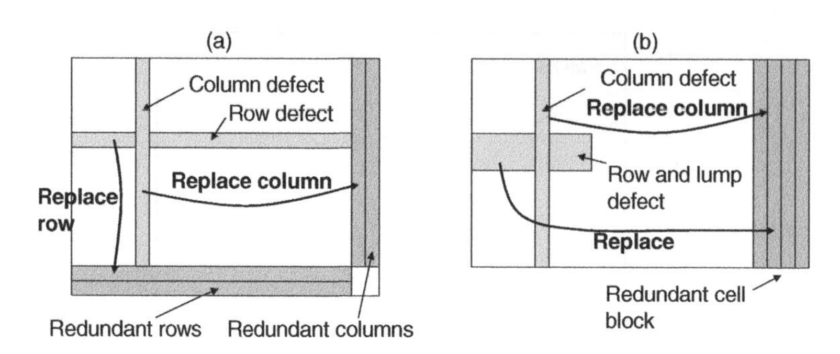

Figure 2.55 Adaptive block redundancy for handling double-stacked SRAM defects: (a) conventional row and column redundancy; and (b) adaptive block redundancy. (Based on K. Sohn *et al.*, (Samsung), ISSCC, February 2008 [45].)

designed to handle the lump-type defects and to improve repair efficiency and flexibility. The adaptive block redundancy added a redundant block to a group of blocks that shared the same I/Os and used that redundant block for replacing both row- and column-type defects. A row repair was treated as a column repair by switching the row address to a column and block address. Figure 2.55 shows a conventional row and column redundancy and the new adaptive block redundancy [45].

The advantages were the following: the ratio of redundant cells for row and column is adjusted easily, the shape of a group of cells for a single repair can be easily modified by address mapping, and the method is easy to implement because a normal cell block is used for the redundant cell array with a partial change of address inputs. This method was found to repair 30% more defects than conventional redundancy methods.

To determine the speed of each chip at the wafer level, a short cycle test (SCT) using BIST was designed. At the rising edge of the external clock, a high-frequency clock of four or eight pulses was generated. A pattern generator produced read and write control signals, addresses, and a write-data-pattern selector. Test results were merged in the data output as compressed data. If all data met the expected data, a signal went high. This test result could also be used for known-good die [45].

2.7.3 *NAND Flash Memory Development Using Double-Stacked S^3 Technology*

At the 30 nm technology node, the fabrication cost of a planar NAND flash device was expected to increase. This increase in cost had been delayed for a generation by using multilevel cell technology to increase the bit density without moving to a more expensive scaled technology. It could potentially be delayed further by using a 3D two-tier stacked technology, which would permit the cost to continue down about a 40% cost curve as shown in Figure 2.56 [46].

The S^3 technology was used for making a 3D NAND flash memory by Samsung in December of 2006 [46]. NAND cell arrays were formed on the ILD and on the bulk silicon to double the memory density without increasing the chip size. The feasibility of the technology was shown by an operational 32-bit NAND flash memory cell string in 63 nm dimensions. The flash was made using a TANOS CT cell technology. A new NAND cell operational method

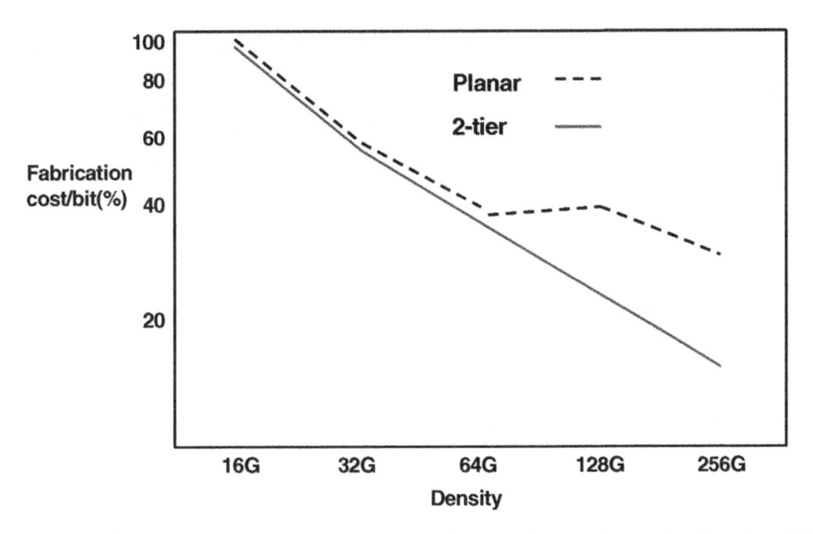

Figure 2.56 Fabrication cost per bit (%) vs. density of NAND flash technology. (Based on S.M. Jung, (Samsung), IEDM, December 2006 [46].)

called source-body tied (SBT) was developed to maximize the advantages of 3D stacked NAND cell structures [46].

The vertical structure of the 3D stacked NAND flash cell string is shown in Figure 2.57 [46]. The second active layer is perfect single-crystal like an SOI layer. The bit-line and common source line (CSL) are formed through the second active layer. The well bias is simultaneously applied to both layers by the CSL using the SBT technique.

A vertical SEM photograph of the fabricated 3D stacked NAND cell string is shown in Figure 2.58 [46]. Shown are the bit-line and CSL penetrating both active layers of single-crystal silicon.

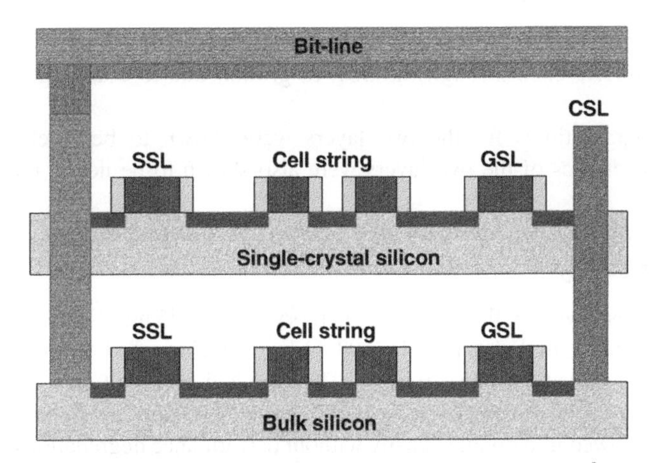

Figure 2.57 Vertical structure of the 3D stacked NAND flash cell string using S^3 technology. (Based on S.M. Jung, (Samsung), IEDM, December 2006 [46].)

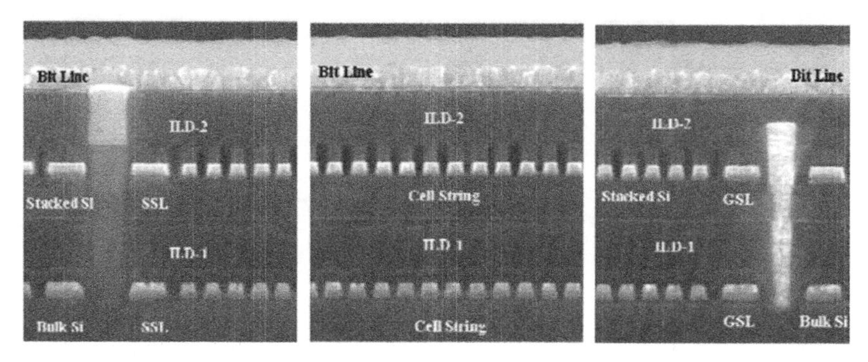

Figure 2.58 Vertical SEM cross-section of 3D NAND cell string. (S.M. Jung *et al.*, (Samsung), IEDM, December 2006 [46], with permission of IEEE.)

The first layer of cell strings was formed on the substrate layer. An ILD was formed on the first layer followed by an SOI single-crystal silicon layer. The upper layer cell strings were formed in the second silicon layer directly above the bottom layer cell strings. The memory cell uses a stack of TANOS CT technology for data storage. The bit-line and CSL lines were patterned simultaneously on both layers and etched through vertically from the upper silicon layer to the bottom active layer. The bit-line hole was filled first with N-doped polysilicon and then with tungsten.

Both cell strings were connected through a single contact hole to the same bit-line. The x-decoders of the upper and lower arrays were laid out separately at the end of the arrays. It required only three additional photolithographic layers to double the density of the NAND cell by stacking a second layer of cell arrays on the ILD. This design and process meant that the upper and lower cell strings had the same electrical characteristics [46].

Erase time was maintained by having the common source of the cell string tied with the body of the string. This permitted the cell strings to be erased by a conventional body bias erase method.

During the program operation, a negative bias could be applied at the SSL gate to reduce leakage current interference through the SSL to ground due to coupling of the grounded SSL gate.

Erase and program times for the two layers were shown to be essentially the same. Endurance characteristics of the two layers were also shown to be nearly the same.

2.7.4 4Gb NAND Flash Memory in 45 nm 3D Double-Stacked S^3 Technology

A 4Gb NAND flash with a floating gate memory cell in 45 nm CMOS technology was described by Samsung in February of 2008 [47]. The chip used 3D double-stacked S^3 technology with a shared bit-line structure. This resulted in a cell size of 0.0021 μm^2/bit per unit feature area. It had a silicon layer dedicated decoder and layer-compensated control schemes, which allowed it to double density without performance degradation compared to the planar NAND flash. Device characteristics included read and cycle access times of 40 μs, typical program time of 800 μs, erase time of 2 ms, and power supply ranging from 2.5 to 3.6 V.

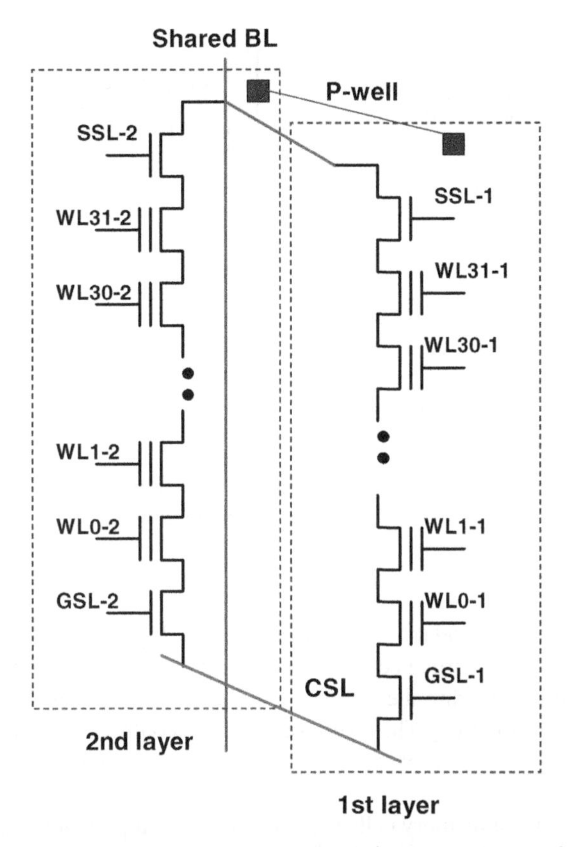

Figure 2.59 Schematic circuit diagram of two NAND chains in two layers of a 3D double-stacked NAND flash. (Based on K.T. Park *et al.*, (Samsung), ISSCC, February 2008 [47].)

The chip used two stacked silicon layers each containing 2Gb of multilevel cell memory array. Each array consisted of 1024 blocks with a page size of (2kB + 64). A 32-cell string with two dummy cells was used with a block size of 256kB. Two word-line decoders were placed with one serving each layer to reduce program and read disturb characteristics and also to reduce word-line loading. A 2kB page buffer, which was shared by the two layers, was located on the bottom side of the memory array. All peripheral circuitry, including the high-voltage-generating circuitry, was formed on the first silicon layer. A schematic circuit diagram of two NAND chains showing the layer placement and shared bit-line and shared CSL line is shown in Figure 2.59 [47].

The same authors discussed the reliability of this technology in January of 2009 [48]. Because the 3D stacked memory cells are formed on different silicon layers, various memory cell characteristics may be different across the silicon layers. A different V_{th} distribution on each silicon layer could cause a widening of the total V_{th} distribution. This could result in a degradation of program performance if using a conventional programming method. Conventional programming uses an incremental step program pulse determined by the fastest cell located at one side of the distribution.

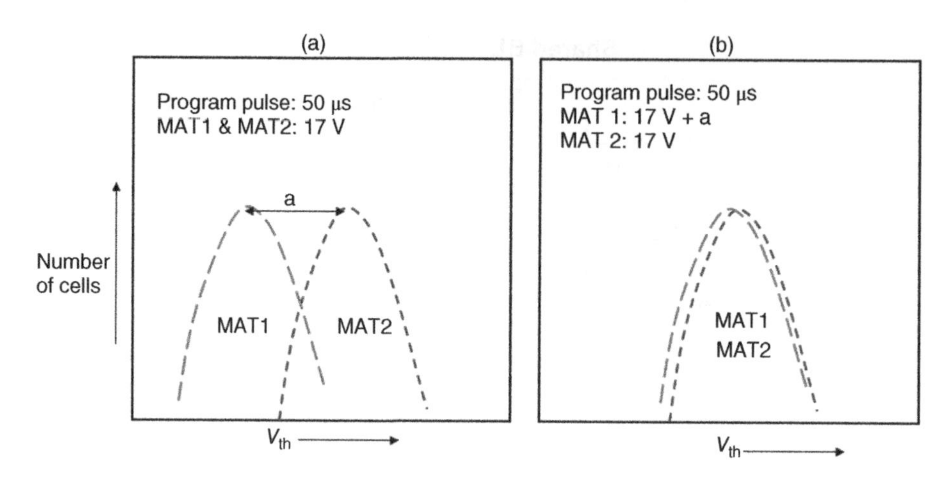

Figure 2.60 Layer-compensated programming method: (a) conventional method; and (b) layer-compensated method. (Based on K.T. Park *et al.* (2009) (Samsung), *IEEE Journal of Solid-State Circuits*, 44(1), 208 [48].)

To reduce performance degradation, a layer-compensated program method was used, which resulted in programming performance comparable to that of the conventional planar device. This method entailed considering the timing difference in the average V_{th} distribution of the two array layers, MAT_1 and MAT_2. The layer-compensated control method added the difference α in the V_{th} distribution to the program pulse for the slower circuit to compensate, as shown in Figure 2.60 [48].

Only one of the arrays of memory cells was formed on the top memory layer. All peripheral circuitry including the high-voltage-generation circuits were formed on the first silicon layer. Figure 2.61(a) shows a schematic cross-section of the array structure of the two-layer S^3 technology with a unit feature indicated that includes a memory transistor in the top layer and

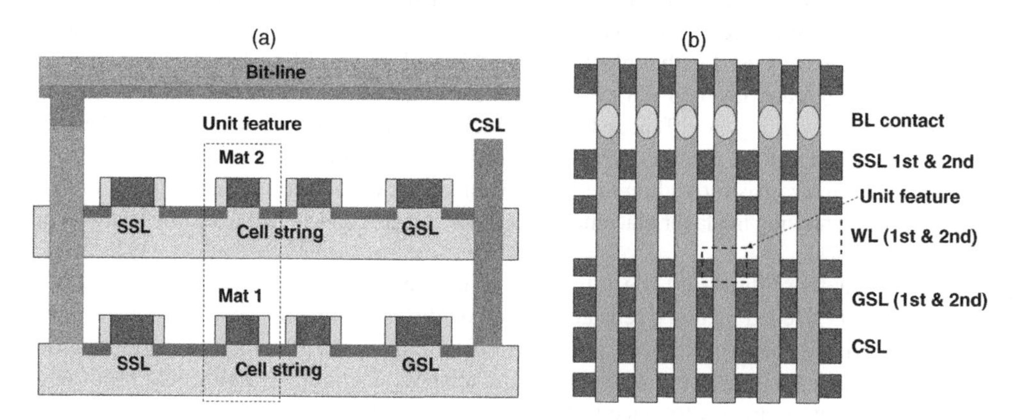

Figure 2.61 Two-layer S^3 technology: (a) schematic cross-section; and (b) top-down layout. (Based on K.T. Park *et al.* (2009) (Samsung), *IEEE Journal of Solid-State Circuits*, 44(1), 208 [48].)

one in the bottom layer. Figure 2.61(b) shows a top view of the layout with a unit feature size indicated that includes a memory transistor in the top layer and one in the bottom layer [46].

Because of the shared BL structure, the page buffer could access both of the layers of memory cells so that the BL loading of the 3D device is nearly the same as the loading of a conventional single-layer NAND flash chip. The extra loading was less than 3% of the total BL loading, so there was no performance penalty. Performance of the erase operation, however, could be degraded by an increased parasitic well capacitance. This degradation is in the range of several hundreds of microseconds, which, during the long, several-microseconds erase time, is also relatively negligible.

A 32-cell string with two dummy cells was used to reduce abnormal disturbances of edge memory cells adjacent to select transistors in the NAND string. These disturbances were caused by hot electron injection induced by band-to-band tunneling at the junction with the select transistor and by boosted channel potential leakage due to word-line-to-word-line capacitive coupling noise. The addition of the dummy transistors and use of proper bias voltage on the dummy cells during operation could reduce these disturbances significantly.

References

1. Xuan, P. *et al.* (December 2003) FinFET SONOS flash memory for embedded applications, (University of California Berkeley, Lawrence Berkley National Laboratory). IEDM.
2. Oh, C.W. *et al.* (December 2004) Damascene gate FinFET SONOS memory implemented on bulk silicon wafer, (Samsung). IEDM.
3. Cho, B.S. *et al.* (June 2005) Hf-silicate inter-poly dielectric technology for sub 70 um body tied FinFET flash memory, (Samsung). VLSI Technology Symposium.
4. Hwang, J.R. *et al.* (December 2005) 20 nm gate bulk-FinFET SONOS flash, (TSMC, National Chiao Tung University). IEDM.
5. Lee, J.J. *et al.* (March 2006) Retention reliability of FinFET SONOS device, (Samsung). IRPS.
6. Sung, S.K. *et al.* (2006) Fully integrated SONOS flash memory cell array with BT (body tied)–FinFET structure, (Samsung). *IEEE Transactions on Nanotechnology*, **5**(3), 74.
7. Sung, S.K. *et al.* (June 2006) SONOS-type FinFET device using p+ poly-Si gate and high-k blocking dielectric integrated on cell array and GSL/SSL for multi-gigabit NAND flash memory, (Samsung). VLSI Technology Symposium.
8. Kim, S. *et al.* (June 2006) Paired FinFET charge trap flash memory for vertical high density storage, (Samsung). VLSI Technology Symposium.
9. Koo, J.M. *et al.* (June 2008) Vertical structure NAND flash array integration with paired FinFET multi-bit scheme for high-density NAND flash memory application, (Samsung). VLSI Technology Symposium.
10. Lue, H.T. *et al.* (December 2009) Understanding STI edge fringing field effect on the scaling of charge-trapping (CT) NAND flash and modeling of incremental step pulse programming (ISSP), (Macronix). IEDM.
11. Kim, W. *et al.* (2010) Arch NAND flash memory array with improved virtual source/drain performance, (Seoul National University). *IEEE Electron Device Letters*, **31**(12), 1374.
12. Ren, J.J. *et al.* (2012) Nonplanar NiSi nanocrystal floating-gate memory based on a triangular-shaped Si nanowire array for extending nanocrystal memory scaling limit (University of California Riverside, Nanjing University). *IEEE Electron Device Letters*, **33**(10), 1390.
13. Lee, S. *et al.* (June 2012) Highly scalable STT-MRAM with 3-dimensional cell structure using in-plane magnetic anisotropy materials, (Samsung). VLSI Technology Symposium.
14. Liu, Y.X. *et al.* (12 September 2011) Variability analysis of scaled poly-Si channel FinFETs and tri-gate flash memories for high density and low cost stacked 3D-memory application (AIST), ESSDERC, p. 203.
15. Liu, Y.X. *et al.* (October 2012) Experimental study of tri-gate SOI-FinFET flash memory, (AIST, Meiji University). SOI Conference.
16. Cho, H., Kapur, P., Kalavade, P., and Saraswat, K.C. (December 2007) Highly scalable vertical double gate NOR flash memory, (Stanford University, Intel). IEDM.

17. Lombardo, S. *et al.* (December 2007) Advantages of the FinFET architecture in SONOS and nanocrystal memory devices, (CNR-IMM, STMicroelectronics, CEA-LETI). IEDM.
18. Choi, S.J. *et al.* (2009) Enhancement of program speed in dopant-segregated Schottky-barrier (DSSB) FinFET SONOS for NAND-type flash memory, (KAIST, ETRI, National Nanofab Center Korea). *IEEE Electron Device Letters*, **30**(1), 78.
19. Choi, S.J. *et al.* (2009) High injection efficiency and low-voltage programming in a dopant-segregated Schottky barrier (DSSB) FinFET SONOS for NOR-type flash memory. *IEEE Electron Device Letters*, **30**(3), 265.
20. Choi, S.J. *et al.* (June 2009) Performance breakthrough in NOR flash memory with dopant-segregated Schottky-barrier (DSSB) SONOS devices, (KAIST). VLSI Technology Symposium.
21. Choi, S.J., Han, J.W., Moon, D.I. *et al.* (2010) P-channel nonvolatile flash memory with a dopant-segregated Schottky-barrier source/drain, (KAIST, ETRI). *IEEE Transactions on Electron Devices*, **57**(8), 1737.
22. Yun, J.G. *et al.* (2009) Independent double-gate fin SONOS flash memory fabricated with sidewall spacer patterning, (Samsung, Seoul National University). *IEEE Transactions on Electron Devices*, **56**(8), 1721.
23. Walker, A.J. (December 2008) Sub-50nm DG-TFT-SONOS—the ideal flash memory for monolithic 3-D integration, (Schiltron). IEDM.
24. Lin, H.C., Lin, Z.M., Chen, W.C., and Huang, T.Y. (2011) Read characteristics of independent double-gate poly-si nanowire SONOS devices, (National Chiao Tung University). *IEEE Transactions on Electron Devices*, **58**(11), 3771.
25. Chen, W.C., Hsu, H.H., Chang, Y.C. *et al.* (June 2010) Investigations of performance enhancement in a poly-Si nanowire FET featuring independent double-gated configuration and its nonvolatile memory applications, (National Chiao Tung University, National Nano Device Laboratories). VLSI Technology Symposium.
26. Chen, H.T., Hsieh, S.I., Lin, C.J., and King, Y.C. (2007) Embedded TFT NAND-type nonvolatile memory in panel, (ITRI, STAR-NTHU). *IEEE Electron Device Letters*, **28**(6), 499.
27. Kim, B. *et al.* (April 2011) Investigation of ultra thin polycrystalline silicon channel for vertical NAND flash, (Samsung). IRPS.
28. Wu, C.Y., Liu, Y.T., Liao, T.C. *et al.* (2011) Novel dielectric-engineered trapping-charge poly-Si TFT memory with a TiN-alumina-nitride-vacuum-silicon structure, (National Chiao Tung University). *IEEE Electron Device Letters*, **32**(8), 1095.
29. Lee, S., Song, E.B., Kim, S. *et al.* (2012) Impact of gate work-function on memory characteristics in $Al_2O_3/HfO_x/$ Al_2O_3/graphene charge-trap memory devices, (Seoul University). *Applied Physics Letters*, **100**(2), 023109.
30. Cho, S. *et al.* (June 2010) Highly scalable vertical bandgap-engineered NAND flash memory, (Seoul National University). Device Research Conference.
31. Cho, S. *et al.* (2011) A charge trap folded NAND flash memory device with band-gap-engineered storage node, (Seoul National University, Stanford University). *IEEE Transactions on Electron Devices*, **58**(2), 288.
32. Hsu, T.H. *et al.* (2007) A high-performance body-tied FinFET bandgap engineered SONOS (BE-SONOS) for NAND-type flash memory, (Macronix). *IEEE Electron Device Letters*, **28**(5), 443.
33. You, H.W. and Cho, W.J. (2012) Nonvolatile poly-Si TFT charge-trap flash memory with engineered tunnel barrier, (Kwangwoon University). *IEEE Electron Device Letters*, **33**(2), 170.
34. Lai, E.K. *et al.* (December 2006) A multi-layer stackable thin-film transistor (TFT) NAND-type flash memory, (Macronix). IEDM.
35. Lue, H.T. *et al.* (December 2005) BE-SONOS: A bandgap engineered SONOS with excellent performance and reliability, (Macronix). IEDM.
36. Wang, S.Y. *et al.* (April 2007) Reliability and processing effects of bandgap engineered SONOS (BE-SONOS) flash memory, (Macronix). IRPS.
37. Lue, H.T. *et al.* (June 2010) A highly scalable 8-layer 3D vertical-gate (VG) TFT NAND flash using junction-free buried channel BE-SONOS device, (Macronix). VLSI Technology Symposium.
38. Hsaio, Y.H. *et al.* (December 2012) Modeling the variability caused by random grain boundary and trap-location induced asymmetrical read behavior for a tight-pitch vertical gate 3D NAND flash memory using double-gate thin-film transistor (TFT) device, (Macronix). IEDM.
39. Shim, W.B. *et al.* (2012) Stacked gated twin-bit (SGTB) SONOS memory device for high-density flash memory, (Seoul National University, Stanford University, Samsung). *IEEE Transactions on Nanotechnology*, **11**(2), 307.
40. Jung, S.M. *et al.* (December 2004) Highly area efficient and cost effective double stacked S3 (stacked single-crystal Si) peripheral CMOS SSTFT and SRAM cell technology for 512Mbit density SRAM, (Samsung). IEDM.

41. Jung, S.M. *et al.* (June 2005) Highly cost effective and high performance 65 nm S^3 (stacked single-crystal Si) SRAM technology with 25 F2, 0.16 μm^2 cell and doubly stacked SSTFT cell transistors for ultra high density and high speed applications. VLSI Technology Symposium.
42. Jung, S.M., Lim, H., Kwak, K.H., and Kim, K. (2010) A 500-MHz DDR high-performance 72-Mb 3-D SRAM fabricated with laser-induced epitaxial c-Si growth technology for a stand-alone and embedded memory application, (Samsung). *IEEE Transactions on Electron Devices*, **57**(2), 474.
43. Jung, S.M. *et al.* (June 2007) High speed and highly cost effective 72 Mbit density S3 SRAM technology with doubly stacked Si layers, peripheral only CoSix layers and tungsten shunt W/L scheme for standalone and embedded memory, (Samsung). VLSI Technology Symposium.
44. Son, Y.H. *et al.* (June 2007) Laser-induced epitaxial growth (LEG) technology for high density 3-D stacked memory with high productivity, (Samsung). VLSI Technology Symposium.
45. Sohn, K. (February 2008) A 100 nm double-stacked 500 MHz 72 Mb separate I/O synchronous SRAM with automatic cell-bias scheme and adaptive block redundancy, (Samsung), ISSCC.
46. Jung, S.M. (December 2006) Three dimensionally stacked NAND flash memory technology using stacking single crystal Si layers on ILD and TANOS structure for beyond 30 nm node, (Samsung). IEDM.
47. Park, K.T. *et al.* (February 2008) A 45 nm 4-Gb 3-dimensional double-stacked multi-level NAND flash memory with shared bit-line structure, (Samsung). ISSCC.
48. Park, K.T. *et al.* (2009) A fully performance compatible 45 nm 4-gigabit three dimensional double-stacked multi-level NAND flash memory with shared bit-line structure, (Samsung). *IEEE Journal of Solid-State Circuits*, **44**(1), 208.

3

Gate-All-Around (GAA) Nanowire for Vertical Memory

3.1 Overview of GAA Nanowire Memories

Vertical gate-all-around (GAA) nanowire (NW) charge-trapping (CT) memories have been developed with both single-crystal silicon channels and with polycrystal silicon channels. Both types of memory and their development are discussed in this chapter. The evolution toward junctionless memory transistors for long GAA NW strings is also covered. Single-crystal nanowire channels can be made from silicon fins or from stacking layers of dielectric alternated with a sacrificial material. Polysilicon NW CT memory stacks have been shown to benefit from radial configuration and also from the small radius of the NW in scaled arrays. Stacked NW horizontal channel CT NAND memory string processes are discussed. Vertical channel NW CT memories both deposited in an etched channel and deposited on the outside of a cylindrical channel using a replacement technology are also covered. This chapter discusses the technology and cell development of these memory devices, and Chapter 4 discusses their product development.

3.2 Single-Crystal Silicon GAA Nanowire CT Memories

3.2.1 Overview of Single-Crystal Silicon GAA CT Memories

The early GAA memories used single-crystal silicon fins formed from the surface of the silicon wafer as the starting point for making a single-crystal silicon wire. This wire was detached from the substrate and oxidized, and then CT and gate materials were added. Initially, vertically stacked twin NWs were processed from a single-crystal silicon fin. Both silicon nitride and silicon nanocrystals were used for the CT layer. Individual SiNWs were also made from a single-crystal silicon fin. When NWs thinner than 10 nm were used, back-tunneling was suppressed and the V_{th} window enlarged. High-κ dielectric blocking layers

Vertical 3D Memory Technologies, First Edition. Betty Prince.

were able to suppress gate electron injection, which improved erase speed, and HfO_2 CT layers improved retention time. TaN metal gate electrodes prevented unwanted back-tunneling of electrons due to their high work function, so erase speed improved.

3.2.2 An Early GAA Nanowire Single-Crystal Silicon CT Memory

The development of an early GAA CT memory using thin twin SiNWs with a silicon nitride trapping layer was described by Samsung in June of 2007 [1]. The memory was programmed using channel hot electron injection (CHEI) and hot hole injection (HHI) mechanisms. The program speed was 1 μs at $V_d = 2$ V, $V_g = 6$ V, and the erase speed was 1 ms at $V_d = 4.5$ V, $V_g = -6$ V using a 2–3 nm NW and 30 nm gate. The effect on threshold voltage shift (ΔV_{th}) and the program/erase (P/E) characteristics were studied.

It was found that as the NW diameter decreased, programming speed increased and the threshold voltage shift increased. This was because the electric field increased as the NW diameter decreased, as shown in Figure 3.1 [1].

A twin SiNW silicon–oxide–nitride–oxide–silicon (SONOS) structure was made by forming twin SiNWs from single-crystal silicon fins. Tunnel oxide (Tox) was then deposited, followed by a SiN trap layer, a blocking oxide, and a gate. A schematic diagram of the two-wire CT memory structure before the gate was deposited is shown in Figure 3.2 [1]. After n+ poly gate deposition, damascene dummy layers were removed. The NW diameter was 3 nm and ONO layer thickness was 3.0 nm, 3.5 nm, and 6.0 nm, respectively.

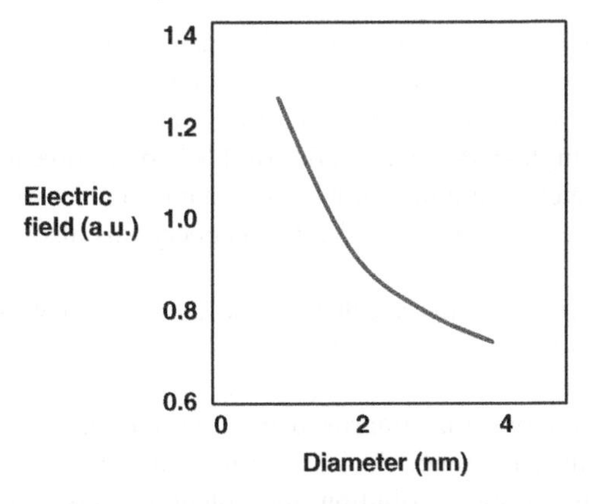

Figure 3.1 Electric field normal to NW surface vs. NW diameter. (Based on S.D. Suk *et al.*, (Samsung), VLSI Technology Symposium, June 2007 [1].)

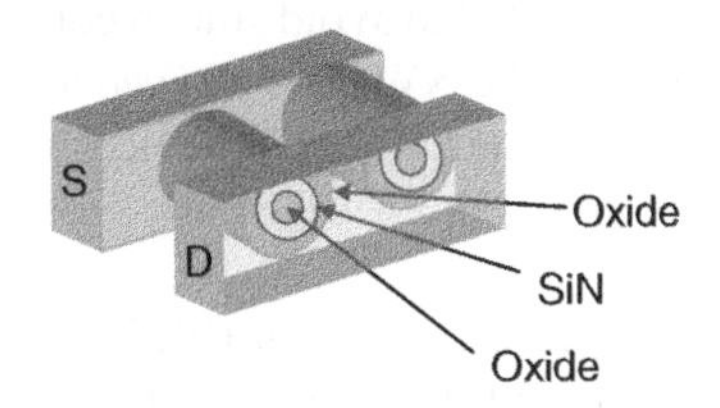

Figure 3.2 Schematic diagram of twin silicon NW charge-trapping memory structure. (Based on S.D. Suk *et al.*, (Samsung), VLSI Technology Symposium, June 2007 [1].)

Table 3.1 Comparison of GAA SiNW SONOS memory with dual-gate FinFET SANOS memory [1].

Structure	Dual-Gate FinFET	GAA SiNW
Program	4.5 V, 100 μs V_g @ 18 V	4.2 V 100 μs @ $V_g = -10$ V, $V_d = 0$ V
Erase	4.5 V, 1 ms @ 18 V	4.2 V, 1 ms @ $V_g = -12$ V, V_s, $V_d = 0$ V

Based on S.D. Suk *et al.*, (Samsung), VLSI Technology Symposium, June 2007 [1].

A comparison was made of the GAA SiNW SONOS memory device with a planar dual-gate FinFET SANOS memory device with a blocking layer of Al_2O_3 [1]. Both cells used Fowler-Nordheim tunneling for P/E. The GAA device had a SiNW diameter of 2–3 nm, a gate length (L_g) of 30 nm, and the stack was a 2 nm *Tox*, 3.5 nm SiN, and 6 nm blocking oxide. The FinFET flash had L_g of 63 nm with stack of 4 nm *Tox*, 5 nm SiN, and 12 nm Al_2O_3 blocking layer.

The GAA SiNW SONOS had a P/E speed and ΔV_{th} similar to the dual gate FinFlash in spite of using lower P/E voltage, as shown in Table 3.1.

The SiN on the GAA SiNW was also thinner. The conclusion was that GAA SiNW SONOS is a candidate for a < 20 nm CT memory.

3.2.3 Vertically Stacked Single-Crystal Silicon Twin Nanowire GAA CT Memories

A vertically stacked CT GAA single-crystal SiNW memory was discussed in December of 2007 by IME-A*STAR and the National University of Singapore (NUS) [2]. The vertically stacked twin NW body was used to reduce the footprint per NW channel. The twin NWs were processed from a single-crystal silicon fin. Both silicon nitride and silicon nanocrystals were used for the CT layer. The transient memory characteristics were fast due to both the NW channel structure and the nanocrystals. The nanocrystals also increased the threshold voltage shift.

A similar GAA CT SiNW memory cell with the twin cell NWs stacked vertically was discussed in May of 2008 by NUS, IME-A*STAR, and the University of Bologna [3]. A tilted-view scanning electron microscope (SEM) image of the stacked NWs is shown in Figure 3.3 [3].

The SiNW memories showed an improved program and erase speed at relatively low voltage compared to planar reference devices. Performance enhancement was studied as a function of the electron energy distribution, the potential energy profile, and the electric field along each layer surrounding the channel. The ONO stack was 4.5 nm, 4.5 nm, and 8 nm, respectively, and deposited sequentially by low-pressure chemical vapor deposition (LPCVD). The tunnel oxide was deposited to avoid silicon consumption in the NW. The gate was 130 nm of deposited polysilicon. The NW cells showed a threshold shift of 2.6 V in 1 μs with a +11 V program pulse on the gate. The NW devices required 1 ms to erase, which is much faster than equivalent planar cells. A schematic cross-section of the SONOS NW memory cell is shown in Figure 3.4 [3].

The GAA cylindrical SiNW field-effect transistor (FET) had optimal electrostatic control with nearly ideal subthreshold slope and only a minor short-channel effect. The cylindrical geometry and inverse logarithmic dependence of the insulator capacitance on its thickness meant the gate length could be scaled with channel diameter without reducing the gate

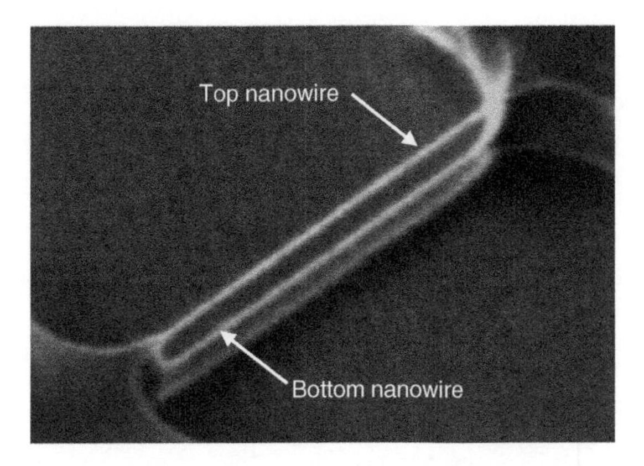

Figure 3.3 Tilted-view SEM image of stacked NWs. (J. Fu *et al.* (2008) (NUS, A*STAR, University of Bologna), *IEEE Electron Device Letters*, 29(5), 518 [3], with permission of IEEE.)

dielectric thickness. Figure 3.5 shows a comparison of P/E characteristics of the vertically stacked twin SiNW CT memory cell with an L_g of 850 nm, SiNW diameter of 5 nm, ONO thickness of 4.5 nm, 4.5 nm, and 8 nm, respectively, with a reference planar device of gate width $(W) = 5\ \mu m$ [3]. The ΔV_{th} memory window of the NW CT memory was significantly larger than that of the reference planar memory. Both were programmed at +11 V and erased at −11 V.

3.2.4 GAA CT NAND Flash String Using One Single-Crystal SiNW

A GAA CT NAND flash string using one SiNW was made in June of 2008 by Samsung [4]. The NW was processed from a single-crystal silicon fin. The drive current was sufficient, at

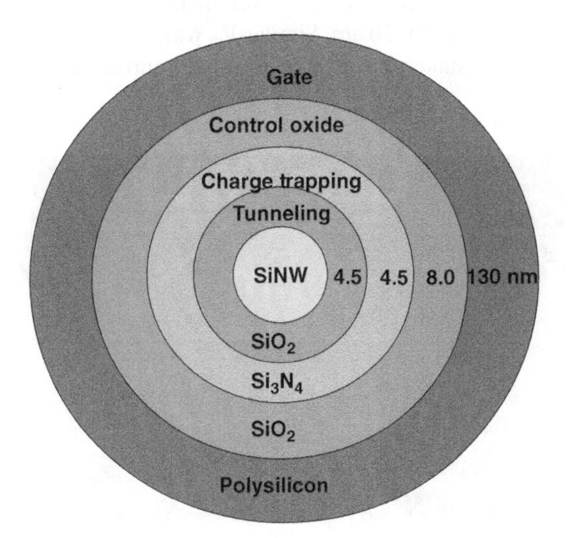

Figure 3.4 Schematic cross-section of SONOS NW memory cell (Based on J. Fu *et al.* (2008) (NUS, A*STAR, University of Bologna), *IEEE Electron Device Letters*, 29(5), 518 [3].)

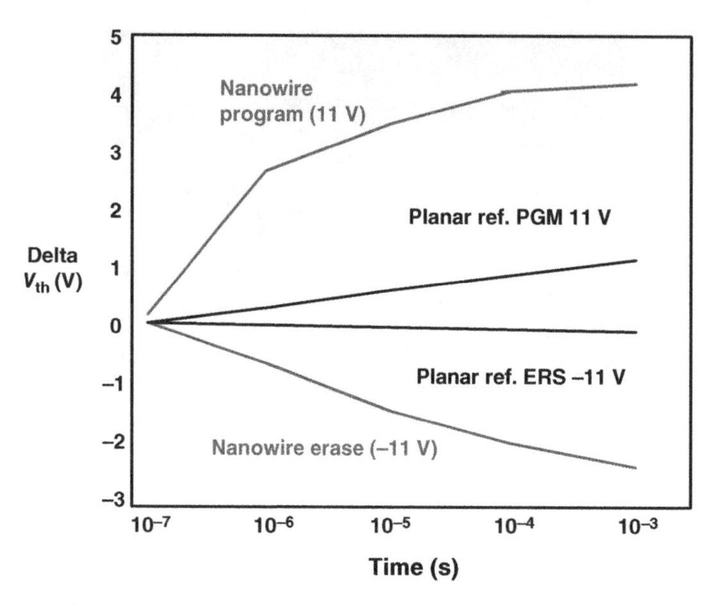

Figure 3.5 Comparison of P/E speed of vertically stacked SiNW SONOS with reference planar device. (Based on J. Fu *et al.* (2008) (NUS, IME-A*STAR, University of Bologna), *IEEE Electron Device Letters*, 29(5), 518 [3].)

over 1 μA for a single 8-bit NAND string with 7 nm SiNW diameter. Fowler-Nordheim tunneling was used for program and erase. A V_{th} window of 4.5 V and a P/E speed of about 10 μs were obtained. Figure 3.6 is a SEM image showing the bit-line of a single NW in an 8-bit SONOS NAND string [4].

A 7 nm wide cylindrical NW was used with ONO thickness of 3.0 nm, 7.8 nm, and 5.8 nm, respectively. The gate width was 45–50 nm. Similar V_{th} was found for 4-bit and 8-bit strings. A large source/drain (S/D) resistance resulted in low read current due to nonoptimized S/D

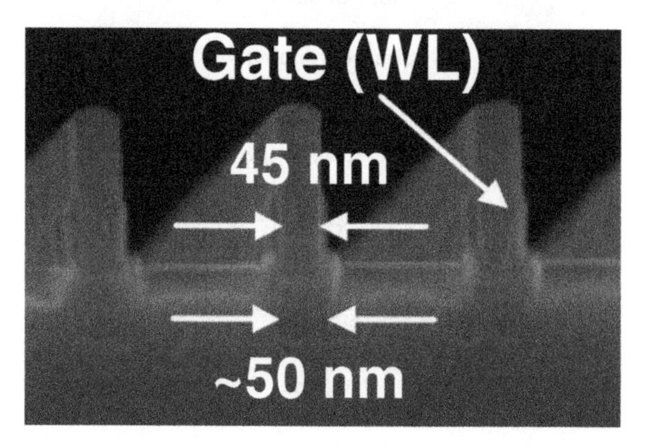

Figure 3.6 SEM image of a single NW in an 8-bit SONOS NAND string. (K.H. Yeo *et al.*, (Samsung), VLSI Technology Symposium, June 2008 [4], with permission of IEEE.)

doping conditions. A V_{th} window of 5.2 V was found at $V_g = 14$ V and P/E time of 100 µs/100 ms in an 8-bit NAND string. P/E speeds with different NW diameters were compared, and the results showed that back-tunneling was suppressed and the V_{th} window was enlarged due to reduced electric fields in the control oxide with 7 nm NW compared to 30 nm and 50 nm NWs.

3.2.5 Single-Crystal SiNW CT Memory with High-κ Dielectric and Metal Gate

A SiNW nonvolatile CT memory cell using a high-κ dielectric and metal gate was discussed in September of 2008 by A*STAR and NUS [5]. The TAHOS (TaN–Al_2O_3–HfO_2–SiO_2–Si) nonvolatile memory cell was made using a top-down method for the GAA architecture. It had faster program and erase speed than a similar SiNW GAA SONOS cell. In the SONOS cell, the erase speed was slower than the program speed due to the large 4.4 eV hole barrier through the tunnel oxide. In the TAHOS technology the increased field in the tunnel oxide improved the program speed. The erase speed was nearly the same as the program speed due to the suppression of gate electron injection by the Al_2O_3 blocking layer, which improved erase speed. The deep trap levels in the HfO_2 CT layer improved retention time. A stable memory window up to 10^4 P/E cycles was shown.

In June of 2009, A*STAR and NUS further discussed this CT SiNW memory that used high-κ dielectrics for the blocking layer and the trapping layer and a metal gate [6]. The CT GAA memory with Al_2O_3–HfO_2–SiO_2 stack and TaN metal gate was made on a single-crystalline SiNW. The use of the high-κ material reduced the electric field in the blocking layer. The TaN metal gate electrode prevented unwanted back-tunneling of electrons due to its higher work function. The memory speed was improved for the TAHOS memory over a similar SONOS memory, as shown in Figure 3.7 [6]. This figure shows the P/E characteristics for both a SONOS and a TAHOS NW CT memory programmed and erased at 12 V. The NW

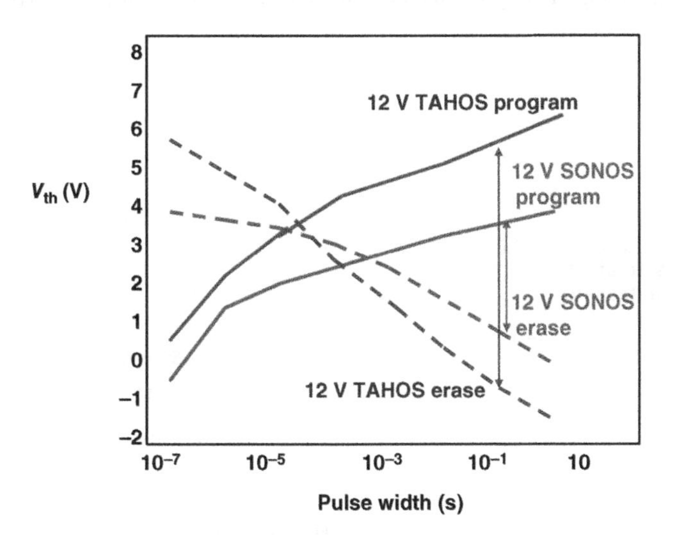

Figure 3.7 12 V program and 12 V erase characteristics for (a) SONOS; and (b) TAHOS NW CT memory with the same GAA NW curvature. (Based on J. Fu *et al.* (2009) (A*STAR, NUS), *IEEE Electron Device Letters*, 30(6), 662 [6].)

configuration of both memories was the same to ensure that the effect of the GAA curvature would not be a factor.

The NW was derived from a p-type silicon-on-insulator (SOI) wafer with a 50 nm SOI on a 150 nm buried oxide (BOX) layer. A narrow fin was defined from the SOI and was connected to wider source and drain pads. Because the SOI was thin, only a single NW was formed from each fin. The wire was kept relatively thick at a diameter of 25–30 nm due to concern about the weight of the TaN. The tunnel layer of 5 nm of SiO2 was deposited by LPCVD. The high-κ dielectrics, 7 nm of HfO_2 as the CT layer, and 10 nm of Al_2O_3 as the blocking oxide layer, used atomic layer deposition (ALD). Annealing at 700 °C for 30 s followed, and 200 nm thick TaN was sputtered [6].

3.2.6 Improvement in Transient V_{th} Shift After Erase in 3D GAA NW SONOS

The origin of transient V_{th} shift after erase in both a planar TANOS device and in a 3D GAA single-crystal SiNW structure using a SONOS gate stack was discussed by KAIST and Hynix in December of 2012 [7]. The main mechanism for this transient V_{th} shift in TANOS devices was found to be hole redistribution in the CT layer.

In the GAA SONOS device, a smaller diameter of SiNW showed better erase efficiency due to the electric field concentration effect on the tunnel oxide. This is illustrated in Figure 3.8, which shows erased saturation voltage vs. NW diameter in GAA SONOS devices [7]. The SiNW was made on the bulk silicon substrate using a deep reactive ion etch (RIE) process.

The GAA SONOS device showed the same transient V_{th} shift after erase that was found in the planar TANOS devices, and the V_{th} shift had the same mechanism of hole redistribution in the trap layer. This V_{th} shift can be reduced by scaling the channel length and diameter of the SiNW GAA SONOS device, possibly due to a compensation effect by charge crowding and lateral charge spreading. This gives the advantage of fast erase to the NW-type 3D CT flash memory.

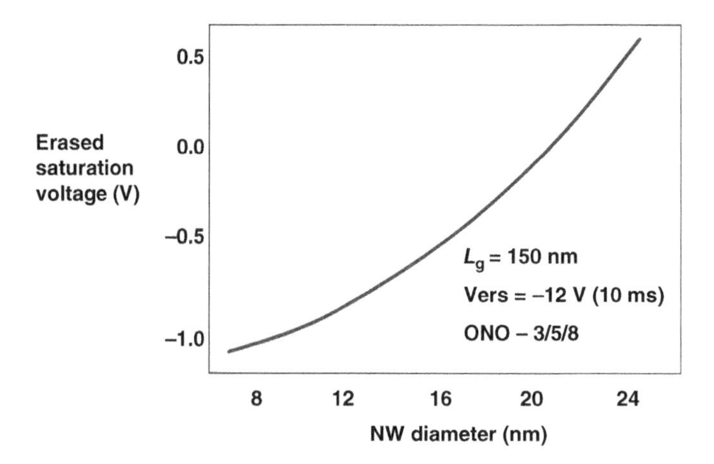

Figure 3.8 Erased saturation voltage vs. NW diameter in GAA-SONOS devices. (Based on J.K. Park *et al.*, (KAIST, Hynix), IEDM, December 2012 [7].)

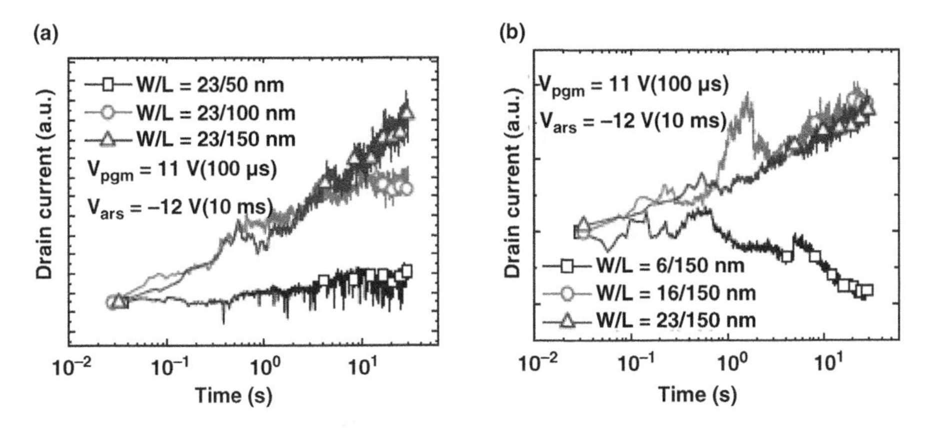

Figure 3.9 Transient I_d of GAA-SONOS devices by (a) channel length; and (b) NW diameter. (J.K. Park *et al.*, (KAIST, Hynix), IEDM, December 2012 [7], with permission of IEEE.)

Figure 3.9 illustrates the transient I_d of GAA SONOS devices with (a) different channel length and (b) different NW diameter [7]. The transient V_{th} shift becomes smaller when the channel length and wire diameter are scaled. Sudden drain current fluctuation in GAA SONOS may be due to single electron effects or random telegraph noise.

A new erase method was proposed to speed up the erase process for the GAA-SONOS devices. Prior to applying a read voltage, a small positive gate pulse is applied right after the main erase pulse. This pulse aided in fast redistribution of holes in the CT layer and reduced the V_{th} transition period after erase.

3.2.7 Semianalytical Model of GAA CT Memories

An analysis of the transient dynamics of GAA CT memories was presented by the Politecnico di Milano and Micron Technology in September of 2011 [8]. This involved solving the Poisson equation in cylindrical coordinates and modifying the Fowler-Nordheim formula for tunneling through cylindrical dielectric layers. Analytical results were confirmed with experimental data.

This model was used to examine the effect of device curvature on both P/E and data retention. The electric field and energy band profile was calculated for the GAA ONO–based CT cell and a planar CT cell with the same thickness of gate dielectrics for $V_g = 12$ V and neutral nitride as a function of position. The calculated program and erase transients at $V_g = +/-12$ V for the GAA CT cell are shown in Figure 3.10 for different radii [8]. In all cases the starting time is 10^{-12} s and $\Delta V_{th} = 0$ for the simulations. As the radius decreases, the program and erase speed increases so that the ΔV_{th} window increases. The calculated retention transients at 85 °C for the GAA CT cell are shown in Figure 3.11 for different energies (E_T) and a 3 nm radius [8].

3.2.8 Nonvolatile GAA Single-Crystal Silicon Nanowire Memory on Bulk Substrate

The fabrication of a 10 nm wide and 50 nm long CT transistor memory using a single-crystal silicon junctionless NW made from a bulk substrate was discussed by KAIST in May of

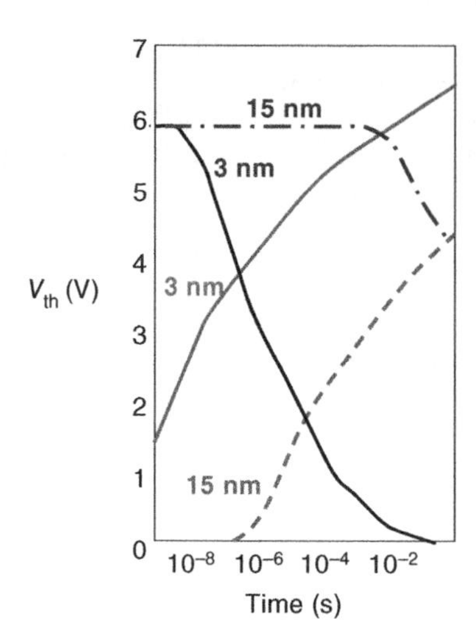

Figure 3.10 Program and erase transients in GAA CT cell by device radius, showing speed increasing as the radius decreases. [8] (Based on S.M. Amoroso *et al.*, (2011) (Politecnico di Milano, Micron Technology), *IEEE Transactions on Electron Devices*, 58(9), 3116 [8].)

2011 [9]. A deep RIE system was used to make a junctionless transistor with a suspended silicon NW channel with a 10 nm width and 50 nm length that was completely separated from the bulk substrate. The process sequence is illustrated in Figure 3.12 [9].

An implantation was followed by a photoresist mask of the NW area and an anisotropic etch. C_4F_3 passivation was deposited and a second isotropic etch done. The result was a detached

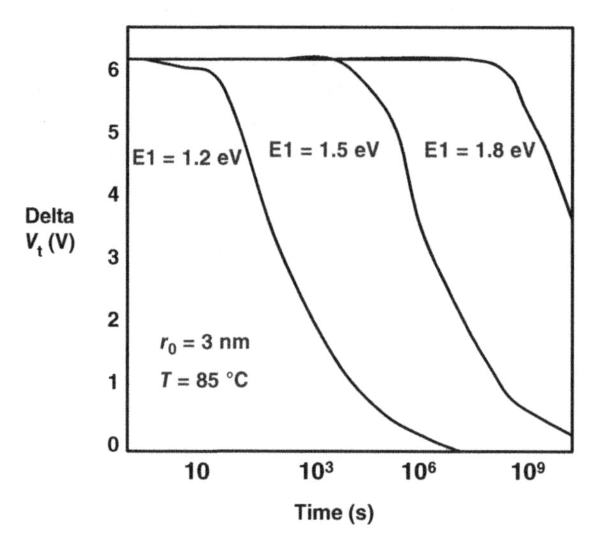

Figure 3.11 Illustration of retention transients at 85 °C for GAA CT MONOS cell for 3 nm radius. (Based on S.M. Amoroso *et al.* (2011) (Politecnico di Milano, Micron Technology), *IEEE Transactions on Electron Devices*, 58(9), 3116 [8].)

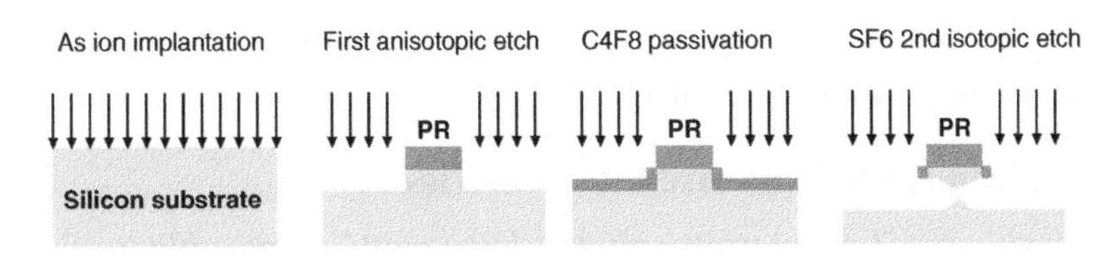

Figure 3.12 Formation sequence of the silicon NW on a bulk substrate. (Based on S.J. Choi *et al.* (2011) (KAIST), *IEEE Electron Device Letters*, 32(5), 602 [9].)

NW completely separated from the bulk substrate. Boron was implanted and sacrificial oxidation used to reduce the width of the SiNW further to 10 nm and to make the channel smooth. An ONO layer was then formed on the SiNW using thermal oxide followed by LPCVD nitride/TEOS oxide and LPCVD polysilicon for the gate. The ONO dimensions were 2.8 nm, 6.2 nm, and 7 nm, respectively. A NW length of 50 nm was then patterned. The gap distance between the SiNW and the bulk substrate was about 250 nm. The junctionless transistor that formed was in depletion mode, so it was normally turned on. This means the ON-state current was determined mainly by the doping concentration of the channel. This means there was no electric field perpendicular to the direction of current flow in the ON state, and the transconductance, g_m, decreased slowly, which could provide the potential for further scaling to below 20 nm.

Program and erase were done using Fowler-Nordheim tunneling. Voltages of 11 to 14 V were used on the gate for programming with the source and drain grounded. For erase, the cells were first programmed with a +14 V pulse for 1 ms followed by negative pulses ranging from −12 to −15 V on the gate. A 5 V programming window resulted, with no degradation of the subthreshold slope. This wide window implied that multilevel operation could be possible.

This same deep RIE fabrication process was used by KAIST in June of 2011 to make a nine-layer CT single-crystal SiNW NAND string on a bulk silicon wafer as illustrated in Figure 3.13 [10]. This was done by iterative plasma etch processes to form and separate

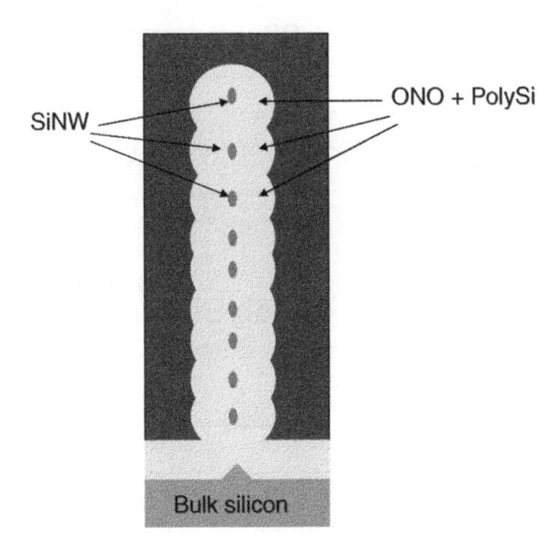

Figure 3.13 Schematic cross-section of nine-layer charge-trapping NW NAND string. (Based on S.J. Choi *et al.*, (KAIST), VLSI Technology Symposium, June 2011 [10].)

each of the nine SiNWs. A transmission electron microscope (TEM) image indicated that the SiNWs were not degraded by the etching steps.

3.3 Polysilicon GAA Nanowire CT Memories

3.3.1 Polysilicon CT Memories with NW Diameter Comparable to Polysilicon Grain Size

Thin-film transistor (TFT) polysilicon CT memories have historically been relatively low speed with poor subthreshold properties due to the grain structure of the polysilicon being small compared to the size of the channel, resulting in significant grain boundary effects. Reducing the cross-sectional dimensions of the polysilicon NW channel to make it comparable to the grain size was found to significantly improve performance. In addition, the GAA structure enhances the gate controllability. These effects make possible polysilicon NW SONOS memories with reasonable electrical characteristics.

A GAA polysilicon NW channel memory was discussed in March of 2009 by A*STAR and NUS [11]. The ONO-based memory cell used a 23 nm diameter NW and showed fast program and erase speed along with improved subthreshold transistor behavior. The memory performance improved as the NW width decreased. The NWs were made by depositing a 60 nm amorphous silicon (a-Si) onto a SiO_2 substrate. A fin was patterned by phase shift mask lithography followed by silicon dry etch. The fins were oxidized to convert them into NWs by steam oxidation. During this step the a-Si crystallized into polysilicon grains of 15 to 30 nm in diameter. The NWs were released by etching followed by deposition of ONO for the tunnel oxide, nitride, and blocking oxide using plasma-enhanced CVD (PECVD) at 400 °C. A 130 nm a-Si was deposited, patterned, and etched to form the gate electrode. Phosphorous was then implanted and activated, followed by contact, metallization, and sintering.

The cross-section of the GAA polysilicon NW device was roughly rectangular with a NW channel width of 23 nm and height of 36 nm, as shown in Figure 3.14 [11]. The channel width was roughly the size of the polysilicon grains.

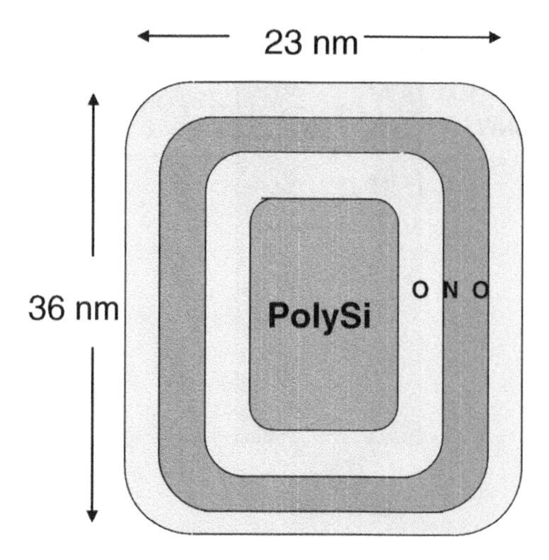

Figure 3.14 Rectangular cross-section of GAA polysilicon NW device. (Based on J. Fu *et al.* (2009) (A*STAR, NUS), *IEEE Electron Device Letters*, 30(3), 246 [11].)

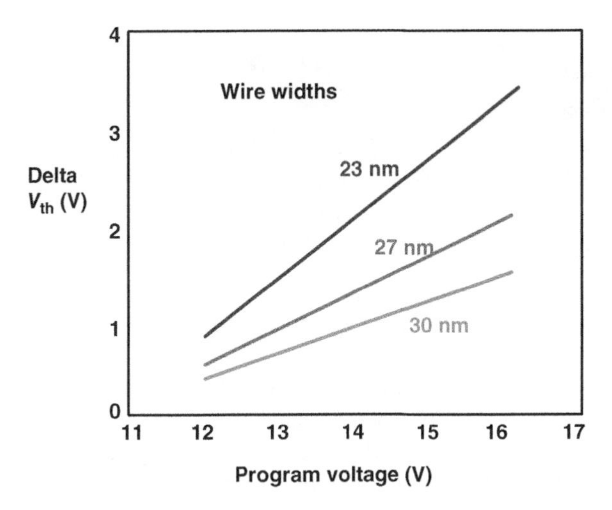

Figure 3.15 V_{th} shift dependence on program voltage for different NW widths. (Based on J. Fu *et al.* (2009) (A*STAR, NUS), *IEEE Electron Device Letters*, 30(3), 246 [11].)

During programming and erase, for a device with L_g of 350 nm, a trend was observed for the smaller-diameter wire to exhibit larger V_{th} shift, as shown in Figure 3.15 [11]. The different NW widths were derived from different fin widths on the wafer.

A low subthreshold slope value of 122 mV/dec was found, which is significantly less than for conventional polysilicon devices. This was attributed to the enhanced gate control in the GAA structure and the reduced number of grain boundaries in the channel region.

Fowler-Nordheim tunneling was used to characterize the cell by grounding the source and drain and stressing the gate. A programming speed of 1 μs at $V_g = 15$ V resulted in a threshold voltage shift of 2.96 V. The programming speed improvement was attributed to the crowding and convergence of electric field lines at the corners of the rectangular channel structure. The erasing time for 2.96 V was 1 ms at $V_g = -16$ V. The slow erase speed was attributed to the higher hole energy barrier at 4.6 eV and to the larger effective mass of the holes compared to the electrons in SiO_2, which results in reduced tunneling of holes.

3.3.2 Various GAA Polysilicon NW Memory Configurations

Several simple and low-cost methods of fabricating NW SONOS devices with multiple-gate configurations were shown by National Chiao Tung University and the National Nano Device Labs of Taiwan in March of 2011 [12]. These configurations included the following: side gate, which was a one-sided gate; omega gate, which was three-sided; and a GAA structure.

The initial device fabrication was the same for all configurations, as shown in Figure 3.16. A silicon substrate had thermal oxide grown, followed by a nitride-TEOS-nitride stack deposited on the thermal oxide, as shown in Figure 3.15(a) [12]. This stack was then patterned and etched. The differential etch rates of SiN and TEOS formed rectangular cavities at the two sides of the TEOS layer, as shown in Figure 3.15(b). Undoped a-Si was then deposited and annealed to form polysilicon, as shown in Figure 3.15(c) [12].

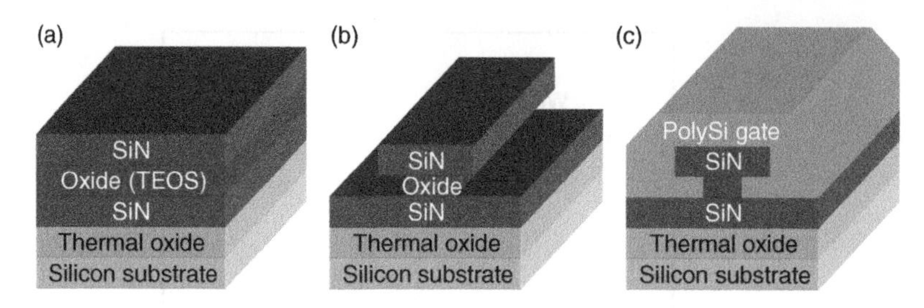

Figure 3.16 Common initial steps in formation of three types of NW structures. (Based on H.H. Hsu *et al.* (2011) (National Chiao Tung University), *IEEE Transactions on Electron Devices*, 58(3) [12].)

Figure 3.17 shows the next steps of the process, which define the gate type [12]. Figure 3.17 (a) shows side-gate (SG) devices in which an anisotropic dry etch removed the polysilicon everywhere except in the cavities shielded by the nitride shelf that acted as a hardmask, leaving one side of the polysilicon NW exposed. Figure 3.17(b) shows the three-sided omega gate in which additional wet etch steps were used to remove the nitride hardmask and dummy TEOS, while a part of the bottom nitride was left to hold up the NWs that then had three sides exposed. Figure 3.17(c) shows the GAA devices in which the bottom nitride was also removed and the NW was left hanging between the source and drain regions, thereby exposing all four sides of the NW. All splits then received on the gate an ONO stack and an n+ polysilicon, which served as gate electrode.

The channel width of the side-gate device is about 20 nm, while the channel widths for the omega gate and GAA devices are 50 nm and 60 nm, respectively, due to the additional sides exposed in the latter two devices.

Fowler-Nordheim tunneling was used for program and erase for all device types. The P/E efficiency of the omega gate and GAA devices was better than for the side-gate devices due to the increase in curvature of corners of the NW channels (a result of more etch steps) and more exposed sides. The GAA device had the fastest P/E speed, showing that improved gate controllability has a significant impact on the NW memory characteristics. The GAA devices were found to have the best transfer and memory characteristics. Drain-induced barrier lowering (DIBL) was negligible for the GAA devices due to the ultrathin channel body and improved gate controllability.

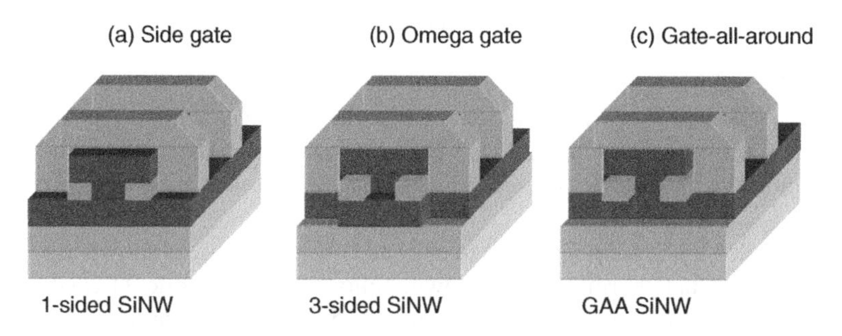

Figure 3.17 Types of NW SONOS devices: (a) side gate; (b) omega gate; and (c) gate-all-around. (Based on H.H. Hsu *et al.* (2011) National Chiao Tung University, *IEEE Transactions on Electron Devices*, 58(3), 641 [12].)

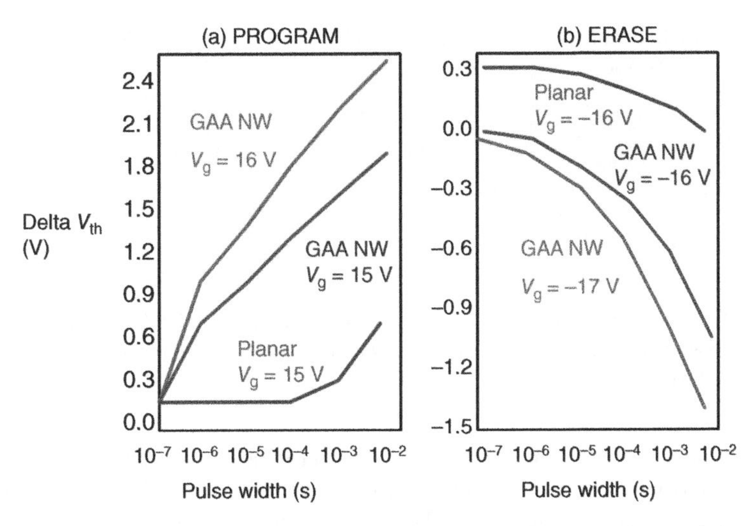

Figure 3.18 GAA multiple NW channel memory (a) program (b) erase characteristics. (Based on P.C. Huang *et al.* (2010) (National Chiao Tung University), *IEEE Electron Device Letters*, 31(3), 216 [13].)

A rectangular polysilicon multiple NW CT TFT memory was discussed in March of 2012 by National Chiao Tung University (NCTU) [13]. The 1 μm long 20-channel polysilicon NWs were made using a spacer patterning technique. Solid-phase crystallization was performed at 600 °C for 24 hours to turn the a-Si into a polycrystalline silicon structure. The corners of the rectangular NW structure resulted in enhanced program speed and a memory window that was enlarged over the characteristics of a planar polysilicon TFT memory device.

The presence of a nonuniform electric field in the channel region during program and erase was confirmed by simulation. The P/E characteristics of the GAA multiple-NW channel memory vs. a conventional planar memory are shown in Figure 3.18 [13]. During erase, the planar devices showed a threshold voltage shift that resulted from gate injection, but the GAA devices were immune to this behavior. The threshold voltage of the planar device was shown to experience a positive shift due to gate injection.

A simulation of the electric field showed that the enhancement in P/E efficiency was a result of the large number of corners where the strongest induced electric field occurred. The polysilicon NW device also showed improved endurance and data retention over planar devices.

3.3.3 Trapping Layer Enhanced Polysilicon NW SONOS

In July of 2011, the NCTU and the National Nanodevice Lab of Taiwan discussed trap layer engineering in a GAA SONOS polysilicon NW device in which silicon nanocrystals were added in the nitride storage layer [14]. The density of nanocrystals was around 2.6×10^{11} cm^{-2}. The polysilicon SONOS NW was rectangular with rounded corners and a width of about 30 nm. A method was devised to incorporate the nanocrystals in various locations in the nitride layer at the blocking layer interface, the middle of the nitride, and the nitride tunnel oxide interface, as shown in Figure 3.19 [14].

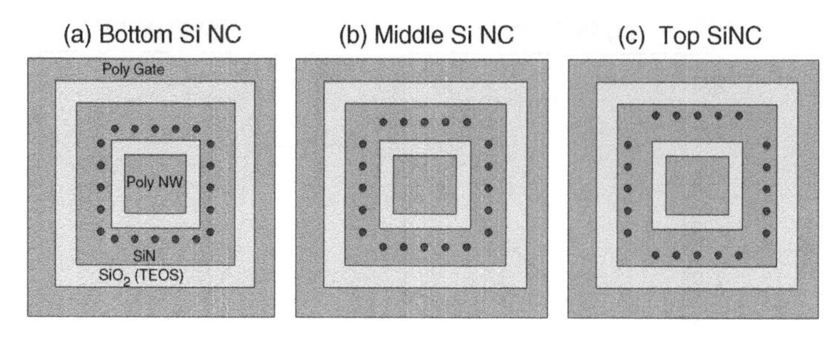

Figure 3.19 Trap layer engineering of the silicon NC location in a SONOS NW. (Based on C.W. Luo *et al.* (2011) (National Chiao Tung University, National Nano Development Laboratories), *IEEE Transactions on Electron Devices*, 58(7), 1879 [14].)

A study was done to find the most effective location for the nanocrystals in order to improve memory characteristics. The best P/E characteristics were found for the bottom nanocrystals because they had a faster V_{th} increase rate. The best retention characteristics, however, were for the middle nanocrystals, where the electrons were harder to detrap. The optimal nanocrystal location was between the middle and bottom interfaces of the nitride layer.

A CT memory with GAA polysilicon NW and a HfAlO trapping layer was discussed in March of 2013 by NCTU [15]. It was demonstrated that the GAA configuration improved the P/E efficiency compared with a planar technology. The study also showed that including Al in the dielectric improved retention and endurance over a device without the Al. The Al was also thought responsible for slowing the recrystallization of the dielectric film. High-κ dielectrics like HfO_2 and Al_2O_3 can be used to replace the nitride CT layer in SONOS memories to improve the program and erase efficiency because the equivalent oxide thickness is reduced and the deeper band structure improves data retention. The GAA structure enhances gate control and tunneling probability through the tunnel oxide due to increase in surface curvature of the wrapper NW channel. In this study, the GAA structure and the high-κ dielectric were combined, and the effect on the characteristics of the resulting CT device was determined.

A cross-section of the device layers of the NW device is shown in Figure 3.20 [15]. The device structure was TAHOS. An 85 nm diameter NW channel had TEOS oxide deposited for the 3 nm tunnel oxide, 16 nm of HfO_2 or HfAlO was used for the CT layer, and 11 nm of Al_2O_3 was used for the blocking layer. The high-κ layers were deposited using ALD. The thickness of the n+ polysilicon source and drain were 100 nm, and the gate electrode used 150 nm thick TiN. The resulting diameter of the NW channel was about 85 nm. After the gate electrode was deposited, the devices were annealed at 600 °C for 20 s. A top-down schematic view of the device tested is shown in Figure 3.21 [15].

The subthreshold swing (SS) for both the HfAlO and HfO_2 GAA NW CT devices was less than 150 mV/dec, and DIBL was negligible. Both factors were due to the very small–radius NW channel and the GAA configuration.

For program operation +16 V and for erase −14 V were applied to the gate with source and drain grounded. The threshold voltage shift as a function of program time is shown in Figure 3.22(a) and as a function of erase time in Figure 3.22(b) [15]. The results for both the HfAlO and the HfO_2 GAA NW devices and for the planar HfO_2 device are shown. V_{th} shifts for programming at 1 μs were 3 V for the GAA HfO_2 device and 2.6 V for the HfAlO device.

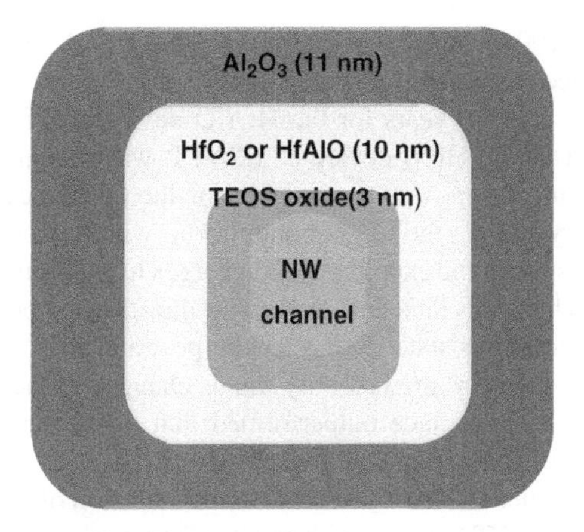

Figure 3.20 Schematic cross-section of the TAHOS NW. (Based on K.H. Lee *et al.* (2013) (National Chiao Tung University), *IEEE Electron Device Letters*, 34, (3) [15].)

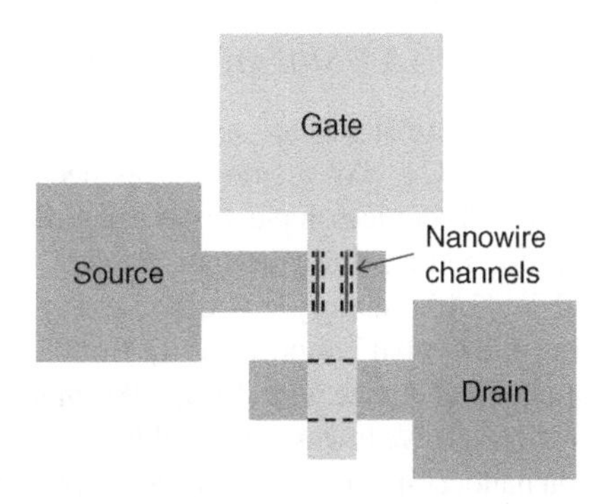

Figure 3.21 Top-down schematic of NW device tested. (Based on K.H. Lee *et al.* (2013) (National Chiao Tung University), *IEEE Electron Device Letters*, 34(3), 393 [15].)

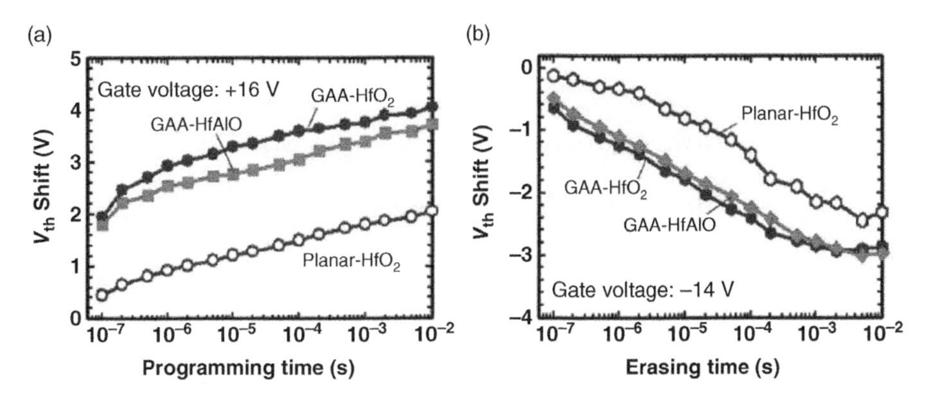

Figure 3.22 V_{th} as a function of (a) program time; and (b) erase time. (K.H. Lee *et al.* (2013) (National Chiao Tung University), *IEEE Electron Device Letters*, 34(3), 393 [15], with permission of IEEE.)

For erase, the V_{th} shifts at 100 μs were 2.4 V for the GAA HfO_2 and 2.3 V for the HfAlO device. V_{th} shifts for the reference planar device were significantly smaller.

The retention window after 10 years for the HfAlO device was found to be about 2.2 V, which is larger than that of the HfO_2 device at 1.2 V. It was thought that the difference in crystallinity of the trapping layers was responsible for the difference in threshold window. Since the HfO_2 is polycrystalline, the CT is nonuniform, with defect density larger near the grain boundaries. This results in the excess trapped charges leaking out faster due to the rise in local electric potential. When Al is included, the recrystallization temperature increases and the dielectric stays in the amorphous state even after the post-anneal.

The HfAlO GAA NW memory showed only minor changes in the transfer characteristics after 10 000 P/E cycles. Its endurance outperformed that of the HfO_2 memory. The GAA HfAlO memory showed a good extrapolated window of 2.2 V after 10 years and only minor shift in the transfer curves after 10^4 P/E cycles due to the inclusion of Al in the HfO_2 layer. The conclusion was that the GAA CT memory with an HfAlO trapping layer is promising for use in 3D high-density flash memory.

3.4 Junctionless GAA CT Nanowire Memories

3.4.1 3D Junctionless Vertical GAA Silicon NW SONOS Memories

Defining junction doping in NWs is difficult and adds to the process complexity. For this reason the use of junctionless GAA CT NW memories has been investigated. Experiments have indicated that junctionless CT devices have lower manufacturing cost, full memory functionality, a high I_{on}–I_{off} ratio, and a robust thermal budget.

A vertical single-crystal SiNW GAA memory was discussed in May of 2011 by A*STAR, Nanyang Technical University, and Global Foundries [16]. Both junction-based and junctionless devices were made. A junctionless device with 20 nm NW channel had a memory window of 3.2 V in 1 ms with P/E at +15 V/−16 V. It was comparable to the performance of a junction-based NW device with a silicon nanocrystal trap layer that achieved a memory window of 2.7 V in 1 ms with P/E at +15 V/−16 V. The process complexity, however, was reduced by the absence of junctions. The junctionless vertical SiNW was thought to be a suitable platform for stacked high-density multilevel cell (MLC) memory applications. A schematic cross-section of the SiNW SONOS device is shown in Figure 3.23(a) without junction and in Figure 3.23(b) with junction [16].

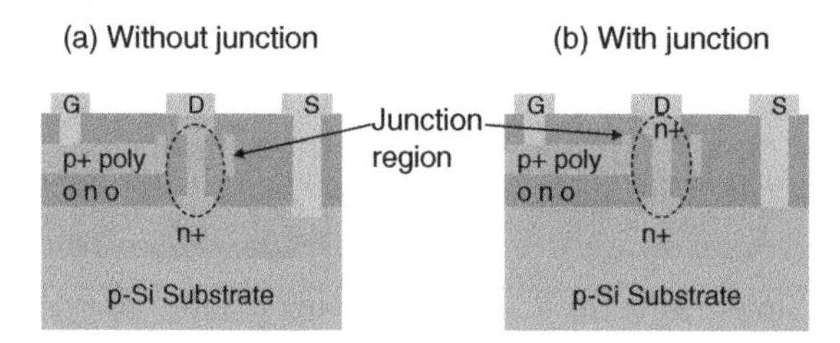

Figure 3.23 Schematic cross-section of silicon NW SONOS (a) without junction; and (b) with junction. (Based on Y. Sun *et al.* (2011) (Nanyang Technical University, A*STAR, GlobalFoundries), *IEEE Transactions on Electron Devices*, 51(5), 1329 [16].)

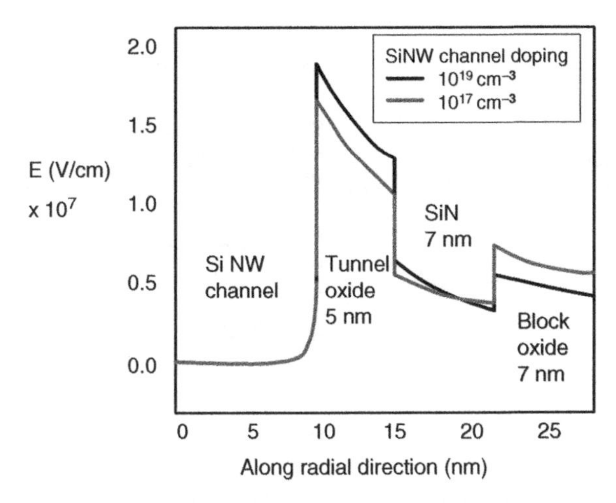

Figure 3.24 Illustration of electric field across the junctionless device vs. channel doping. (Based on Y. Sun *et al.*, (A*STAR, Nanyang Technical University, GlobalFoundries), IMW, 22 May 2011 [17].)

The same group reported further on this junctionless device in May of 2011 [17]. Devices with $L_g = 120$ nm were used with various wire diameters and dopant concentrations.

The I_d–V_g characteristics of the junction and junctionless devices were comparable with subthreshold swing <70 mV/dec and OFF current $<1 \times 10^{-12}$ A. The junctionless device was normally OFF due to the high work function of the p+ gate polysilicon relative to the n-channel, which depleted the channel electrons. The V_{th} increased as the channel diameter decreased due to the deep depletion in the narrow wires. Channel doping had an impact on the P/E speeds, with programming being faster for high doping and erase being faster for low doping.

A simulation showed that the electric field across the tunnel oxide became larger with increased channel doping, while the electric field across the blocking oxide became smaller using the same gate voltage and a 20 nm diameter NW, as shown in Figure 3.24 [17].

The programming speed with high-channel doping was enhanced by the large vertical electric field at the channel to tunnel oxide interface, the depressed electric field in the blocking oxide, and the high electron density at the surface of the heavily doped channel. The erase, however, was slower with high-channel doping due to the low hole density in the channel and the increased hole tunneling barrier height. High-temperature data retention at 85 °C for the junctionless SONOS with high-channel doping of 1×10^{19} cm^{-3} showed less V_{th} degradation compared to the lightly doped junctionless SONOS. The memory window showed little degradation up to 10^3 s.

Direct trap-to-band tunneling from the nitride traps to the channel conduction band tended to be the main discharge mechanism and, with increased channel doping, the energy states close to the channel conduction band tended to be occupied by electrons, which reduced the probability of further electron injection from the trap layer. Endurance was 10^5 cycles at 85 °C with little degradation of the P/E window.

For a vertical silicon NW array, the junctionless SONOS cell can be achieved by vertically stacking memory cells along a wire. Due to reduced processing, the junctionless SONOS has lower manufacturing cost yet retains full memory functionality, a high I_{on}/I_{off} ratio, and a robust

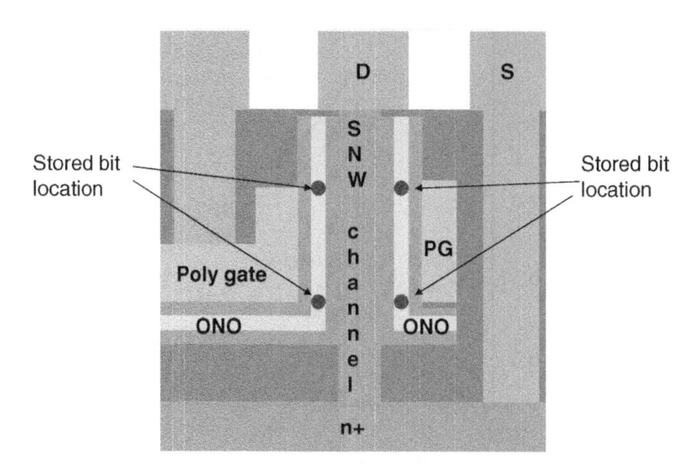

Figure 3.25 Schematic cross-section of 2-bit storage vertical NW CT memory device. (Based on Y. Sun (2011) (Nanyang Technical University, A*STAR, GlobalFoundries), *IEEE Electron Device Letters*, 32, (6), 725 [18].)

thermal budget, making it promising for highly scaled 3D multilevel integration of NAND Flash.

Further development with this junctionless SONOS vertical SiNW GAA memory was discussed in June of 2011 by Nanyang Technical University, Global Foundries, and A*STAR [18]. This device had two physical storage nodes per cell. The two physical bits per cell were evaluated for the second bit effect, and P/E speeds, endurance, and data retention were obtained. The relaxed channel length in the vertical structure permitted more tolerance for overcoming scaling-related reliability issues, and the absence of junctions along with the multiple storage nodes reduced the cost.

A schematic cross-section of the two-physical-bit storage device is shown in Figure 3.25 [18]. The charges were separately stored in the nitride layers at the top and the bottom of the vertical wire channel. The wire diameter was about 20 nm, and the gate length was 120 nm. The bottom of the wire was the source. The ONO thickness was 5 nm, 7 nm, and 7 nm, respectively.

The I_d–V_g characteristics of the junction less SONOS were comparable with the junctioned SONOS. These properties were attributed to the scaled wire channel with the added gate control of the GAA structure. The p+ gate resulted in a normally “off” behavior, where the high work function of the gate poly relative to the n-channel depleted the channel electrons. The increase in V_{th} with reduction in channel diameter resulted from deep depletion in the narrow wires. The aggressive scaling of the cell made conventional CHE injection unsuitable for mass data storage due to its high power consumption, so a low-power P/E method was used, where Fowler-Nordheim channel erase helped raise the V_{th}. The programming was done through band-to-band HHI. To read the top bit, a source bias was applied to reduce the channel potential near the bottom bit. To read the bottom bit, a drain bias was applied to reduce the channel potential near the top bit. Operation biases are shown in Table 3.2 [18].

Retention characteristics of the junctionless device were studied, and it was found that the V_{th} window was maintained after 10^5 s at 85 °C for both virgin devices and 10^4 cycled devices. Endurance was shown up to 10^5 cycles with no V_{th} shift observed.

Table 3.2 Operation conditions of a junctionless vertical SiNW-based CT memory with 2-bit storage.

	Top bit			Bottom bit		
	V_g	V_d	V_s	V_g	V_d	V_s
Program	−5 V	5 V	0 V	−5 V	0 V	5 V
Erase	15 V	−3 V	−3 V	15 V	−3 V	−3 V
Retention	—	0 V	1.6 V	—	1.6 V	0 V

Based on Y. Sun *et al.* (2011) (Nanyang Technical University, A*STAR, GlobalFoundries), *IEEE Electron Device Letters*, 32, (6), 725 [18].

3.4.2 Junctionless GAA SONOS Silicon Nanowire on Bulk Substrate for 3D NAND Stack

Junctionless NAND flash NW CT memories with a virtual S/D induced as an inversion layer by the gate-fringing field have been reported for stacked and vertical channel devices, as shown in Figure 3.26 [10]. These devices have lightly doped S/D junctions and are used because implantation is difficult for a 3D stackable process.

There is an issue, however, with loss of read current due to the high S/D resistance that is induced by an inadequate fringing field. The floating body potential can also be an issue. To fix the floating body potential during erase operations, a sufficient number of holes must be generated by band-to-band tunneling from the S/D junctions. This means the S/D junctions must be heavily doped and uniform.

In June of 2011, KAIST discussed a junctionless GAA SONOS NAND flash device with a homogeneously n+ doped SiNW on a bulk substrate, which is illustrated in Figure 3.27 [10].

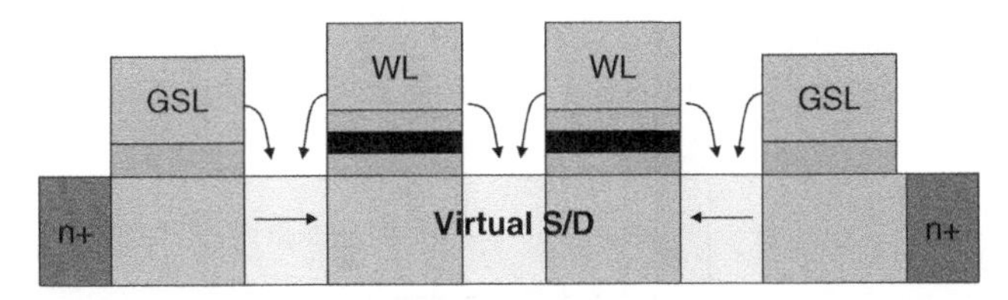

Figure 3.26 Junctionless NAND flash polysilicon NW CT memory with virtual source/drain. (Based on S.J. Choi *et al.*, (KAIST), VLSI Technology Symposium, June 2011 [10].)

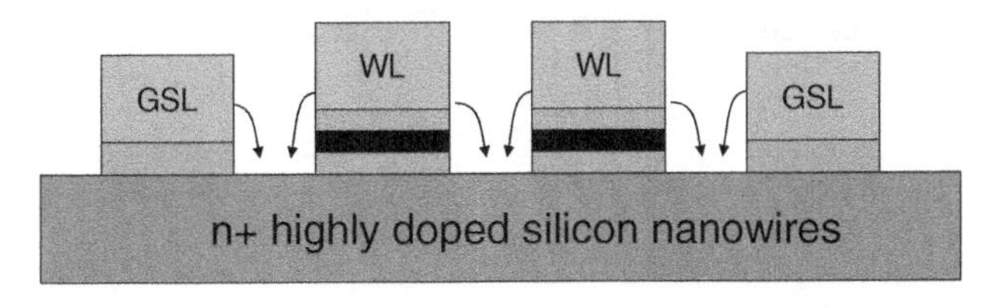

Figure 3.27 Junctionless "all-around-gate" SONOS NAND flash on silicon NW. (Based on S.J. Choi *et al.*, (KAIST), VLSI Technology Symposium, June 2011 [10].)

The SiNW diameter was 4 nm, and the gate length was 20 nm. A deep RIE process was used to form the SiNW. The junctionless SONOS device had a read current >10 μA, a V_{th} window >6.5 V, a narrowed distribution of the erased V_{th}, and endurance of 10^5 cycles. The process was used to implement nine layers of single-crystal SiNW vertically integrated for a 3D NAND device.

3.4.3 Modeling Erase in Cylindrical Junctionless CT Arrays

A new erase saturation issue in cylindrical CT cells integrated along junctionless NAND strings was modeled by the Politecnico di Milano and IFM-CNR in December of 2012 [19]. The effect is a result of not properly inducing an inversion layer in the intercell regions of the string during read when a positive charge is stored in the cells.

A parametric analysis indicated that the erase saturation issue improved when narrow intercell regions or large substrate radii were used. The length of the intercell regions, however, has a significant effect on other string parameters including the string's immunity to lateral charge migration during data retention This results in a tradeoff in the design of the intercell regions of the string. This tradeoff is relatively easy for large-substrate radii but becomes more difficult for small-substrate radii. Parameters for the simulated cylindrical junctionless NAND string are shown in Figure 3.28 [19].

Simulations showed that for negative Q_s, string conduction represented by V_{th} was correctly set by channel inversion in the selected cell and that intercell regions were correctly inverted by the positive word-line bias. For positive Q_s, however, the low bias applied to the selected word-line was sufficient to strongly invert the selected cell channel but not the adjacent intercell regions. This meant that V_{th}, in this case, was controlled by these regions regardless of Q_s.

This nonlinearity of V_{th} for Q_s in the positive region is displayed in Figure 3.29, which shows ΔV_{th} as a function of Q_s for two V_{pass} values applied to the unselected word-lines [19]. This result confirms that the inversion of the intercell regions by the word-line bias is critical for a cylindrical memory due to fringing fields being weaker in the cylindrical geometry than in the planar geometry.

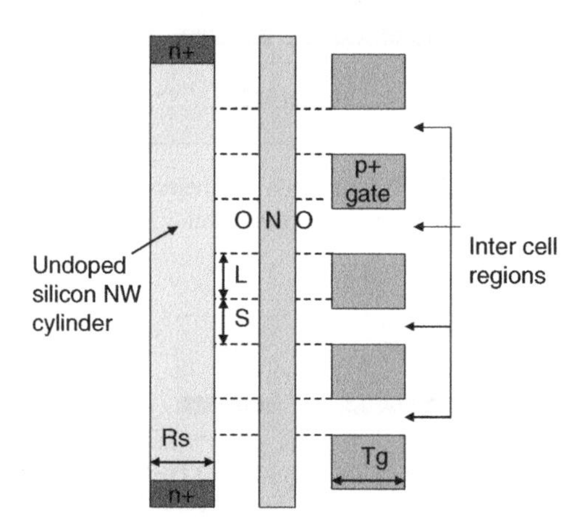

Figure 3.28 Parameters of simulated cylindrical junctionless NAND string. (Based on A. Maconi *et al.*, (Politecnico di Milano, IFN-CNR), IEDM, December 2012 [19].)

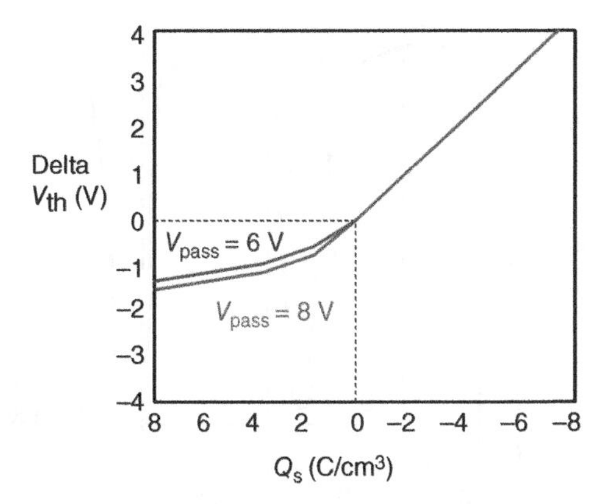

Figure 3.29 ΔV_{th} as a function of Q_S. (Based on A. Maconi *et al.*, (Politecnico di Milano, IFN-CNR), IEDM, December 2012 [19].)

Figure 3.30 illustrates electron concentration along the string [19]. Going from a cylinder radius (R_s) of 50 nm to 5 nm resulted in an increase in electron concentration under the gate (*L*) while leading to a reduction in minimum value of the electron concentration in the intercell regions (*S*).

This reduction increased the bias value necessary to reach the threshold condition of 100 nA flowing through the string, which resulted in a higher cell V_{th} sensed by the read operation [19].

Because the new erase saturation effect was a result of the intercell regions becoming more dominant on string conduction as the positive charge in the cells was increased, a solution to the issue was reducing the intercell regions *S*. Figure 3.31 illustrates the potential for the word-lines to better control the inversion of the intercell regions when these regions are shorter. This figure illustrates electron concentration along the string for $R_s = 100$ and $R_s = 50$ [19].

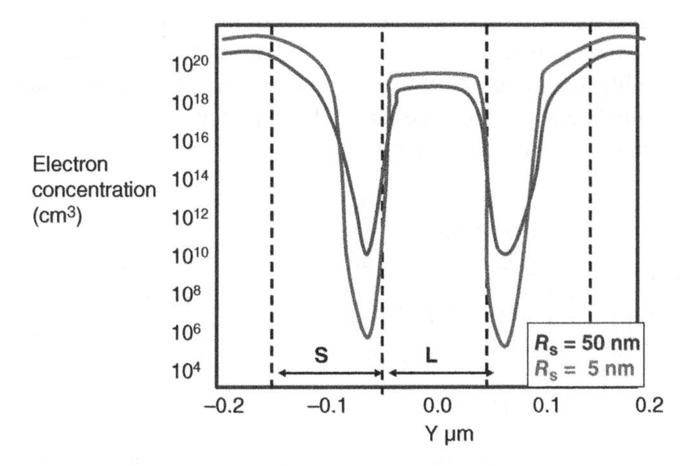

Figure 3.30 Illustration of electron concentration along the NAND string for different string radii. (Based on A. Maconi *et al.*, (Politecnico di Milano, IFN-CNR), IEDM, December 2012 [19].)

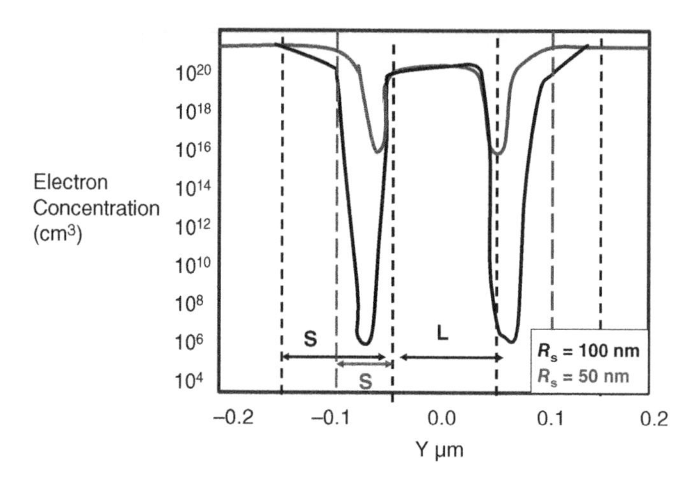

Figure 3.31 Illustration of electron concentration along the string with reduced intercell region *S* for R_s of 100 nm and 50 nm. (Based on A. Maconi, (Politecnico di Milano, IFN-CNR), IEDM, December 2012 [19].)

The potential for relieving erase saturation by reducing the length of the intercell region (*S*) is limited by the effect of lateral electron migration during data retention, which requires *S* to be a minimum size for minimizing cell-to-cell crosstalk.

In July of 2013, the Politecnico di Milano and CNR Milano further discussed their detailed simulation analysis of the erase performance of a junctionless CT memory array in cylindrical geometry [20]. The simulation in this study showed that a saturation of the erased threshold voltage occured as a result of incomplete inversion of the intercell regions when a positive charge was stored in the cells.

Figure 3.32 shows a schematic of the device electrostatics at threshold voltage when a positive charge is stored in the cell by the read operation [20]. A bottleneck for string conduction is the inversion of the intercell regions close to the selected cell, because all the other substrate regions are strongly inverted by V_{pass} or by the positive stored change. The erase saturation issue was studied as a function of string and cell parameters. This showed lower erase capability for large cell-to-cell separation, small substrate radius, and small equivalent oxide thickness for the gate stack. It adds new constraints to the design of cylindrical junctionless memory technology.

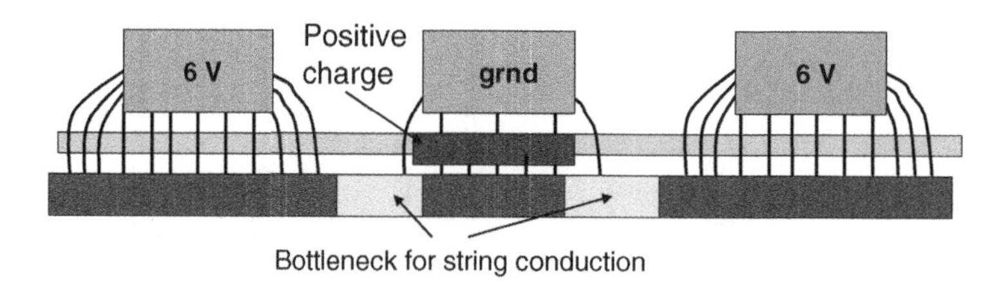

Figure 3.32 Schematic of device electrostatics at *Vth* with positive read charge stored in cell. (Based on A. Maconi *et al.* (2013) (Politecnico di Milano, Con. Naz. de Ricerche Milano), *IEEE Transactions on Electron Devices*, 60(7), 2203 [20].)

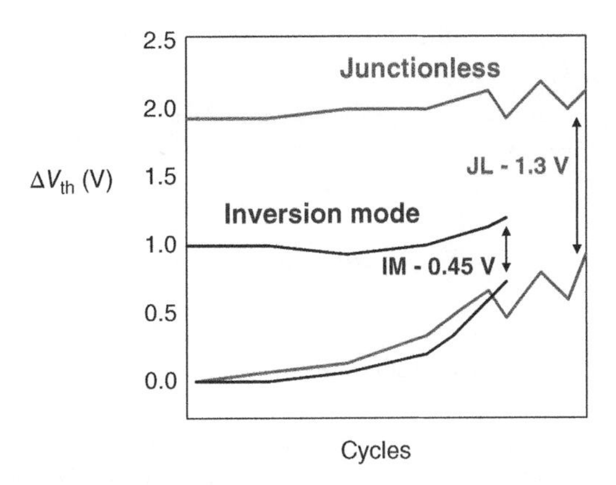

Figure 3.33 Endurance characteristics of junctionless and inversion mode NAND flash. (C.Y. Chen (2013) (NTHU, National Nano Development Laboratories), *IEEE Electron Device Letters,* 34(8), 993 [21].)

3.4.4 HfO_2–Si_3N_4 Trap Layer in Junctionless Polycrystal GAA Memory Storage

A junctionless polycrystalline-based flash memory that used an HfO_2–Si_3N_4 (HN) stacked trapping layer for memory storage was discussed in August of 2013 by National Tsing Hua University (NTHU) and the National Nano Device Labs of Taiwan [21]. The slow erase speed of junctionless devices was an issue, and bandgap engineering was found to improve operational and reliability characteristics. By properly arranging the bandgaps of different materials or by controlling the dielectric formation processes, electrons and holes could be injected more efficiently during P/E operations, while long escape paths could be maintained to suppress electron leakage during retention.

The effects of the HN stacked trapping layer on junctionless and inversion-mode memory devices were studied and compared. The HN stacked trapping layer was proposed to improve P/E speed due to the lower conduction band level and the larger trap density of the HfO_2 compared with the Si_3N_4.

The junctionless memory showed faster programming speed than the inversion mode memory due to its heavily doped n-channel. The junctionless devices had comparable erase speeds due to the more effective electron detrapping. In addition, the junctionless memory with the HN trapping layer showed better retention characteristics and maintained a larger window after 10^5 P/E cycles. Both the junctionless and inversion mode devices were programmed at 15 V for 20 μs and erased at −15 V for 200 μs. The junctionless memory kept a 1.3 V window after 10^5 P/E cycles compared to a 0.45 V window remaining for the inversion mode part after 10^4 cycles, as shown in Figure 3.33 [21].

3.5 3D Stacked Horizontal Nanowire Single-Crystal Silicon Memory

3D NW NAND flash memories can be made with horizontal bit-lines or with vertical bit-lines. These memories can be made with single-crystal silicon or with polysilicon substrates. NAND flash memories with horizontal bit-lines using single-crystal silicon are discussed in this section.

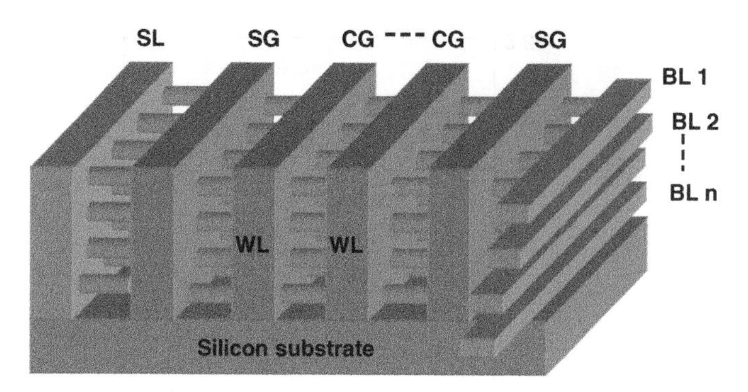

Figure 3.34 Concept of 3D crystalline silicon memory with horizontal NW bit-lines with common vertical gates. (Based on T. Ernest *et al.*, (CEA-LETI, Minetec), IEDM, December 2008 [23].)

3.5.1 Process for 3D Stacked Horizontal NW Single-Crystal Silicon Memory

An experiment with vertically stacking the NW memories using SiGe as a sacrificial layer was conducted by A*STAR, ITT, and Silterra in November of 2008 [22]. The differential oxidation rate between Si and SiGe yielded well-separated devices with higher drive current. Low drive current had been an issue with NW devices [22].

A stacked SiNW memory process for advanced 3D integration was discussed in December of 2008 by CEA-LETI and Minetec [23]. Standard microelectronic equipment was used to build up alternating layers of SiGe–Si superlattice epitaxy, which was etched into fins and then subjected to selective SiGe etching to form NWs. It was found possible to stack up to 19 layers of SiGe–Si, and the use of selective epitaxy allowed the number of stacked silicon layers to be determined locally. A 3D crystalline NW memory matrix with common gates for word-line control was proposed, as shown in Figure 3.34 [23].

Continuing to examine this idea, a multiple-level horizontal stacked crystalline silicon GAA NW CT memory technology was discussed in December of 2009 by CEA-LETI, MINATEC, IMEP-LAHC, and INPG-MINATEC [24]. The fabrication steps for a three-level single-crystal SiNW CT memory are illustrated in Figure 3.35 [24]. Fabrication consisted of building up a selective epitaxy of alternate layers of Si and SiGe on an SOI substrate, defining a fin, then performing anisotropic dry etching and SiGe dry isotropic etching to isolate the SiNWs. This is followed by ONO deposition to form the tunnel oxide, trapping nitride layer followed by a SiN selective etch and barrier oxide. The polysilicon N+ gate deposition followed.

The completed device had a 600 nm gate length and three 6 nm diameter GAA SONOS NW cells. Programming and erase were by Fowler Nordheim tunneling. A programming window up to 8 V was found with no subthreshold slope degradation. No erase saturation was observed up to −18 V, which showed that the introduction of a metal gate or high-κ interpoly oxide is not a requirement for CT NWs. Retention was good after 10^4 cycles.

The same group discussed the technology further in May of 2010, presenting an electrical characterization of the technology [25]. Detailed simulations of the write, erase, and retention characteristics using uniform Fowler-Nordheim tunneling were discussed. Program and erase

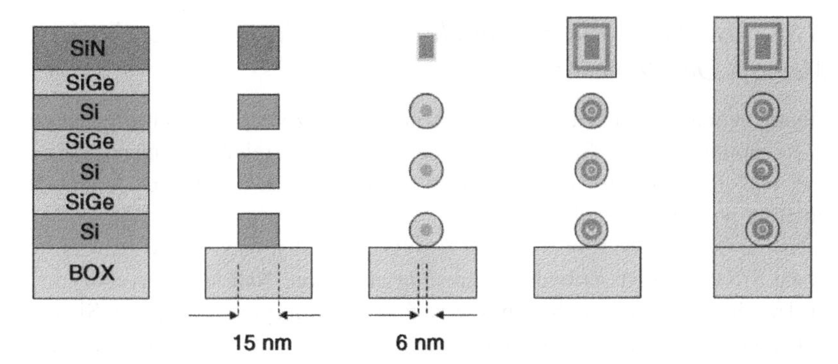

Figure 3.35 Fabrication sequence of three-level GAA single-crystal CT NW memory. (Based on A Hubert *et al.*, (CEA-LETI, MINATEC, IMEP-LAHC, INPG-MINATEC), IEDM, December 2009 [25].)

speeds were found to differ for cylindrical NW memories and for planar memories. Cylindrical SONOS NW devices initially program within 1 μs for high voltages such as 14 to 18 V, and after that the speed slows down. Planar devices, however, show slow initial programming followed by higher programming speed. A similar behavior occurs for erase, with a less than 1 μs erase occurring in cylindrical SONOS cells but not planar cells even with a high erase voltage. An illustration of this effect is shown in Figure 3.36 [25].

Cylindrical devices show no trace of erase saturation effect even without metal gates. The performance of the cylindrical devices was found better than planar devices due to the thicker 6 nm tunnel oxide compared with 2 nm for planar. The faster dynamics of the cylindrical CT device could be explained as an electrostatic effect of the higher electric field in the tunnel oxide of the cylindrical cell compared to the planar cells and the lower field in the barrier oxide of the cylindrical cells than in planar cells.

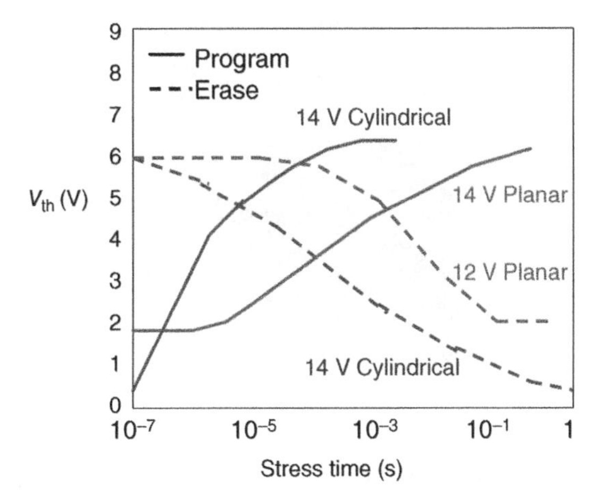

Figure 3.36 Program and erase at 14 V for cylindrical SiNW ONO devices with $R = 3$ nm and ONO = 6/5/8 nm, and for planar devices at 14 V program and 12 V erase with ONO = 2.2/5.0/8.6 nm. (Based in E. Nowak *et al.*, (CEA-LETI, MINATEC, IMEP-LAHC, INPG-MINATEC), IMW, May 2010) [25].)

3.5.2 A Stacked Horizontal NW Single-Crystal Silicon NAND Flash Memory Development

The development of a 3D stacked single-crystal silicon horizontal NW memory was described, including its epitaxial SiGe–Si process. Operational and reliability characteristics were determined, and a new electrical layer selection method was developed. A unit 3D block design structure was also described.

A 3D flash array, intended for use as a NAND flash, that used many stacked horizontal single-crystal SiNWs was described by Samsung and Seoul National University in April of 2011 [26]. The single-crystal NW bit-lines were formed from epitaxially grown SiGe–Si layers on a Si substrate. By stacking the bit-lines, the memory density could be increased without scaling the lateral dimensions. The device structure and fabrication included electrical isolation of the stacked NWs. Simulations indicated that NAND flash memory operation was possible. An array of these devices with stacked NW GAA memory devices on the bit-lines were made, and memory characteristics such as program and erase were measured. It was called the stacked array (STAR) architecture. A schematic of the 3D stacked NW memory is shown in Figure 3.37 [26].

The bit-line STAR had single-crystal silicon channels and a GAA structure, and the unit cell had a GAA structure with a cylindrical channel. If the S/D doping process was precisely controlled, a body contact region could be made for the block erase operation. The bit-lines and word-lines were perpendicular, and multiple stacked bit-lines shared one word-line.

Numerical simulations of operational characteristics were performed for the 3D single-crystal NW STAR CT array. The I_d–V_g characteristics and the electric field on the ONO dielectrics showed better transistor performance and P/E characteristics than a comparable planar memory device. Endurance characteristics for the NW array are illustrated in Figure 3.38 (a), and data retention characteristics are shown in Figure 3.38(b) [26]. Endurance for program of 10 V in 100 μs and erase of −9 V in 500 μs showed an initial 3.4 V memory window and a

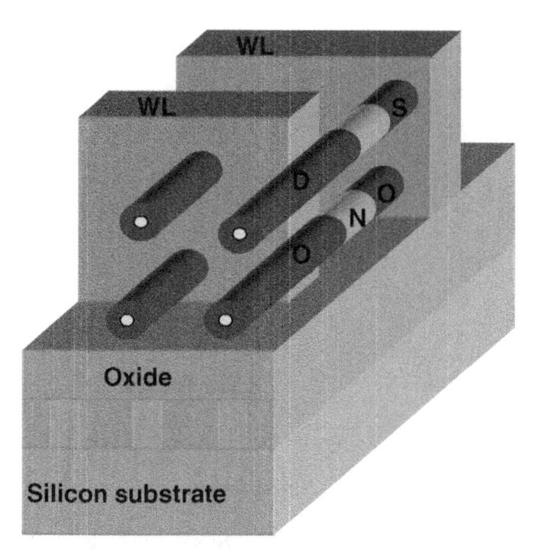

Figure 3.37 Schematic of 3D stacked NW flash array. (Based on J.G. Yun *et al.* (2011) (Samsung, Seoul National University), *IEEE Transactions on Electron Devices*, 58(4), 1006 [26].)

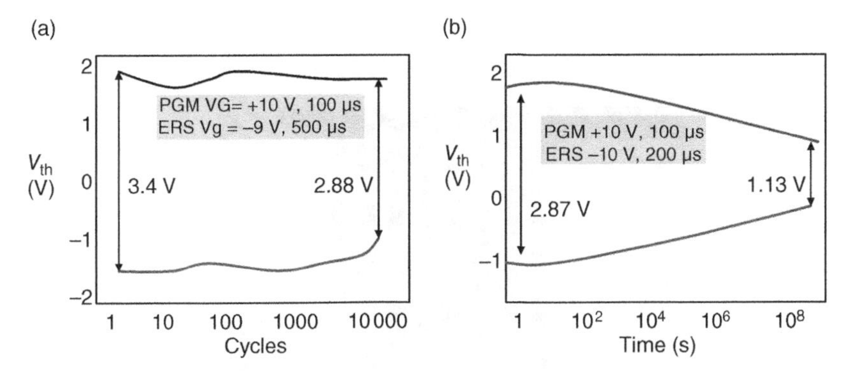

Figure 3.38 Simulated endurance and data retention characteristics of the NW "STAR" memory. (Based on J.G. Yun *et al.* (2011) (Samsung, Seoul National University), *IEEE Transactions on Electron Devices*, 58(4), 1006 [26].)

window of about 2.88 V after 10^4 cycles. Data retention at room temperature showed about a 1.13 V memory window after 10^8 seconds.

Numerical simulations of the flash memory operation were developed, and various memory characteristics, such as program and erase select gate operation, were measured. The I_d–V_g curves and the electric field on the ONO dielectrics showed better transistor performance and P/E characteristics than a conventional planar device.

A new electrical layer-selection method in a bit-line stacked 3D NAND STAR memory array was discussed by Samsung and Seoul National University in July of 2011 [27]. A schematic of the proposed 3D NAND flash memory with horizontal bit-lines is shown in Figure 3.38. The figure shows the source select lines, the horizontal word-lines, and the staircase body [27].

Selecting a bit-line in a stack of horizontal bit-lines required a new selection method. The stacked layers were selected by using multiple source select lines (SSLs) with a unique configuration of erased cells in each layer, as indicated in Figure 3.39 [27]. The number of SSLs required is minimal; for example, to select 252 stacked bit-lines, only 10 SSLs are required. A single vertical stack of source bit-lines was contacted in the interconnect layer above the array. Ground select lines (GSL) were contacted using the terraced body. An etch-through spacer technique was developed to form the terraced body for a vertical contact process. Word-lines were perpendicular to the bit-lines.

For READ, some current degradation, due to the many SSL transistors in a string, was expected, compared with that of a planar array. If the single-crystal silicon channel was compared with a polysilicon channel, however, the current was enhanced. There was some concern with hot-carrier-induced soft programming due to disturb in the SSL transistors, but this could be suppressed by inserting dummy word-lines.

For a word-line stacked array, bit-lines are formed after stacking word-lines. As a result, channels are polycrystal, which can degrade electrical characteristics. A stacked array using single-crystal channels can be made by epitaxial growth of Si–SiGe layers on a Si substrate. In this case, there is a need to distinguish stacked bit-lines during operation. Forming metal contacts on each stacked layer is difficult. Lithography and the etch process of stacked layers were complex. A trim and etch process using photoresist could have been used, but it required many operations.

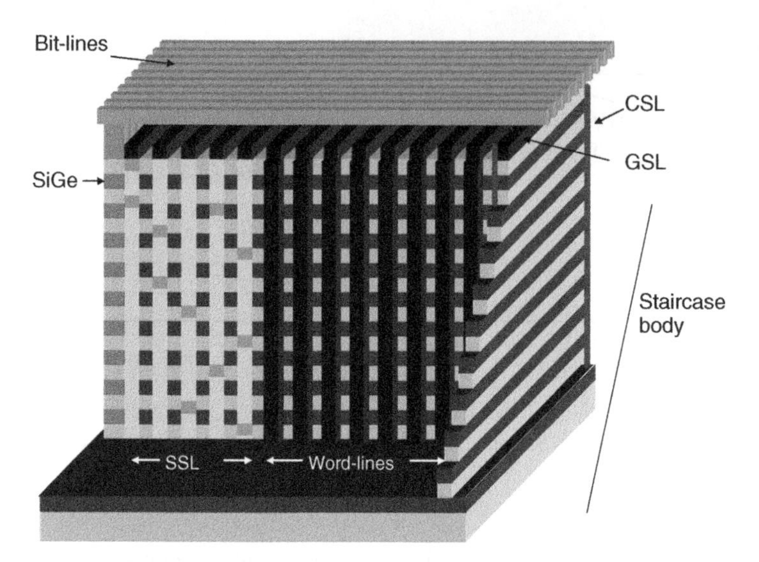

Figure 3.39 Schematic of 3D NAND flash memory with horizontal bit-lines. (Based on J.G. Yun *et al.* (2011) (Samsung, Seoul National University), *IEEE Transactions on Electron Devices*, 58(7), 1892 [27].)

A new layer selection by erase method was proposed to distinguish stacked bit-lines, and a simple contact process was developed using an "etch-through-spacer" technique. Laterally distributed lines were selected by the bit-line wires, and vertically stacked lines were selected by the operation of SSLs. For the initial erase operation, the bodies of each stacked layer needed to be contacted. But with a common source line next to a ground select line, there is no space to contact the body for the erase operation. Terraced body regions were therefore directly contacted for the layer-by-layer erase operation. Program and read disturbance were expected to increase compared to planar NAND flash arrays and to be dependent on the number of stacked layers.

The 3D STAR single-crystal stacked NAND flash memory was discussed further in January of 2012 by Samsung and Seoul National University [28]. This NAND flash technology was targeted at the high-density and low-cost smart phone and tablet market. The single-crystalline STAR architecture was proposed as a 3D unit structure, which took into consideration the structure and operation of the full array. The total number of cells on one word-line in the channel-stack type of NAND had no limitation. The conclusion was that the channel-stack type of NAND architecture had a smaller unit cell size, better performance, and more extendability than the vertical channel type of architecture. The STAR unit component was used to design a 3D block and a full chip architecture. An example of the unit building structure is shown in Figure 3.40 [28]. Full array operation methods were considered for a terabit NAND flash memory capacity.

To evaluate the performance of the STAR architecture, a 3D TCAD simulation was used. The advantages found for the STAR architecture over the vertical gate were as follows: current drivability, subthreshold swing, and the single-crystal channel, which provided uniform distribution of cell performance due to the absence of defects at the grain boundaries. Another

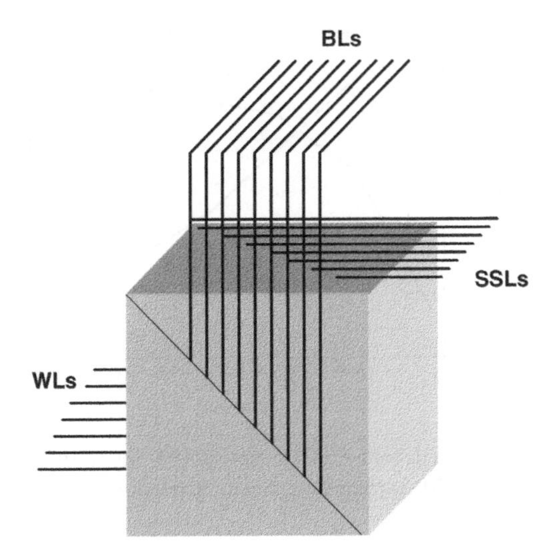

Figure 3.40 Schematic of unit building block in the STAR NAND flash structure. (Based on Y. Kim *et al.* (2012) (Seoul National University, Samsung), *IEEE Transactions on Electron Devices*, 50(1), 35 [28].)

advantage of the channel stack is that the read current drivability is independent of the number of layers stacked.

A 3D NAND flash memory prototype using a nitride-based CT single-crystal channel stacked GAA array was discussed by Samsung and Seoul National University in August of 2013 [29]. This 3D NAND flash architecture had four levels of stacked horizontal single-crystal silicon NW channels (CSTAR). This CSTAR NAND flash memory was fabricated and its operations verified.

The CSTAR architecture was made of four different devices with two gates controlled by word-lines, an SSL, and an S/D contact to each layer for layer selection. A schematic view of a CSTAR device and array is shown in Figure 3.41 [29].

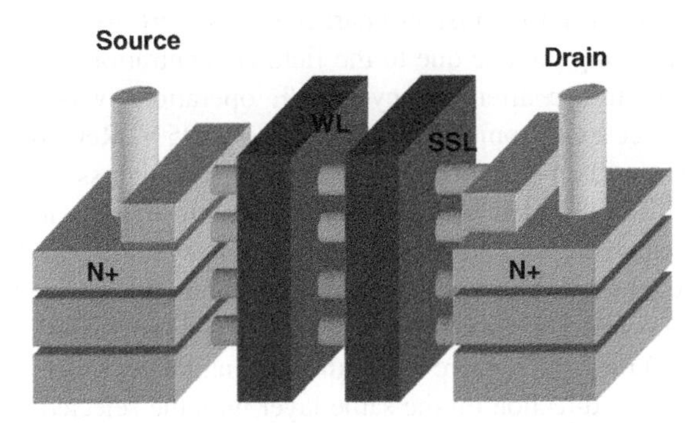

Figure 3.41 Schematic of four-level crystalline silicon C-STAR NAND flash architecture. (Based on Y. Kim *et al.* (2013) (Samsung, Seoul National University), *IEEE Electron Device Letters*, 34(8), 990 [29].)

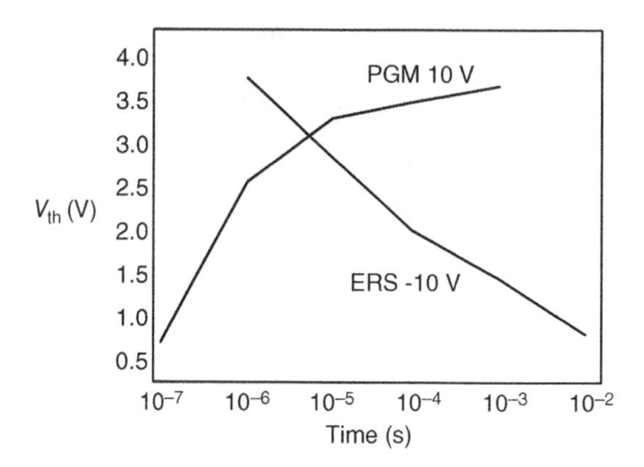

Figure 3.42 Program and erase speeds for a CSTAR array at 10 V for 100 μs for program and −10 V for 1 ms for erase. (Based on Y. Kim (2013) (Samsung, Seoul National University), *IEEE Electron Device Letters*, 34(8), 990 [29].)

During process integration of the CSTAR flash memory, the silicon germanium selective etching was a critical step. To make the single-crystal SiNW channels, an epitaxial growth was done of multiple Si–SiGe layers on the Si wafer. After active channel definition, the sacrificial SiGe layers were removed to isolate the stacked silicon channels from each other. Due to the slope of the etching profile, the diameters of the different-layer SiNW channels were different. The diameters of these channels were as follows: first layer: 13 nm; second layer: 32 nm; third layer: 45 nm; and fourth layer: 55 nm. Because the electric field is inversely proportional to the radius of the cylindrical channel, program speed variation could be an issue between the layers. It was expected that, over time, better equipment and improved etching processes would be required to obtain a more vertical sidewall slope in deep trench etches.

The CSTAR device and array were characterized. Program operation involved applying a word-line voltage of 10 V for 100 μs. For the erase operation, −10 V for 1 ms was applied. A large V_{th} window resulted, as illustrated in Figure 3.42 [29].

Because the device channel is single-crystal silicon, a good subthreshold swing of about 120 mV/decade was shown. Fast transient characteristics were found in the P/E operation even at a low 10 V operating voltage due to the field concentration effect of the cylindrical channel geometry. Endurance after 10^4 cycle P/E operations was a 2.84 V_{th} window, showing only a 3.7% decrease from the initial value of 2.95 V. Retention of the cells after 10^3 cycles at 85 °C showed an extrapolated accelerated charge loss of 25% after 10 years.

The array operation is illustrated in the schematic circuit diagram in Figure 3.42, which shows three levels of channels with three channels on each level [29]. Each channel had a bit-line contact, an SSLx select device, a WLy word-line, and a GSL device to control CSL and body contact. In the array operation, a cell was selected. In addition to the inhibited cells used in operating planar NAND flash, there were two additional inhibit cells used on the diagonal on a different layer and in the *y* direction on the same layer with the selected cell, as illustrated in Figure 3.43 [29]. To prevent an inhibited cell from being programmed, the PROGRAM INHIBIT operation by self-boosting was used as in the conventional planar NAND array. The Selective program operation in a specific layer was successfully verified.

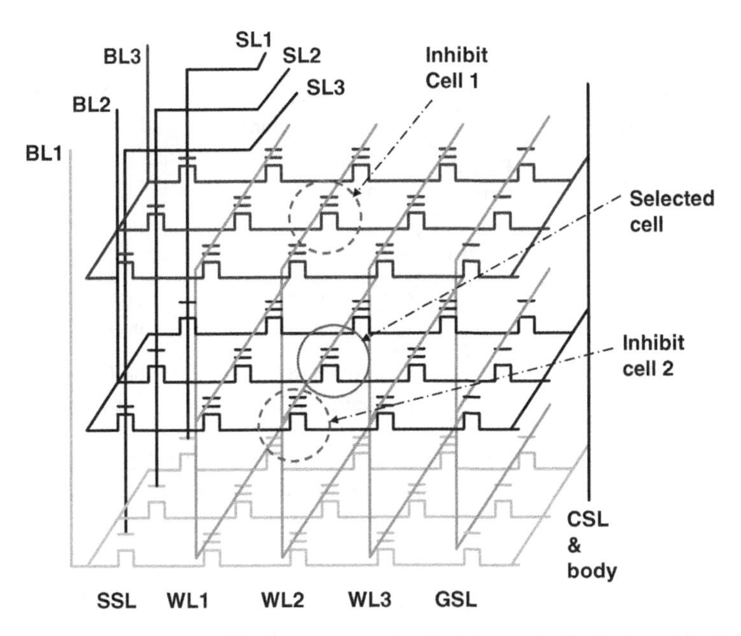

Figure 3.43 Schematic circuit diagram of array showing 3D inhibited cells during cell select. Inhibit cell 1 is the neighboring cell on the adjacent word-line in the *Z* direction, and inhibit cell 2 is the neighboring cell on the adjacent word-line in the *Y* direction. (Based on Y. Kim *et al.* (2013) (Samsung, Seoul National University), *IEEE Electron Device Letters,* 34(8), 990 [29].)

The sloped etch profile made the diameter of each NW channel different. Because the electric field was inversely proportional to the radius of the cylindrical channel, program speed variation could be an issue. To reduce this variation, the operating speed in the bottom layer with the widest channel dimension needed to be checked and minimum operating times determined for managing all the layers. This 3D stacking technology was thought to be promising for the terabit era.

3.6 Vertical Single-Crystal GAA CT Nanowire Flash Technology

3.6.1 Overview of Vertical Flash Using GAA SONOS Nanowire Technology

Single-crystal GAA CT NW flash memories start with a single-crystal NW. This section discusses NWs made from a vertical silicon fin on a silicon substrate.

3.6.2 Vertical Single-Crystal Silicon 3D Flash Using GAA SONOS Nanowire

A complementary metal–oxide–silicon (CMOS)-compatible GAA CT flash memory that used a vertical silicon pillar for the NW was reported by A*STAR in August of 2009 [30]. The schematic process flow is illustrated in Figure 3.44 [30]. In Figure 3.44(a) the single-crystal NW originated as a tall thin silicon fin with a diameter down to 50 nm. ONO was deposited

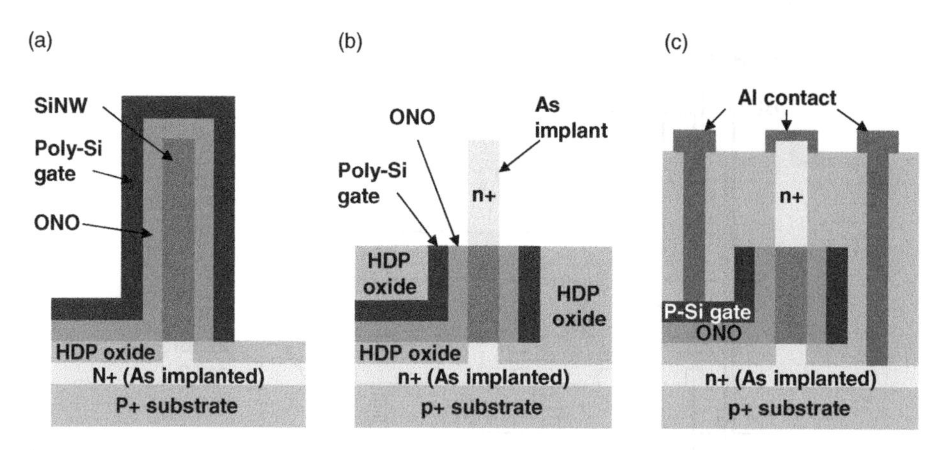

Figure 3.44 Schematic process flow for vertical silicon nanowire SONOS flash memory. (Based on M. Chen *et al.* (2009) (A*STAR, Nanyang Technical University), *IEEE Electron Device Letters*, 30(8), 879 [30].)

followed by a polysilicon gate and a high-density plasma oxide. In Figure 3.44(b) the top of the NW was exposed and implanted with As. Figure 3.44(c) shows that SiO_2 was deposited and aluminum contacts formed. The NW diameter was 50 nm, and the memory gate length was 150 nm. The unoptimized SONOS vertical gate stack had a SiO_2–Si_3N_4–SiO_2 thickness of about 5 nm, 5 nm, and 6 nm, respectively, and showed well-behaved memory characteristics for P/E window, data retention, and endurance.

Retention characteristics at 85 °C showed no degradation after 10^8 s after programming at $V_g = 16$ V or 18 V for 1 ms and erase at $V_g = -16$ V or -18 V with $V_d = 5$ V for 1 ms. Endurance was done at $V_g = 18$ V program for 10 ms and erase at $V_g = -18$ V with $V_d = 5$ V for 10 ms. The device did not show any P/E window narrowing after 10^4 cycles.

In May of 2010, Nanyang Technical University and A*STAR reported further on their vertical SiNW CT memory [31]. The 50 nm channel width was reduced to a 20 nm diagonal, somewhat-square geometry. A top-down method was used for manufacture. This device had a significant P/E speed performance improvement due to enhancement of the electric field of the tunneling dielectric near the corners of the square, which had a smaller radius of curvature because of the smaller diameter. Comparison of the program and erase characteristics for both the 20 nm and 50 nm devices are shown in Figure 3.45 [31]. The improvement in V_{th} window for the 20 nm silicon NW vertical CT device was significant.

Retention and endurance were also improved for the 20 nm sample over the larger 50 nm device.

The performance improvements were due to the enhanced field in the tunneling oxide and also a reduction in the field in the blocking oxide because of its reduced radius of curvature. Fowler-Nordheim tunneling was used for both program and erase. L_g was 150 nm. Endurance showed no degradation up to 10^4 cycles at 25 °C. Multibit programming capability was indicated.

3.6.3 Fabrication of Two Independent GAA FETs on a Vertical SiNW

The fabrication of two independently controlled GAA metal–oxide–silicon FETs (MOSFETs) on a single vertical single-crystal SiNW using a CMOS process technology was discussed in

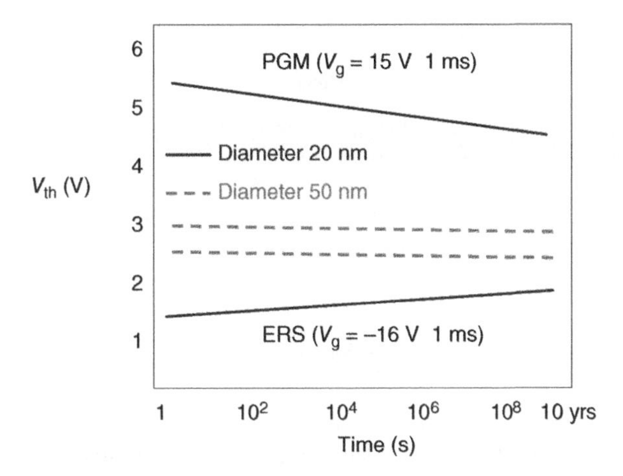

Figure 3.45 Program and erase characteristics of the 20 nm vs. 50 nm SiNW vertical CT device. (Based on Y. Sun *et al.* (2010) (Nanyang Technical University, A*STAR), *IEEE Electron Device Letters*, 31(5), 390 [31].)

November of 2011 by A*STAR and the University of California, Santa Barbara [32]. The second gate is stacked vertically on top of the first gate, maintaining the same footprint. A dielectric isolated the two gates. This structure had potential for stacked CT memory as well as for CMOS logic. A schematic diagram of the two stacked gates is shown in Figure 3.46 [32].

A NAND string using two-level stacked junctionless GAA SONOS cells made on a vertical SiNW was discussed in December of 2011 by A*STAR, Nanyang Technological University, Peking University, and Global Foundries [33]. The cell string had a footprint of $6F^2$. The stacked SiNW memory cells had dimensions down to 30 nm. These cells had well-behaved

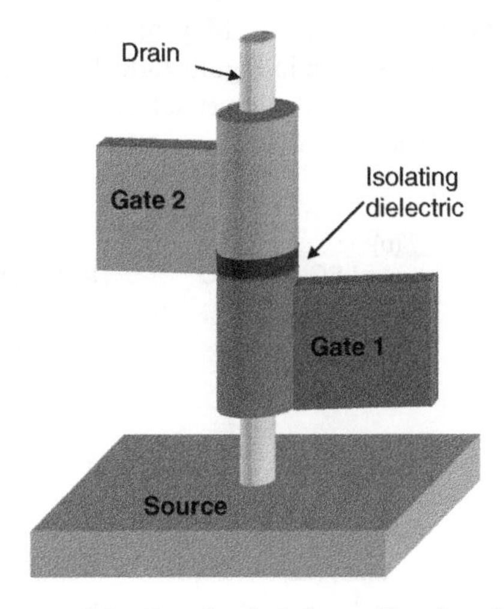

Figure 3.46 Schematic Diagram of the Two Stacked Gates (Based on X. Li *et al.* (2011) (A*STAR, University of California Santa Barbara), *IEEE Electron Device Letters*, 32(11), 1492 [32].)

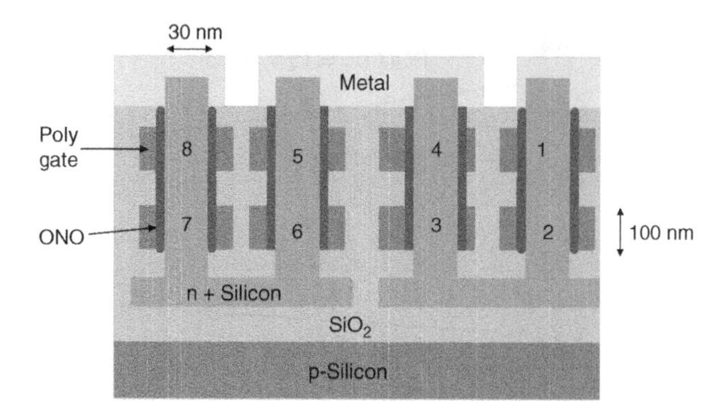

Figure 3.47 Schematic cross-section of an eight-cell junctionless GAA CT NAND string. (Based on Y. Sun *et al.*, (A*STAR, Nanyang Technical University, Peking University, GlobalFoundries), IEDM, December 2011 [33].)

memory characteristics, including program and erase speed, endurance, data retention, and program disturb. A vertically stacked device structure improved the lateral bit density, and the junctionless channel reduced process complexity and cost. The device had a very low thermal budget. A two-level junctionless channel GAA CT device was built on a vertical SiNW with $6F^2$ footprint and integrated into a NAND string. An eight-cell NAND flash string is illustrated in the schematic cross-section in Figure 3.47 [33]. The SSL transistors can be formed on the peripheral region. The wire diameter was 30 nm, and $L_g = 100$ nm. Wire channel doping was $10^{19}\,cm^{-3}$. Subthreshold slope was 84 mV/dec with $I_{off} = 1 \times 10^{-12}$ A.

3.6.4 Vertical 3D Silicon Nanowire CT NAND Array

An architecture to expand the number of vertical SiNW junctionless stacked charge trapping devices in a NAND array for 3D applications was discussed by A*STAR, Nanyang Technical University, and GlobalFoundries in May of 2011 [17]. The formation process concept for the vertical NW array is shown in Figure 3.48 [17]. The vertical SiNW array was formed and isolated, then a gate stack for the lower select gate was formed and etched. This process resulted in the formation shown in the cross-section in Figure 3.48(a). The gate stack

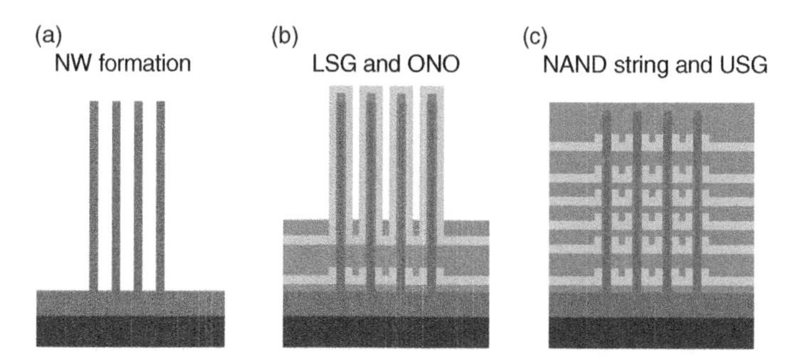

Figure 3.48 Vertical silicon NW junctionlesss stacked CT NAND string array. (Based on Y. Sun *et al.*, (A*STAR, Nanyang Technical University, GlobalFoundries), IMW, 22 May 2011 [17].)

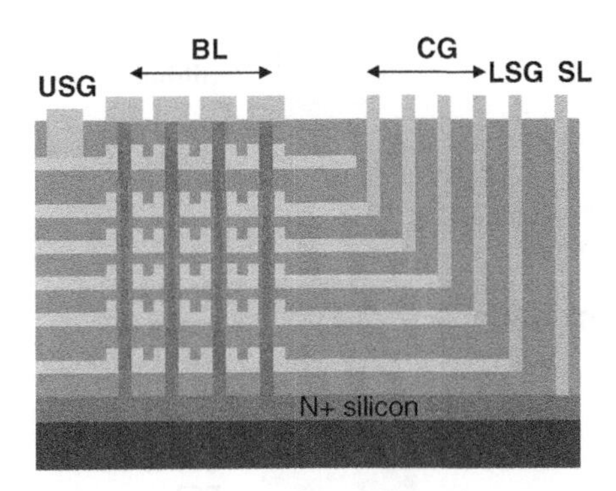

Figure 3.49 Completed 3D vertical silicon nanowire CT NAND string. (Based on Y. Sun *et al.*, (A*STAR, Nanyang Technical University, GlobalFoundries), IMW, 22 May 2011 [17].)

of the first control gate was then deposited along with high-density plasma isolation, and the ONO layer was formed, resulting in the cross-section shown in Figure 3.48(b). Once the ONO layer was formed, it could be shared by all the memory cells along each wire. The gate stack of the first control gate was then etched, and subsequent control gates and high-density plasma isolation layers in the NAND string were formed. The array was completed with the formation of the upper select gates, as shown in Figure 3.48(c) [17].

Access contacts to each of the independent control gates for the NAND string were next formed using a stair-step structure along with bit-line contacts, upper and lower select gate contacts, and source line contact. The completed device is shown in Figure 3.49 [17].

3.7 Vertical Channel Polysilicon GAA CT Memory

Vertical GAA flash memory string arrays using a single polysilicon NW punched hole channel have also been made. Characterization and modeling of these devices and some variations are discussed.

3.7.1 Multiple Vertical GAA Flash Cells Stacked Using Polysilicon NW Channel

The bit-cost scalable (BiCS) NAND flash memory was discussed by Toshiba in December of 2007 [34]. This device consisted of multiple vertical GAA flash cells stacked with a single polysilicon NW channel. To form the vertical strings of flash cells, multiple layers of polysilicon gates interspersed with dielectric spacers were stacked. Holes were etched in this stack, and the GAA NW memory devices were built up from the outside of the hole to the inside, forming a SONS FET memory device that substituted SiN for the gate oxide. A schematic cross-section of the resulting vertical NAND string is illustrated in Figure 3.50. The polysilicon layers acted as gates for the NAND memory string transistors as well as for the upper and lower select gates [34].

In this process, which was punched then plugged, the order of forming the body, gate dielectric, and gate electrode of the vertical FET was reversed from the conventional FET. This gate-first process exposed the gate dielectric to a dilute hydrofluoric acid (HF) clean before the polysilicon deposition. Rather than using an engineered SiO_2 gate dielectric to withstand the etch, an SiN gate dielectric was used.

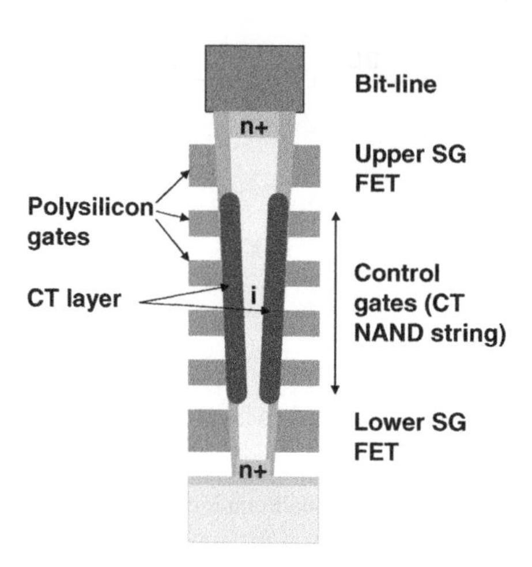

Figure 3.50 Schematic cross-section of vertical channel GAA flash NAND string. (Based on Y. Fukuzumi *et al.*, (Toshiba), IEDM, December 2007 [34].)

Because the subthreshold characteristics of polysilicon FETs were dependent on the trap density at the grain boundary, controllability was improved if the silicon in the device body was made thinner than the depletion width to reduce the volume of poly and reduce the total number of traps. A dielectric filler was used inside the thin polysilicon body. The concentric silicon shell made V_{th} less sensitive to trap density fluctuation and was expected to provide better controllability. It was thought that with continued scaling the diameter of the plug would eventually be small enough to suppress V_{th} variation. The device had reasonable operating window after 10 years. This vertical GAA FET structure is illustrated in Figure 3.51 [34].

A vertical SONS FET was proposed for the memory element. In this case the surface of the gate dielectric was SiN, which was durable against the dilute HF treatment, unlike the SiO2 used in conventional SONOS. The memory FETs worked in depletion mode with the body polysilicon being either undoped or uniformly lightly n-doped to form a junctionless memory transistor string. The bottom of the string was connected to the common source diffusion formed on the silicon substrate. For erase, a hole current was generated by gate-induced drain leakage (GIDL) near the lower select gate and used to raise the body potential [34].

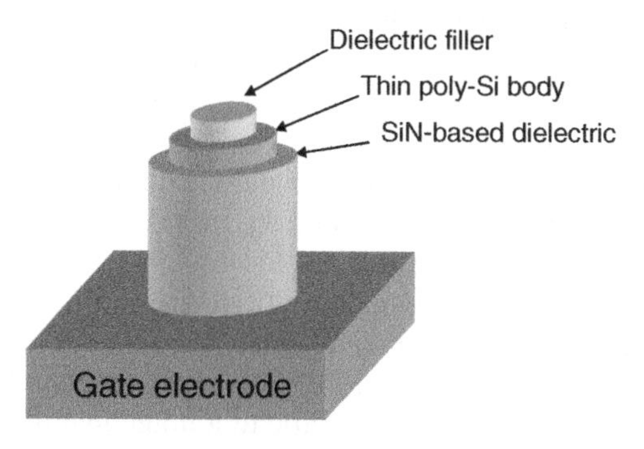

Figure 3.51 Schematic of vertical FET structure. (Based on Y. Fukuzumi *et al.*, (Toshiba), IEDM, December 2007 [34].)

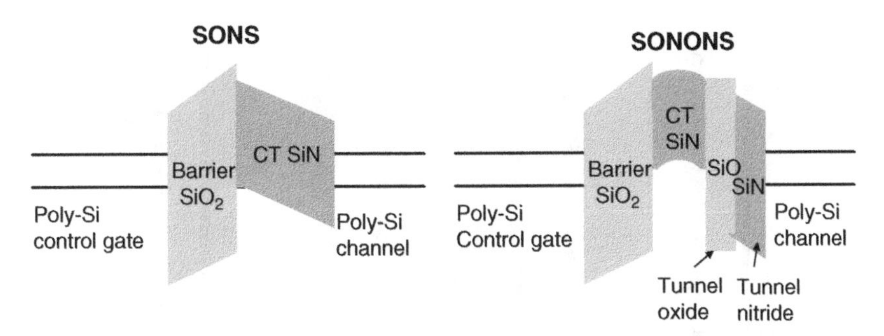

Figure 3.52 Schematic illustration of energy barriers for (a) SONS; and (b) SONONS films. (Based on Y. Komori *et al.*, (Toshiba) IEDM. December 2008 [35].)

An improvement in the memory FET was discussed by Toshiba in December of 2008 [35]. The SONS FET process was replaced with a SONONS FET process. The ON film continued to be required for process integration, but the retention and disturb characteristics were a concern due to the lack of a tunnel barrier film. An ONON film stack provided a better CT film stack in terms of disturb and reliability concerns because it included the tunnel oxide layer covered by a tunnel nitride film to withstand the dilute HF etch. A schematic illustration of the energy barriers for the SONS and the SONONS films is shown in Figure 3.52 [35].

The 10-year 85 °C data retention of the SONONS FET had a wider V_{th} window of 2.7 V compared to that of the SONS FET, which was 1.1 V. It was confirmed that the erase operation was induced by a hole current generated by GIDL. The boost efficiency of the pillars and the ONON film stack effectively suppressed disturb on unselected pillars. The BiCS flash was found to be capable of terabit array density without issues of disturb [35].

A silicided control gate for the vertical channel NAND string was discussed by Toshiba in December of 2009. This was intended to reduce the resistance–capacitance (RC) delay caused by the high resistivity of the doped polysilicon control gate. In this process, a slit is formed between the vertical NAND strings, and all gates are silicided simultaneously. A metal film is deposited by CVD on the sidewall of the control gate exposed by the slit as well as on the memory control gates. The slit divides the control gate stack into two comb-shaped gate stacks, as shown in Figure 3.53 [36]. There was no issue with the work function difference between a partially silicided structure and fully silicided structure [36].

3.7.2 V_{th} Shift Characteristics of Vertical GAA SONOS and/or TANOS Nonvolatile Memory

Development of a model for GAA SONOS/TANOS flash memory was discussed by Seoul National University in May of 2010 [37]. The BiCS flash memory technology was used for the model. An illustration of a cross-section of the simulated structure used in the model is shown in Figure 3.54 [37]. The simulation was performed with 2D cylindrical formulation.

The V_{th} shift of the GAA SONOS and/or TANOS device was modeled as a function of the dielectric layer thickness and the radius of the silicon. TANOS is made by substituting a titanium gate for the silicon gate and Al_2O_3 for the SiO_2 blocking oxide in SONOS. The analytical expression obtained indicated that as the radius of the silicon decreased, the V_{th} shift decreased due to increasing capacitance in dielectric layers with the same trapped

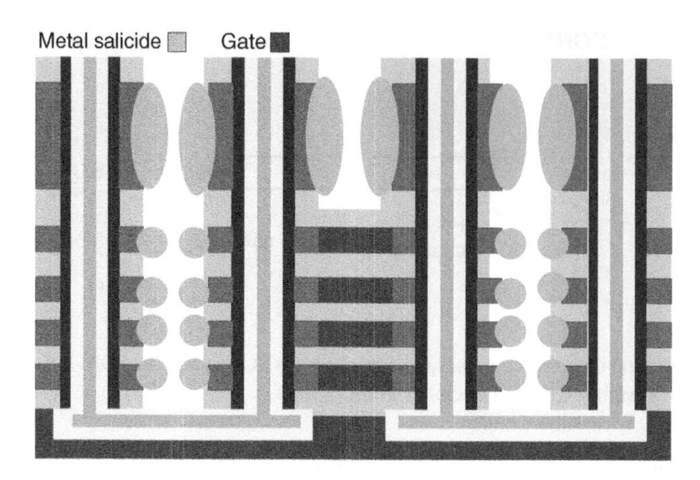

Figure 3.53 Illustration of salicided gate structure in slit between vertical NAND strings. (Based on M. Ishiduki *et al.*, (Toshiba), IEDM, December 2009 [36].)

charge density. The threshold shift is nearly linearly dependant on bottom and top oxide thickness. A larger V_{th} shift could be obtained by increasing the nitride thickness and the top oxide thickness. For the same thickness of dielectric layers, the program speed was faster with TANOS than SONOS, but when the capacitance of the dielectric layers is the same, programming is more efficient for SONOS due to the lower gate work function of SONOS. The maximum V_{th} shift of TANOS was lower than for SONOS due to the high dielectric constant of the TANOS blocking layer.

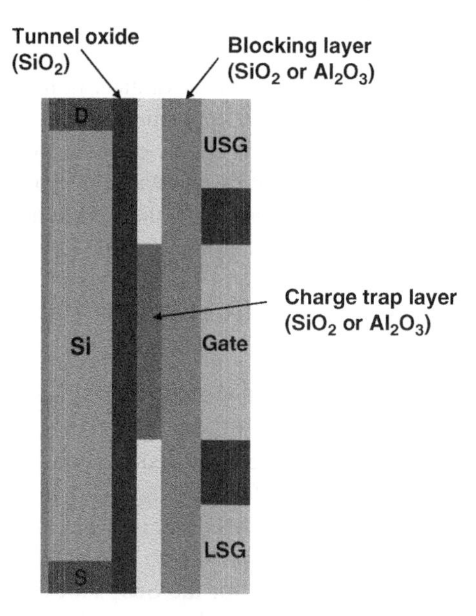

Figure 3.54 Illustration of 2D cross-section of simulated cylindrical BiCS flash memory structure. (Based on J. Ji *et al.*, (Seoul National University), IMW, 16 May 2010 [37].)

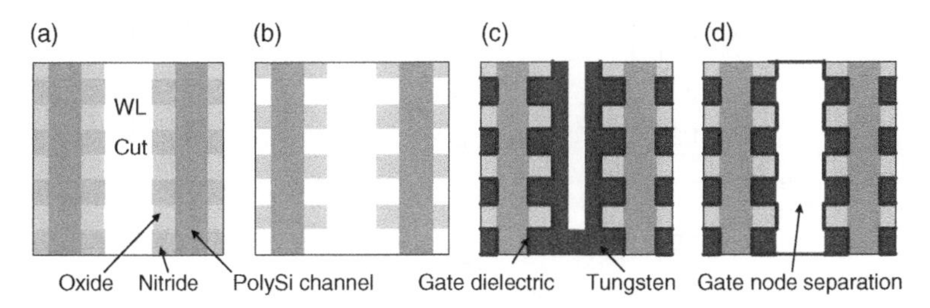

Figure 3.55 Process sequence for gate replacement for the TCAT NAND flash. (J. Jang *et al.*, (Samsung), VLSI Technology Symposium, June 2009 [39].)

3.7.3 GAA Vertical Pipe CT Gate Replacement Technology

A vertical polysilicon plug GAA process with a tungsten metal gate using a gate replacement process was discussed in June of 2009 by Samsung [38]. The process was initially called the terrabit cell array transistor (TCAT). After a damascene tungsten replacement process was added in June of 2010, it was called the vertical NAND (VNAND) [39].

Figure 3.55 illustrates the process sequence for gate replacement [39]. The damascene tungsten (W) replacement process begins with a multilayer stack with alternating oxide and nitride layers. Vertical holes are etched through this stack and filled with rows of vertical polysilicon plugs. In Figure 3.55(a) a line-type word-line cut is dry-etched through the stack between each row array of channel polysilicon plugs. In Figure 3.55(b) the gate replacement process to define the metal gate SONOS structure is initiated by removing the sacrificial nitride layer using wet etching. In Figure 3.55(c) the nitride is replaced by a deposited ONO gate dielectric followed by a tungsten metal layer in a "gate-last" process. In Figure 3.55(d) gate node separation is achieved by partially etching the tungsten.

In June of 2010, Samsung discussed moving from this gate replacement process on the GAA TCAT to a damascene tungsten gate replacement process that permitted scaling the technology. The new device was called the VNAND [38]. A bird's eye view of the GAA CT memory structure is shown in Figure 3.56 for a 120 nm channel and a scaled 40 nm channel structure [38].

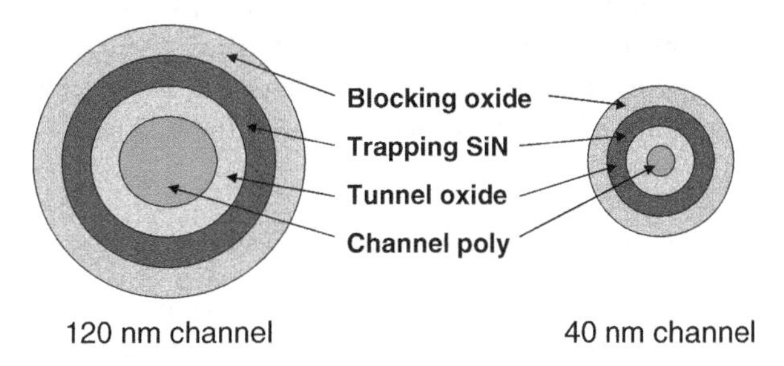

Figure 3.56 VNAND GAA charge-trapping memory structure. (Based on W-S Cho *et al.*, (Samsung), VLSI Technology Symposium, June 2010 [38].)

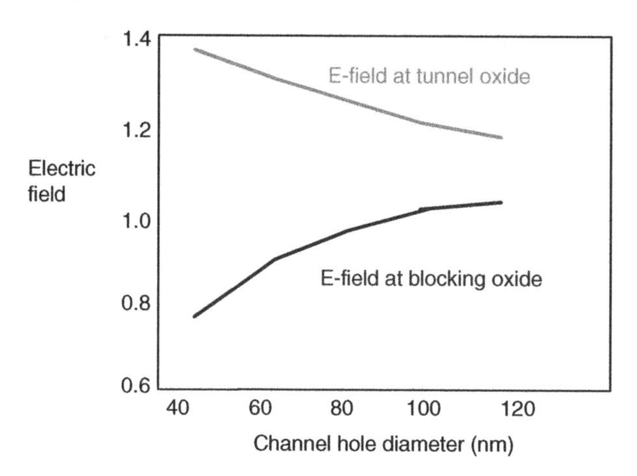

Figure 3.57 Electric field In tunnel and blocking layer vs. channel hole size. (Based on W-S Cho *et al.*, (Samsung), VLSI Technology Symposium, June 2010 [38].)

The electric field induced in the tunnel layer and the blocking layer as a function of the channel hole size is shown in Figure 3.57 [38]. As the radius of the GAA structure decreased, the tunnel oxide electric field increased, and the blocking oxide electric field decreased.

3.7.4 Bilayer Poly Channel Vertical Flash for 3D SONOS NAND

A vertical cylindrical cell with a 22 nm diameter bilayer polysilicon channel for use in a 3D NAND flash memory was discussed by IMEC and ASM in November of 2011 [40]. The minimum lateral cell area was $4F^2$, and there were no pipeline connections. The smallest functional cells had a memory hole diameter of 45 nm, which resulted in a 22 nm channel. This was thought equivalent to an 11 nm planar cell technology in the case of 16 stacked cells. A schematic cross-section of the 3D vertical cylindrical NAND cell is shown in Figure 3.58. A close-up view of the memory device stack is also shown [40].

A thin a-Si layer was used inside the memory hole along with an ONO gate stack that formed the blocking oxide, silicon nitride, and tunnel oxide. The a-Si layer protected the tunnel oxide during opening of the gate stack at the bottom of the memory hole and then served as the first layer of the bilayer polysilicon channel. After ONO etch, undoped silicon was deposited. The bottom junction was activated by anneal, and the top junction was implanted.

A 100 µs program time at +16 V and 1 ms and erase at −14 V were used. With a 22 nm channel diameter, a 5 V memory window was obtained. The erased state was stable for all tested conditions. The programmed state retention showed a weak temperature activation attributed to tunneling as the charge-loss mechanism. A more than 4 V P/E window was maintained after 10 000 cycles.

3.7.5 3D Vertical Pipe CT Low-Resistance (CoSi) Word-Line NAND Flash

The use of a CVD cobalt silicide for a low-resistance multilayered word-line electrode in a 3D vertical pipe NAND flash was described by Hynix in May of 2011 [41]. Plasma vapor deposition (PVD) is primarily used in planar architectures for thin-film scaled processes. but in 3D structures it is unable to form silicide in a scaled deep trench. CVD was found to have step

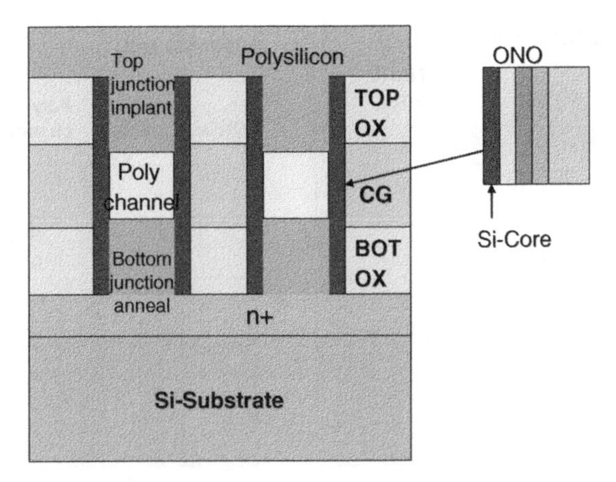

Figure 3.58 Vertical cylindrical SONOS cell. (Based on G. van den Bosch *et al.*, (IMEC, ASM), *IEEE Electron Device Letters,* 32(11), 1501 [40].)

coverage of about 97% with a trench aspect ratio of 15:1. The CoSi was formed using *in-situ* first anneal without a capping metal.

The resistivity of CVD CoSi was shown to be close to that of PVD CoSi, and the thermal stability is comparable up to 900 °C. Material properties discussed included Co–Si ratio, interlayer thickness, and surface roughness. A schematic cross-section of part of the channel and word-lines of the 3D vertical pipe NAND flash using CVD CoSi is shown in Figure 3.59 [41].

The structure was formed by depositing eight pairs of oxide/polysilicon layers. The vertical polysilicon SONOS pipe transistor strings were formed, and then line-like deep trenches were formed between the vertical pipes. CVD CoSi was formed on the drain and source contacts and showed similar electrical properties to PVD CoSi. The conclusion was that CVD Co is suitable for forming the CoSi word-line of a 3D vertical pipe NAND Flash.

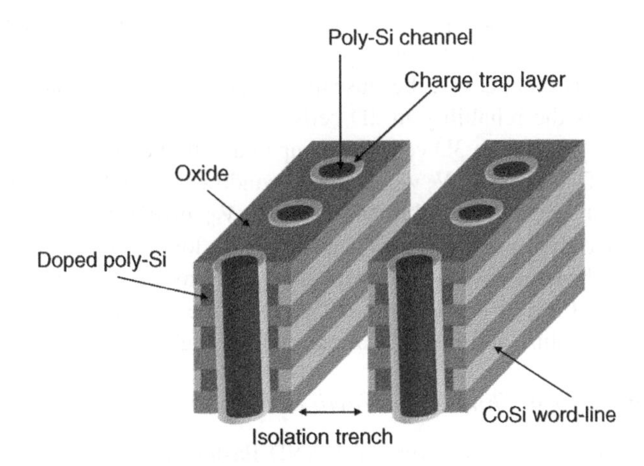

Figure 3.59 Schematic cross-section of 3D vertical pipe NAND flash with Co silicide word-line. (Based on M.S. Kim *et al.*, (Hynix), IITC/MAM, 8 May 2011 [41].)

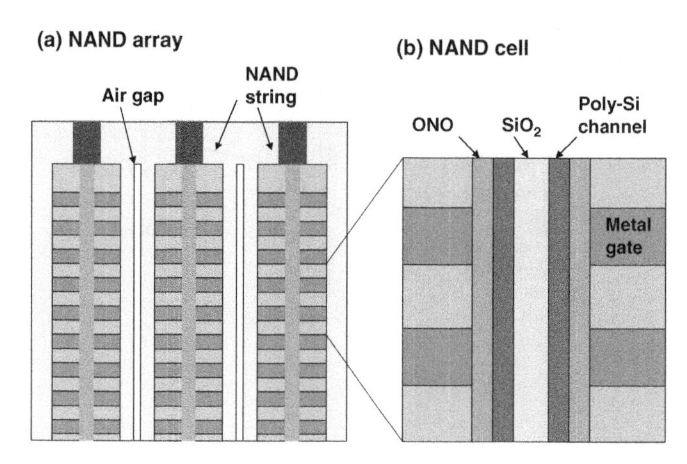

Figure 3.60 Schematic cross-section of (a) NAND array; and (b) individual cell. (Based on E.S. Choi, S.K Park, (Hynix), IEDM, December 2012 [42].)

3.7.6 Vertical Channel CT 3D NAND Flash Cell

A vertical channel 3D NAND flash cell using a CT mechanism that was intended to minimize both stack height and word-line resistance was discussed by Hynix in December of 2012 [42]. Characteristics of the CT NAND cell were discussed, including V_{th} distribution, disturbance, and reliability. These were compared with a similarly structured floating gate (FG) cell. A schematic cross-section of the vertical channel NAND array and one of the individual cells is shown in Figure 3.60 [42].

The stack height of this technology was reduced by inserting the ONO layer in the plug. The word-line height is reduced by using low-resistance tungsten. WL height was also reduced by using a merged cell array architecture. Minimizing the total WL stack height was investigated because multistack etching technology is a key issue in implementing 3D technologies. This was a gate-last process because the blocking oxide of the ONO stack is refreshed at the metal gate replacement step. The TiN gate MANOS cell technology was applied without gate direct patterning, because this produces etching damage at the gate edge and seriously degrades the reliability in 2D cells.

The cell V_{th} vs. P/E bias of the 3D cell was compared with the 2D planar FG cell. The V_{th} window for the 3D CT cell was −5 V, while the V_{th} window for the 2D FG cell was 4 V. This was judged sufficient for 2- and 3-bit MLCs. Cell V_{th} distribution is critical to avoid interference. The interference of the CT 3D cell was very low at 30 nm gate spacing [42].

Another issue was the temperature-dependent charge-loss mechanism. The Arrhenius curve does not work for the CT cell because the nitride loses the stored charge by band-to-band tunneling at low temperature and by thermal emission at high temperature [42].

3.7.7 Read Sensing for Thin-Body Vertical NAND

A sensing method for thin-body CT vertical NAND flash devices was discussed in August of 2011 by Stanford University and Seoul National University [43]. Without increasing the number of read operations, the P/E states of a memory cell could be identified with precision

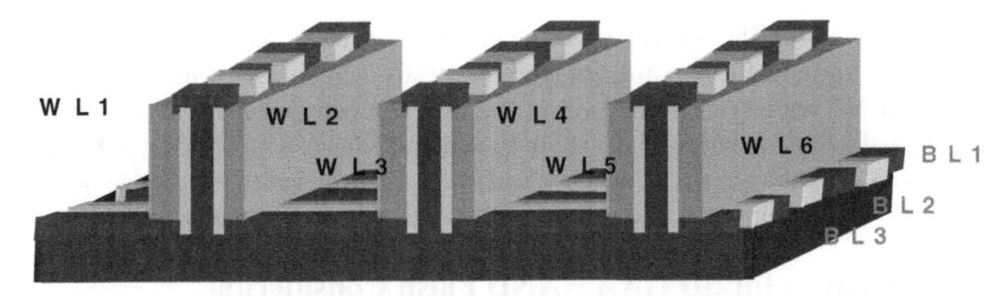

Figure 3.61 One level of a CT vertical channel thin body NAND flash array. (Based on S. Cho, B.G. Park, (Stanford University, Seoul National University), *IEEE Transactions on Electron Devices*, 58(8), 2814 [43].)

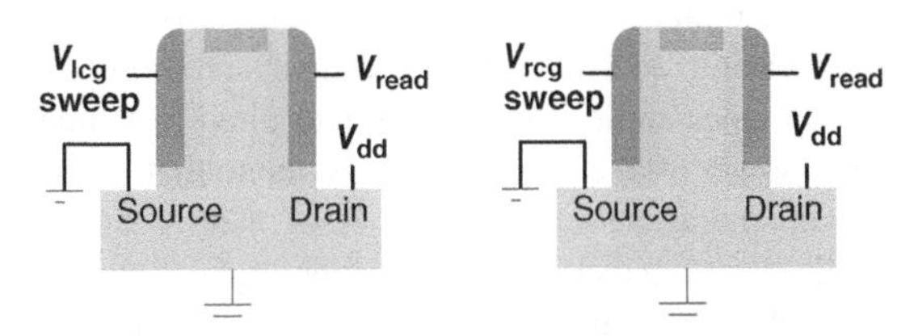

Figure 3.62 Schematic view of sensing procedure. (Based on S. Cho, B.G. Park, (Stanford University, Seoul National University), *IEEE Transactions on Electron Devices*, 58(8), 2814 [43].)

even when electrical interference existed between cells having a thin vertical channel in common. A schematic of the CT vertical channel thin-body NAND flash array is shown in Figure 3.61 [43]. Only one level of the vertical stack is shown.

By folding bit-lines and using sidewall word-lines, a $2F^2$ cell is expected. Higher V_{pass} and higher channel doping was used to suppress cell-to-cell electrical interference. Higher self-boosting efficiency was required for programming inhibition of an unselected word-line in MLC operation [43].

The sensing scheme involved double-sensing per 2 bits and was validated for a device with channel thickness of 80 nm. This method can also be used as a reference for establishing a smart sensing scheme for MLC NAND flash. A schematic view of the sensing procedures is shown in the schematic cross-sections in Figure 3.62. Read operations are shown on (a) the left cell and (b) the right cell. There is no contact on the top n+ region between control gates. The storage dielectric is ONON (20 nm, 20 nm, 60 nm, 60 nm, respectively).

The new sensing scheme was done by switching read from one side to the other by applying (V_{lcg}, V_{rcg}) = (LOW,HIGH) and (HIGH,LOW) alternatively. This improved the accuracy in the read operations. If intermediate voltages are introduced at multiple (V_{lcg}, V_{rcg}) points, it could be evolved into reading methodologies for MLC operations in NAND flash memories and work even with electrical interference.

3.8 Graphene Channel Nonvolatile Memory with Al_2O_3–HfO_x–Al_2O_3 Storage Layer

The functionality of graphene-based nonvolatile memory devices using a single-layer graphene channel and an Al_2O_3–HfO_x–Al_2O_3 charge storage layer was discussed in January of 2012 by

Seoul University [44]. The impact of the gate material work function on the memory characteristics was studied using different types of metals including Ti, with a work function of 4.3 eV, and Ni, with a work function of 5.2 eV. The ambipolar carrier conduction of graphene caused an enlargement of the memory window, which was 4.5 V for a Ti gate and 9.1 V for a Ni gate. The increase in memory window was attributed to the change in the flat-band condition and to suppression of electron back-injection in the gate stack.

3.9 Cost Analysis for 3D GAA NAND Flash Considering Channel Slope

A cost analysis for vertical cylindrical channel 3D NAND flash as a function of the number of device layers and the slope of the channel with the vertical was discussed in November of 2013 by Schiltron [45]. It was shown that, for a sloped channel, there is a minimum in die cost after which costs rise with increasing device layers. The analysis showed that there is a small parameter space defined by taper angle and vertical gate pitch in which the vertical channel architecture can maintain a lower total chip cost. This is due to that fact that any nonzero taper angle results in a large cell pitch at the top of the stack. The conclusion was that vertical processes using a deep hole or trench etch must have taper angle of close to zero degrees, or the total cost will be more than obtained in using lithography-intensive approaches to minimize cell areas on all layers.

A result of the model used is shown in Figure 3.63 [45], which plots die cost for a 256Gb NAND flash die as a function of the number of device layers using a vertical channel with taper angle as the parameter. The vertical gate pitch is maintained at 40 nm.

For taper angles of 1 degree or more, there is a clear minimum in die cost as a function of number of device layers. This means that to be cost effective, an etch technique must be used that results in less than a 1-degree taper.

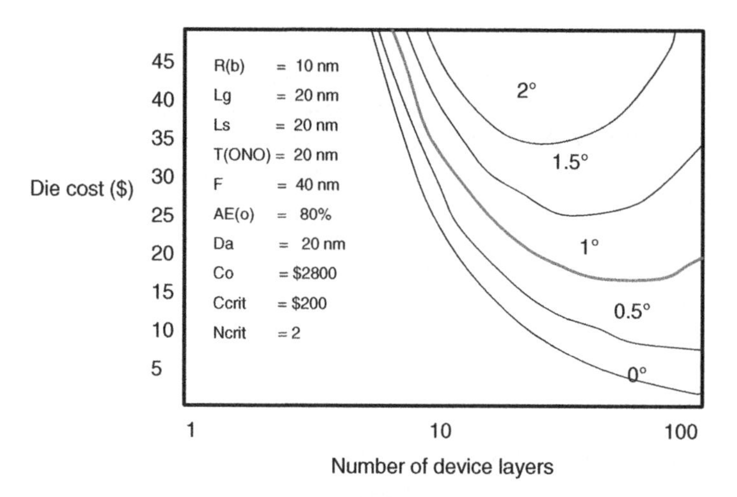

Figure 3.63 Die cost vs. number of device layers as a function of taper angle. (Based on A. J. Walker, (Schiltron), *IEEE Transactions on Semiconductor Manufacturing*, 26(4), 619 [45], with permission of Schiltron.)

References

1. Suk, S.D. *et al.* (June 2007) Gate-all-around twin silicon nanowire SONOS memory, (Samsung). VLSI Technology Symposium, p. 142.
2. Fu, J. *et al.* (December 2007) Trap layer engineered gate-all-around vertically stacked twin Si-nanowire nonvolatile memory, (IME Singapore, NUS). IEDM.
3. Fu, J. *et al.* (2008) Si-nanowire based gate-all-around nonvolatile SONOS memory cell, (NUS, IME-A*STAR, University of Bologna). *IEEE Electron Device Letters*, **29**(5), 518.
4. Yeo, K.H. *et al.* (June 2008) Gate-all-around single silicon nanowire MOSFET with 7 nm width for SONOS NAND flash memory, (Samsung). VLSI Technology Symposium.
5. Fu, J., Singh, N., Yang, B. *et al.* (September 15, 2008) Si-nanowire TAHOS (TaN/Al2O3/HfO2/SiO2/Si) nonvolatile memory cell, (A*STAR, NUS). ESSDIRC.
6. Fu, J., Singh, N., Zhu, C. *et al.* (2009) Integration of high-k dielectrics and metal gate on gate-all-around Si-nanowire-based architecture for high-speed nonvolatile charge-trapping memory, (A*STAR, NUS). *IEEE Electron Device Letters.*, **30**(6), 662.
7. Park, J.K. *et al.* (December 2012) Origin of transient V_{th} shift after erase and its impact on 2D/3D structure charge trap flash memory cell operations, (KAIST, Hynix). IEDM.
8. Amoroso, S.M., Compagnoni, C.M., Mauri, A. *et al.* (2011) Semi-analytical model for the transient operation of gate-all-around charge-trap memories, (Politecnico di Milano, Micron Technology). *IEEE Transactions on Electron Devices*, **58**(9), 3116.
9. Choi, S.J. *et al.* (2011) Nonvolatile memory by all-around-gate junctionless transistor composed of silicon nanowire on bulk substrate, (KAIST). *IEEE Electron Device Letters*, **32**(5), 602.
10. Choi, S.J., Moon, D.I., Duate, J.P. *et al.* (June 2011) A novel junctionless all-around-gate SONOS device with a quantum nanowire on a bulk substrate for 3D stack NAND flash memory, (KAIST). VLSI Technology Symposium.
11. Fu, J., Jiang, Y., Singh, N. *et al.* (2009) Polycrystalline Si nanowire SONOS nonvolatile memory cell fabricated on a gate-all-around (GAA) channel architecture, (A*STAR, NUS). *IEEE Electron Device Letters*, **30**(3), 246.
12. Hsu, H.H., Lin, H.C., Luo, C.W. *et al.* (2011) Impacts of multiple-gated configuration on the characteristics of poly-Si nanowire SONOS devices, (National Chiao Tung University). *IEEE Transactions on Electron Devices*, **58** (3), 641.
13. Huang, P.C., Chen, L.A., and Sheu, J.T. (2010) Electric-field enhancement of a gate-all-around nanowire thin-film transistor memory, (National Chiao Tung University). *IEEE Electron Device Letters*, **31**(3), 216.
14. Luo, C.W., Lin, H.C., Lee, K.H. *et al.* (2011) Impacts of nanocrystal location on the operation of trap-layer-engineered poly-Si nanowired gate-all-around SONOS memory devices, (National Chiao Tung University, National Nano Device Laboratory). *IEEE Transactions on Electron Devices*, **58**(7), 1879.
15. Lee, K.H., Lin, H.C., and Huang, T.Y. (2013) A novel charge-trapping-type memory with gate-all-around poly-si nanowire and HfAlO trapping layer, (National Chiao Tung University). *IEEE Electron Device Letters*, **34**(3), 393.
16. Sun, Y. *et al.* (2011) Vertical-Si-nanowire-based nonvolatile memory devices with improved performance and reduced process complexity, (Nanyang Technical University, A*STAR, GlobalFoundries). *IEEE Transactions on Electron Devices*, **51**(5), 1329.
17. Sun, Y. *et al.* (May 22, 2011) Junction-less stackable SONOS memory realized on vertical-Si-nanowire for 3-D application, (A*STAR, Nanyang Technical University, GlobalFoundries). IMW.
18. Sun, Y., Yu, H.Y., Singh, N. *et al.* (2011) Junctionless vertical-Si-nanowire-hannel-based SONOS memory with 2-bit storage per cell, (Nanyang Technical University, A*STAR, GlobalFoundries). *IEEE Electron Device Letters*, **32**(6), 725.
19. Maconi, A., Compagnoni, C.M., Spinelli, A.S., and Lacaita, A.L. (December 2012) A new erase saturation issue in cylindrical junction-less charge-trap memory arrays, (Politecnico di Milano, IFN-CNR). IEDM.
20. Maconi, A., Compagnoni, C.M., Spinelli, A.S., and Lacaita, A.L. (2013) New erase constraint for the junction-less charge-trap memory array in cylindrical geometry, (Politecnico di Milano, Con. Naz. de Ricerche Milano). *IEEE Transactions on Electron Devices*, **60**(7), 2203.
21. Chen, C.Y., Chang-Liao, K.S., Wu, K.T., and Wang, T.K. (2013) Improved erasing speed in junctionless flash memory device by HfO2/Si3N4 stacked trapping layer, (NTHU, Naional. Nano Device Laboatories). *IEEE Electron Device Letters*, **34**(8), 993.
22. Singh, N. *et al.* (2008) Si, SiGe nanowire devices by top-down technology and their applications, (A*STAR, IIT, Silterra). *IEEE Transactions on Electron Devices*, **55**(11), 3107.

23. Ernest, T. *et al.* (December 2008) Novel Si-based nanowire devices: Will they serve ultimate MOSFET's scaling or ultimate hybrid integration?, (CEA-LETI, Minetec). IEDM.
24. Hubert, A. (December 2009) A stacked SONOS technology, up to 4 levels and 6 µm crystalline nanowires, with gate-all-around or independent gates (Φ-Flash), suitable for full 3D integration, (CEA-LETI, MINATEC, IMEP-LAHC, INPG-MINATEC), IEDM.
25. Nowak, E. (May 16, 2010) In-depth analysis of 3D silicon nanowire SONOS memory characteristics by TCAD simulations, (EA-LETI MINATEC, IMPE-LAHC, INP-MINATEC). IMW.
26. Yun, J.G. *et al.* (2011) Single-crystalline Si stacked array (STAR) NAND flash memory, (Samsung, Seoul National University). *IEEE Transactions on Electron Devices*, **58**(4), 1006.
27. Yun, J.G., Park, S.H., and Park, B.G. (2011) Layer selection by erase (LASER) with an etch-through-spacer technique in a bit-line stacked 3-D NAND flash memory array, (Samsung, Seoul National University). *IEEE Transactions on Electron Devices*, **58**(7), 1892.
28. Kim, Y. *et al.* (2012) Three-dimensional NAND flash architecture design based on single-crystalline stacked array, (Seoul National University, Samsung). *IEEE Transactions on Electron Devices*, **50**(1), 35.
29. Kim, Y., Kang, M., Park, S.H., and Park, B.G. (August 2013) Three-dimensional NAND flash memory based on single-crystalline channel stacked array, (Samsung, Seoul National University). *IEEE Electron Device Letters*, **34** (8), 990.
30. Chen, M. *et al.* (2009) Vertical-Si-nanowire SONOS memory for ultrahigh-density application, (A*STAR, Nanyang Technical University). *IEEE Electron Device Letters*, **30**(8), 879.
31. Sun, Y., Yu, H.Y., Singh, N. *et al.* (2010) Multibit programmable flash memory realized on vertical Si nanowire channel, (Nanyang Technical University, A*STAR). *IEEE Electron Device Letters*, **31**(5), 390.
32. Li, X., Chen, Z., Shen, N. *et al.* (2011) Vertically stacked and independently controlled twin-gate MOSFETs on a single Si nanowire, (A*STAR, University of California Santa Barbara). *IEEE Electron Device Letters*, **32**(11), 1492.
33. Sun, Y. (December 2011) Demonstration of memory string with stacked junction-less SONOS realized on vertical silicon nanowire, (A*STAR, Nanyang Technical University, Peking University, GlobalFoundries). IEDM.
34. Fukuzumi, Y. *et al.* (December 2007) Optimal integation and characteristics of vertical array devices for ultra-high density, bit-cost scalable flash memory, (Toshiba). IEDM.
35. Komori, Y. *et al.* (December 2008) Disturbless flash memory due to high boost efficiency on BiCS structure and optimal memory film stack for ultra high density storage device, (Toshiba). IEDM.
36. Ishiduki, M. *et al.* (December 2009) Optimal device structure for pipe-shaped BiCS flash memory for ultra high density storage device with excellent performance and reliability, (Toshiba). IEDM.
37. Ji, J., Park, B.G., Lee, J.H., and Shin, H. (May 16, 2010) A comparative study of the program efficiency of gate all around SONOS and TANOS flash memory, (Seoul National University). IMW.
38. Cho, W.-S. *et al.* (June 2010) Highly reliable vertical NAND technology with biconcave shaped storage layer and leakage controllable offset structure, (Samsung). VLSI Technology Symposium.
39. Jang, J. *et al.* (June 2009) Vertical cell array using TCAT (terabit cell array transistor) technology for ultra high density NAND flash memory, (Samsung). VLSI Technology Symposium.
40. van den Bosch, G. *et al.* (2011) Highly scaled vertical cylindrical SONOS cell with bilayer polysilicon channel for 3-D NAND flash memory, (IMEC, ASM). *IEEE Electron Device Letters*, **32**(11), 1501.
41. Kim, M.S. *et al.* (May 8, 2011) CVD-cobalt for low resistance word line electrode of 3D NAND flash memory, (Hynix). IITC/MAM, p. 1.
42. Choi, E.S. and Park, S.K. (December 2012) Device considerations for high density and highly reliable 3D NAND flash cell in near future, (Hynix). IEDM.
43. Cho, S. and Park, B.G. (2011) A novel sensing scheme for reliable read operation of ultrathin-body vertical NAND flash memory devices, (Stanford University, Seoul National University). *IEEE Transactions on Electron Devices*, **58**(8), 2814.
44. Lee, S., Song, E.B., Kim, S. *et al.* (2012) Impact of gate work-function on memory characteristics in $Al_2O_3/HfO_x/Al_2O_3$/graphene charge-trap memory devices, (Seoul National University). *Applied Physics Letters*, **100**(2), 023109.
45. Walker, A.J. (2013) A rigorous 3-D NAND flash cost analysis, (Schiltron). *IEEE Transactions on Semiconductor Manufacturing*, **26**(4), 619.

4

Vertical NAND Flash

4.1 Overview of 3D Vertical NAND Trends

4.1.1 3D Nonvolatile Memory Overview

There is a strong potential today for 3D nonvolatile memory to replace planar memory for 10 nm nodes and beyond [1]. One candidate for bringing in the 3D era appears to be the NAND flash memory, and its basic operational modes are expected to be similar to those in the planar device. Currently, the basic types of 3D vertical memory are the vertical channel gate-all-around (GAA) NAND flash strings, the vertical gate devices with horizontal thin-film transistor (TFT) double-gate NAND flash strings, and the earlier stacked planar NAND structures discussed in Chapter 2. Both charge-trapping (CT) and floating gate (FG) 3D devices have been investigated. Some characteristics expected with the vertical channel and vertical gate CT NAND flash devices are shown in Table 4.1 [1].

Various issues are expected in 3D NAND flash configurations. Program disturbs may increase in 3D as a result of the proximity of cells in the z direction, and an increase is expected in various program inhibit conditions on unselected strings as a result of increased number of pages on the same word-line. Junction capacitance is expected to be significantly reduced by thin channels, and junctionless source/drain (S/D) structures should help suppress program disturbances. The vertical channel NAND is expected to be less dependent on lithography than planar NAND because thickness of materials rather than lithography will determine the bit-line (BL) and word-line (WL) pitches. Vertical gate NAND is still somewhat dependent on lithography because WLs are patterned, but BL pitch is determined by the thickness of the channel, WL, and storage layer. Basic operations are expected to be similar to those of planar devices. Primary differences are expected to be an increase in program disturbance in the z direction, which may require different program inhibit conditions on various unselected strings. Because 3D NAND flash has reduced junction capacitance due to having thin channels and junctionless S/D, program disturbance can be suppressed. A comparison of planar and two types of 3D cell arrays is shown in Table 4.2 [1].

Vertical 3D Memory Technologies, First Edition. Betty Prince.

Table 4.1 Comparison of 2D and Two Early 3D NAND Flash Features.

Dimension	2D Planar	3D Vertical Channel	3D Vertical Gate
Gate Structure	Stacked	Gate-all-around	Dual-gate TFT
Unit Cell Size	$4F^2$	$6F^2$	$4F^2$
Scaling Issue	Lithography	Film thickness	Lithography/film thickness
Coupling	Strong	Low	Moderate
Coupling Direction	Word-line + bit-line	Vertical	Bit-line + vertical

Based on J. Choi, K.S. Seol, (Samsung), VLSI Technology Symposium, June 2011 [1].

Table 4.2 Comparison of Planar and 3D NAND Operations.

	2D	3D	3D
3D Cell Array Type	Planar	Vertical channel	Vertical gate
WL Direction	Horizontal	Horizontal	Vertical
BL Direction	Horizontal	Vertical	Horizontal
Program/Erase	FN	FN	FN
Program Disturb	Channel coupling	NOP	NOP, vertical coupling
Stacks for 1x nm	1	16–64	0–32
Lithography Factor	Strong	Weak	Moderate

Based on J. Choi, K.S. Seol, (Samsung), VLSI Technology Symposium, June 2011 [1].

4.1.2 Architectures of Various 3D NAND Flash Arrays

Several 3D NAND flash array architectures that have been studied for potential products are discussed in this chapter. These include 3D CT devices with vertical pipe channel structures (e.g., P-BiCS, VRAT, TCAT, and VSAT) and with horizontal channel and vertical gate structures (e.g., vertical gate NAND and vertical gate BE-SONOS). These architectures are discussed in this chapter along with 3D vertical channel FG structures (e.g., ESCG, S-SCG, DC-SF, and SCP).

A comparison of the vertical CT architectures was given by Macronix in May of 2010 using technology computer-aided design (TCAD) simulations [2]. The structure and critical dimensions of the vertical channel TCAT, P-BICS, and VSAT architectures are illustrated in Figure 4.1 [2]. The TCAT, P-BiCS, and VSAT each have a horizontal pitch of 3F and vertical pitch of 2F, resulting in a $6F^2$ cell area.

The vertical gate NAND flash architecture with stacked horizontal channels is discussed later in this chapter, and an illustration with critical dimensions is shown in Figure 4.2 [2]. The critical cell area of the horizontal channel architecture is $2F \times 2F$.

In June of 2007, Toshiba discussed a cost comparison of the 3D planar NAND flash with stacked single-crystal polysilicon layers, which was discussed in Chapter 2, and the 3D vertical channel bit cost scalable (BiCS) NAND flash, which is discussed in this chapter [3]. The results of this study are shown in Figure 4.3.

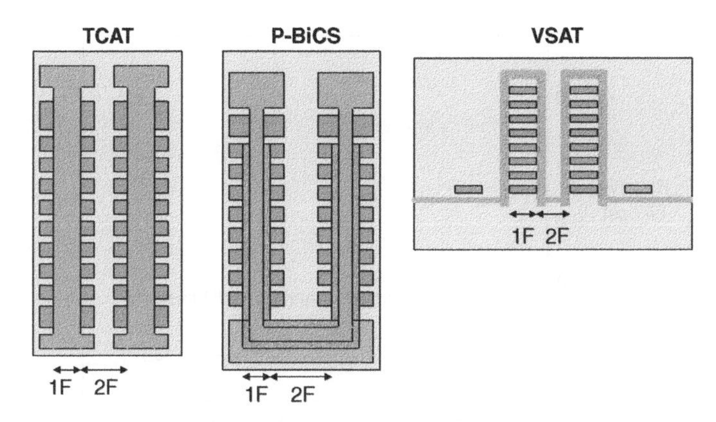

Figure 4.1 Vertical channel CT NAND flash architectures. (Based on Y.H. Hsiao *et al.*, (Macronix), IMW, 16 May 2010 [2].)

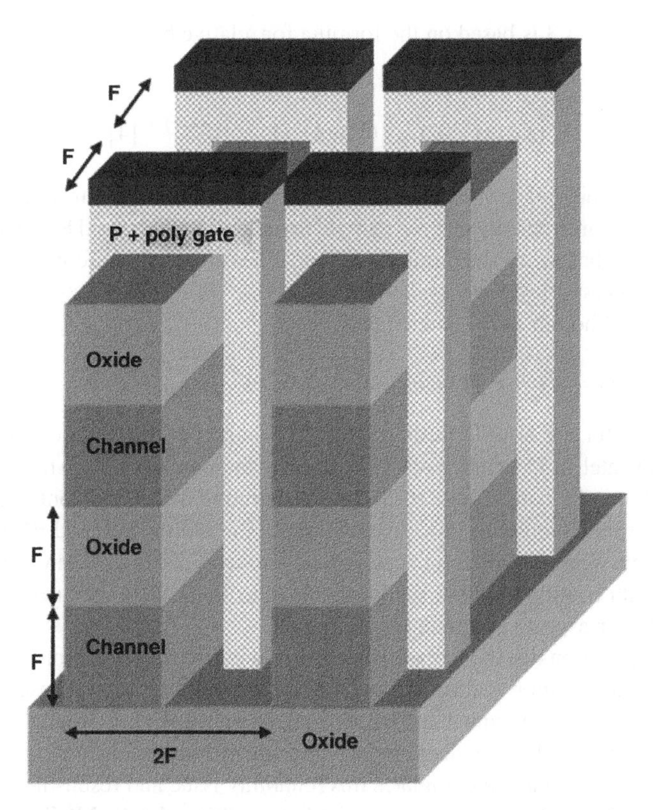

Figure 4.2 Stacked horizontal channel CT architectures. (Based on Y.H. Hsiao *et al.*, (Macronix), IMW, 16 May 2010 [2].)

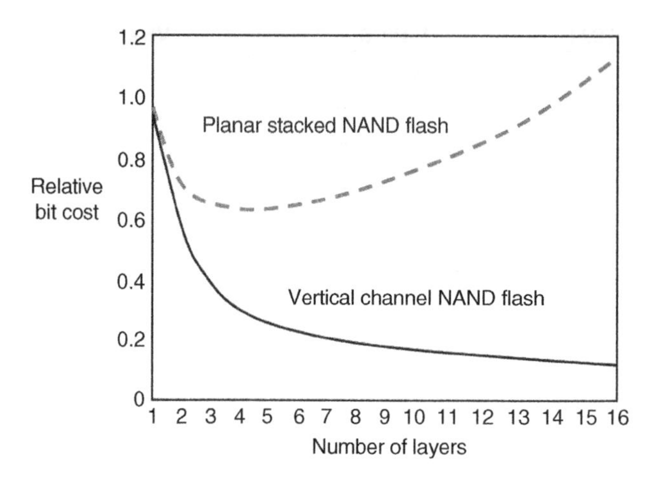

Figure 4.3 Cost comparison of 3D stacked NAND and BiCS 3D NAND flash. (Based on H. Tanaka *et al.*, (Toshiba), VLSI Technology Symposium, June 2007 [3].)

The chart in Figure 4.3 is based on the formula for relative bit cost of the two architectures, which is:

$$(1/n)(C_c + C_{sl})[(1 + A)/(1 - Y)]^{(n-1)} \quad [4]$$

Where n is the number of stacked layers, C_c is the cost for the common part, C_{sl} is the cost per single layer, A is the area penalty rate for a single layer, and Y is the yield loss per single layer. It appears to indicate that, when the number of layers goes above about four, the relative bit cost of the 3D single-crystal layer stack also begins to rise while the bit cost of the BiCS stack continues to fall, although at a slower rate.

4.1.3 Scaling Trends for 2D and 3D NAND Cells

Scaling directions for the NAND flash cell for 20 nm and beyond were discussed by Micron Technology and Intel in December of 2012 [4]. For the conventional wrapped floating gate (FG) cell, which has the control gate (CG) wrapped around the FG, to continue scaling below 20 nm, the aspect ratio would become greater than 10 in both the WL and BL directions. Two possible directions were identified for scaling the wrapped FG cell. These were to return to the planar FG cell with continued lateral scaling or move to 3D NAND cell arrays. The planar cell would have a lower aspect ratio than the wrap cell.

Reliability degradation due to increased electric field is also a scaling limitation. Both FG and CG widths are reduced upon scaling the cell, which creates sharp tips. These tips increase the electric field locally and cause unwanted electron injection from FG to CG and from CG to active area. During program and erase electrons are trapped and accumulate, which limits the cell-cycling capability. A planar cell eliminates this reliability issue and results in increased cycling capability. A large threshold window is required for multilevel-cell (MLC) capability. As the wrapped cell is scaled, a larger threshold voltage (V_{th}) window is needed to compensate for the increased program V_{th} distribution and the cell-to-cell interference. The planar cell can reduce oxide stress because there is a smaller cell-to-cell interference, and cycling capability is increased.

A CT cell was considered, as it has the advantage of scalability and reduction of interference. The nitride CT cell has, however, poorer program/erase (P/E) characteristics than the FG cell, which reduces the possibility of using MLC storage. In addition, a poor programming slope tends to degrade the program disturb window. The planar FG cell has better P/E window and program slope along with better cycling endurance.

It was concluded that the planar FG cell has the best option for sub 20 nm NAND flash until the cell is scaled to the point where cell noise, statistical fluctuation, and data retention become limiting. Beyond this point, where scaling the planar FG NAND flash cell becomes unfeasible, the 3D NAND architectures need to be considered.

Current development on 3D architectures falls into two broad categories: the vertical channel 3D NAND and the vertical gate (horizontal channel) 3D NAND. Because the channel width can be wider in both architectures, the cell size can be larger even though the effective lateral area is reduced by the stacking of multiple tiers of cells [4].

A schematic illustration of the (a) vertical channel (string) and the (b) horizontal channel (string) vertical 3D NAND, which is also called the vertical gate, are shown in Figure 4.4 [4].

The following analysis maintains constant both the cell stack thickness of the vertical channel device and the tier thickness of the horizontal channel vertical gate device. Several factors are involved in analyzing the 3D vertical NAND flash [4]. One factor is effective area feature size, which is defined as the square root of cell area divided by the number of tiers, all divided by 2. This is considered an indicator of cost. Another factor is cell physical feature size, which is defined as the square root of the channel width times the channel length and is an indicator of reliability. String conductivity, which is the channel width divided by the string length, is used as an indicator of cell current.

For the vertical channel and horizontal channel 3D NAND flash, a comparison of the effective area feature size and the physical feature size is illustrated in Figure 4.5 for 20, 30, 40 and 50 nm process features and 16 and 64 vertical (3D) tiers. The figure indicates that vertical channel 3D NAND can have a larger physical cell size, while the horizontal channel 3D NAND can have fewer tiers for the same effective area feature size [4].

The cell architecture and process parameters should be defined so that physical cell feature size and the effective area feature size are well balanced. The wider channel width and shorter

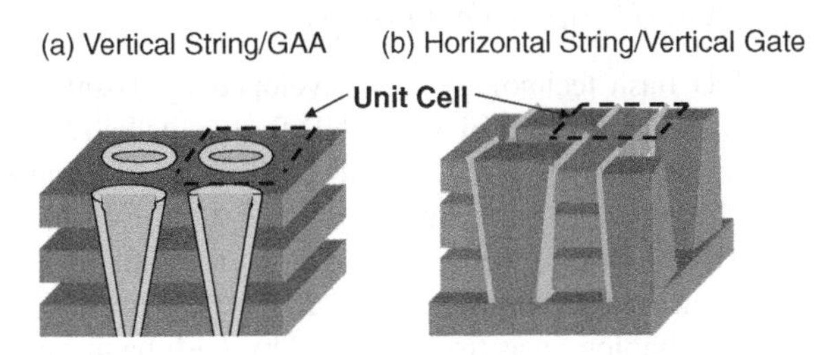

Figure 4.4 Schematics of (a) vertical channel 3D NAND string; and (b) vertical gate 3D NAND string showing unit cell. (Based on A. Goda and K. Parat, (Micron, Intel), IEDM, December 2012 [4].)

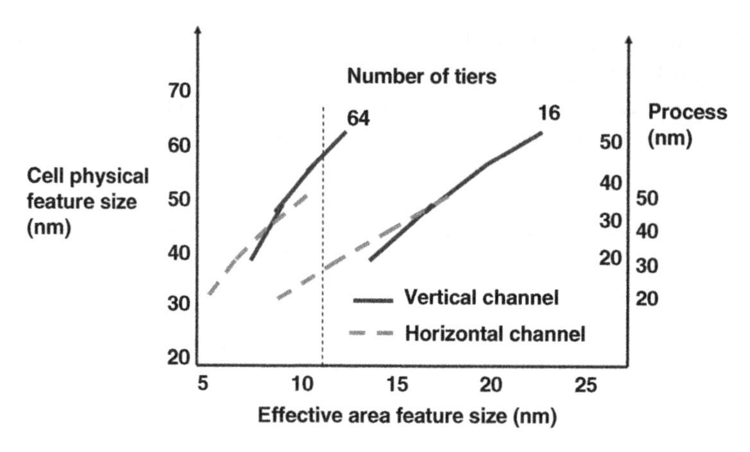

Figure 4.5 Effective area feature size vs. physical feature size for 3D NAND flash for 16 and 64 tiers. (Based on A. Goda, K. Parat, (Micron), IEDM, December 2012 [4].)

string length of the vertical channel NAND tends to enhance string conductance, which improves the sense operation. The polysilicon channel mobility needs to maintain this advantage. This mobility requirement becomes more critical as the number of tiers increases. For the horizontal channel NAND, the string conductance lacks the GAA structure, so the string conductance is more difficult to determine. The number of tiers can, however, be increased without affecting the string conductance. The taper of the vertical pillar is important to obtaining good cell area scaling because the effective area scaling benefit from the tier stack is reduced when taper results in a larger cell footprint. Cell noise can increase in the vertical structure. Cell noise includes interference, random telegraph signal noise, and statistical fluctuations. All of these factors can reduce the physical cell size advantage of 3D NAND. The conclusion was that the 2D NAND should be scaled as far as possible, and then when the 3D NAND is undertaken, the key to cell scaling is defining the cell architecture to maximize physical cell size while minimizing effective cell area. Tight threshold voltage placement is important.

4.2 Vertical Channel (Pipe) CT NAND Flash Technology

4.2.1 BiCS CT Pipe NAND Flash Technology

The BiCS CT pipe NAND flash technology was developed by Toshiba. In July of 2013, Toshiba announced it was moving to vertical 3D NAND flash in its Fab 5 in Japan, which was planned for completion in the summer of 2014, with facilities and equipment installation to begin later. The facility is expected to be capable of running the BiCS process for 3D NAND flash memories. In late 2012, a 16-layer prototype BiCS device with 50 nm diameter vertical channel was announced with production expected in 2015 [5].

The BiCS NAND flash technology was first discussed by Toshiba in June of 2007. The 3D pipe array structure was intended to increase NAND flash density and was expected to be ready simultaneously with the 32 nm technology node. In the BiCS technology, rather than building transistor layers up vertically one at a time, an entire stack of electrode plates are punched through in one operation. A plug consisting of a string of CT NAND flash transistors is built up in the via holes that are created. This technology permits the number of NAND transistors to be increased without increasing the number of process steps. Figure 4.6 shows a schematic cross-section of the CT NAND flash memory string with vertical CT cells and cross-sections of the vertical field effect transistors (FETs) [3].

In this process, after the via holes are punched through the stack, the walls are lightly doped. A vertical n-FET is deposited in the via to form the lower select gate (LSG). This is followed by a GAA CT oxide–nitride–oxide (ONO)-based technology, which is deposited in reverse order along the via walls, forming a series of junctionless CT NAND cells. Finally, another vertical n-FET is deposited near the top of the plug to form the upper select gate (USG). The device density increases with the stack height, but the number of process steps remains the same. This stacked multiple-layer CT memory array has a constant number of critical lithography steps regardless of the number of stacked layers used [3].

The fabrication sequence of the BiCS NAND flash is as follows. The LSG transistor, memory string, and USG transistor are made separately. A stack of alternate dielectric and p+ polysilicon layers is formed. The holes through the gate layers for the transistor channel and memory plug are punched in a single lithography operation, and then low-pressure chemical vapor deposition (LPCVD) tetraethylorthosilicate (TEOS) for the transistor and ONO film for the memory are deposited. The bottoms of the dielectric films are removed by reactive ion etching (RIE), and the remaining hole is plugged by amorphous silicon. Arsenic is implanted

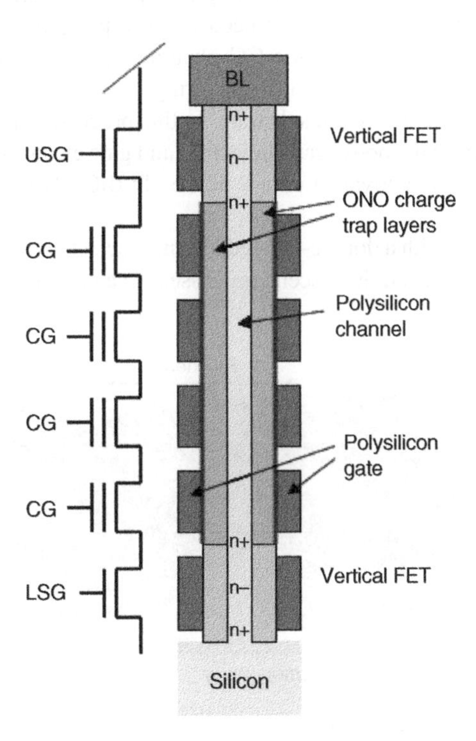

Figure 4.6 Schematic cross-section of the vertical CT cell string and vertical SG FETs. (Based on H. Tanaka *et al.*, (Toshiba), VLSI Technology Symposium, June 2007 [3].)

and activated for the drain and source of the upper device. The ONO is deposited as 5 nm of LPCVD TEOS film as the top blocking oxide, 11 nm of LP-CVD silicon nitride (Si_3N_4) film for the trapping layer, and 2.5 nm of LPCVD TEOS film as the tunnel oxide [3].

The edges of the CG are processed into a stair-like structure by repeating the RIE and resist slimming. To minimize disturb, an entire stack of CG and lower select line are etched with a slit, which separates a block of memory plugs from each other. Only the USG is cut into line pattern to work as a row address selector, while the LSG is common to the pillars. Via holes and BLs are made on the array and peripheral circuit simultaneously. The resulting stacked BiCS NAND flash array is shown in Figure 4.7 [3].

I_d–V_g characteristics of the select gate FET were determined. Subthreshold slope was 250 m/dec. I_{off} was 30 pA, on-current was 2.4 μA. Endurance to 10^5 cycles showed a V_{th} shift of less than 0.5 V after 10^4 cycles. After 1000 cycles, 2.5 V of the threshold window remained after 10 years [3].

The Toshiba BiCS NAND flash with its stacked vertical array flash memory was discussed by Toshiba again in December of 2007. In 90 nm technology this device had a $6F^2$ cell array, which is illustrated schematically in the top-down diagram in Figure 4.8 [6].

When manufacturing this technology, entire stacks of electrode plates were punched through and plugged with polysilicon and other materials at one time. A series of vertical FETs are formed in the punched hole, which act as a NAND string. The vertical GAA FET and the silicon–oxide–nitride–silicon SONS FET memory elements were described in Chapter 3. The memory FETs form a junctionless channel working in depletion mode because they are either undoped or lightly n-doped. This avoids the need to form p-n junctions within the plug. A single bit is accessed at the intersection of a CG plate and a vertical NAND string, which is selected by a BL and a USG. The bottom of the string is connected to the common source diffusion, which is formed on the silicon substrate. In this process, which is punched and then plugged, the order of forming the body, gate dielectric, and gate electrode is reversed compared to the conventional FET. The equivalent circuit of the 3D BiCS NAND flash array is shown in Figure 4.9 [6].

A $4F^2$ cell array structure with a double-layered alternate select gate was also described. This technology was based on RIE and damascene processes, so a critical lithography step was not

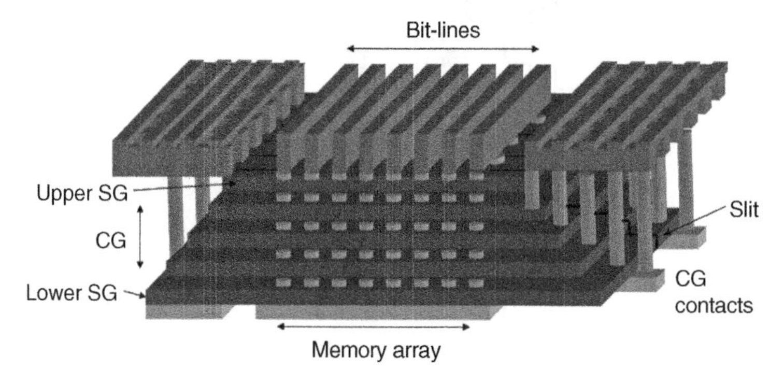

Figure 4.7 3D schematic of stacked BiCS NAND flash array. (Based on H. Tanaka, *et al.*, (Toshiba), VLSI Technology Symposium, June 2007 [3].)

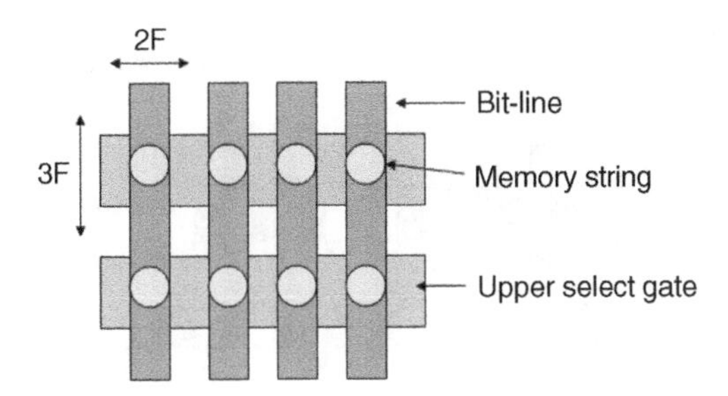

Figure 4.8 Top-down view of $6F^2$ cell array of the BiCS flash memory. (Based on Fukuzumi, Y., *et al.*, (Toshiba), IEDM, December 2007 [6].)

required. The 8-terabit capacity BiCS flash was expected to maintain a bit-cost superiority over 3D layers of stacked horizontal NAND flash [6].

A BiCS NAND flash memory with floating pillars and an oxide–nitride–oxide–nitride (ONON) structure was discussed by Toshiba in December of 2008 [7]. Program and erase operation on the NAND string of the BiCS flash memory was achieved, and it was forecast that this structure could be used for a terabit memory density. Operationally, a single bit of the BiCS flash is accessed at the intersection of a CG plate selected by an USG and a NAND string pillar selected by a BL. The bottom of the pillars is connected to the common source diffusion on the silicon substrate.

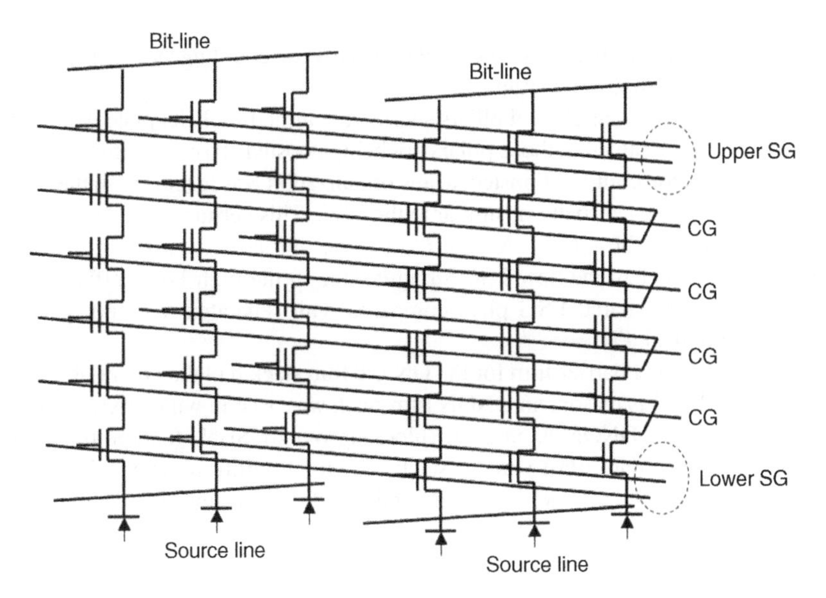

Figure 4.9 Equivalent circuit of 3D BiCS NAND flash. (Based on Y. Fukuzumi, *et al.*, (Toshiba), IEDM, December 2007 [6].)

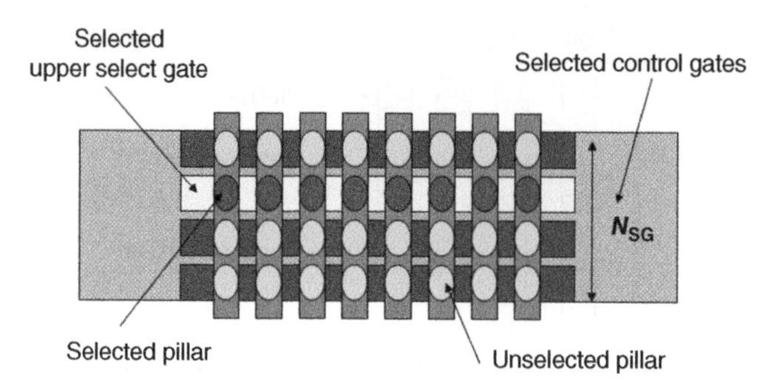

Figure 4.10 Top-down view of BiCS flash memory in program or read operation. (Based on Y. Komori *et al.*, (Toshiba), IEDM, December 2008 [7].)

A new erase operation was needed because the pillars of NAND string are not connected directly to a p-well. Erase was done by raising the potential voltage of the pillars of the NAND string by injecting holes generated by gate-induced drain leakage (GIDL) at the junction-side edge of the select gate. They found that V_{th} increased with maximum electric field, showing that the erase operation is caused by hole current generated by GIDL.

Another concern was disturbance of unselected pillars during program or read operation. CGs were shared by several rows of pillars, which also share a USG, as illustrated in Figure 4.10 [7]. The selected USG is shared by selected pillars. The selected CGs are shared by both selected and unselected pillars. This means program/read operation on pillars selected by a USG can result in stress on unselected pillars.

Despite a large number of stress cycles on unselected pillars, the V_{th} shift after stress was suppressed by high boost efficiency due to the absence of coupling between pillars of the NAND string and the silicon substrate.

During a read operation, unselected pillars sharing the turned-on LSG plate were connected to the source line. This resulted in those pillars not being boosted. To ensure that the unselected pillars in a read operation could be boosted, a new organization was formed with the LSG plates changed to a "line-and-space" pattern, similar to the USG. This permitted the USG and LSG to be turned on and off synchronously. As a result, the V_{th} shift after read disturb on unselected pillars was much smaller in the new LSG structure due to the high boost efficiency.

The configuration of shared LSG plate compared with separated line-and-space LSG is illustrated in Figure 4.11 [7].

The absence of a tunnel barrier film for the ON CT film stack had the potential for retention and disturb issues. For this reason, an ONON film stack was used in which oxide covered by a SiN film was used. The ONON is an SiO_2 (blocking oxide)–Si_3N_4 (charge-trapping)–SiO_2 (tunnel oxide)–Si_3N_4 (tunnel film) vertical stack. The tunnel film in the SONONS structure inhibited release of trapped charge in the SiN and was found to improve data retention and disturb characteristics [7].

4.2.2 *Pipe-Shaped BiCS (P-BiCS) NAND Flash Technology*

A pipe-shaped BiCS (P-BiCS) flash memory in 60 nm technology was discussed by Toshiba in June of 2009 [8]. The 3D stacked 1Gb cell/layer pipe-shaped flash memory used a 16-cell

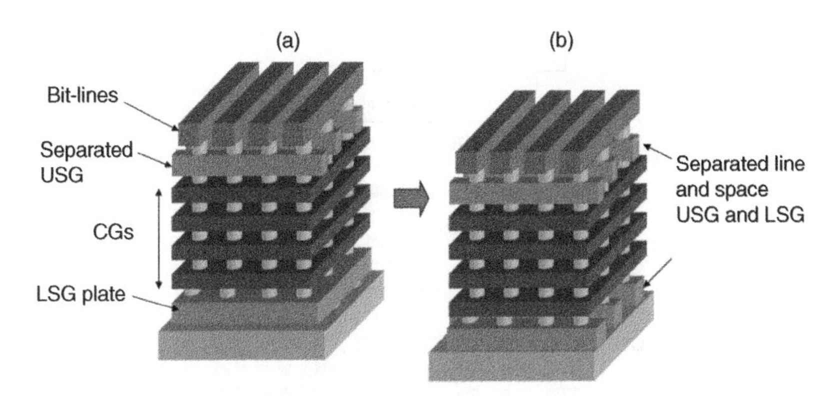

Figure 4.11 Illustration of shared LSG plate changed to separate line and space LSG. (Based on Y. Komori *et al.*, (Toshiba), IEDM, December 2008 [7].)

string. The effective bit cell size was 0.00082 μm². A branched CG configuration and a new erase operation were used for this memory. The target application was terabit storage devices. The BiCS cell array had multiple stacked CG plates and polysilicon pillars formed through the CG plates. The intersections of the CG plates and polysilicon pillars formed SONOS memory cells. In the original configuration of this cell, the source line was diffused in the silicon substrate and had a high resistance that degraded the read margin. This high-resistance issue was solved in the P-BiCS cell by using a folded pipe structure for the cell string with the source line formed in a metal layer, as shown in Figure 4.12 [8].

The cell string consists of two pillars with a pipe connecting the pillars at the bottom. The drain and source select gates now lie over the CG plates. The configuration of the CGs and row decoders was modified to accommodate this new structure. In the original BiCS memory, the CGs were shared by several neighboring rows of cell strings to reduce area. In the new P-BiCS flash, the CG is configured with fork-shaped plates with four branches, where each branch controls two pages of cells. The block is formed by 16 vertically stacked pairs of CGs, with each pair of CG plates arranged in a staggered layout. Select gates are individually separated for the selectivity of cell strings [8].

A further refinement on the P-BiCS vertical NAND flash with 16 stacked layers and MLC operation in 60 nm technology was discussed by Toshiba in June of 2009 [9]. The BiCS flash uses Fowler-Nordheim (FN) tunneling for P/E operations. The FN tunneling originates from the strong curvature effect of the small pipe radius. The P/E characteristics are dependent on the hole diameter, as shown in Figure 4.13 [9], where the 60 nm diameter hole can be seen to have a wider P/E V_{th} window than the larger diameter holes. The effective cell area per bit is 0.00082 μm².

Device characteristics of the 32Gb P-BiCS NAND flash included a 60 nm complementary metal–oxide semiconductor (CMOS) process, cell size per bit of 0.00082 μm², memory hole diameter of 66 nm, 16-layer CG stack, and 32 NAND gate string length. The back-end-of-line metallization was three layers of metal including Al–Cu–W. Characteristics of the device included 10-year data retention with no degradation and a V_{th} shift of less than 0.3 V after 100 000 read cycles at 7.5 V.

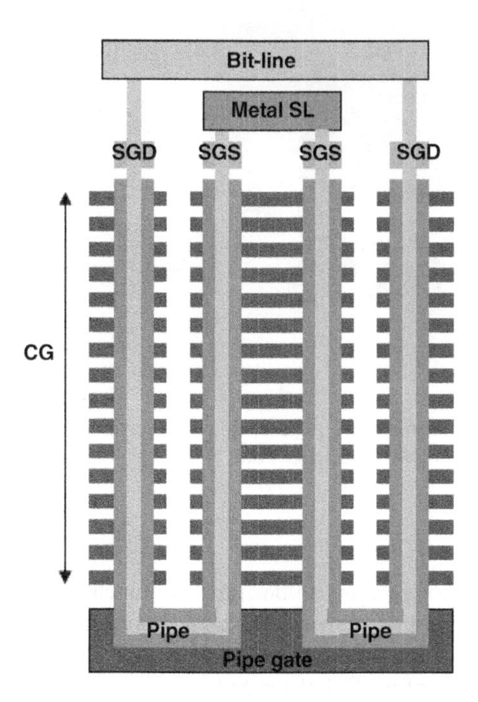

Figure 4.12 P-BiCS SONOS NAND flash cell array with vertical pillars. (Based on T. Maeda *et al.*, (Toshiba), VLSI Circuits Symposium, June 2009 [8].)

Issues addressed by this part included minimizing damage from dilute hydrofluoric acid (HF) treatment for "gate-first" process by using Si_3N_4-based tunnel film, improving the immunity to read disturb and data retention in order to operate it as an MLC, improving clamping of the high-resistance source line during read operation, and improving control of the diffusion profile of the LSG transistor. The device characteristics of the improved process indicate the potential for MLC operation [9].

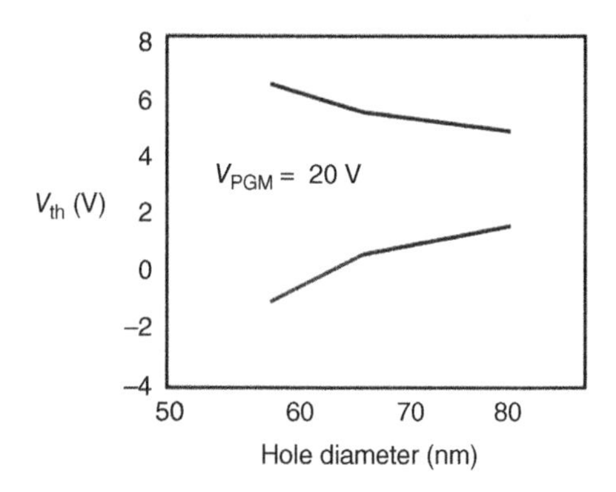

Figure 4.13 V_{th} (V) vs. hole diameter for 20 V P/E voltage for P-BiCS NAND flash. (Based on R. Katsumata *et al.*, (Toshiba), VLSI Technology Symposium, June 2009 [9].)

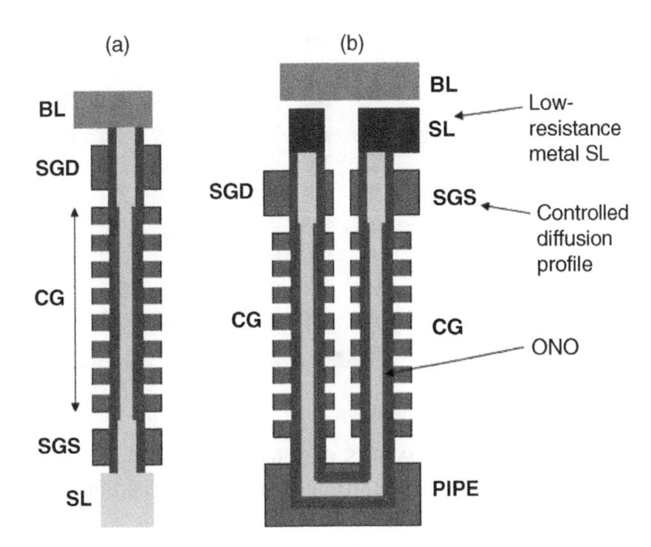

Figure 4.14 Illustration of the (a) straight vertical; and (b) U-shaped NAND string. (Based on M. Ishiduki *et al.*, (Toshiba), IEDM, December 2009 [10].)

The process technology for a U-shaped P-BiCS flash memory were discussed by Toshiba in December of 2009 [10]. An asymmetric S/D profile was used for the select gate and metal salicide for the CG. The select line was low-resistance metal wiring. This resulted in good performance and reliability. The straight-shaped vertical NAND string was changed to a U shape, which improved data retention and widened the V_{th} window due to less process damage on the tunnel oxide during fabrication. Illustrations of the straight vertical NAND string and the U-shaped NAND string are shown in Figure 4.14 [10].

A gate-first deposition process showed good endurance during the P/E sequence. Figure 4.15(a) illustrates the electrode–polysilicon stack, and Figure 4.15(b) shows the

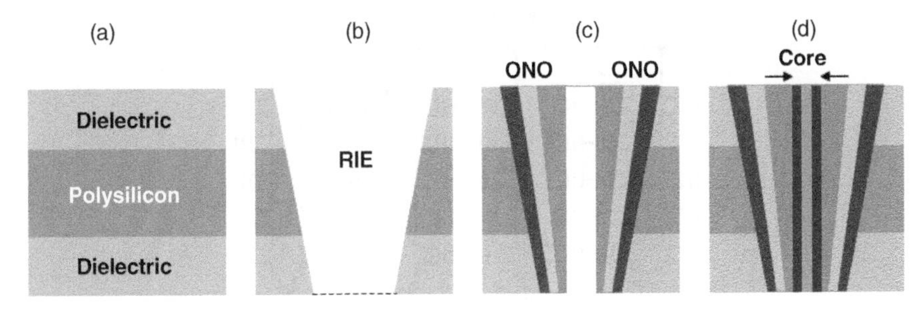

Figure 4.15 Illustration of gate-first deposition process for BiCS flash memory (a) electrode stack; (b) punch through of electrode stack using RIE; (c) ONO film deposited on walls of memory hole; (d) deposition of body polysilicon and core filler. (Based on M. Ishiduki, *et al.*,(Toshiba), IEDM, December 2009 [10].)

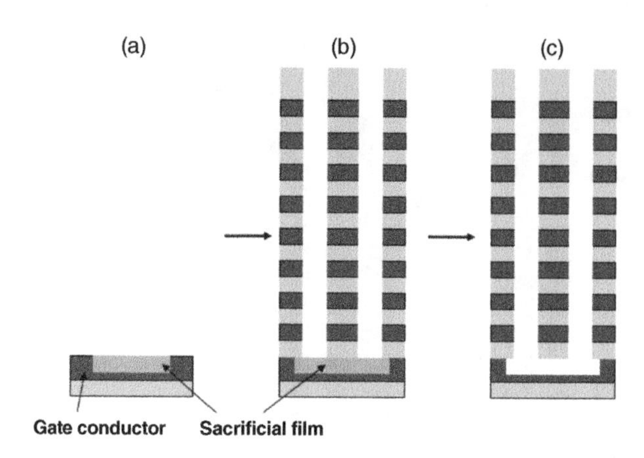

Figure 4.16 Pipe connection process for P-BiCS flash memory showing (a) deposition of sacrificial film; (b) memory hole punch using RIE; and (c) removal of the sacrificial film. (Based on M. Ishiduki *et al.*, (Toshiba), IEDM, December 2009 [10].)

stack punched through using RIE to form the memory hole. Figure 4.15(c) shows an ONO film deposited on the walls of the memory hole, and Figure 4.15(d) shows deposition of the body polysilicon and the core filler [10].

The pipe connection process, which used a sacrificial film, was added to the straight-shaped BiCS, as shown in the schematic cross-section for the P-BiCS flash memory in Figure 4.16 [10].

The BL contacts of the BiCS and P-BiCS architectures are shown in the schematic circuit diagram in Figure 4.17 [10]. This diagram shows one 8-cell planar NAND string and four 8-cell vertical NAND strings using the P-BiCs structure. The BL of the P-BiCS has a large number of

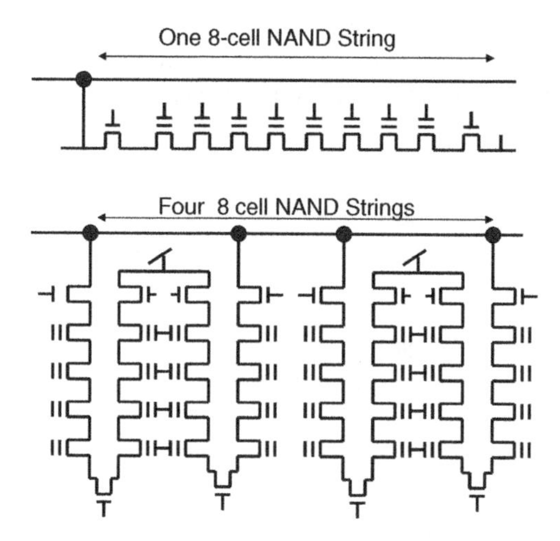

Figure 4.17 Circuit schematic of the (a) BiCS; and (b) P-BiCS NAND string showing both number of bit-line contacts and lateral density increase. (Based on M. Ishiduki *et al.*, (Toshiba), IEDM, December 2009 [10].)

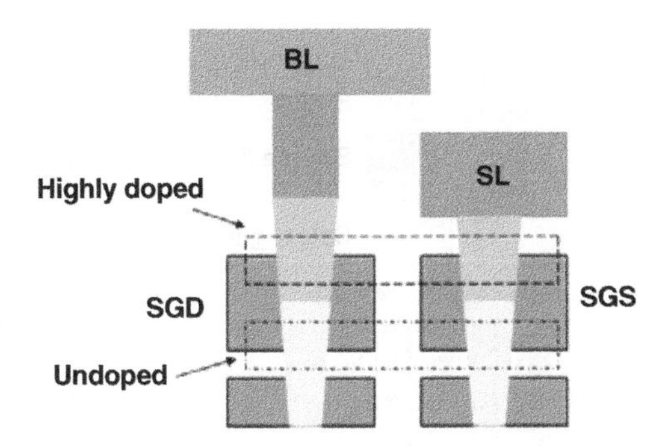

Figure 4.18 Schematic illustration of asymmetric S/D structure of vertical SG transistor. (Based on M. Ishiduki *et al.*, (Toshiba), IEDM, December 2009 [10].)

contacts on a single NAND String. This means the leakage current of the unselected NAND string from BL to source line must be decreased to avoid summing to a larger leakage current and to get a good ON–OFF ratio for the read operation.

An asymmetric source and drain structure was used for the vertical select gate of the P-BiCS flash. Using this structure, holes are effectively generated by GIDL current when source line bias is boosted for the erase operation, while leakage current is suppressed when source line bias is 0 V and the channel pillar is biased slightly positive. An illustration of the asymmetric S/D structure of the select gate (SG) transistor is shown in Figure 4.18 [10].

Metal salicide deposited with a CVD process was used for the CG of the P-BiCS to lower the resistivity of the doped polysilicon gate. Two orders of magnitude improvement in the sheet resistance was attained. As a result the resistance–capacitance (RC) delay was reduced by an order of magnitude. A test array of the P-BiCS NAND flash array with 1GB of cells per plane was made, and its functionality was confirmed for multilevel data storage [10].

In June of 2010, Toshiba reviewed developments on its P-BiCS flash 3D vertical technology [11]. This process reduces chip cost by stacking the memory array vertically using a punch and plug process. They have shown a pipe-shaped BiCS flash with U-shaped NAND string and MLC operation. The change from a vertical NAND string with a straight-shaped to a U-shaped pipe improved data retention and increased the V_{th} window because of less process damage on the tunnel oxide during processing. Functionality was shown using a 32Gb test chip with 16 stacked layers and MLC operation in 60 nm technology.

An optimal device structure to solve known issues and improve reliability was shown. The number of contacts per BL was much larger than for a planar NAND device, so the leakage current of the unselected NAND string needed to be reduced to avoid summing up a large leakage in the unselected strings. An asymmetric S/D structure for the select gate was used in which holes were generated by GIDL current when the source line bias was high during erase, but leakage current was suppressed when the source line bias was 0 V and the channel pillar was biased slightly positive. Another improvement was that metal salicide was used with CVD processing for the CG of the P-BiCS.

In April of 2012, Toshiba discussed its roadmap for 2013, which included development of BiCS for NAND and also cross-point arrays [12]. The P-BiCS NAND flash was described as a U-shaped NAND string with a back gate at the bottom of the U to reduce parasitic resistance of the bottom part of the U-shaped pipe. There was no diffusion between the CGs. An undoped

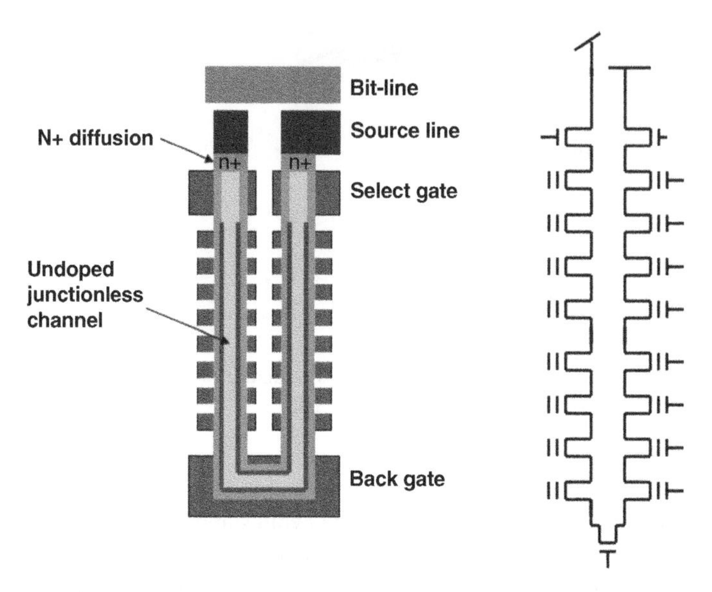

Figure 4.19 Illustration of P-BICS NAND flash (a) cross-sectional structure; and (b) schematic circuit diagram. (Based on J. Ohshima, (Toshiba), GSA/SEMATECH Memory Conference, 16 April 2012 [12].)

polysilicon channel was used for the memory transistors. The select gate used an asymmetric S/D structure to reduce OFF current. An illustration of the P-BiCS NAND flash structure is shown in Figure 4.19 [12].

Scaling and device design for a 3D stackable BiCS NAND flash were discussed in May of 2013 by the University of Tokyo [13]. CG length (L_g) and spacing (L_{space}) were studied because they could be separately varied in 3D NAND and affect the cell area and electrical characteristics.

A 3D device simulation was used to determine the expected characteristics of L_g and L_{space}. It was shown that $L_g = L_{space} = 20$ nm is possible for BiCS-type 3D NAND with a 90 nm diameter hole. Programming voltage can be reduced from 20 to 17 V. L_g and L_{space} need to be the same in order to handle the tradeoff between memory window and disturbance. If the number of stacked layers is 18 with a layer pitch of 40 nm, the effective cell size of the 3D NAND corresponds to that of 15 nm planar NAND technology.

In 3D stackable NAND, the number of stacked layers (N_{layer}) is increased to reduce the bit cost instead of shrinking the planar cell area. New scaling and design methodologies are required for 3D NAND because they differ significantly from those of planar NAND. A schematic cross-section of the BiCS 3D NAND cell array is shown in Figure 4.20 [13].

An issue with the 3D NAND is the decrease in cell density in the planar direction. The BiCS hole must by filled with about 20 nm of ONO film. Because the ONO film is not aggressively scaled to maintain the memory window and reliability, the diameter of the BiCS hole is not scalable. For this reason N_{layer} should be increased with a finite taper angle "θ" in the BiCS hole, as shown. Because the minimal line-and-space pattern is not required for the CG formation in 3D NAND, the parameters L_{cg}, L_g and L_{space} can be chosen separately. This design flexibility is unique to the 3D NAND for L_g and L_{space}. The effective cell area (A_{eff}) of the 3D

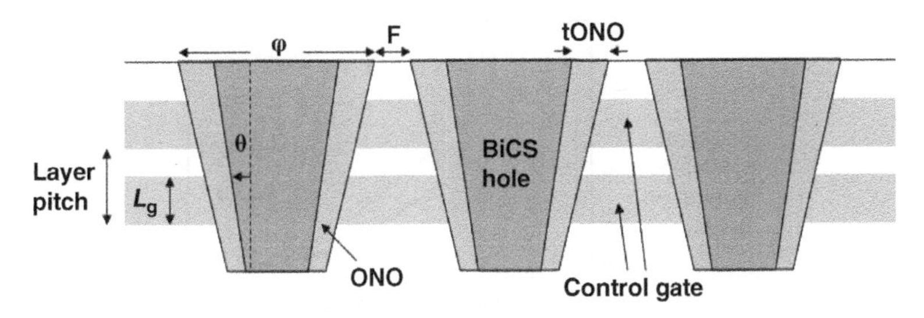

Figure 4.20 Illustration of cell cross-section of BiCS hole. (Based on Y. Yanagihara *et al.*, (University of Tokyo), IMW, 20 May 2013 [13].)

NAND was studied, where $A_{eff} = A/N_{layer}$ where A is the cell area in each layer. An expression for A_{eff} was derived as follows.

$$A_{eff} = \{2R_B + 2N_{layer}(L_g + L_{space})\tan\theta + 2t_{ONO} + F\}^2 / N_{layer}$$

Where R_b = the bottom radius of the BiCS hole, T_{ono} is the thickness of the ONO layer, and F is the feature size, which is the spacing between BiCS holes. A_{eff} of the 3D NAND depends on the cell pitch. Increasing the number of layers becomes less effective for A_{eff} reduction because the BiCS hole pitch increases with the total height of the 3D NAND in the presence of the BiCS hole taper, as shown in Figure 4.21 [13].

The required minimum number of layers (N_{layer}) to achieve an A_{eff} smaller than the planar NAND cell is shown in Figure 4.22 for electrical characteristics achievable by top cell diameter $\varphi = 90$ nm [13].

To achieve effective cell size equivalent to 15 nm planar NAND technology, a 40 nm layer pitch with 18 layers is needed. This study explores the L_g and L_{space} design window for 40 nm layer pitch. 3D device simulations were done to study their impact on the electrical parameters.

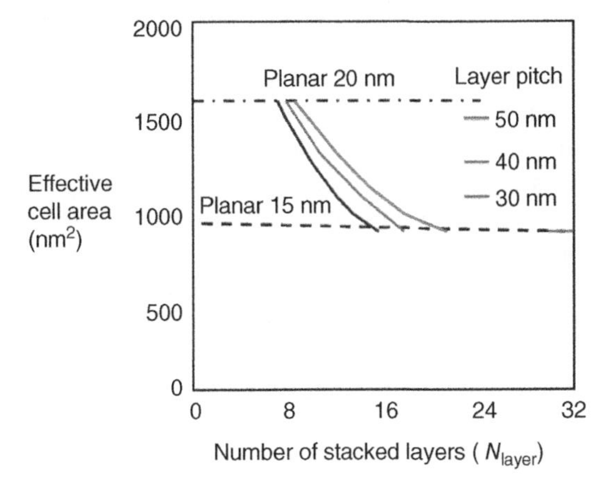

Figure 4.21 Effective cell area vs. number of stacked layers with BiCS hole taper shown between an equivalent planar 15 nm and 20 nm cell areas. (Based on Y. Yanagihara *et al.*, (University of Tokyo), IMW, 20 May 2013 [13].)

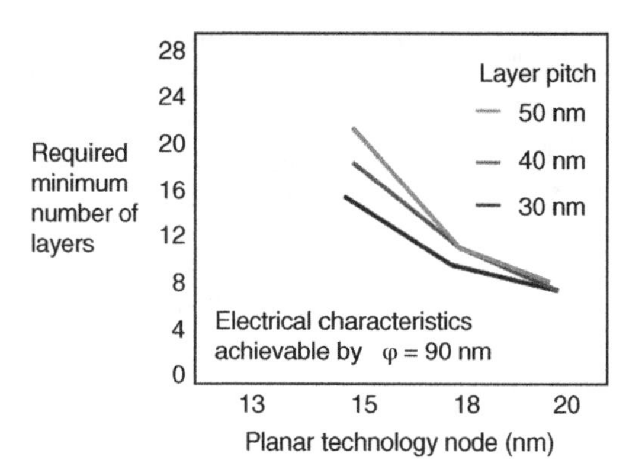

Figure 4.22 Required minimum N_{layer} vs. planar technology node by layer pitch. (Based on Y. Yanagihara *et al.*, (University of Tokyo), IMW, 20 May 2013 [13].)

Results showed that 3D NAND has good on-current (I_{on}) and subthreshold slope (SS) compared with the planar NAND. Program voltage (V_{pgm}) was reduced from 20 to 17 V. V_{th} roll-off and V_{th} shift, however, became worse than that of planar NAND below $L_g = L_{space} = 20$ nm in 3D NAND. This was because coupling with the neighboring cell became stronger than in planar NAND. Having the same L_g and L_{space} is better for a large memory window and small disturbance. The conclusion was that 3D BiCS-type NAND with effective cell size of 15 nm compared to planar NAND is possible with $L_g = L_{space} = 20$ nm and 40 nm layer pitch if a 90 nm diameter hole is used and the number of stacked layers is 18.

The BiCS technology was reviewed by Toshiba in June of 2013 [14]. The BiCS technology reduces bit cost by using a punch and plug process to vertically stack memory arrays. While the technology is initially being used to stack NAND flash memory, the BiCS technology was expected to be used for other memory devices as well. The BiCS process involves a stack, a punch, and a plug. Multiple layers of plate electrodes and dielectric films are stacked to isolate the various plate electrodes. The multilayer stack is then punched through from top to bottom, with only one lithography step being used. The inside of the holes is then covered with the memory film, and the holes are plugged with vertical pillar electrodes. The memory cells are found at the intersection of the plate electrodes and pillar electrodes. An advantage of using this processing method is that there is no increase in critical lithography, etching, or planarization when layers are added to increase memory density. The basic concept of BiCS is shown in Figure 4.23 [14]. A top-down view cross-section of the GAA memory plug is shown in Figure 4.24 [14].

The study claims that the relative bit cost of the BiCS technology compared to the cross-point array process is as shown in Figure 4.25. As the number of layers increases, the BiCS technology is expected to cost less per bit than the cross-point array technology [14].

3D NAND flash memories have several critical issues that need to be addressed before they can go into scalable mass production. These include WL RC delay, WL and hole formation, data retention, and improvement in cell current. The polysilicon TFT process used in the BiCS flash had improved mobility. The process also has a large electric field contrast due to the difference of channel curvature radius, which permits the blocking oxide to be pure SiO_2. The SONOS process has good data retention due to the use of bandgap engineering [14].

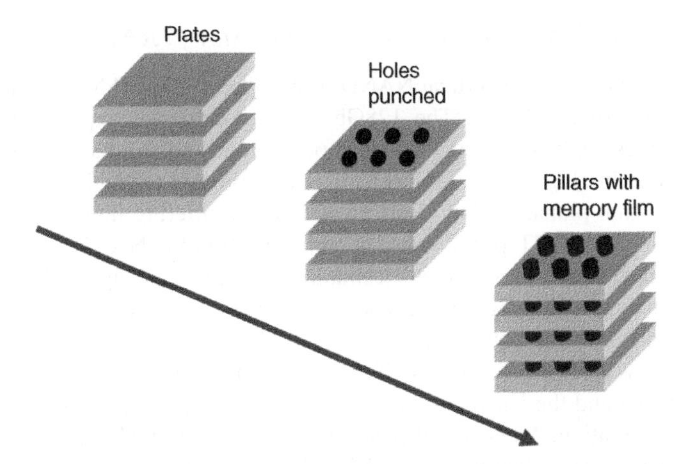

Figure 4.23 Basic concept of BiCS technology. (Based on A. Nitayama and H. Aochi, (Toshiba), VLSI Technology Symposium, June 2013 [14].)

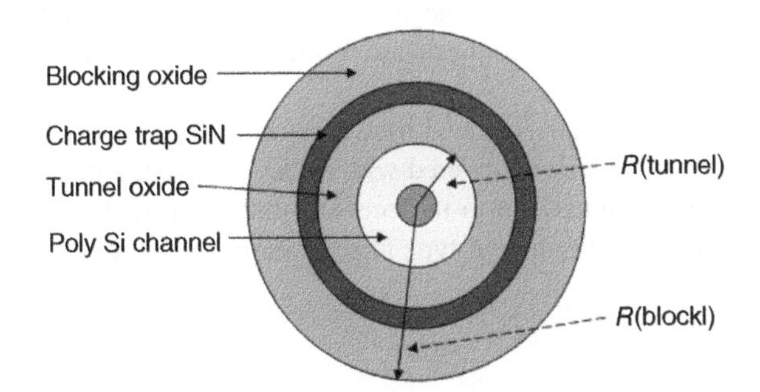

Figure 4.24 Schematic of GAA memory plug cross-section. (Based on A. Nitayama and H. Aochi, (Toshiba), VLSI Technology Symposium, June 2013 [14].)

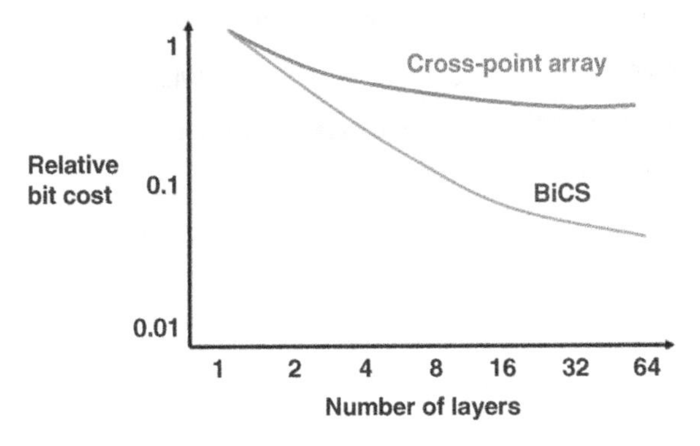

Figure 4.25 Bit cost of BiCS and cross-point array technology by number of layers. (Based on A. Nitayama and H. Aochi, (Toshiba), VLSI Technology Symposium, June 2013 [14].)

4.2.3 Vertical CT Vertical Recess Array Transistor (VRAT) Technology

Mass production of an early 3D vertical NAND flash called the V-NAND was announced by Samsung in August of 2013 [15]. The 128Gb capacity chip was expected to be used in consumer electronics applications such as industrial storage and solid-state drives (SSDs). It used both 3D CT technology and a vertical interconnect process technology to provide twice the scaling of a 20 nm planar NAND flash. In addition to offering higher reliability, it was expected to double the WRITE performance of a planar 10 nm FG NAND flash memory. The device used a vertical interconnect process technology that can stack up to 24 layers vertically using a special etching technology that connects the layers electronically by punching holes from the highest layer to the bottom [15].

The vertical recess array transistor (VRAT) 3D NAND flash memory was discussed in June of 2008 by Samsung and the University of California, Los Angeles [16]. The 3D technology used planarized integration. Another higher-density method of designing 3D NAND flash was also described, which was called the zigzag VRAT (Z-VRAT). A schematic cross-section of the VRAT is shown in Figure 4.26 [16]. The dashed line shows the path the current takes past the four vertical ONO-based CT stacked flash transistors.

The VRAT structure increases memory density through stacking multiple layers. Fully depleted channels permit keeping the gate length unscaled to help suppress subthreshold leakage current (SLC). Vertical interconnects are formed at the same time as the VRAT array, which helps reduce cost.

The VRAT consists of vertically chained transistors and interconnects. Multiple stacks of nitride and undercut oxide layers are covered with a thin, fully depleted polysilicon channel. The channel length is maintained due to its recessed shape, while the double-gate structure increases the drive current. A tilt-angle n-type implantation is used to dope the channels [16].

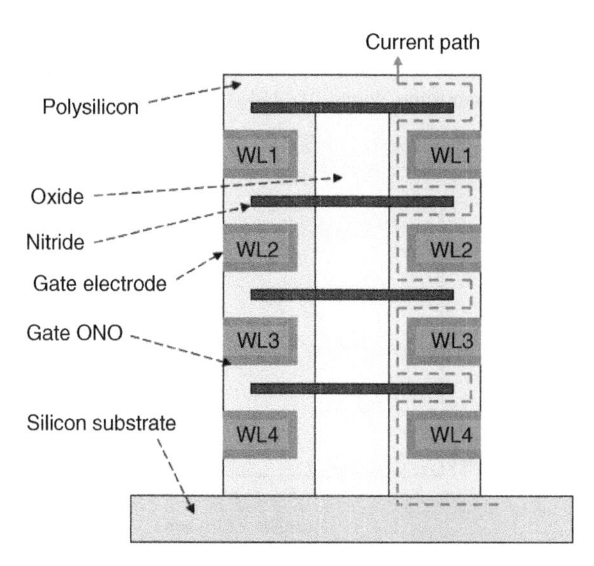

Figure 4.26 Schematic of four transistor VRAT NAND Chain (Based on J. Kim *et al.*, (University of California Los Angeles, Samsung), VLSI Technology Symposium, June 2008 [16].)

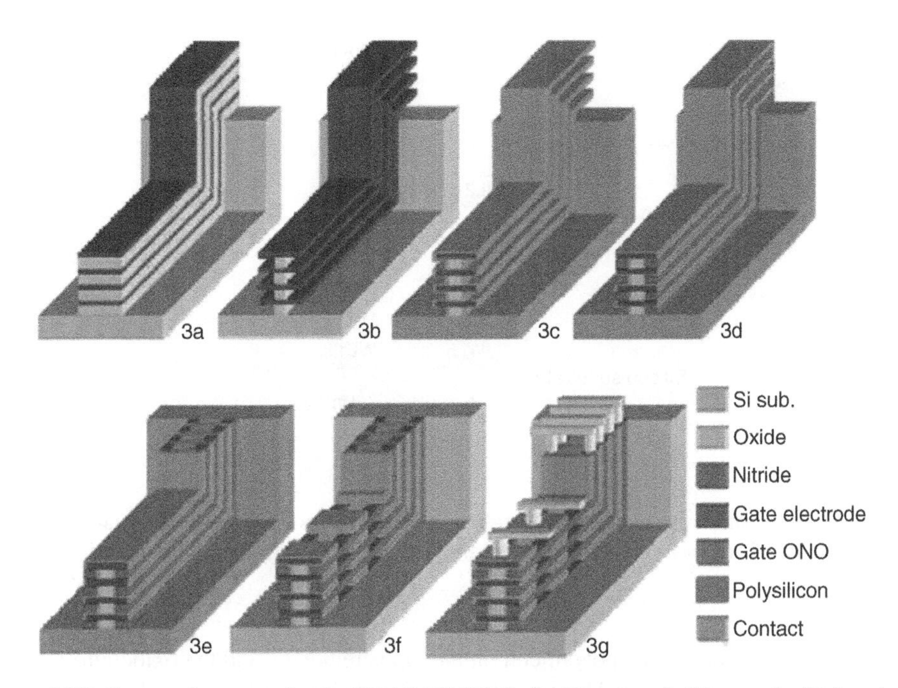

Figure 4.27 Process Sequence for the VRAT NAND flash. (Based on J. Kim *et al.*, (University of California Los Angeles, Samsung), VLSI Technology Symposium, June 2008 [16], with permission of IEEE.)

The process sequence is illustrated in Figure 4.27 [16]. In this sequence, multiple alternating layers of oxide and nitride films are deposited partially sloping up a silicon mesa. These are patterned lithographically and etched to form the active region (Figure 4.27(a)). These oxide layers are then selectively etched with a wet process to form an undercut (Figure 4.27(b)). The polysilicon is then deposited to form the channel material of the active transistors (Figure 4.27(c)). A nitride CT material is deposited in the undercut between the tunnel oxide and the control oxide to form the ONO followed by the WL electrodes. An etch-back removes the WL electrode material on the sidewall, leaving the WL electrodes in the undercut space but isolated from one another (Figure 4.27(d)). Planarization using CMP exposes the gate electrodes in a horizontal plane on the slope (Figure 4.27e). N-type implant is done at a tilt angle to form the S/D, and each string is isolated (Figure 4.27(f)). Contacts are then completed (Figure 4.27(g)) [16].

4.2.4 Z-VRAT CT Memory Technology

It was estimated that memory density of 128Gb could be obtained with the VRAT in 50 nm technology. A further increase in memory density can be obtained potentially by splitting a VRAT into two narrow mesas through the middle, which doubles the memory density. A Z-VRAT with eight transistors in a vertical stack was simulated, and a zigzag current path was shown along the polysilicon channel. Samsung estimated that a NAND flash memory density as high as 256Gb could be obtained using the Z-VRAT in 50 nm technology with 16 stacks. A schematic cross-section of the Z-VRAT with a six-transistor vertical string is shown in Figure 4.28 [16].

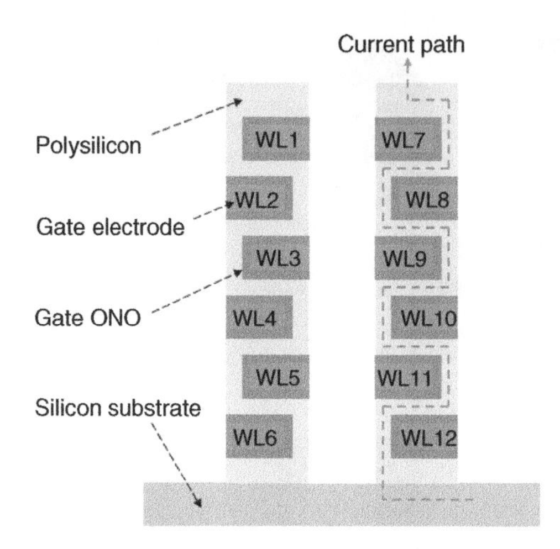

Figure 4.28 Schematic cross-section of Z-VRAT with six transistors in vertical string. (Based on J. Kim *et al.*, (University of California Los Angeles, Samsung), VLSI Technology Symposium, June 2008 [16].)

In the Z-VRAT process, the polysilicon on the top is removed, and the oxide/nitride layers are etched using a wet process. The gate electrode is filled, planarization is done to form the contacts, and finally the contacts and metal connections are formed. An illustration of the 3D Z-VRAT technology is shown in Figure 4.29 [16].

A comparison of possible memory density at the 50 nm node was done for the planar, VRAT, and Z-VRAT array architectures and is shown in Table 4.3 [16].

Figure 4.29 Illustration of 3D Z-VRAT technology. (Based on J. Kim *et al.*, (University of California Los Angeles, Samsung), VLSI Technology Symposium, June 2008 [16].

Table 4.3 Comparison of Memory Density at 50 nm for Planar, VRAT, and Z-VRAT Arrays.

	Planar	VRAT		Z-VRAT	
Density (Gb)	16	64	128	128	256
Layers	1	8	16	8	16
MLC/SLC	MLC	SLC	SLC	SLC	SLC

Based on J. Kim *et al.*, (UCLA, Samsung), VLSI Technology Symposium, June 2008 [16].

4.2.5 Vertical NAND Chains—VSAT with "PIPE" Process

A vertical stacked array transistor (VSAT) that was combined with a PIPE process (Planarized Integration on the same PlanE) was discussed by Samsung in June of 2009. The full PIPE method of vertical interconnection was thought to be an improvement over the "stair-like" PIPE connection method used previously for cost effectiveness, measured in lateral area of silicon and in simplicity of manufacture. A schematic diagram of the two interconnection methods is shown in Figure 4.30. The vertical interconnection method is deposited on a sloping surface and planarized to expose the pipes at the surface [17].

The 24-transistor VSAT NAND string is formed using three vertical stacks with eight devices per stack. The BL contacts four devices on one side of the stack and four devices on the other side of the stack. Devices on opposite sides of the stack share a common WL. A schematic diagram of the VSAT and PIPE is shown in Figure 4.31 [17]. This method of forming vertical NAND chains decreased the off-current level in the polysilicon channel transistor by five orders of magnitude by using a thin 20 nm body and a double-gate-in-series structure. Hydrogen annealing was used to improve the subthreshold swing and mobility of the polysilicon channel transistor. The VSAT + PIPE NAND technology was thought to be an improvement over the previously shown VRAT + PIPE 3D NAND structure, which created an undercut space and filled it with electrodes. Improvements in subthreshold performance and channel mobility also resulted from using a thinner polysilicon channel [17].

An equivalent circuit diagram showing the VSAT device with three vertical strings is shown in Figure 4.32 [17]. This device was developed at the 100 nm node. Storage capacity was expected to reach 128GB with 16 multiple layers on the 50 nm node [17].

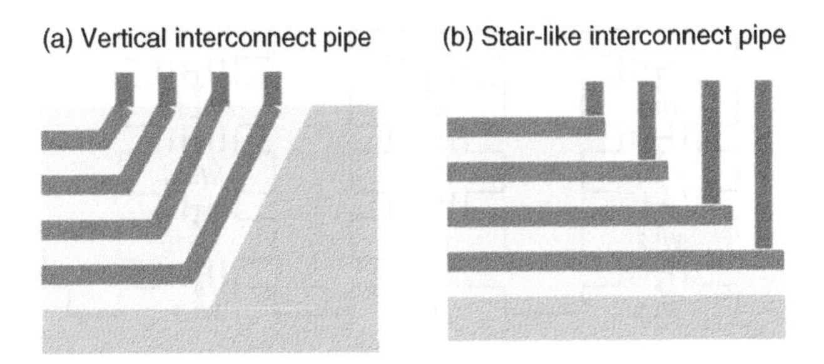

Figure 4.30 Vertical interconnection "PIPE" vs. "PIPE stairs." (Based on J. Kim *et al.*, Samsung, VLSI Technology Symposium, June 2009 [17].)

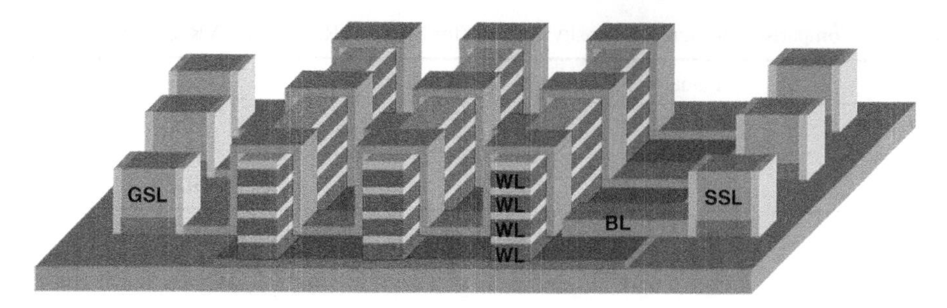

Figure 4.31 Schematic cross-section of VSAT + PIPE technology along with bit-line. (Based on J. Kim *et al.*, (Samsung), VLSI Technology Symposium, June 2009, Figure 3 [17].)

4.2.6 Vertical CT PIPE NAND Flash with Damascene Metal Gate TCAT/VNAND

A damascene tungsten (W) metal gate CT cell in a 3D vertical NAND flash string was proposed by Samsung in June of 2009 [18]. It was called a terabit cell array transistor (TCAT) and was made by a gate replacement process. Bulk erase operation of the cell was demonstrated. A schematic cross-section of the Samsung TCAT vertical NAND flash string structure with details of the selection transistors is shown in Figure 4.33 [18].

The cell string has six NAND cell transistors with a string select (SSL) transistor on top and a ground select line (GSL) transistor on the bottom. The SSL and GSL transistors are formed at the same time as the cell transistors are. The TCAT vertical NAND flash technology with metal gate and bulk erase operation was projected as a potential technology for terabit memory capacity. A 32-string array was characterized. The subthreshold slope was about 320 mV/dec and on–off ratio was $>10^6$. A good threshold distribution was attributed to the GAA body effect. Program disturb was suppressed by V_{th} adjustment of select transistors [18].

Samsung has indicated that its TCAT process solves several of the issues with the earlier Toshiba BiCS stacked NAND string technology. This included a potential difficulty in etching the metal–oxide multilayer simultaneously and a concern that a circuit change might be necessary to apply negative bias on the WL during erase to avoid potential GIDL erase.

The process sequence is as follows: initially an oxide–sacrificial nitride multilayer stack is deposited, the channel hole is etched and filled with polysilicon, gate pads are etched to form a

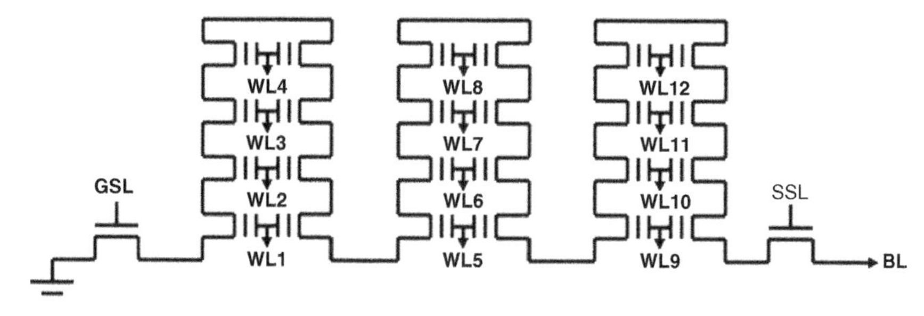

Figure 4.32 Equivalent circuit diagram of VSAT NAND flash with three vertical strings. [17] (Based on J. Kim *et al.*, (Samsung), VLSI Technology Symposium, June 2009, Figure 6 [17].)

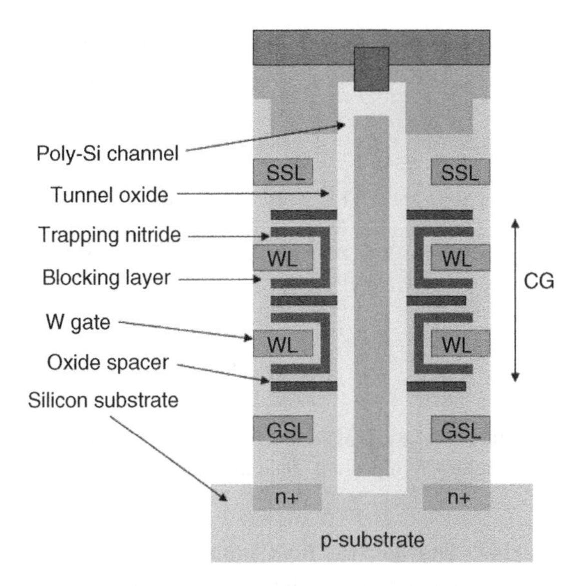

Figure 4.33 Schematic cross-section of Samsung vertical NAND flash string structure (TCAT). (Based on J. Jang *et al.*, (Samsung), VLSI Technology Symposium, June 2009 [18].)

pyramid-like stack that provides access for contacts to the WLs, the WL cut is dry etched to separate the vertical device strings and the sacrificial nitride is removed by wet etching, a gate dielectric (ONO) is deposited, and a common source line (CSL) implant is done followed by deposition of the tungsten WLs. Contacts to the WL steps are formed in the back-end-of-the-line process [18].

The TCAT NAND flash memory is capable of bulk erase operation because the channel poly plug in the TCAT structure is connected to the silicon substrate so that conventional bulk erase operation can be performed and the NAND string can be operated without a significant change to the conventional peripheral circuitry used in the planar flash. Erase saturation at about 1 V was an issue but was expected to improve with a channel hole diameter of 45 nm, which will cause the electric field in the tunnel layer to become stronger. A 120 nm hole diameter was used in this test device. Endurance was 10^4 cycles, and data retention at 85 °C showed a V_{th} window of 5.6 V after 10 years. This was expected to permit MLC operation [18].

Further efforts on its 3D NAND TCAT technology with a 90 nm channel hole were discussed by Samsung in June of 2010 [19]. The TCAT was improved by the addition of damascened metal gates, which provided low resistance. A new tungsten fill metal process was adopted to reduce WL resistance by increasing the portion of low-resistivity tungsten in the metal gate. The resistance was reduced by more than 70% over the conventional process without changing the dimension of the gate. Lowering the WL resistance improves the performance of NAND devices in a given chip area or permits reducing chip size without affecting performance.

Lateral charge spreading was decreased by a biconcave structure that could be achieved by using a damascened gate structure. A controlled offset between the BL contact and the select transistor was also used. A suppressed disturbance property was achieved by using this offset, which reduces the leakage current through the select transistor. A comparison of the trap layer structure of the (a) BiCS 3D NAND and (b) concave TCAT VNAND device is

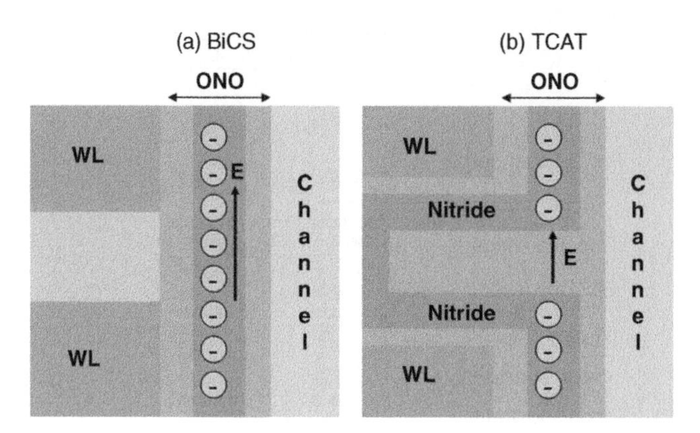

Figure 4.34 Trap layer structure of (a) BiCS 3D NAND; and (b) TCAT 3D NAND. (Based on W.S. Cho *et al.*, (Samsung), VLSI Technology Symposium, June 2010 [19].)

shown in Figure 4.34 [19]. The electric field induced between adjacent cells that are programmed in different states tends to accelerate charge losses at high temperature. The biconcave TCAT structure helps prevent lateral charge losses.

With scaling from 120 to 40 nm channel width, the electric field induced in the tunnel layer is increased, while the field in the blocking layer is reduced as the diameter of the channel hole is scaled. For a 90 nm channel width, a V_{th} window of about 7 V was achieved using $V_{pgm} = 18$ V and $V_{ers} = -18$ V, as shown by the P/E characteristics in Figure 4.35 [19]. Data retention after 10 years showed less than a 0.5 V shift in V_{th} window.

Reliability of the TCAT VNAND CT Flash arrays was simulated by Samsung in June of 2012 [20]. Two intrinsic variation sources of the cell threshold voltage induced by traps in the polysilicon channel material in a 3D VNAND flash using CT were described. These variation sources were random trap fluctuations (RTF) and random telegraph noise (RTN). RTN was shown to be enhanced by the polysilicon material used for the channel. A model was developed

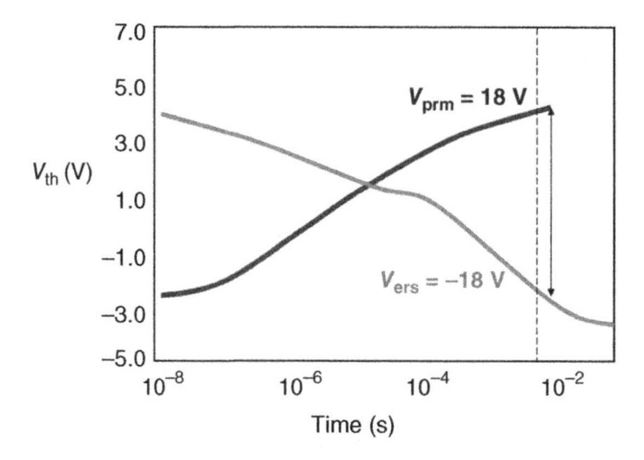

Figure 4.35 Program and erase characteristics of a 90 nm channel width VNAND. (Based on W.S. Cho *et al.*, (Samsung), VLSI Technology Symposium, June 2010 [19].)

that explained the asymmetric RTN distribution observed after endurance testing. The model enabled prediction of V_{th} distribution for VNAND devices during MLC operation.

RTF is due to fluctuations of a trap location inside the polysilicon channel. These traps were shown to follow a Poisson statistic, and it was shown that trap density was the correct metric for evaluating the electrical performance of the polysilicon channel. RTN follows an exponential distribution due to RTF in polysilicon. The RTN energy distribution indicated that most of the RTN traps were present at the Fermi level. It was found that cycling the cell induced a generation of RTN traps during programming. The model used was found suitable for the complex 3D structure and permitted structure optimization and statistical prediction for the vertical NAND structure. It also modeled accurately the MLC distribution, permitting prediction of the ECC requirements for the VNAND [20].

4.2.7 3D NAND Flash SB-CAT Stack

In December of 2011, Seoul National University discussed a 3D stacked NAND Flash which used a common gate structure and a shield layer. A trench structure was used instead of a through-hole structure. The 3D structure is called a Square wave-shaped BL cell array transistor (SB-CAT). A 2-D schematic is shown in Figure 4.36 [21].

The trench structure permitted V_{th} variation in cells of a NAND cell string to be reduced and the number of stacked control-gate electrodes in a gate stack was increased. In the trench between adjacent CG stacks, gate O/N/O stack, poly-Si body, backside oxide and shield layer

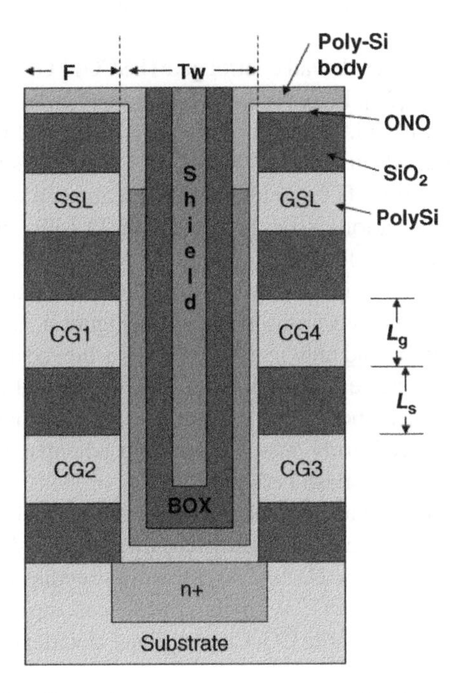

Figure 4.36 3-D stacked NAND flash using trench structure with shield. (Based on M.K. Jeong *et al.* (2011) (Seoul National University), *IEEE Transactions on Electron Devices*, 58(12), 4212 [21].)

were formed. The 3D stacked cell strings had three layers of vertically stacked CGs. V_{th} was controlled by applying bias to the shield layer. Reasonable cycling and retention and good pass-gate properties were shown in the bottom of the trench. The n+ region in the bottom of the cell can count as a CG in the bottom cell. Two layers of CG's were used in the structure [21].

The SB-CAT structure, by using common CG and shield layer, assures an effective cell area of 4F2/n. A cell string is made with three layers of electrodes which includes two selection devices, SSL and GSL, and four cells. Using the trench for isolating adjacent stacks means the stacking of electrodes is more expandable than a conventional through-hole-type structure. The cell string showed retention window of 0.84 V after 10 years and cycling with V_{th} shift of 1.5 V after 5×10^3 P/E cycles. The cell in the bottom of the trench was confirmed to work well so that both vertical surfaces could be connected successfully by turning on the bottom cell. By controlling the bias of the trench shield, the cell threshold voltage shift could be controlled for both program and erase states [21].

A study of the ID-VGS characteristics of a 3D stacked NAND flash string with a common gate and a shield layer was discussed in May of 2010 by Seoul National University [22]. Body cross-talk was eliminated completely. The proposed common gate and shield layer structures were thought to be promising for future vertical NAND flash string memories.

4.3 3D FG NAND Flash Cell Arrays

4.3.1 3D FG NAND with Extended Sidewall Control Gate

Several efforts have been made to use FG NAND flash cells in vertical channel arrays. This technology development is an attempt to extend the conventional FG structure into the 3D architectures.

Arrays made with the extended sidewall control gate (ESCG) FG cell were discussed in May of 2010 by Tohoku University [23]. In this cell, the S/D region has an electrically inverted layer caused by an extended sidewall CG. The surface electron density of the S/D is higher than that of a conventional FG cell with diffused S/D. The result is a low-resistive S/D. The ESCG cell uses enhancement mode operation. The cylindrical FG structure cell with the extended sidewall CG was intended to overcome various issues possible with CT cells, such as charge spreading, which can reduce the V_{th} window for MLC operation.

A 4-bit vertical FG NAND cell array using the ESCG cell along with an equivalent schematic circuit diagram of the cell array is shown in Figure 4.37 [23]. Normal flash cell operation with fast programming and high read current was achieved as a result of the increased coupling ratio and the low-resistive S/D technique. The 3D vertical flash memory cell array with the new S/D technique had about 50% less interference with neighboring cells than the planar FG NAND cell. This cell array was considered a potential candidate for a fast and reliable terabit 3D vertical NAND flash cell array.

The interference effects of direct and indirect coupling paths of neighboring cells for 3D FG NAND cell arrays with an ESCG cell structure were discussed in June of 2011 by Tohoku University and JST-CREST [24]. The 3D NAND cell array structure with ESCG was shown to fully suppress the interference effects. With the ESCG cell structure, the FG is formed cylindrically and fully surrounded by the CG. Conventional electrical S/D techniques can used with the ESCG structure.

Capacitive coupling effects between adjacent FGs are illustrated in Figure 4.38. Figure 4.38(a) illustrates the coupling capacitance between adjacent FGs for the conventional

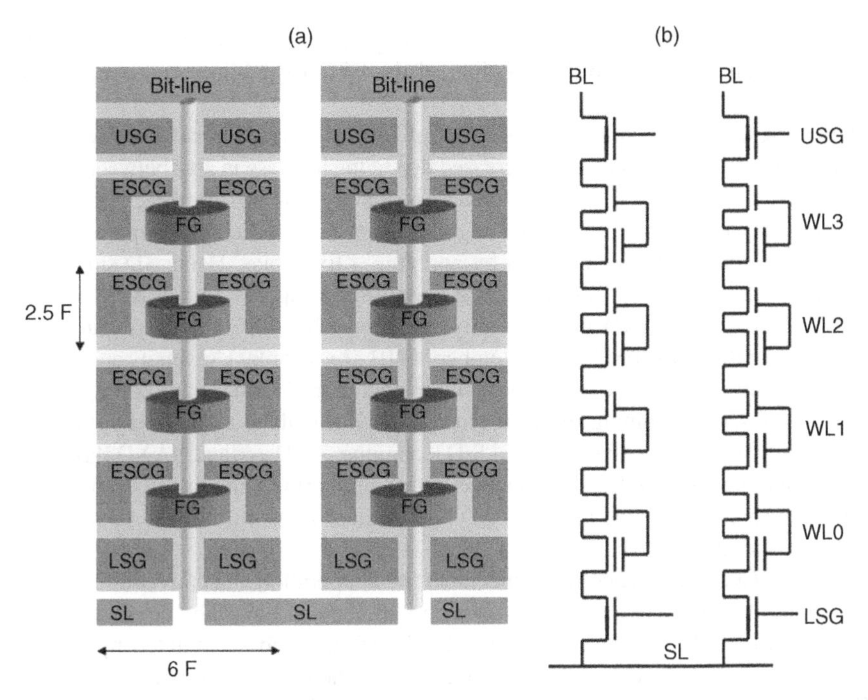

Figure 4.37 Schematics of (a) a 4-bit vertical FG NAND cell array with ESCG; and (b) an equivalent circuit diagram. (Based on M.S. Seo, S.K. Park, and T. Endoh, (Tohoku University), IMW, 16 May 2010 [23].)

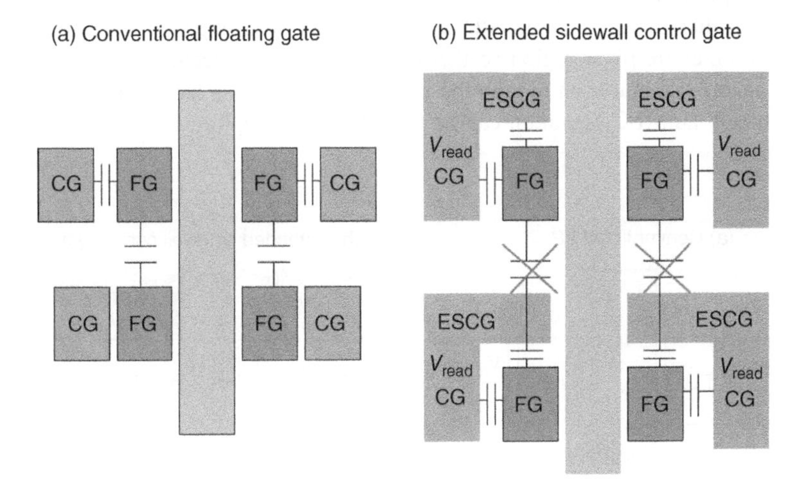

Figure 4.38 Illustration of capacitive coupling effects of (a) conventional FG cell; and (b) ESCG cell. (Based on M.S. Seo and T. Endoh, (Tohoku University, JST-CREST), VLSI Technology Symposium, June 2011 [24].)

FG structure, and Figure 4.38(b) illustrates the coupling capacitance between adjacent FGs for the ESCG cell [24]. Due to the coupling of the ESCG shielding structure, the FG–FG capacitance is suppressed.

With the ESCG cell structure, the FG–FG coupling capacitance was sufficiently suppressed, and the CG coupling capacitance was increased due to the increase in area between the CG and the FG. For the ESCG cell, the interference is mainly due to the direct coupling from neighboring FGs to the ESCG channel region. In order to efficiently suppress the direct coupling effect, the ESCG is controlled by applying a fixed bias. Using 3D simulation, the interference of the optimized ESCG cell was found to be suppressed below 50 mV.

The 3D vertical FG NAND flash memory cell array with ESCG was compared to a 3D vertical FG cell without the ESCG in September of 2011 by Tohoku University and Hynix [25]. The ESCG structure allowed enhancement mode operation. Normal flash cell operation was shown to have fast programming and good read current due to the increase in coupling ratio and the use of an ESCG low-resistive electrical S/D technique. A 3D vertical NAND flash cell array with ESCG had about 50% less interference with neighboring cells than a conventional 3D vertical FG NAND without the ESCG. A 3D NAND flash array with ESCG showed fast read and program and good reliability.

Using the ESCG structure, both enhancement mode operation and conventional bulk erase were achieved. Figure 4.39 shows a cross-sectional view of both the conventional FG and the proposed ESCG [25]. The FG is formed cylindrically and surrounded by the CG. An electrical S/D is formed in the junctionless channel when a positive bias is applied to the ESCG structure. At the surface of the pillar, the electrically inverted electron density of the ESCG cell was found to be higher than that of the conventional FG cell with a diffused S/D by about an order of magnitude.

Cross-sectional views of the conventional FG and ESCG cell during bulk erase are shown in Figure 4.40 [25]. The diffusion S/D junctions of the conventional FG device may change the silicon pillar potential when the bulk erase bias propagates into the channel. With the ESCG cell, when the positive erase bias is applied in the channel region, holes accumulate at the surface of the pillar region, and no S/D junction exists. This means the erase potential is transferred to the entire pillar region so that normal bulk erase occurs.

The conclusion was that even though the ESCG FG cell size is larger than the vertical CT NAND flash cell, an MLC operation can be done without coupling effects with neighboring

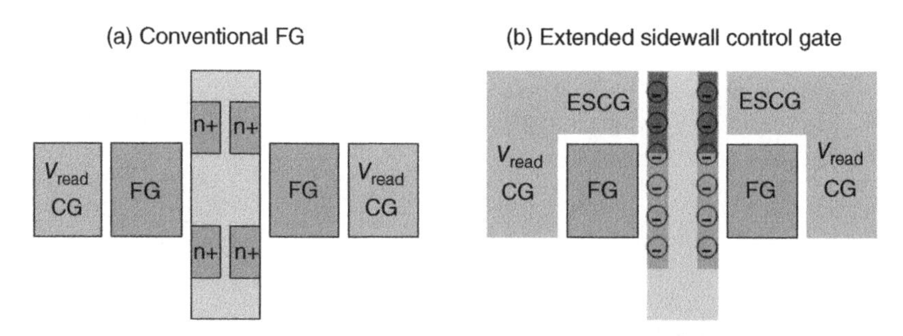

Figure 4.39 Cross-section of (a) conventional FG cell; and (b) ESCG cell. (Based on M.S. Seo, S.K. Park, and T. Endoh, (Tohoku University, Hynix), *IEEE Transactions on Electron Devices*, 58(9) [25].)

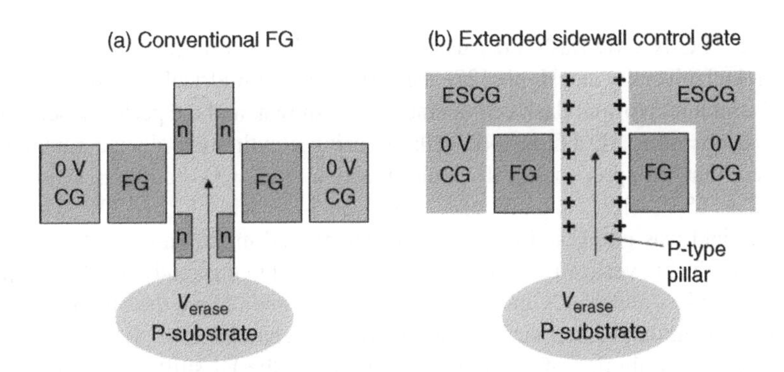

Figure 4.40 Schematic cross-section during bulk erase of (a) a conventional FG cell; and (b) an ESCG cell. (Based on M.S. Seo, S.K. Park, and T. Endoh, (Tohoku University, Hynix), *IEEE Transactions on Electron Development*, 58(9) [25].)

cells. This MLC operation can potentially improve the bit cost and provide superior reliability characteristics [25].

4.3.2 3D FG NAND with Separated-Sidewall Control Gate

A 3D vertical polysilicon pillar FG NAND flash memory cell array using a separated-sidewall control gate (S-SCG) was proposed by Tohoku University and Hynix in May of 2011 [26]. This cell has a cylindrical FG with a line-type CG and an S-SCG structure.

With the S-SCG NAND cell array, the same operating conditions can be achieved as in the conventional planar FG NAND. Bulk erase is realized by using an electrically inverted S/D technique. To prevent direct interference and disturbance issues, the voltage levels in the SCG regions were controlled to 8 V in program, 0 V in erase and 3.5 V in read operation. The S-SCG cell array fully suppressed both the interference effects and the disturbance problems by removing the direct coupling effect in the same cell string. The cell was thought to have good potential for a terabit 3D vertical NAND flash cell array with MLC operation.

Figure 4.41 illustrates the structure of various vertical channel FG cells including the conventional FG cell, the ESCG cell, and the S-SCG cell [26].

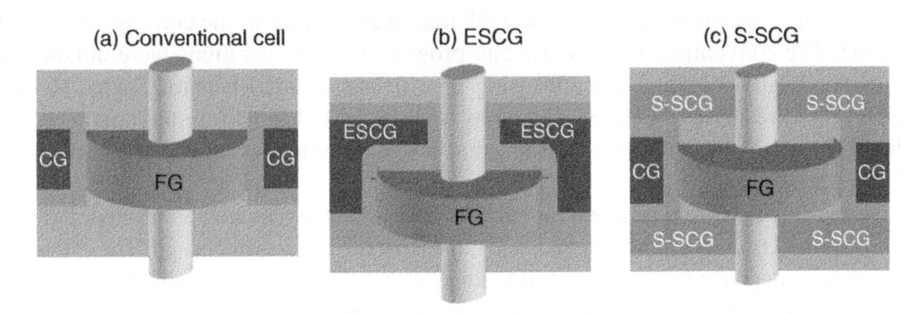

Figure 4.41 Various types of FG vertical channel cell structures: (a) conventional FG; (b) ESCG; and (c) S-SCG. (Based on M.S. Seo *et al.*, (Tohoku University, Hynix), IMW, 22 May 2011 [26].)

A 3D vertical FG NAND flash array using the S-SCG was discussed further in August of 2012 by Tohoku University and Hynix [27]. This cell used a cylindrical FG with a line-type CG with S-SCG structure. Normal flash cell operation was shown, and the performance was found better than various other 3D FG NAND cells by using a cylindrical device simulation. The study showed that the proposed cell could reach a high CG coupling ratio. The cell programmed at 15 V with $V_{th}=4$ V and erased at 14 V with $V_{th}=-3$ V. The retention mode electric field was good, and there was sufficient read-mode on-current margin.

The benefit of the 3D vertically stacked S-SCG addressed in this study was MLC operation. This operation could significantly increase the bit density and reduce the bit cost of the array, compensating for the larger cell size of the FG vertical channel cells. Key factors in MLC operation were the distribution of the programmed V_{th} and the V_{th} shift. All of the 3D vertical channel NAND arrays have a bit cost that is inversely proportional to the number of stacked cells. MLC operation is an issue with the CT NAND cell, which has difficulties with charge spreading and with retention characteristics. The distribution of the programmed V_{th} is affected by interference and disturbance issues. Other issues are direct coupling effects from neighboring cells in the same cell string.

Process complexity was a significant factor in stacking cells, and the scaling was limited by high-voltage cell operations, which could limit scaling in sub–10 nm technology due to the fixed dielectric thickness of about 20 nm, which included tunnel oxide and interpoly dielectric layers. High program voltage was a critical issue because it was applied directly to the sidewall CG. This study showed the superior cell performance of the S-SCG NAND cell including low-voltage P/E operation, good retention mode electric field, fully suppressed interference, direct disturb issues, and sufficient read-mode on-current margin.

The ESCG structure discussed previously suppressed the interference coupling effect so that high CG coupling capacitance could be obtained and the S/D structure could electrically invert the pillar at the same time. Direct disturb to neighboring passing cells, however, remained an issue as a result of the high coupling capacitance in the ESCG. This issue was addressed by the S-SCG structure. A schematic cross-sectional view of the S-SCG array is shown in Figure 4.42 [27].

In a 20 nm–type technology, the S-SCG horizontal feature size was about 24 F^2 ($6F \times 4F$), and the vertical height was 2F, resulting in a cell volume of about 48 F^3. This height could be achieved by using a thin, prestacked SCG. The common SCG method was applied in order to minimize the number of contacts, as shown in the equivalent circuit of the cell arrays in Figure 4.43 [27].

The operating conditions of the S-SCG cell array were similar to the conventional planar FG NAND. The conventional bulk erase and program inhibit operations were obtained by using an electrically inverted S/D and using an efficient channel-boosting ratio. Medium voltage levels were applied to the SCG regions, such as 11 V during program operation. The V_{scg} during the read operation was minimized to 1 V in order to increase the initial V_{th} when the FG had no charge.

The process included two different layers. The polysilicon layers were deposited, the pillar and FG regions were cylindrically etched, and then barrier oxide was deposited. The cylindrical FG region was deposited using a unidirectional etch process, and then tunnel oxide and polysilicon pillar were deposited. The CG was then self-aligned by using an etching ratio and a silicide process. The self-aligned process used the difference of the etch ratio between the two deposited layers [27].

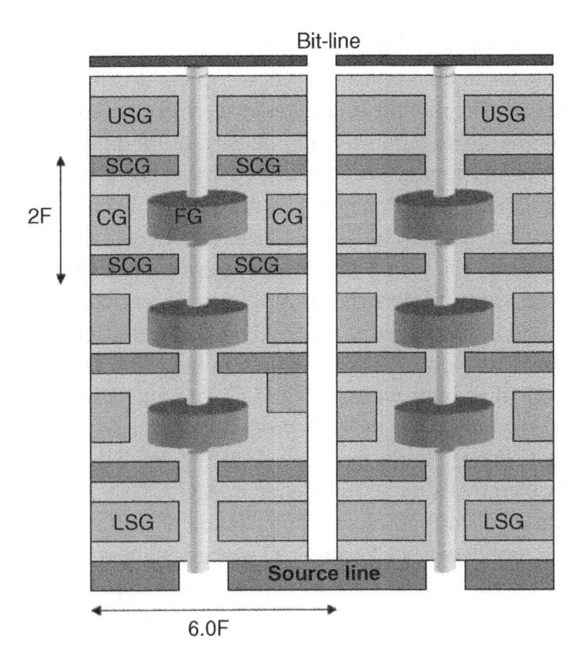

Figure 4.42 S-SCG array cross-section. (M.S. Seo, B.H. Lee, S.K. Park, and T. Endoh (2012) (Tohoku University, Hynix), *IEEE Transactions on Electron Development*, 59(8), 2018 [27].)

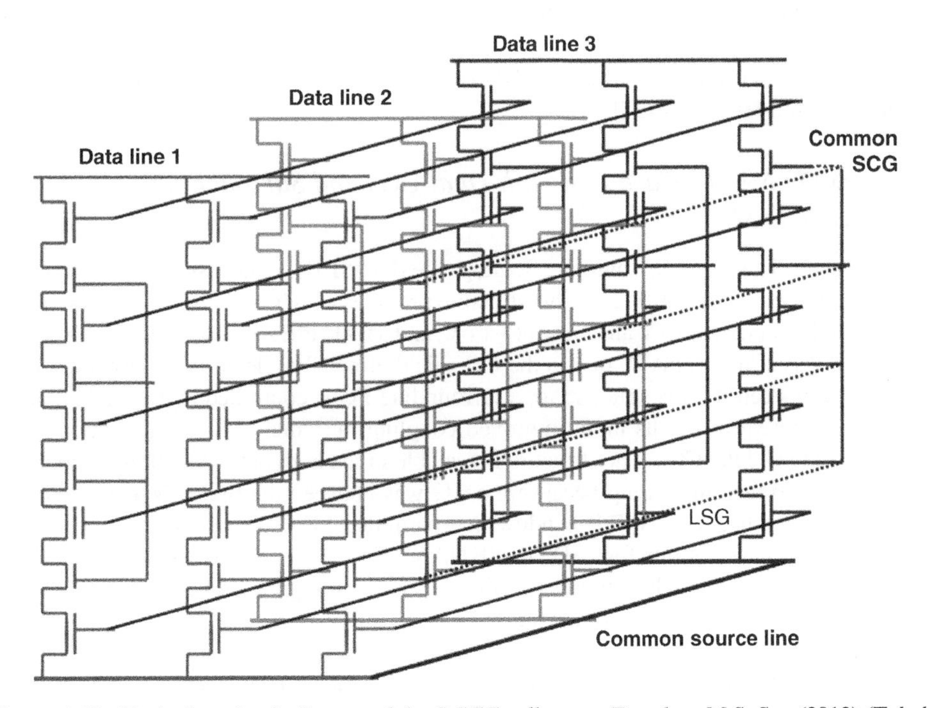

Figure 4.43 Equivalent circuit diagram of the S-SCG cell array. (Based on M.S. Seo (2012) (Tohoku University, Hynix), *IEEE Transactions on Electron Development*, 59(8), 2018 [27].)

To evaluate the characteristics of the S-SCG cell, 2D device simulations were used together with cylindrical coordinates in the vertical direction. An eight-cell NAND string was used to simulate an unselected cell. The mobility coefficients were decreased using a constant mobility model to indicate mobility degradation by trap density in the polysilicon pillar region. The amount of charge in the FG region was controlled by using a nonlocal tunneling model. Cylindrical device simulations were done on a 30 nm cell to obtain FG coupling capacitance, P/E voltage, disturbance and interference issues, read on-current, and retention mode E-field. In the S-SCG cell model it was found that the program voltage improved as the V_{scg} was increased, and the erase voltage improved as V_{scg} was decreased. Even if only half of the conventional program voltage is applied to the SCG node, the S-SCG NAND cell still had low-voltage P/E operations as a result of its excellent coupling ratio [27].

Disturb becomes more of an issue with the high CG coupling ratio used in many 3D vertical NAND cell arrays—particularly when the high program voltage is directly applied to the SCG. In addition, the high SCG voltage limits the vertical scaling. With the S-SCG cell, however, the V_{scg} level can be controlled by using an independent SCG. The V_{read} or V_{pass} and V_{scg} can be combined to prevent direct disturb problems. To maintain reliability of the interpoly dielectric (IPD), the voltage difference between CG and SCG was maintained below 9 V. This means V_{scg} can be increased to 11 V with V_{pass} at 2 V during program to improve the program efficiency. V_{read} can also be 10 V with V_{scg} at 1 V during the read operation to reduce external resistance of the neighboring cells [27].

The feasibility of having an MLC and the number of stacked cells obtainable compared with conventional 3D cells was addressed. The S-SCG cell fully suppressed both the interference effect and the direct disturbance issue. It had a low E-field in retention mode, which indicated good potential for highly reliable MLC operation with an FG cell. The vertical cell height was decreased compared to other vertical NAND cells by using a predeposited thin S-SCG and applying medium-level bias [27].

4.3.3 3D FG NAND Flash Cell with Dual CGs and Surrounding FG (DC-SF)

A 3D dual control gate with surrounding floating gate (DC-SF) NAND flash cell was discussed by Hynix in December of 2010 [28]. This structure provides a high coupling ratio, low-voltage operation with program at 15 V and erase at −11 V, and a wide P/E window of 9.2 V. FG-to-FG interference was a low 12 mV/V due to the CG shield effect. The device was multibit with 2 bits/cell or 4 bits/cell. In this structure, one surrounding FG is controlled by two neighboring CGs. This arrangement results in an improved coupling ratio due to the larger surface area between the FG and the CGs. The structure is scalable in the horizontal direction. There is no FG-to-FG interference in a string because the CG was between the two FGs and acted as a shield. The device had a 9.2 V P/E window that permitted MLC operation.

A method for read operation for a 3D DC-SF NAND flash memory was discussed in May of 2011 by Hynix [29]. A single-cell cross-section is shown in Figure 4.44(a), and a schematic cross-section of an array is shown in Figure 4.44(b).

Based on the model used, the selected cell V_{th} was increased by the neighbor cell's high V_{th} because the neighbor cell does not have enough $V_{pass\text{-}read}$ to be a pass transistor in a conventional read operation. To prevent this neighboring-cell high V_{th} effect, a higher $V_{pass\text{-}read}$ is applied to the CG of the neighbor cell, and a lower $V_{pass\text{-}read}$ is applied to the

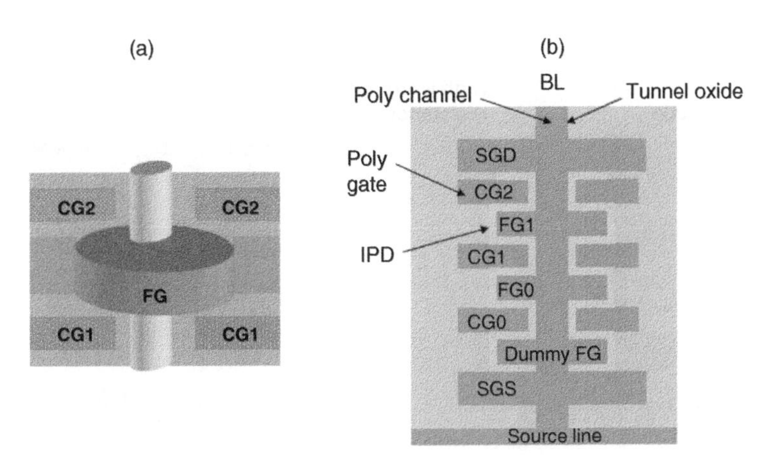

Figure 4.44 3D dual control gate with surrounding FG NAND string. (Based on H.S. Yoo *et al.*, (Hynix), IMW, 22 May 2011 [29].)

CG of the next neighbor cell to compensate for the FG potential of the neighbor cell. For read operation of an MLC, the $V_{pass\text{-}read}$ modulation must be decreased in correspondence to the selected cell read voltage. Using this new read scheme, a stable read operation was achieved for 2 bits/cell and 3 bits/cell operation [29].

A 3D NAND flash memory that used a metal control gate last (MCGL) process for a DC-SF cell was discussed by Hynix in June of 2012 [30]. This process corrected several issues identified in previous DC-SF cells. The MCGL process used a low-resistive tungsten metal WL with high-κ IPD, resulting in low damage on the tunnel oxide/IPD and a preferable FG shape. A conventional bulk erase could be used due to the direct connection between channel poly and the p-well by the channel contact holes. It was projected that up to 3-bit MLC configurations could be made using this technology, which would permit 512Gb 3D NAND flash memories.

In the previous DC-SF cells, there were several problems identified: high WL resistance of the poly gate, damage on IPD ONO by FG separation, FG field confinement at the FG edge during programming, and issues caused by GIDL erase. This study proposed a new MCGL process that was intended to solve these issues.

The MCGL process avoided some of the drawbacks of the conventional DC-SF process, as illustrated in Figure 4.45. Figure 4.45(a) shows the conventional DC-SF process that used ONO IPD and poly CGs. This process suffered from IPD damage, lack of FG field confinement at corners, and high WL resistance. The new MCGL process is illustrated in Figure 4.45(b). It has straight tunnel oxide, which avoids corners, and uses a high-κ IPD to resist damage and a low-resistance tungsten WL [30].

The MCGL process had a shorter distance between substrate and LSG, which produced a high saturation current by reducing the parasitic resistance in the string. The P/E window was 8 V. Neighbor cell program disturbance was reduced because V_{pass} for the neighbor cell was decreased during program. The cell-to-cell FG-to-FG interference was 30 mV/V due to shielding effects of the tungsten WL. For program and erase cycles, the V_{th} shift after cycling was less than 1.3 V after 1000 cycles. Data retention testing showed a shift of 60 mV after a 250 °C 120-minute bake, which is comparable to planar FG NAND. To avoid source line bounce issues during read, the SL contact was made 1 contact per 128 strings [30].

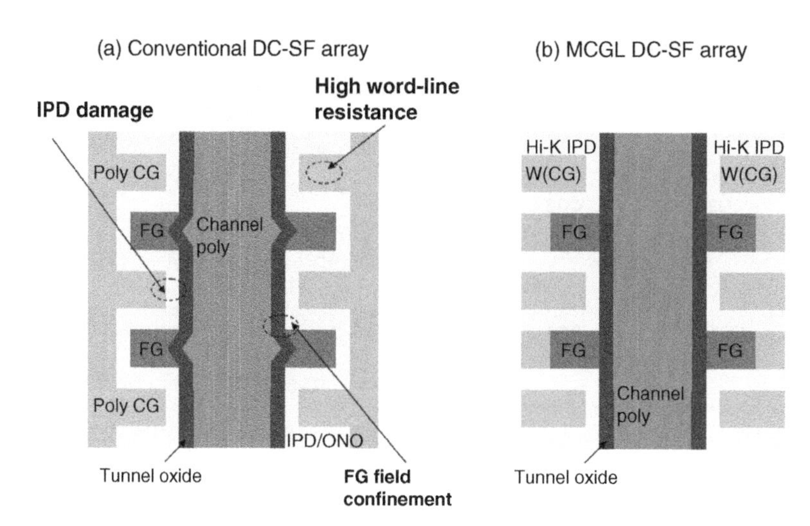

Figure 4.45 Schematic cross-section of (a) conventional DC-SF process showing issues with this process; and (b) MCGL process showing improvement in these issues. (Based on Y. Noh *et al.*, (Hynix), VLSI Technology Symposium, June 2012 [30].)

The MCGL process developed to improve performance and reliability of the DC-SF cell was discussed further in April of 2013 by Hynix and Hiroshima University [31]. Read and program operation schemes were developed. In the read operation, both higher and lower $V_{pass\text{-}read}$ were alternately applied to unselected CGs to compensate for lower FG potential to a pass transistor. In the program scheme, the optimized V_{pass} was applied to a neighboring WL or the selected WL to prevent program disturb and charge loss through IPD. The use of the MCGL process and new read and program methods permitted a higher-performance, more reliable DC-SF cell for use in 3D NAND flash memories [31].

The conventional read operation for the DC-SF NAND flash string involved applying the read voltage (V_r) to the two neighboring CGs of the selected FG and applying $V_{pass\text{-}read}$ to unselected CGs. V_r was a predetermined value for cell read. The MLC read operation is defined as shown in Figure 4.46 [31].

A new read operation was developed for the DC-SF NAND string. For each V_r for PV1, PV2, and PV3, $V_{pass\text{-}read2}$ and $V_{pass\text{-}read1}$ should be different voltages to compensate for the neighboring FG1 potential. This new read operation is illustrated in the schematic circuit

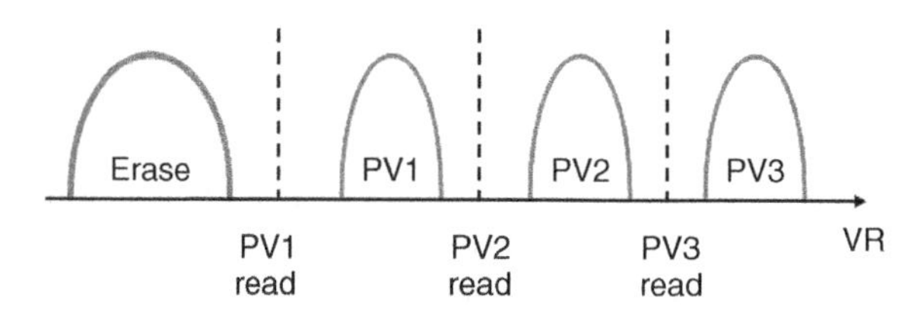

Figure 4.46 Illustration of levels for MLC read operation. (Based on S. Aritome *et al.* (2013) (Hiroshima University, Hynix), *IEEE Transactions on Electron Devices*, 60(4), 1327 [31].)

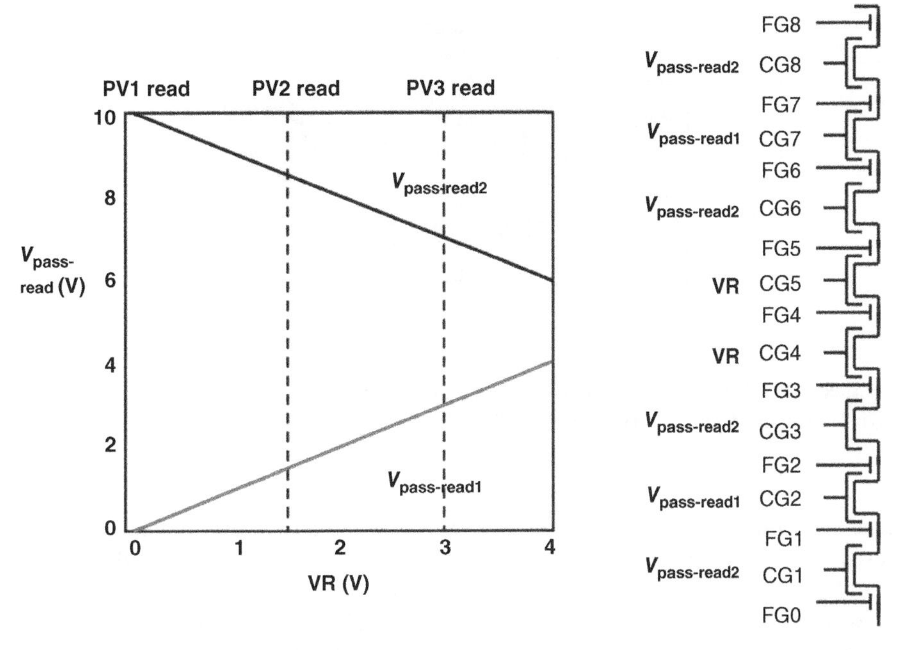

Figure 4.47 Read of the DC-SF NAND string: (a) V_{pass} vs. $V_{R(V)}$; and (b) schematic circuit diagram of $V_{pass\text{-}read}$ vs. V_R. (Based on S. Aritome *et al.*, (Hiroshima University, Hynix), *IEEE Transactions on Electron Devices*, 60(4), 1327 [31].)

diagram in Figure 4.47(b), where $V_{pass\text{-}read1}$ and $V_{pass\text{-}read2}$ have a dependence on V_r as shown the graph in Figure 4.47a [31]. The operating voltages for the new read method for the DC-SF NAND string for various values of V_r were specified.

The programming method for a DC-SF cell needed to be optimized to avoid program disturb issues. There are two inhibit modes for the DC-SF cell: the electron injection mode and the charge loss mode. The electron injection mode had a weak electron injection stress caused by a high field in the tunnel oxide due to the FG coupled with two CGs. This mode became an issue in the erase state (at lower V_{th}) because of the high field of the tunnel oxide. The charge loss mode is unique to the DC-SF cell in that electrons in the FG are ejected to the CG by the high field in the IPD. This mode becomes an issue with high cell V_{th}. To minimize program disturb issues, V_{pass_n-2} and V_{pass_n+2} must be optimized. P/E cycling showed V_{th} shift after cycling is less than 1.3 V after 1000 cycles. Data retention was found comparable with conventional planar FG NAND flash.

4.3.4 3D Vertical FG NAND with Sidewall Control Pillar

A 3D vertical FG NAND flash cell array using a sidewall control pillar (SCP) was discussed in May of 2012 by Tohoku University and Hynix [32]. This cell had a cylindrical FG and a SCP with a line-type CG, as shown in Figure 4.48 [32]. The SCP structure is connected to the NAND string's polysilicon pillar on both sides of each FG. A tight vertical space length is achieved by using a self-aligned process for deposition of the polysilicon pillar. The SCP NAND cell has the same operating conditions as the DC-SF NAND cell. To compensate for the

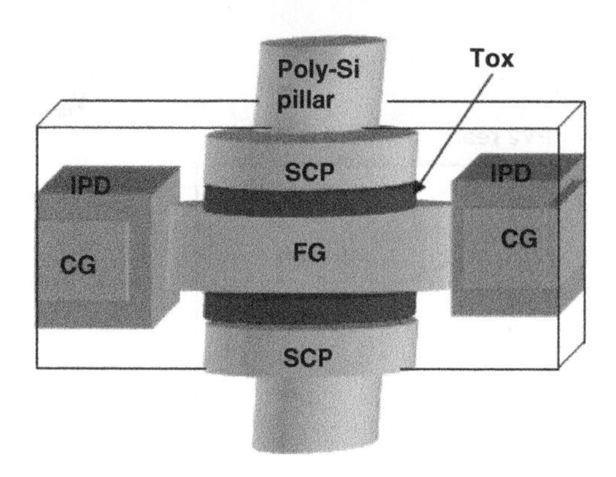

Figure 4.48 3D FG NAND cell with SCP. (Based on M.S. Seo, J.M. Choi, S.-K. Park, and T. Endoh, (Tohoku University, Hynix), IMW, 20 May 2012 [32].)

increase in channel capacitance, the FG width was decreased by 15 nm and a high-κ material was used for the IPD. The cell programmed at 18 V with $V_{th} = 4$ V and erased at 17 V with $V_{th} = -3$ V, which was comparable to conventional FG NAND cells. The same vertical scaling found with a CT 3D NAND cell was achieved for sub–20 nm technology. This cell was expected to have potential for use in a terabit 3D vertical NAND cell.

Vertical scaling of 3D FG NAND cell arrays was limited by the interference with neighboring cells. As a result several 3D vertical FG NAND cells with an SCG that reduces the interference issue of the FG NAND cell have been shown. These vertical FG NAND cells have such features as ESCG, DC-SF, and separated SCG. To suppress the interference coupling effect of the FG-type cell, the SCG is used. High program speed is achieved due to the increased coupling ratio of the SCG. Cells with FG NAND SCGs, however, have issues in both neighboring passing cell disturb and reliability of the IPD next to the SCG because high operating voltage is applied to the SCG during P/E operations. The SCG NAND cell has difficulties scaling vertically below 30 nm because it is limited by SCG and IPD thickness.

The operation of the 3D vertical FG NAND cell with SCP was compared with 3D FG NAND cells with the SCG structure, including the ESCG, DC-SF, and S-SCG. 3D cylindrical device simulations of an eight-cell NAND string were performed to determine the characteristics of the SCP NAND cell. The simulation assumed FN tunneling for the FG-type NAND cell. For the SCP NAND cell, two kinds of interference effects were found to exist. These are conventional indirect coupling between FGs and direct coupling from FG to SCP. These effects depend on parameters such as the length and width of the SCP (L_{scp}, W_{scp}) and the width of the FG (W_{fg}). The read current and interference effects for the cell as a function of W_{scp}/W_{fg} are shown in Figure 4.49(a), where the I_{read} and interference ΔV_{th} depend on the ratio of W_{scp}/W_{fg}, where $W_{fg} = 30$ nm. This ratio is minimized at 60%. If L_{scp} is decreased, interference worsens due to the coupling effect with the neighboring FG while the I_{read} improves due to the channel resistance of the SCP. Corner effects were also shown to exist and be optimized. The I_{read} and interference ΔV_{th} dependence on L_{scp} are shown in Figure 4.49(b) [32].

Two types of interference effects were found: conventional indirect coupling between FGs and direct coupling from FG to SCP. As the ratio of W_{scp} to W_{fg} is increased, the indirect coupling effect was suppressed, but the direct coupling effect became more critical due to the

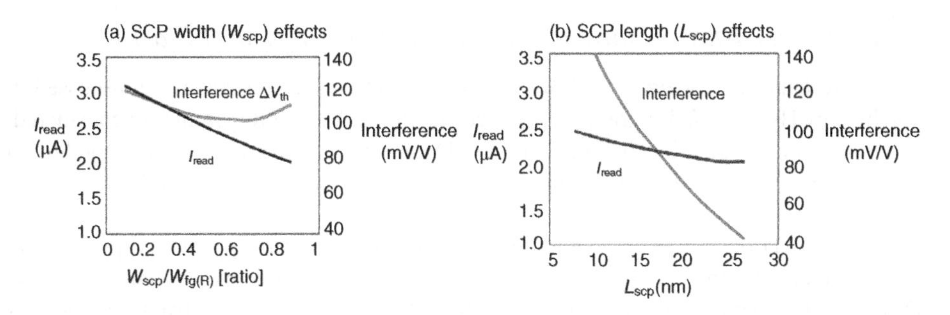

Figure 4.49 (a) SCP width (W_{scp}) effects on I_{read} and interference; (b) I_{read} and interference ΔV_{th} dependence on SCP length (L_{scp}). (Based on M.S. Seo *et al.*, (Tohoku University, Hynix), IMW, 20 May 2012 [32].)

influence of the channel potential in the SCP. As a result the interference effect of the cell was minimized at the 60% ratio of W_{scp} and W_{fg}. If the W_{fg} was decreased, the interference effect improved due to suppression of the coupling capacitance with the neighboring FG, and the I_{read} also improved. To evaluate the characteristics of the cell, 3D cylindrical device simulations were performed vertically on the eight-cell NAND string. The geometric effects of the cell were confirmed. The FG NAND flash memory cell operation using the SCP was shown using simulation to have good scalability and cell performance. The potential for MLC operation was shown to have sufficient interference margin and good reliability characteristics.

The basic cell size of the SCP NAND FG cell is about 60% larger than that of the CT NAND cell. If a 2-bit MLC is used, then the effective cell size of the SCP FG NAND cell is about the same size as a CT NAND cell. A 3-bit MLC could also be used with correspondingly smaller effective cell size.

4.3.5 Trap Characterization in 3D Vertical Channel NAND Flash

A characterization of traps in a 3D vertical channel NAND flash was discussed in December of 2012 by Seoul National University and Hynix [33]. Trap density was extracted in this 3D VC flash. The trap density extracted by the conductance method was $1\text{–}2 \times 10^{12}\ \text{cm}^{-2}\ \text{eV}^{-1}$ in E_c–E_t of 0.15–0.35 eV. Simulation results were compatible with the measurements. The effects of P/E cycling stress on 1/f noise was also studied, and the position was defined as a trap-generating RTN by considering cylindrical coordinates and pass cell resistance in the 3D stacked NAND flash cell. The traps of grain boundaries in vertical polysilicon channels are of interest for characterization because these traps can induce high leakage degradation of the subthreshold swing and degrade mobility. RTN can cause fluctuations in the read current. This study involved the characterization of traps due to the polysilicon grain boundary and characterization of the RTN.

The 3D stacked NAND flash had a vertical polysilicon body, gate dielectric stack with Si_3N_4 layer and a virtual S/D. The cell string was similar to that of the p-BiCS structure. Gate length, space between vertical WLs, and tube diameter are less than 100 nm. Normalized noise power of the 3D stack devices was analyzed, and a method to extract a trap-generating RTN was developed and verified [33].

4.3.6 Program Disturb Characteristics of 3D Vertical NAND Flash

Program disturbance characteristics of 3D vertical NAND flash cell arrays were discussed in May of 2013 by Hynix [34]. Program disturbance in both the *X* and *Y* directions are considered. Characteristics are compared with 2D planar NAND flash cells. A new program method for 3D NAND flash was proposed to determine the disturbance characteristics for MLC NAND flash arrays.

Most 3D NAND flash cells have adopted an undoped or uniformly n-doped polysilicon channel, which makes the subthreshold swing of the select transistor worse than in the case of 2D NAND flash using single-crystal silicon. In addition, the number of strings per block is increased over that of 2D NAND flash, which increases the "number of program" stresses. There is a new program disturb mode that is worse than conventional program disturb. While there is only an *x* program disturbance in 2D NAND flash, an unselected string in 3D NAND flash has potential program disturb on axis *x*, *y*, and *xy*, as illustrated in Figure 4.50 [34].

Each mode has a different backbias level of drain selection line (DSL) transistor, with each V_{bl} and V_{dsl}. Program *Y* disturbance mode has the lowest backbias level, and so can expect the worst cutoff characteristics. Program *X* disturbance is defined at $V_{bl} = V_{cc}$ and V_{dsl}, which is the condition for a conventional 2D planar NAND flash cell. Program *XY* disturbance has a higher DSL backbias level than the 2D case, which means a smaller disturbance V_{th} shift than the other modes. The select transistor with the poly channel shows worse swing characteristics than the 2D select transistor with single-crystal silicon, so the select transistor cutoff characteristics in program disturb mode are worse for a 3D vertical cell than a 2D planar cell [34].

To improve the inherent program disturbance of a 3D cell, a high DSL V_{th} compared to 2D should be used to suppress the leakage current. DSL vias of the selected BL during

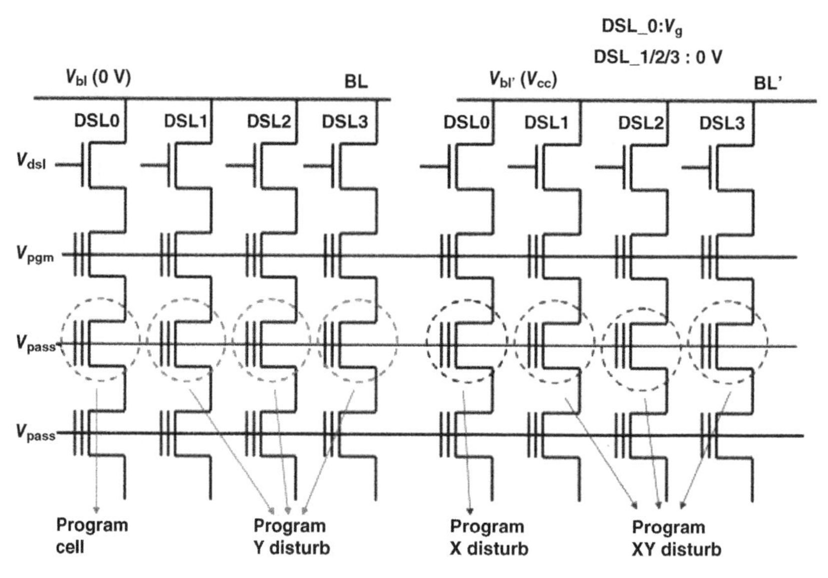

Figure 4.50 3D NAND flash cell program disturb mode. (Based on K.S. Shim *et al.*, (Hynix), IMW, 20 May 2013 [34].)

programming should also be increased by the same amount to prevent slow programming. In the 3D cell structure DSL V_{th} can be controlled by channel V_{th} implant. However, the vertical channel topology could add extra nonuniformity to the DSL V_{th}, unless process control for uniform cell topology is maintained. In addition, negative V_{dsl} needs to be applied to improve cutoff characteristics in program Y disturb mode. If high negative V_{dsl} is applied, GIDL current could be increased in the overlap region between the DSL transistor and the boosting channel and could make the boosting level lower. This means there is an optimum negative DSL bias. Finally, the edge WL's program disturb characteristics are much worse than those of other WLs. Because a 3D cell has a floating body structure, the SSL channel potential is raised by the program bias of the edge WL during program. The channel boosting level of the edge WL becomes lower than other WLs by the DIBL current, and dummy WLs are needed between the select transistors and main WLs to prevent DIBL current [34].

4.4 3D Stacked NAND Flash with Lateral BL Layers and Vertical Gate

4.4.1 Introduction to Horizontal BL and Vertical Gate NAND Flash

3D stacked NAND flash with layered horizontal BLs and vertical gates is an alternative structure to vertical BL pipes with through-holes in which tunneling oxide, polysilicon body, and blocking dielectric are formed sequentially. An attribute of the stacked horizontal BL architecture is that this is the orientation of the BLs in the conventional planar NAND flash.

Issues for vertical pipe structures include a limitation in the number of vertical bit cells and in the large V_{th} distribution in a BL because the through-hole diameter becomes narrower at the bottom as the number of cells increase. Another issue with vertical pipe devices is that metal gate structures tend to have poor retention characteristics during high-temperature operation from stress due to the high thermal expansion coefficient. It is possible that a stacked structure will eliminate some of these issues. Vertical pipe structures also have a pitch scaling issue, which makes a $4F^2$ cell a difficult proposition. Vertical gate architectures do not have the pitch scaling issue, so a $4F^2$ cell is possible; however, the cell select/decode is more difficult.

3D TCAD simulations on various 3D NAND flash memory architectures, including the vertical channel P-BiCS, TCAT, and VSAT as well as the horizontal channel vertical gate memory, were discussed by Macronix in May of 2010 [2]. All structures have an X,Y lateral scaling limitation due to the minimal 20 nm ONO thickness and 10 nm poly channel thickness that can not be scaled further. The characteristics of these technologies are summarized in Table 4.4 [2].

Table 4.4 Characteristics of Various 3D NAND Flash Technologies.

	P-BICS	TCAT	VSAT	VG
Device Structure	GAA	GAA	Planar	Double gate
Cell Size (xy plane)	$>4F^2$	$>4F^2$	$>4F^2$	$4F^2$
Gate Process	Gate first	Gate last	Gate first	Gate last
Direction of Current	Vertical	Vertical	Mixed	Horizontal

Based on Y. H. Hsiao *et al.*, (Macronix), IEEE IMW, 16 May 2010 [2].

Of these architectures, the vertical gate memory was found to have the better X-direction scalability and no penalty for increasing the number of layers in the vertical Z direction because the channel current flows horizontally. Simulations showed that a buried-channel junction-free NAND improved the read current for all 3D NAND arrays. In the vertical pipe structures, NAND string read current was shown to decrease with increasing vertical layers because the string length was being increased. The read current in the vertical gate structures was independent of Z layer number because each layer was independent. A vertical gate stacked NAND flash with horizontal BLs is possible with GAA as well as double-gate structures. Interference in the Z direction was also examined, and both vertical channel and vertical gate memories were shown to suffer from it. A new Z-direction scaling limitation was indicated [2].

4.4.2 A 3D Vertical Gate NAND Flash Process and Device Considerations

A 3D stacked NAND flash memory with lateral BL layers and vertical gates was discussed in June of 2012 by Seoul National University [35]. A schematic view of the process sequence for fabricating a 3D stacked NAND flash memory device is shown in Figure 4.51.

The fabrication sequence consisted of depositing 40 nm of SiO_2 and 100 nm of polysilicon alternately on the silicon substrate followed by boron ion implantation doping with three different energies to dope the three polysilicon layers. The trench was formed using dry etching with a hardmask for active layers. The three stacked layers of polysilicon form BLs, which were separated by SiO_2. The side of the polysilicon was used as a channel. A tunneling oxide–nitride–blocking oxide (ONO) stack was formed by consecutive deposition. After the ONO was formed, a doped polysilicon was deposited for the WL and to fill the trench. The doped polysilicon was then etched to pattern the WLs. The side n+ regions were then formed by ion implant except in the regions overlapped by the WLs. To form the n+ regions, arsenic ions were implanted, tilted by 30 degrees with two rotations, forming the n+ regions on the sides of the etched polysilicon layers.

Anneal of the implanted ions was then done. The blocking oxide and nitride layers between WLs were then selectively removed using a wet etch to separate the CT layers. Contact holes at the bottom and center BLs were formed by etching the top BL layer first and then the center BL near the contact holes for the bottom BL. n+ doping was done at both ends of the BL body to give contact regions for BL wiring. A layer of oxide was deposited, and contact holes were etched. The metal stack Ti–TiN–Al–TiN was sputtered and etched for a metal pad.

The P/E characteristics are shown in Figure 4.52 [35]. This shows the BL current vs. CG bias of the 3D stacked NAND flash. 19 V was used for programming and −18 V for erase. The ΔV_{th} was 2.4 V when the BL current (I_{bl}) was 10 nA. Because the channel was formed on the side of the etched polysilicon body, the device characteristics suffered.

Retention characteristics showed minimal charge loss up to 10^4 s. The V_{th} for the programmed state tended to decrease from 2.44 V initially to 2.1 V after 10^4 s and to 1.84 V after 10^6 s. For endurance, the shift in V_{th} was 1.58 V after 10^4 cycles. To improve the quality of the side channel, a chemical dry etch was used after etching the BL stack to remove etch damage. As a result the current–voltage characteristics improved slightly.

The effect of variation on the vertical profile of gate dimensions in 3D stacked NAND flash with multiple stacked layers was discussed in June of 2012 by Seoul National University [36]. In 3D stacked NAND flash, the number of stacked layers increases to increase the memory

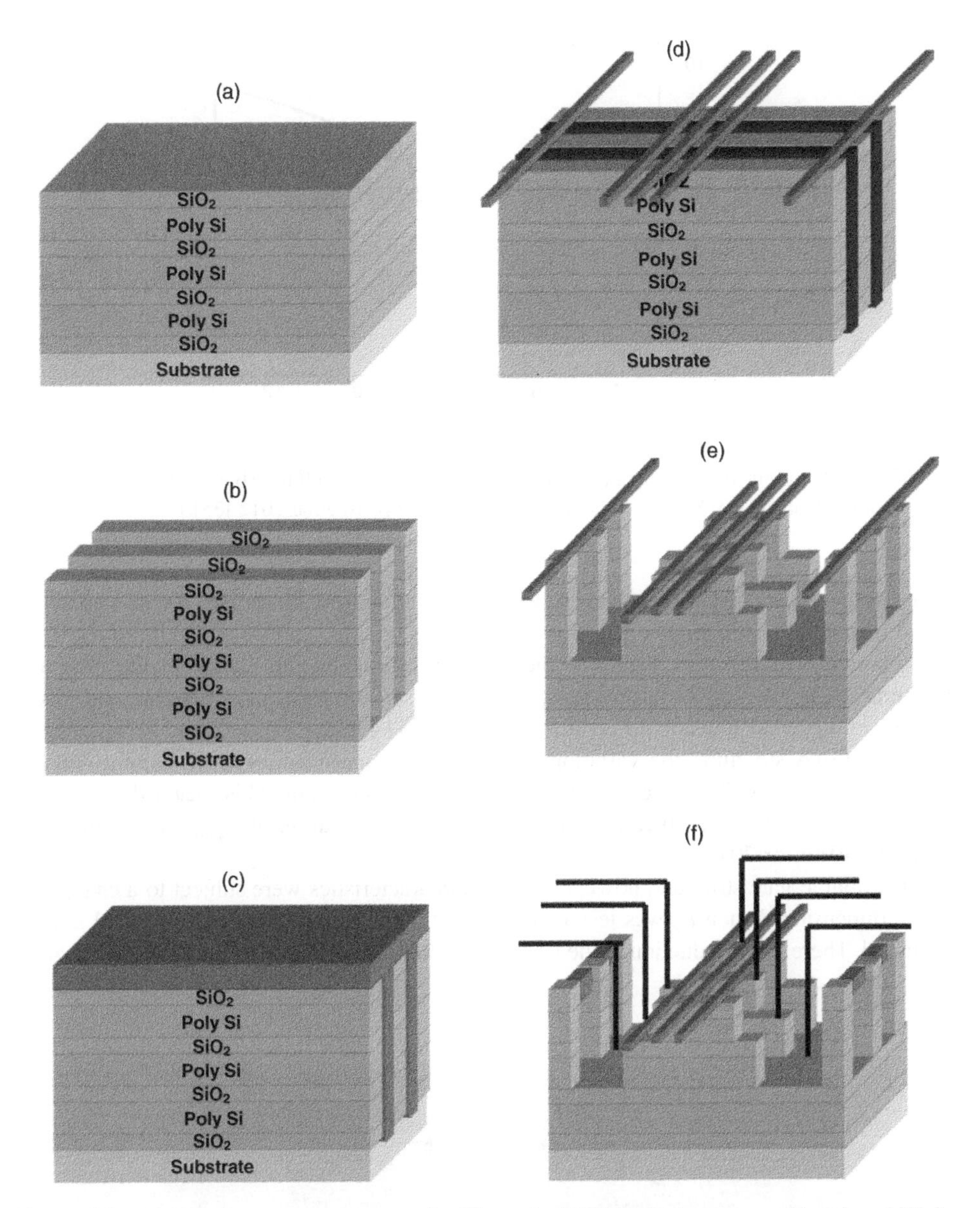

Figure 4.51 Schematic process sequence of a 3D stacked NAND flash memory. (Based on J.W. Lee *et al.*, (Seoul National University), Silicon Nanoelectronics Workshop, 10 June 2012 [35].)

capacity of the storage. If the height of the stacked array is increased, it becomes more important to have a good vertical etch profile because WL gate dimensions are affected.

The effect of variation of gate dimensions on the program characteristics in 3D NAND flash was investigated using a TCAD simulation. The cell characteristics with different 3D structures were compared, including GAA and double-gate CT NAND flash memory. A 3D stacked NAND flash memory is illustrated in Figure 4.53 [36].

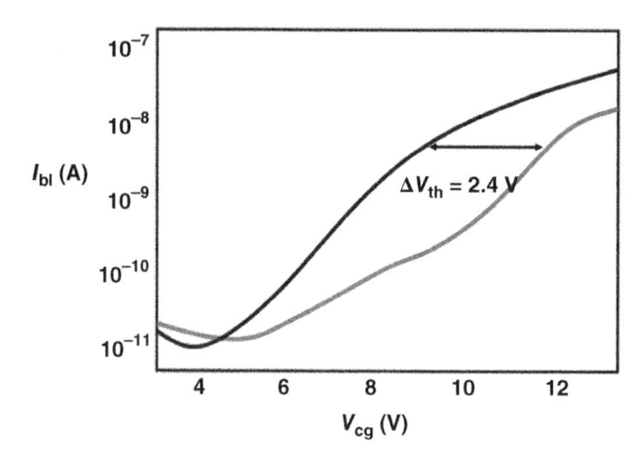

Figure 4.52 BL current vs. control gate bias of the 3D stacked NAND flash. (Based on J.W. Lee *et al.*, (Seoul National University), Silicon Nanoelectronics Workshop, 10 June 2012 [35].)

In an array with multiple stacked layers, a slight deviation from 90 degrees on the etch slope resulted in a significant change of the dimensions between the gates on different layers. This study reported how the dimensions of the gates, in particular the L_g, affect the program characteristics in 3D NAND flash arrays. The change in the gap in double-gate and GAA structures on different layers is illustrated in Figure 4.54 [36].

With the GAA structure, the variation of L_g did not appear to affect the current-voltage characteristics initially. The level of BL current remained constant. This meant the program characteristics of the GAA structure were not changed by variation of L_{gap} whether the gate length was 40 nm or 70 nm.

For the double-gate structure, however, the cell characteristics were subject to a change in the gate dimension. When L_g was less than 50 nm, the program characteristics could not be determined. There was a reduction in the level of BL current as L_g decreased. This was because

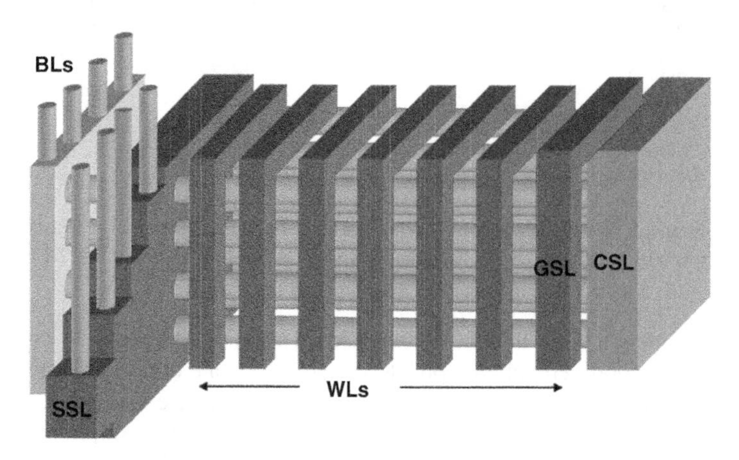

Figure 4.53 3D schematic of 3D stacked NAND flash memory. (Based on J.Y. Seo *et al.*, (Seoul National University), Silicon Nanoelectronics Workshop, 10 June 2012 [36].)

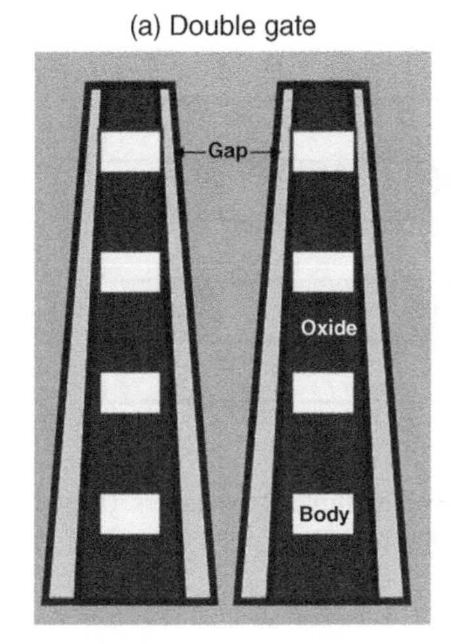

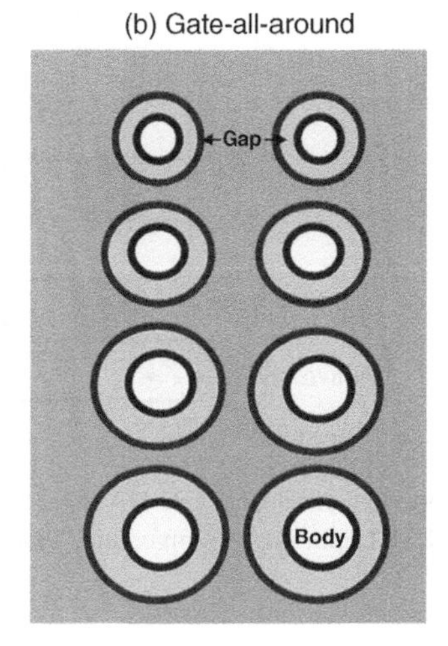

Figure 4.54 Schematic illustration of a (a) double-gate; and (b) GAA structure. (Based on J.Y. Seo *et al.*, (Seoul National University), Silicon Nanoelectronics Workshop, 10 June 2012 [36].)

the fringing field was not sufficient to induce the inversion layer between gates as the distance between gates got longer. This led to low electron density in the virtual S/D area. The result was that the double-gate NAND flash cells located in the upper layer could not guarantee stable program characteristics. To guarantee stable cell characteristics, the gap between WL gates had to be smaller than 50 nm. It was found that the maximum height of total stacked layers for the double-gate NAND as a function of the etch slope was the equal to height when the WL gap is 50 nm at the top of the gates and 10 nm at the bottom of the gates.

The conclusion was that the etch slope limits the total height of the stacked layer when double-gate NAND architectures were used. In addition, the NAND flash memory cells with GAA architecture were less sensitive to variation of the gate dimensions than cells featuring a double-gate structure. The GAA architecture appeared to be more suitable for very high-density NAND flash memory cells.

4.4.3 Vertical Gate NAND Flash Integration with Eight Active Layers

The integration of a vertical gate NAND flash array with multiple (eight) active layers was discussed in June of 2009 by Samsung [37]. Stable operation of program, erase, and read was confirmed. There was no accumulation of program disturb with increased numbers of vertical layers due to the vertical block architecture. A top view of the $4F^2$ cell is shown in Figure 4.55 [37].

One of the issues that vertical NAND flash strings had was increased program disturb and channel resistance as the number of WLs between top BL and bottom CSL increased. In this study, vertical gates and horizontal multiple active layers were introduced to solve some of these issues [37].

The formation of the vertical gate NAND used the following sequence:

1. The source line, BL, and CSL were formed.
2. Then multiple active layers (ONO) were deposited with ion implants and active patterning.

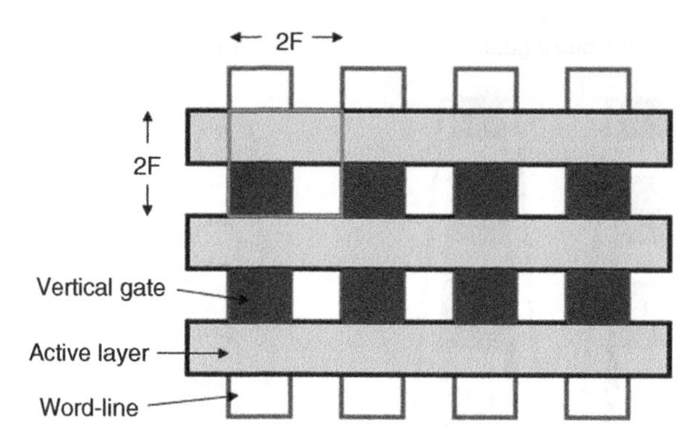

Figure 4.55 Top view of vertical gate NAND flash with multiple active layers. (Based on W.J. Kim *et al.*, Samsung, VLSI Technology Symposium, June 2009 [37].)

3. The CT layer was deposited over the patterned active layers.
4. The vertical gates were formed.
5. Vertical plugs of drain current (DC), source, and body bias (VBB) were plugged [37].

Multiple SSL's were used to select data from a chosen layer out of multiple layers because the vertical gate NAND used common BL and common WL between multiple active layers. The number of process steps for the vertical gate NAND with multiple active layers was comparable to the one for planar NAND. P/E cycling to 1000 cycles and retention to 10^9 s were confirmed. A sketch of the cross-section of the vertical gate NAND along the WL direction is illustrated in Figure 4.56, which shows the eight active layers and the vertical gates [37].

A schematic circuit diagram of the vertical gate NAND array is shown in Figure 4.57 [37]. NAND strings were horizontal in eight layers. WLs and gates were vertical. Source and body

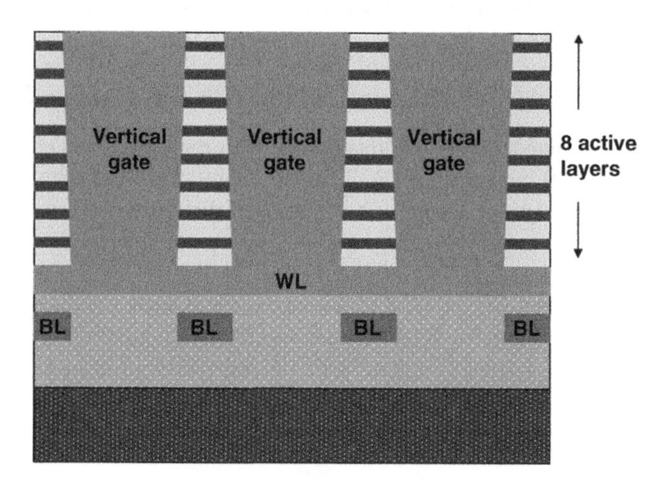

Figure 4.56 Sketch of a SEM of the cross-section of vertical gate NAND along word-line direction. (Based on W.J. Kim *et al.*, (Samsung), VLSI Technology Symposium, June 2009 [37].)

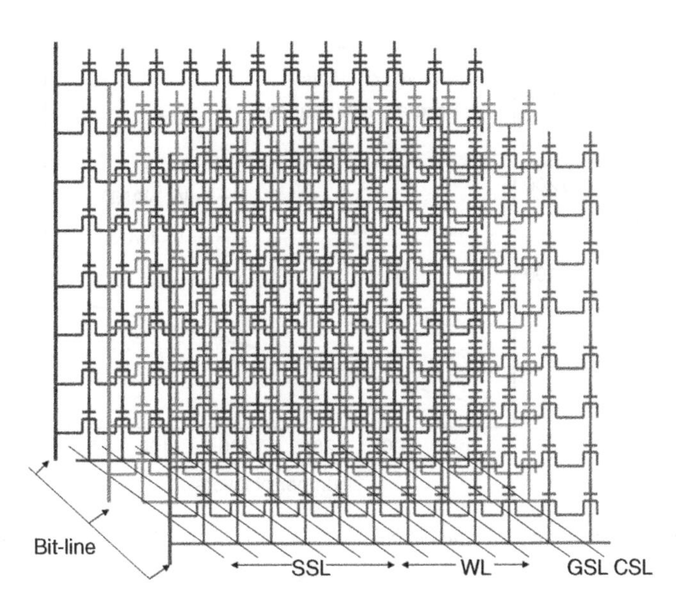

Figure 4.57 Schematic circuit diagram of eight-layer vertical gate NAND array. (Based on W.J. Kim *et al.*, (Samsung), VLSI Technology Symposium, June 2009 [37].)

were tied to CSL. Each BL contained multiple active strings, a common vertical gate, and vertical plugs.

Program and body erase were done, and a 3.7 V program window resulted. The channel of the selected layer was boosted with a channel precharge, while channels of the nonselected layers were boosted without channel precharge during program inhibit operation. No distinguishable difference in program disturbance was found between the channel with and without precharge when there was a program window of 2.8 V. P/E endurance up to 1000 cycles was found.

A vertical block strategy was implemented for the vertical gate NAND, which reduced program disturbance because each vertical block used its own WL during program operation. Factors considered to cause complications in vertical channel NAND devices included number of trimmed WL contacts, program disturbance, and channel resistance due to increasing numbers of WLs between the BL and CSL. The channel resistance of both vertical channel NAND and vertical gate NAND was expected to increase with the number of layers at about the same rate. The number of WL contacts and the PGM disturbance were expected to be significantly different and are shown in Figure 4.58, which illustrates this advantage of the vertical gate NAND over the vertical channel NAND. The factor $x1$ is based on planar NAND with 32 WLs [37].

4.4.4 3D Stacked CT TFT Bandgap-Engineered SONOS NAND Flash Memory

An eight-layer 3D vertical gate TFT bandgap-engineered (BE) CT NAND flash was discussed in June of 2010 by Macronix [38]. A buried channel n-type well device was used to improve the read current of TFT NAND. The buried channel permitted a junction-free structure, which is

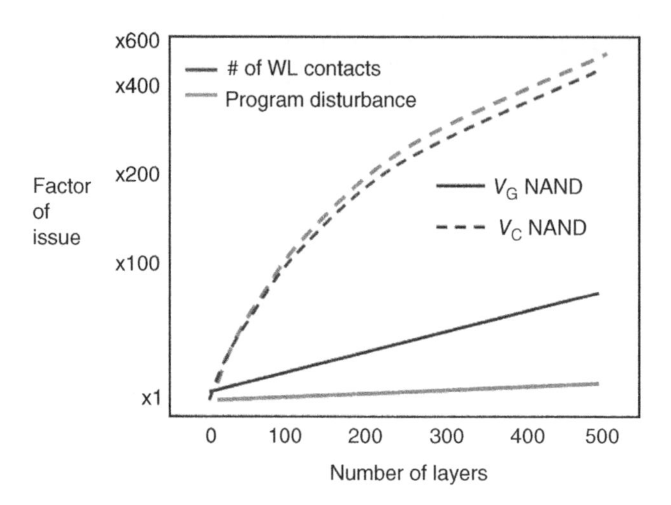

Figure 4.58 Advantages of vertical gate NAND over vertical channel NAND in program disturbance. (Based on W.J. Kim *et al.*, (Samsung), VLSI Technology Symposium, June 2009 [37].)

important for 3D stackable devices. A 6 V disturb-free memory window was achieved. The Z interference between adjacent vertical layers was studied [38].

A 3D array architecture of the vertical gate CT BE-SONOS is shown in Figure 4.59 [38]. The layout resembled the conventional NAND. The WLs were in the vertical plane, and the BLs were in the horizontal plane. Array efficiency is similar to that of the conventional NAND. An n-type depletion mode buried channel was used. The BE CT technology was discussed in Chapter 2 [38].

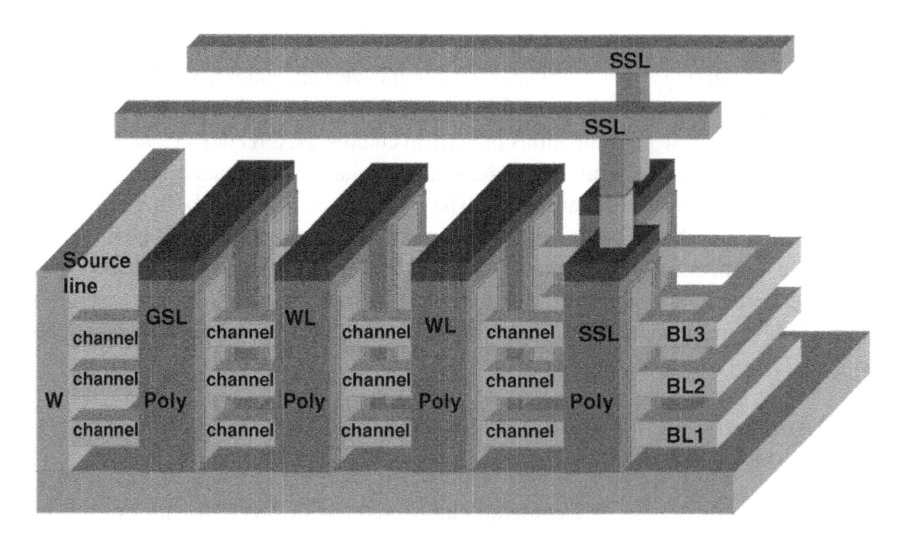

Figure 4.59 3D array architecture of vertical gate charge-trapping BE-SONOS. (Based on H.T. Lue *et al.*, (Macronix), VLSI Technology Symposium, June 2010 [38].)

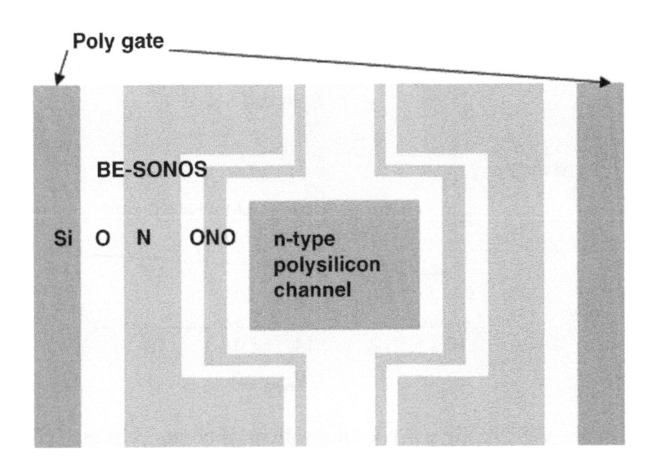

Figure 4.60 Schematic cross-section of p-poly gate and BE-SONOS memory structure. (Based on H.T. Lue *et al.*, (Macronix), VLSI Technology Symposium, June 2010 [38].)

Each cell was a double-gate junctionless TFT BE CT device. A schematic cross-section of the BL in the *X* direction showing the BE CT device is shown in Figure 4.60 [38]. It shows the p-poly gate, the n-type polysilicon buried channel, and the BE-SONOS layer structure including the ONONO. The poly channel thickness was about 18 nm. The half-pitch of the structure was 75 nm technology.

A proposed scaling to a half-pitch of 25 nm in both the *X* and *Y* directions was proposed. This resulted in a vertical pitch of 60 nm, an ONO of 5 nm, 6 nm, and 6 nm, respectively, and a V_{pass} of 7 V. Array characteristics were determined. The read current of a 32-WL TFT NAND array was determined for various channel dopings. The buried-channel n-type well device showed improved read current over the surface channel p-type. A > 7 V memory window was found. The FN erase is different from the conventional substrate erase due to the floating body cell. In this case the BLs and source lines are raised together to provide an efficient FN erase on each WL.

Self-boosting programming was used for array operation with V_{pgm} applied to cell A while neighboring cells B, C, D, and E remained in the erased state. Cell A was the programmed cell with the corresponding BL and source line grounded. In all three dimensions, neighboring cells B, C, D, and E were inhibited. Figure 4.61 shows that during programming of cell A, the neighboring cells could be sufficiently inhibited [38]. The disturb-free window was nearly 6 V. Due to the junction-free channel and the floating body structure, the channel potential was mainly boosted by the V_{pgm}.

Z interference that comes from the fringing field induced potential change due to the floating body. It depended on the thickness of the buried oxide between adjacent layers. When the buried oxide thickness was greater than 40 nm, the *Z* interference was smaller than 200 mV when cell A was programmed with a 4 V V_{th} shift.

The various layers showed equally good endurance. Because the vertical gate NAND array was a "gate-last" process, there was no tunnel oxide damage. After 10^4 P/E cycles, the V_{th} window remained at about 4 V. Program was at 20 V for 200 μs and erase was at −13 V for 100 ms. Retention, after 150 °C bake for 10^5 s, retained a V_{th} window of about 4 V.

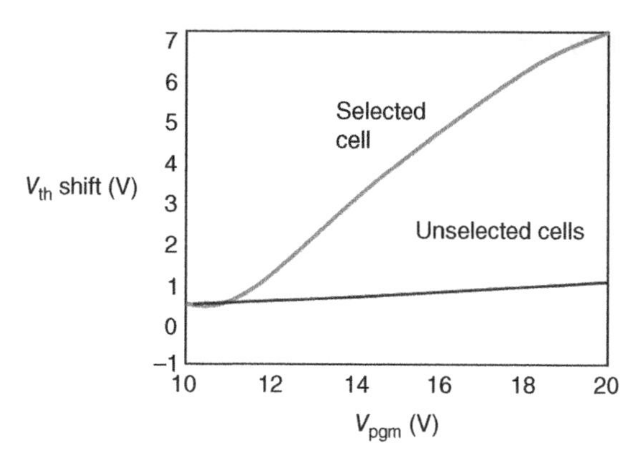

Figure 4.61 Programming cell "A" by self-boosting with neighboring cells in 3D inhibited. (Based on H.T. Lue *et al.*, (Macronix), VLSI Technology Symposium, June 2010 [38].)

4.4.5 Horizontal Channel Vertical Gate 3D NAND Flash with PN Diode Decoding

In June of 2011 Macronix discussed a vertical gate 3D NAND architecture with p-type silicon n-type silicon junction (PN) diodes self-aligned at the source side of the vertical gate [39]. The BE-SONOS junctionless cell technology was used. A PN diode decoding method eliminated the need to make multiple SSL transistors inside the array, which enabled a symmetrical, scalable array structure as shown in Figure 4.62 [39]. A three-step programming pulse waveform was used to implement the program-inhibit method of programming. This method

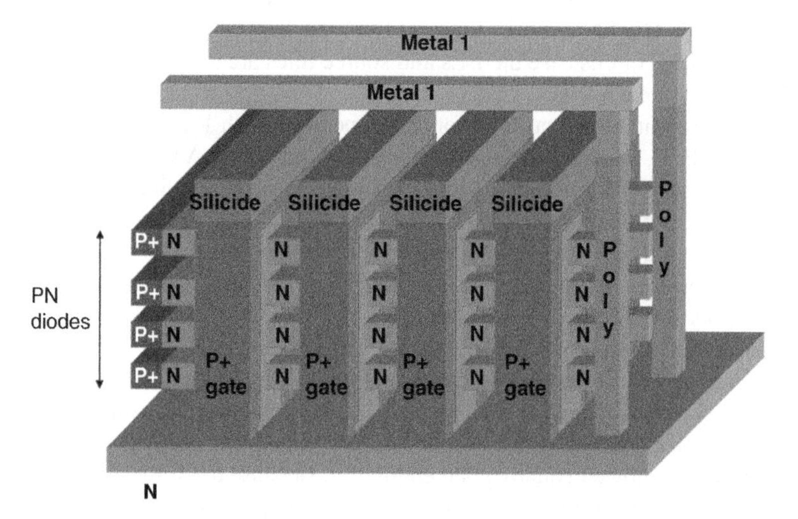

Figure 4.62 Vertical gate 3D NAND with self-aligned PN diodes. (Based on C.H. Hung *et al.*, (Macronix), VLSI Technology Symposium, June 2011 [39].)

took into account the fact that the PN diodes could prevent leakage of the self-boosted channel potential. A greater than 5 V disturb-free program window was shown.

Vertical gate architectures do not have the pitch-scaling issue that the vertical channel architectures have, so a $4F^2$ cell is possible. However, the cell select/decode is more difficult. The solution proposed here was PN diode decoding, which eliminated the need to make multiple SSL gates in one block and resulted in a simple, scalable array.

For PN diode decoding, PN diodes were formed at the source side of the vertical gate NAND. The ON–OFF ratio of the diode was greater than six orders of magnitude, and the reverse current had leakage below 10 pA. Source lines of each memory layer were separately decoded, while WL, BL, SSL, and GSL were common vertically for the multilayer stacks. There was only one SSL and one GSL per block. A three-step programming pulse waveform was used to implement the program inhibit technique. This method used the PN diodes to prevent leakage of the self-boosted channel potential. A greater than 5 V V_{th} window was found free of program disturb. The PN diode decoding technique resulted in robust program disturb immunity without using a complex SSL design, and it had scalable pitch.

4.4.6 3D Vertical Gate BE-SONOS NAND Program Inhibit with Multiple Island Gate Decoding

In May of 2012, Macronix studied the program inhibit performance of their 3D vertical gate NAND flash, which uses a multiple island gate SSL decode. In this architecture, the array efficiency is improved by sharing WLs in the vertical direction and BLs in the lateral direction, as shown in Figure 4.63 [40].

For correct decoding of the array, every channel BL had its own island gate SSL control device. Many channel BLs were grouped together in one unit, and staircase BL contacts were formed to decode the various memory layers. A page operation occurred naturally by selecting

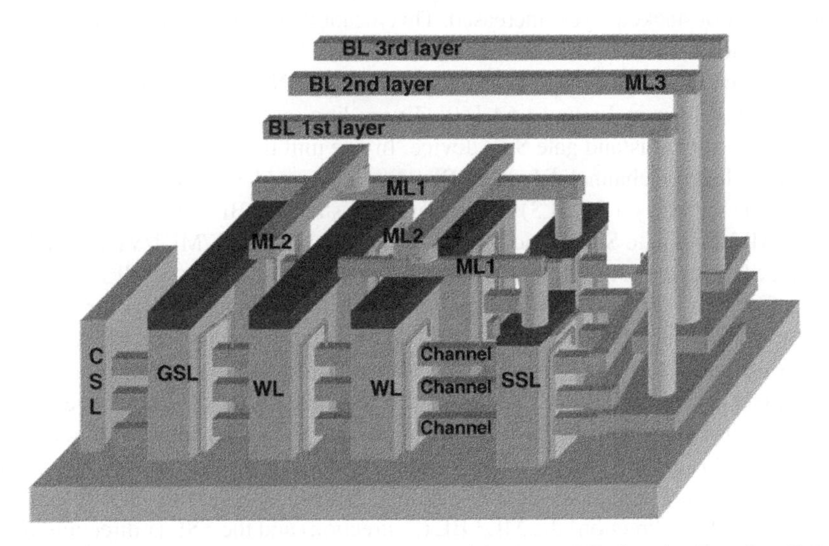

Figure 4.63 3D vertical gate NAND flash with multiple island gate SSL decode. (Based on K.P. Chang *et al.*, (Macronix), IMW, 20 May 2012 [40].)

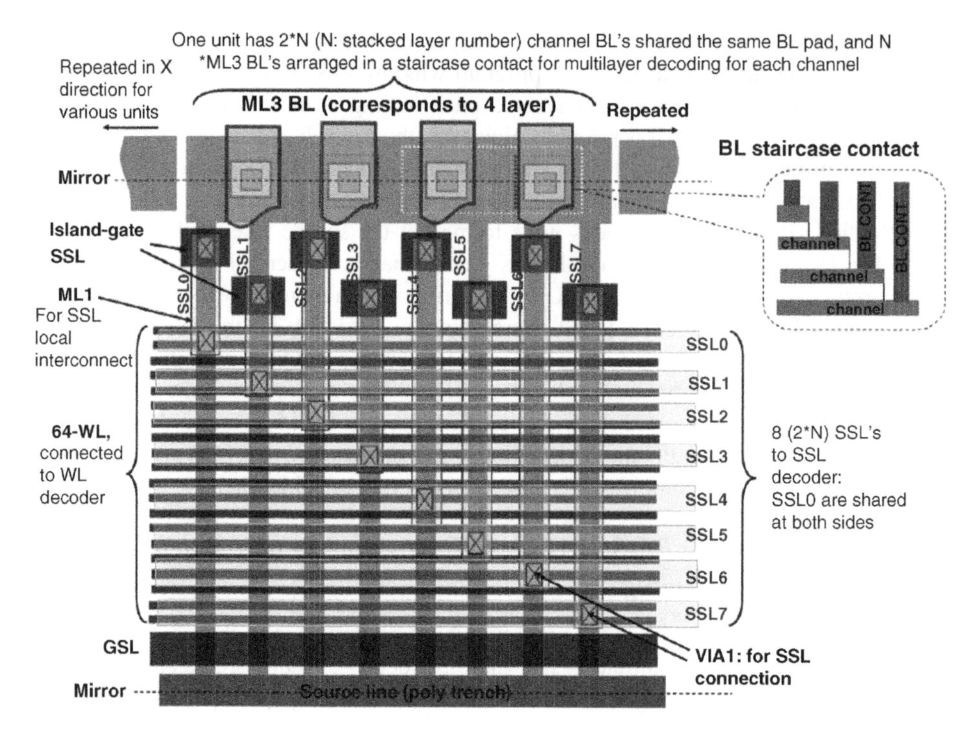

Figure 4.64 Top-view schematic of 3D vertical gate NAND with word-lines of the vertical layers shared. (K.P. Chang *et al.*, (Macronix), IMW, 20 May 2012 [40], with permission of IEEE.)

each island gate SSL device. Due to the multiple SSL devices, the architecture had more pages when the number of stacked layers increased. This meant that program inhibit stress was larger than in conventional 2D NAND.

A top-view schematic diagram of the 3D vertical gate NAND with WLs of the vertical layers shared is shown in Figure 4.64 [40]. Every horizontal channel BL was separately decoded by one vertical island gate SSL device. In one unit that contained $2 \times N$ (where N is the number of layers) channel BLs, all BLs were grouped together for each layer and connected to the metal layer 3 (ML3) BLs through the staircase BL contacts formed at the BL pad region. All island gate SSL devices were connected by CONT/ML1/VIA1/ML2 routing toward the SSL decoder. A CSL was used to share the source lines of all memory layers.

Looking from the top down, the 2D layout schematic in Figure 4.64 is for $N=4$ stacks. Every $2 \times N = 8$ channel BLs are grouped into one unit that share the same BL pad. In the BL pad, a staircase contact is made where each contact corresponds to one memory layer. The staircase contact is connected by ML3 BLs toward the page buffer for memory sensing. Each channel BL has its own island gate SSL for selection. The SSLs are connected through CONT/ML1/VIA1/ML2 toward the SSL decoder. A common source is made to connect source lines of all memory layers. Each device is a double-gate TFT BE-SONOS device and is selected by the intercept of the WL (Y direction) the ML3 BL (Z direction) and the SSL (x direction).The chip has 16Kb BLs and four memory layers. Because the ML3 BL has double X pitch, the total number of ML3 BL is 16Kbit/2 or 8 Kb, and the total unit number is 2Kb. Every unit has eight pages, where each page is defined by the selection of one SSL. To select one page for each WL, a total of 2Kb SSL devices is selected in many parallel units. For all-BL sensing for the four layers, the total number of selected devices are 2Kb × 4 or 8 Kb, which defines the page size [40].

To study the program disturb property during page operation, a two-layer device was made. It was found that the program disturb stress increased linearly as the number of memory layers increased, due to the greater number of pages. Program disturb testing results projected that the device had program disturb immunity sufficient to support more than 32 stacked memory layers.

4.4.7 3D Vertical Gate NAND Flash BL Decoding and Page Operation

The BL decoding and page operation of a 37.5 nm half-pitch 3D BE-SONOS NAND flash with 16 stacking layers was discussed in June of 2012 by Macronix [41]. This array was expected to be cost competitive with a 20 nm 2D NAND flash. A 3D vertical gate NAND was shown using a self-aligned IDG SSL transistor method of decoding. The IDG SSL improved program inhibit and read selection without increasing the cell size.

The decoding method was more difficult for vertical gate devices than for vertical channel ones because the BLs were horizontal and difficult to decode. The decoding method affected the design of the page operation as well as the cell size and array overhead.

This study proposed a self-aligned IDG SSL decoding technique for 3D vertical gate NAND. A 30 nm node range half-pitch 3D NAND was demonstrated, and good program inhibit performance was shown. A schematic of the IDG decoded 3D vertical gate NAND flash array structure is shown in Figure 4.65 [41]. The memory was a double-gate TFT CT device made by the intercept of source lines and BLs. The intercepts of the WL and BL can be seen in the upper structure. The BL contacts formed a staircase with the various memory layers, connecting the corresponding BLs to ML3. The SSLs were divided into a poly gate between the channel BLs and connected through ML1 and ML2 to the SSL decoder. A CSL was made that connects all source lines of every memory layer.

The IDG decoded 3D vertical gate had a layout similar to that of a conventional NAND. It did not require additional area because it was self-aligned and the pitch was scalable. A top-down view of the layout schematic of the IDG decoded array is shown in Figure 4.66 [41]. This was an eight- layer stack. It had conventional WLs and BLs with SSLs divided into 16 groups that corresponded to 16 pages for each WL. The SSLs were connected through ML1 and ML2 to the SSL decoder. The BL pad in each unit shares $2N$ ($16 = 2 \times 8$ layers) channel BLs. Every unit had $N(8)$ different BL contacts that corresponded to various memory layers. When the number of memory stacks was increased, the layer efficiency remained unchanged but the page number of each unit and the BL pad layout did change.

Each memory device was selected by a WL, an ML3 BL, and the page that corresponded to the sandwich of two SSLs. Unselected adjacent pages were inhibited by the IDG SSL operation through the use of an inhibit bias on the other SSL gate. For the IDG SSL operation, a turn-on voltage $V_{ssl} = +2$ V was applied to the two SSLs that sandwiched the selected channel BL. The adjacent unselected channel BL shared one side of the SSL, so a negative inhibit bias $V_{inhibit} = -7$ V had to be applied to the other side of the IDG SSL to turn it off [41].

The I_d–V_g characteristics of the 30 nm range vertical gate TFT device showed good subthreshold behavior, which was due to good gate control capability in a narrow-width

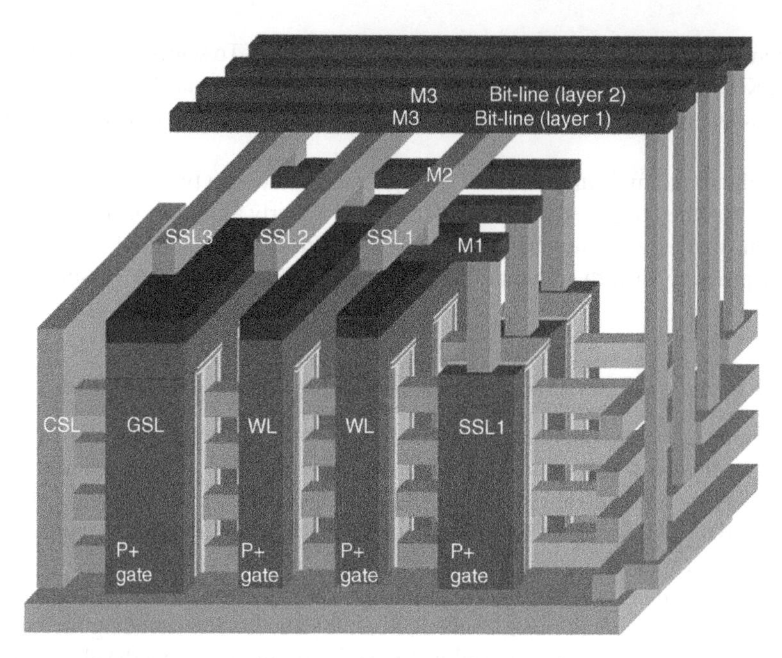

Figure 4.65 Schematic of IDG decoded 3D vertical gate NAND flash array structure. (Based on C.P. Chen *et al.*, (Macronix), VLSI Technology Symposium, June 2012 [41], with permission of Macronix.)

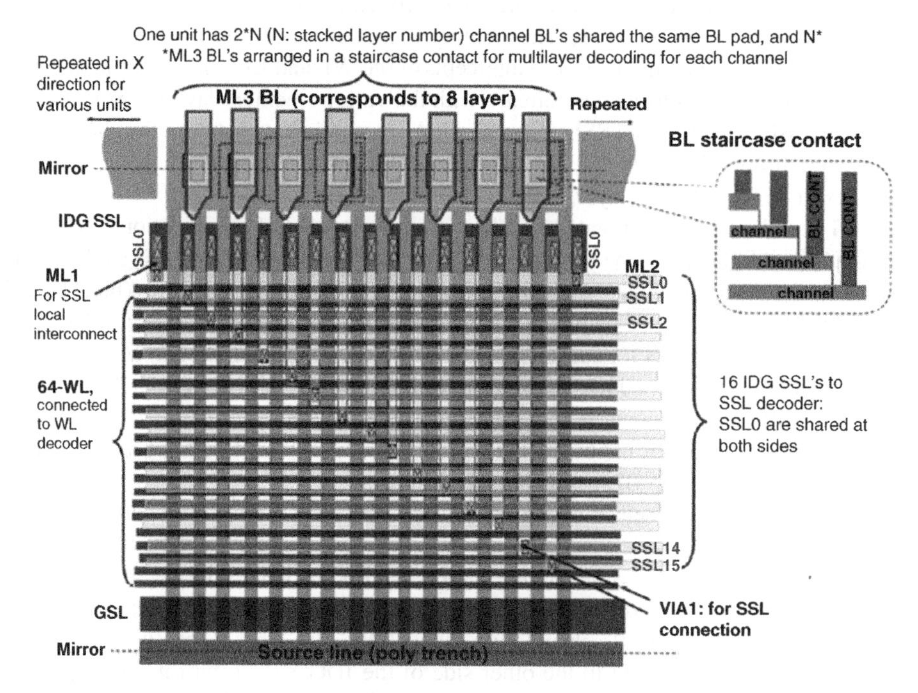

Figure 4.66 Top-down view of layout schematic of ICG decoded array. (C.P. Chen *et al.*, (Macronix), VLSI Technology Symposium, June 2012 [41], with permission of IEEE.)

double-gate device. The I_{dsat} of the array was over 150 nA, which gave sufficient current for memory sensing. The memory window was still well separated after program disturb stressing.

4.4.8 An Eight-Layer Vertical Gate 3D NAND Architecture with Split-Page BL

An eight-layer 3D horizontal channel stacked TFT NAND flash using BE CT devices with 64-WL NAND string with 63% array core efficiency was discussed in December of 2012 by Macronix [42]. The process had a WL half-pitch of 37.5 nm and a BL half-pitch of 75 nm. It was thought that an eight-layer stack device with this pitch could provide a technology at a lower cost than a conventional sub–20 nm 2D planar NAND. It was expected that this technique would permit NAND flash scaling below 15 nm node.

This architecture had two key features: the first was a split-page layout, with even and odd BLs and pages twisted in the opposite direction. This permitted island-gate SSL devices. The other key feature was a BL contact method that minimized the number of process steps. Metal interconnections were laid out in double pitch, which created a larger process window for BL pitch scaling. The 3D vertical gate architecture used WL and BL patterning so that the lateral pitch could be scaled in a similar manner to the 2D NAND. The decoding method for 3D NAND was, however, more difficult than for the vertical channel NAND; the BLs were horizontal and parallel to the multilayers and could, therefore, not be simply connected to metal BLs, as could be done with 2D NAND.

The vertical gate architecture used many rows of normally-on SSL devices to decode the BLs within the NAND string. As the stacked layer increased, the required number of rows of SSLs increased, which reduced the array efficiency. To compensate for this, an island gate SSL device was proposed in order to separate channel BLs for decoding. This permitted the array efficiency to be constant as the stack layers increased. However, making the island gate SSL within each channel BL was difficult with scaled BL pitch.

To improve scalability, a new architecture with a twisted layout was proposed in this study.

NAND strings were divided into even and odd pages so that string current flowed in opposite directions. This permitted the island gate SSL to be laid out at double the pitch of the BL. A top view of this split-page vertical gate architecture is shown in Figure 4.67 [42]. The island gate SSL devices were split into even and odd pages, and GSLs also had even and odd pages in the opposite direction, which provided a twisted BL layout. Each island gate SSL corresponded to one page during page program or read. In the BL direction, each BL pad grouped a total of $2 \times N$ channel BLs, where each even or odd had N SSL devices per page. In the array, MI1 and MI2 connected the SSL gates to the decoder. Page operation selected one SSL in one unit and in parallel selected all units together. Staircase BL contacts in the BL pad connect to different memory layers. These contacts had double the pitch of the channel BL, which improved the process window. A poly plug was used for the staircase BL contact and source contact.

A new method that used only three masks to define the contact for an eight-layer stack was implemented for the staircase contacts. This technique allowed 2^M contacts using M masks. This staircase BL contact method of forming pages used the binary sum of M lithography and etching steps to achieve 2^M contacts. This made it possible to minimize the number of etching steps by using multiple staircase contacts on each layer, which minimized the incremental layer

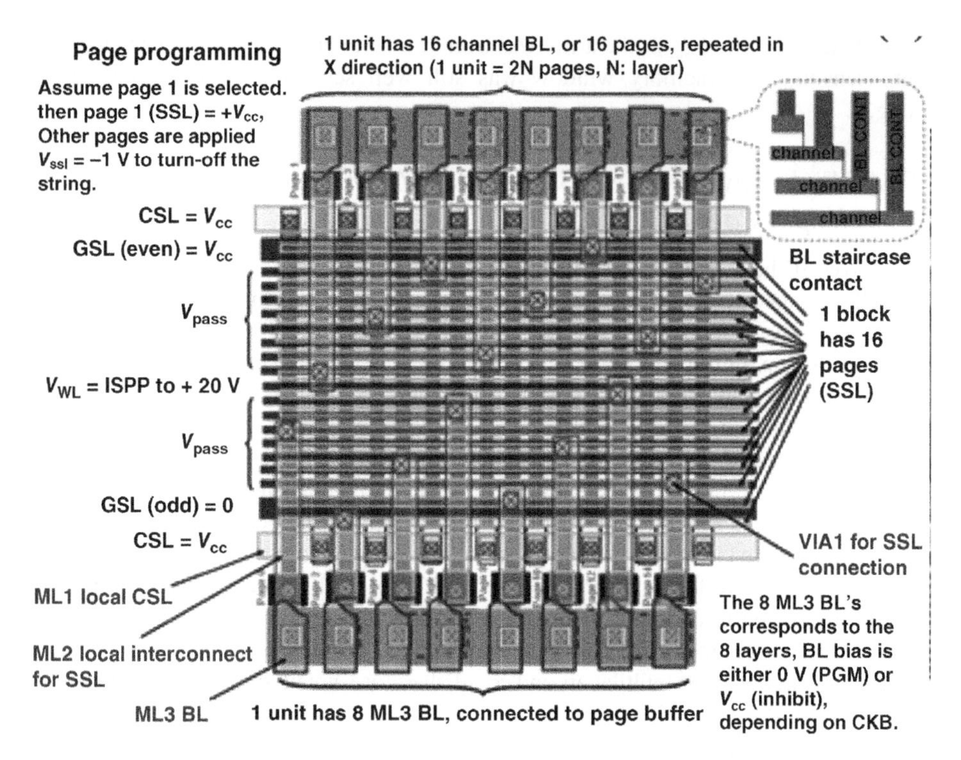

Figure 4.67 Top-down view of split-page vertical gate architecture. (S.H. Chen *et al.*, (Macronix), IEDM, December 2012 [42], with permission of IEEE.)

cost. In this example three masks can be used to carry out eight-layer contacts; that is, random summation of 0,1,2,4 can generate 0,1,2,3,4,5,6,7. This means that to double the memory layer, just one more mask is needed. This is illustrated in Figure 4.68 [42].

It was shown that each of the eight memory layers could be programmed with a memory window greater than 6 V. The schematic circuit diagram in Figure 4.69 shows the program inhibit method used for programming [42]. To program a page, the SSL for a page (page 0) was taken to V_{cc} while other SSLs had a slightly negative voltage applied to guarantee they were turned off. GSL (even) was turned off while GSL (odd) was turned on. V_{cc} was applied to CSL to inhibit. Page 2 (SSL2) was inhibited by floating the entire NAND string. Pages 1 and 3 were inhibited by precharging the channel using CSL and GSL.

Block erase used +13 V on CSL and BL, keeping all WLs at 0 V. A +6 V was applied to the SSL and GSL to reduce GIDL with minimum disturb to SSL and GSL. All memory layers and all WLs could be erased together. In this 3D vertical gate device, every WL had 2^N pages, so to program one WL, every page had to endure a total number of program (NOP) stresses = $2N-1$. Stress capability was to sustain 64 = NOP stresses. The vertical gate NAND had planar ONO, which minimized the V_{pass} disturb. The conclusion was that an eight-layer 3D vertical gate NAND could provide a terrabit memory at 25 nm half-pitch using 32 stacked layers [42].

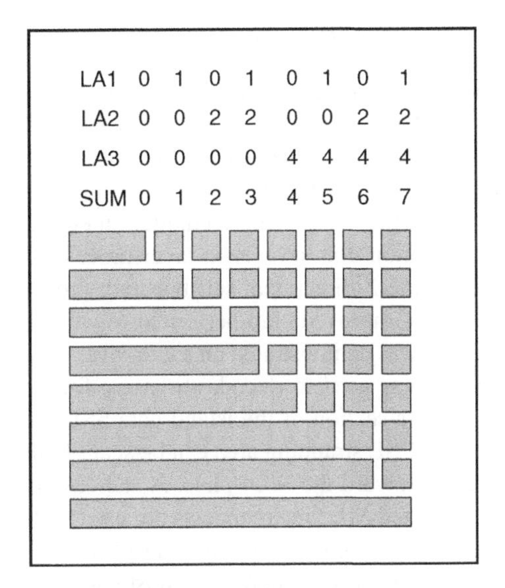

Figure 4.68 Minimal incremental layer cost method for staircase bit-line contacts. Random summation of 0,1,2, and 4 can make 0, 1, 2, 3, 4, 5, 6 and 7. To double the memory layer requires just one additional mask. (Based on S.H. Chen *et al.*, (Macronix), IEDM, December 2012 [42].)

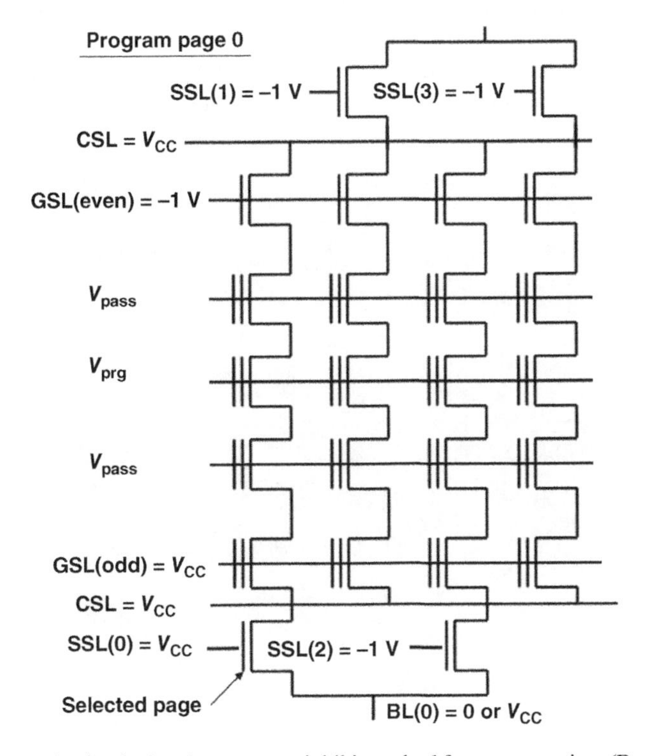

Figure 4.69 Schematic circuit showing program inhibit method for programming. (Based on S.H. Chen, *et al.*, (Macronix), IEDM, December 2012 [42].)

4.4.9 Various Innovations for 3D Stackable Vertical Gate

In December of 2012, Macronix discussed several design innovations for its 3D vertical gate NAND flash technology [43]. These included a “shift-BL scramble” intended to average the BL capacitances, which would provide uniform BL capacitances for various memory layers, optimized read waveforms intended to reduce the hot carrier–induced read disturb during page read, a reverse read with a multiple V_{th} sensing technique intended to compensate different memory layers for the V_{th} variation due to layer-to-layer process difference, and a program inhibit method for minimizing the Z-directional self-boosting program disturb present in 3D stacked memory. The result was improved chip performance with good memory window for SLC and MLC operations. This was demonstrated on a two-layer 3D vertical gate NAND chip.

An island gate SSL device was used to decode the array for different BLs. A scanning electron microscope (SEM) cross-section of the island gate SSL used to decode the array is shown in Figure 4.70 [43]. The CSL strapping area can be seen. In the CSL strapped area, the local ML1 CSL was connected toward the ML4 plate. A staircase BL contact was made and connected to the BL2 BL. ML1 and ML2 interconnects are used to decode the SSLs. An ML3 plate was used as the CSL to supply current during read [43].

The 3D vertical gate stacked flash is considered the most pitch-scalable structure among all of the 3D NAND flash architectures. The decoding method is, however, more complex. To improve the decoding options, an island gate SSL was used to separate the BLs, and each memory layer was connected to its own metal BLs for sensing. The process window for the island gate SSLs was improved by splitting the even and odd pages to opposite sides of the array [43].

Each NAND string had 64 WLs connected in series with channel current flowing horizontally. The gate was shared vertically. Each memory cell was selected by the WL,

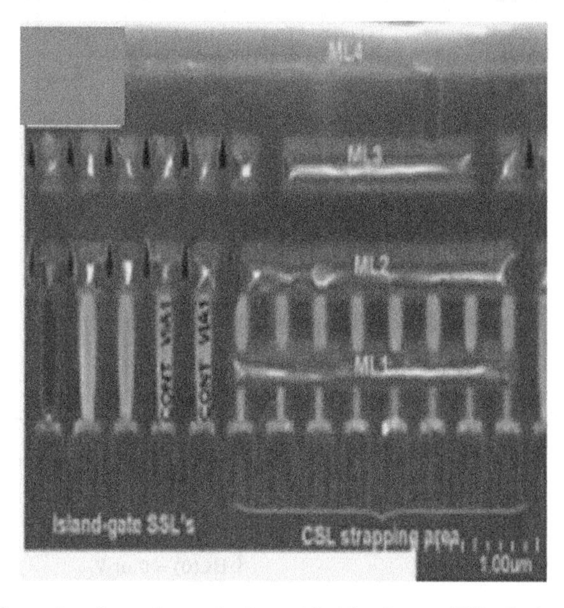

Figure 4.70 3D schematic of two-layer device with island gate SSL device. (C.H. Hung *et al.*, (Macronix), IEDM, December 2012 [43], with permission of IEEE.)

BL M3, and SSL, which was an island gate SSL that separated the BL channel. In the BL direction, one unit had four ($2 \times N$, where N = number of layers) pages, divided as even and odd pairs located on opposite sides. Each BL pad grouped 2 (= N) channel BLs in one side. Staircase BL contacts were made at the BL pad. The ML3 BL was connected to the page buffer for sensing the NAND string. ML1 and ML2 were used to connect SSL devices and the decoder. For page read and write, many SSLs from different units were selected together for high throughput. Four metal layers were used. M4 was the global CSL supporting a large current during page read [43].

The two-layer device was a 512Mb MLC test chip with core memory area of 2.3 mm × 2.6 mm. This area included the page buffer, WL, and SSL decoder and driver. Page size was 512 Bytes. More layers could be stacked with this configuration. The stack height of the array was significantly higher than those of the peripheral CMOS devices, which made planarization difficult. For this reason, the 3D vertical gate array was put in a deep trench in the silicon surface, and the peripheral CMOS devices were made on the original silicon surface, as shown in the schematic cross-section in Figure 4.71 [43].

This CMOS-last process had the advantage of avoiding thermal budget issues on the peripheral CMOS. As a result, the periphery had similar characteristics to those of a 2D array and also had a high thermal budget, which could improve characteristics of the polysilicon TFT devices and optimize the ONO in the BE-SONOS.

An issue was that the ML3 global BL was connected to the different memory layers in each unit. Because of the parasitic capacitance of the BL pad, the capacitance was different for each layer. This led to about a 10% layer-to-layer capacitive difference. This effect was averaged out by using a "shift-BL scramble" design in which each ML3 BL was connected to different

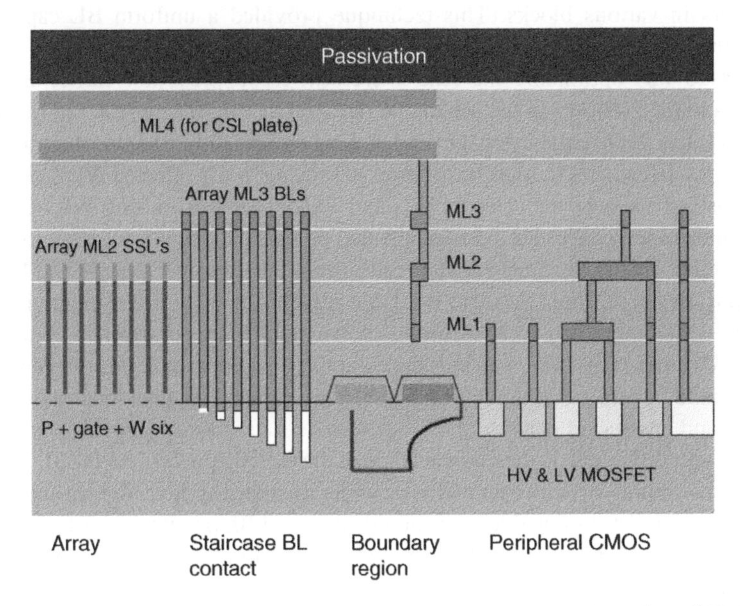

Figure 4.71 Schematic cross-section of 512Mb MCL test chip process. (Based on C.H. Hung *et al.*, (Macronix), IEDM, December 2012 [43].)

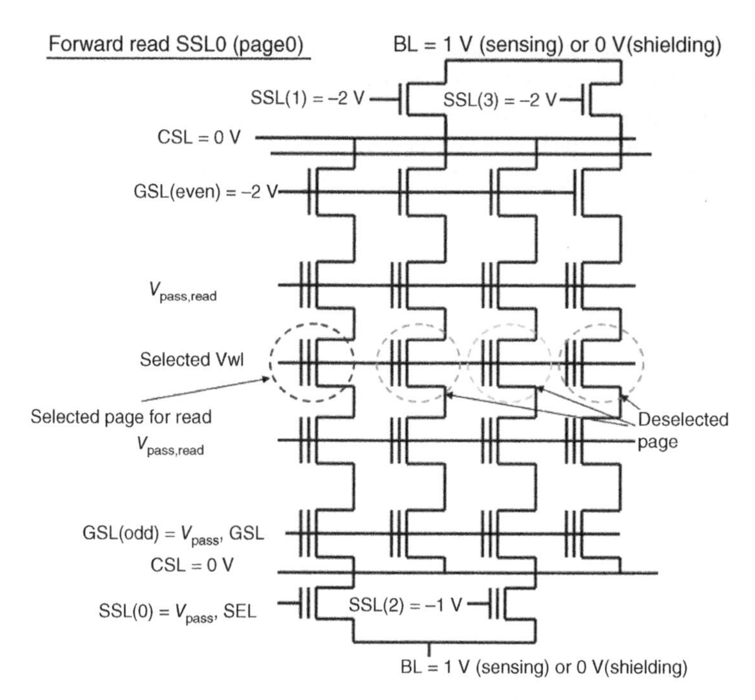

Figure 4.72 Schematic circuit diagram illustrating suppression of hot-carrier-induced read disturb. (Based on C.H. Hung *et al.*, (Macronix), IEDM, December 2012 [43].)

memory layers in various blocks. This technique provided a uniform BL capacitance for voltage sensing.

Another issue was optimizing the read waveform to suppress hot carrier–induced read disturb. An example is shown in the schematic circuit diagram in Figure 4.72 [43]. In this 3D array, each WL has $2 \times N$ pages. A page read is done by selecting SSL0 while turning off the unselected pages. Because SSL must be off prior to sensing, if the selected WL is negative, then local self-boosting between the selected WL and SSL is induced during WL setup for both selected and unselected pages. The boosted channel potential induces hot carrier injection after many read cycles. This is illustrated in the read-timing diagram in Figure 4.73 [43]. Because there are many pages sharing the same WL, this read disturb can be a serious issue.

A solution for minimizing this read disturb is shown in the timing diagram in Figure 4.74 [43]. A "precharge channel to ground" was introduced at the rising phase of WL, while a "discharge channel to ground" was done at the falling phase. This avoided channel local self-boosting for both selected and unselected pages.

A layer-to-layer difference in the vertical slope of the etching process for the 3D NAND flash occurred. The bottom-layer device tended to be wider than the top-layer device and hence had a higher V_{th} and wider distribution due to the nonideal sloped BL profile. It was found possible to optimize the reverse read method so that the top- and bottom-layer V_{th} distribution could be tuned to being nearly the same.

A programming method was used to minimize Z-directional disturb. As an example, for page 0 programming, page 2 was inhibited by floating the NAND string without any precharging of

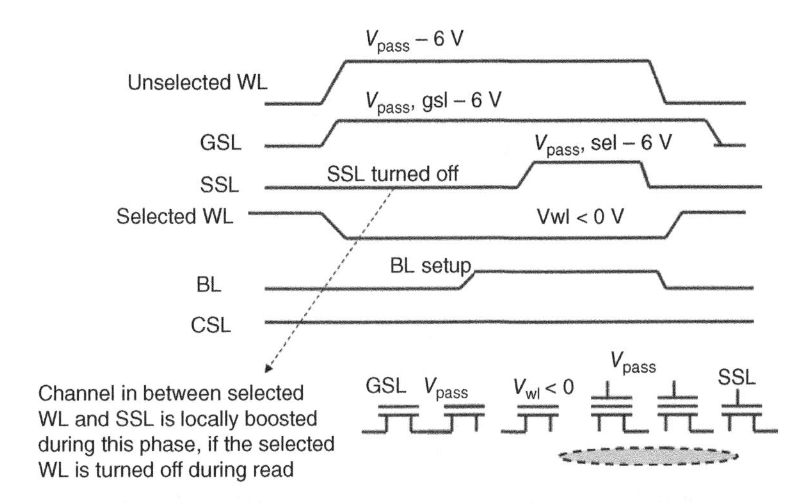

Figure 4.73 Read timing diagram of 3D vertical gate NAND flash showing read disturb. (Based on C.H. Hung, *et al.*, (Macronix), IEDM, December 2012 [43].)

the BL. Meanwhile, pages 1 and 3 were inhibited by precharging the CSL. A typical single-page checkerboard programming was used, where a large memory window of greater than 4.5 V was found within four pages of one WL.

Figure 4.75 shows the results of program disturb for a checkerboard when an adjacent memory layer was programmed [43]. Programming with and without *Z*-disturb showed

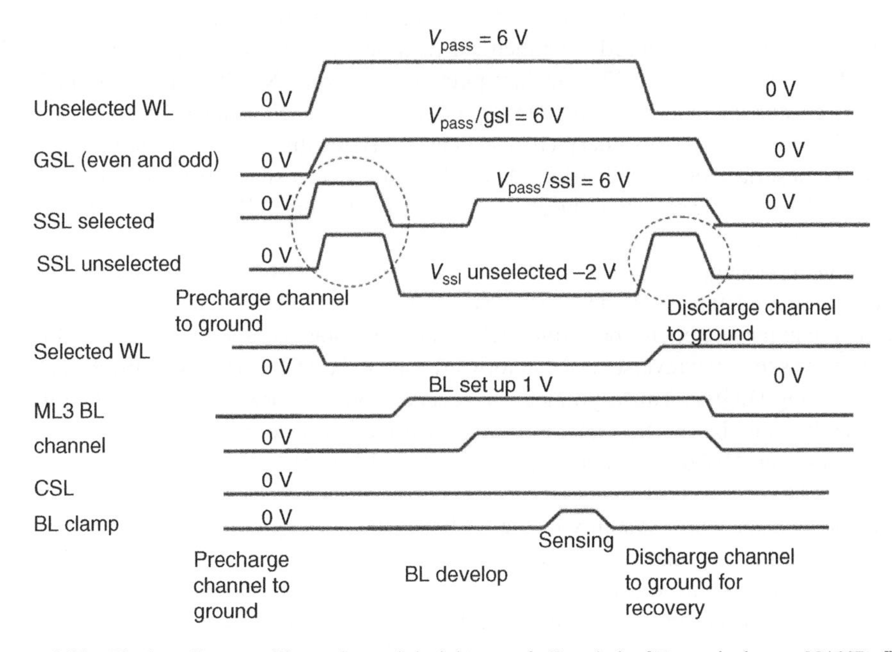

Figure 4.74 Timing diagram illustrating minimizing read disturb in 3D vertical gate NAND flash. (Based on C.H. Hung, *et al.*, Macronix, IEDM, December 2012 [43].)

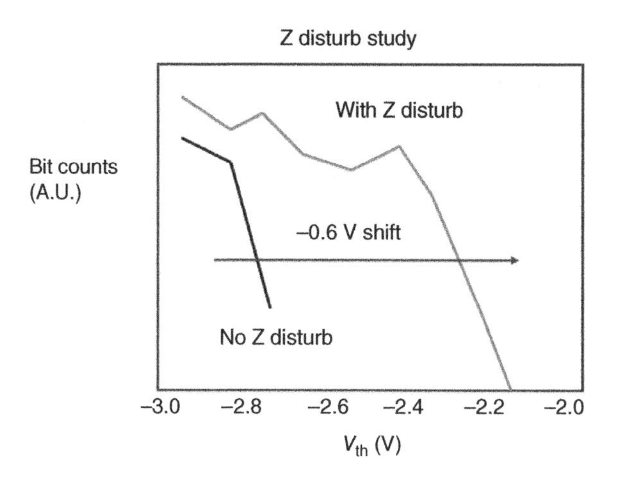

Figure 4.75 Program disturb with checkerboard pattern when adjacent memory layer is programmed. (Based on C.H. Hung, *et al.*, (Macronix), IEDM, December 2012 [43].)

Z-disturb causes about a 0.6 V higher V_{th} shift. The physical reason for the shift was that when adjacent vertical channels were at ground (i.e., programmed), the vertical capacitance coupling could reduce the boosted potential.

4.4.10 Variability Considerations in 2D Vertical Gate 3D NAND Flash

In December of 2012, Macronix discussed modeling the variability caused by random grain boundaries and the trap-induced asymmetrical read behavior for a vertical double gate 3D NAND TFT device [44]. A 37.5 nm half-pitch vertical gate NAND flash was made and characterized, and the random grain boundary effect was modeled using TCAD simulation. In the model used, the grain boundary created interface states that resulted in large local band bending and a surface potential barrier. Major physical mechanisms considered that affect the subthreshold behavior are gate-induced grain barrier lowering (GIGBL) and drain-induced grain barrier lowering (DIGBL).

Using the modeling, the impact of BL and WL critical dimensions of the double-gate TFT device were studied. It was found that narrower BL and larger WL critical dimensions were the most important parameters for providing tight V_{th} distribution and good memory window. An asymmetry was found of reverse read and forward read of the TFT device. This mechanism was explained using DIGBL. This asymmetry of reverse read and forward read could be used to determine the grain barrier trap lateral location and the interface trap density.

The polysilicon TFT device was needed to stack multilayer memory cells. Concerns in this case included large variability due to the random grain boundary traps and low mobility of the polysilicon channel. The variability of the TFT was the primary concern because an uncontrolled V_{th} distribution could cause the device to fail. This is particularly a concern with MLC devices. The low mobility was less critical because the large page size could compensate for a slower read latency. For this reason, the main focus of the study was on the variability model.

Each memory cell was a double-gate TFT BE CT device. A 37.5 mm half-pitch WL device was made. Critical WL dimension was 25 nm. A 64-transistor NAND string was used with an outer SSL–GSL select transistor. The critical dimension in the BL direction was about 30 nm. Several thermal annealing steps were done to increase the grain size and improve the polysilicon channel property. The channel was made of undoped polysilicon in order to reduce the doping fluctuation. The array was junction free with no N+ S/D. The fringe field of the adjacent pass gate bias induced a virtual inversion junction during read operation. The string I_{dsat} was greater than 100 nA, which was sufficient for NAND sensing. It was found that narrower BL critical dimensions greatly improved the subthreshold slope behavior and gave a tighter V_{th} distribution. The transconductance and I_{dsat} were not found degraded for narrower BLs because the inversion channel was induced primarily at the sidewall of the BL so that effective channel width was not proportional to the BL critical dimension [44].

It was estimated that the polysilicon grain sizes ranged between 20 and 50 nm. The TCAD simulation represented the grain boundary traps outside and beneath the selected gate. The subthreshold degradation and V_{th} for both cases were explored. In the case of grain boundary traps outside the selected gate, smaller degradation was found than for the fresh state. For traps beneath the gate, there was significant subthreshold degradation and higher V_{th} [44].

The impact of grain boundary traps in the TFT subthreshold current is shown in Figure 4.76 [44]. The insertion of grain boundaries lowers the surface potential and is the same as creating a higher surface potential barrier because the electron is negatively charged. It requires higher gate bias to reduce the barrier. In the example in Figure 4.76, the grain barrier was assumed to be at the gate center, where the barrier potential is lowered. $V_g = 7$ V was sufficient to reduce the barrier. This phenomenon is called GIGBL.

The drain bias also caused a drain-induced grain barrier lowering (DIGBL) effect, as shown in Figure 4.77, where the larger drain bias helped lower the grain boundary–induced surface potential barrier [44].

The variability of 3D TFT NAND devices was accurately modeled using classical transport 3D TCAD with added grain boundary trap effects. The asymmetry of reverse and forward read

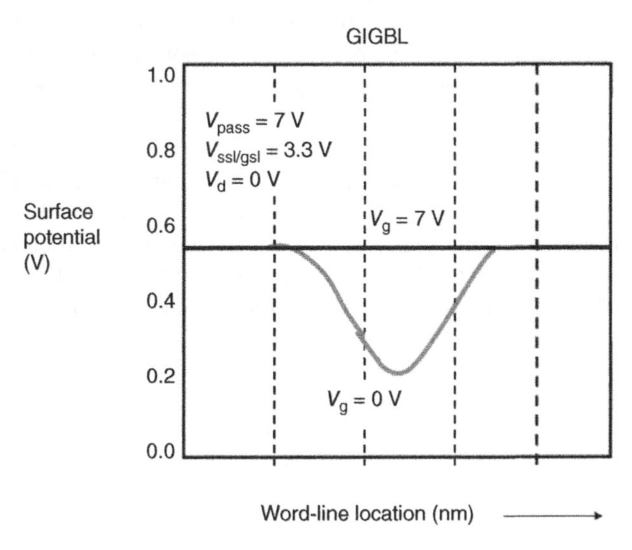

Figure 4.76 Impact of grain boundary traps in TFT subthreshold current. (Based on Y.H. Hsiao *et al.*, (Macronix), IEDM, December 2012 [44].)

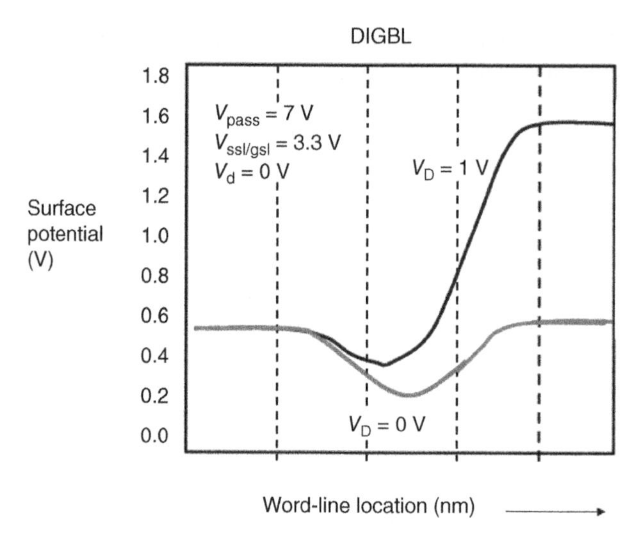

Figure 4.77 Illustration of DIGBL effect. (Based on Y.H. Hsiao, *et al.*, Macronix, IEDM, December 2012 [44].)

was modeled, and a method for profiling the grain boundary trap lateral location and the interface trap density was suggested [44].

4.4.11 An Etching Technology for Vertical Multilayers for 3D Vertical Gate NAND Flash

An etching technology for preparation of smooth vertical multilayer stacked SiO_2 and polysilicon BLs for use in 3D vertical gate NAND flash was discussed by Macronix in May of 2013 [45]. The shape evolution profile required was from tapered to less than 10 nm critical dimension difference between the bottom and top polysilicon layers. This ideal etch profile was done by an etch–trim–etch process sequence. A test chip was made with seven pairs of SiO_2 and polysilicon layers. Initially, an organic antireflective film was deposited on top of the SiON. Photolithography followed using a 193 nm ArF scanner. The etch process consisted of various etch steps carried out using a commercial conductor etching chamber. A layer-to-layer etch sequence was then used; it consisted of etching the SiO_2 and polysilicon layers separately, using different etching plasmas. The critical dimension difference between the eighth (top) polysilicon layer and the first (bottom) polysilicon layer was 42 nm and 77 nm, or a 35 nm difference. This was considered too tapered for use in a 3D vertical gate NAND flash cell. Electrical operation depends on the critical dimension difference between bottom and top polysilicon layers being less than 15 nm.

An etch sequence following the layer-to-layer etch was established. It consisted of an isotropic trim step and a high-energetic CF_x ion bombardment etch step. The resulting critical dimension difference between the bottom and top polysilicon layers was less than 10 nm. The amorphous carbon hardmask was found sufficient even after the energetic bombardment step. The resulting profile was smooth and vertical within 10 nm top to bottom. An illustration of the initial film stack is shown in Figure 4.78 [45]. The stack consisted of seven layers of SiO_2 alternated with polysilicon topped by a thick SiO_2 layer and an amorphous carbon hardmask.

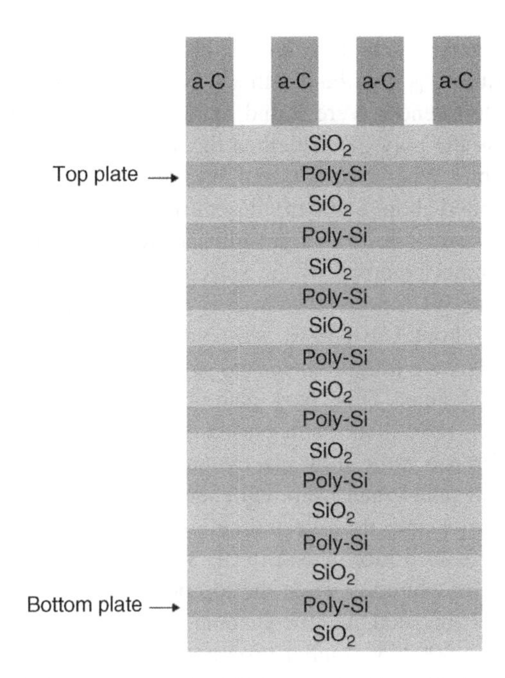

Figure 4.78 Smooth vertical multilayer stacked SiO_2 and polysilicon bit-line stack. (Based on. Z. Yang *et al.*, (Macronix), ASMC, 14 May 2013 [45].)

4.4.12 Interference, Disturb, and Programming Algorithms for MLC Vertical Gate NAND

In June of 2013, Macronix discussed MLC programming in 3D vertical gate NAND flash [46]. 2D FG NAND flash suffered from interference and disturb as the technology scaled, but much of the interference and disturb from scaling was alleviated in the 3D flash because the pitch was generally larger. The 3D NAND flash, however, suffered from a new source of disturb and interference as a result of operating in three dimensions.

A split-page 3D vertical gate NAND flash test chip was used to study the MLC programming algorithm. The memory array consisted of two layers of double-gate polysilicon TFT BE-SONOS CT devices. These devices had 37.5 nm half-pitch WL and a 64-WL NAND string. A 512Mb MLC test chip with 4Kb page size was used to develop the programming algorithms. Programming the split-page 3D vertical gate NAND architecture involved programming the pages. To program page 0, $+V_{cc}$ was applied to SSL0 while other unselected SSLs were turned off. A BL pad was used to group several horizontal channels together. A staircase BL contact was made at the BL pad for memory layer decoding. For FN incremental step pulse programming (ISPP), the programming V_{th} distribution is an indicator of device intrinsic performance. Using polysilicon TFT device improvements, the intrinsic V_{th} sigma (standard deviation) was made smaller than 600 mV, and the distribution was uniformly shifted without erratic tail bits even after cycling. The advantage of a CT device being free of erratic tail bits was that it guaranteed there was no issue with overprogramming. This meant that the required ISPP bias range for various program-verify (PV) states could be estimated in order to minimize the number of programming cycles.

RTN is an intrinsic property of the device that limits the PV distribution. The PV distribution showed a finite V_{th} distribution that was consistent with the RTN distribution. WL and vertical (*Z*) interferences were found to be significant, but BL interference was found negligible because the poly gate shielded the adjacent BLs. Other diagonal interferences were also negligible. When two adjacent WLs in the same layer were programmed, WL interference contributed about a 400 mV V_{th} shift. The WL interference was caused by the junction-free NAND structure with tight-pitched WL because the adjacent pass-gate voltage affected the channel potential of the selected gate. The grain boundary interface trap could also make the WL interference worse. When the adjacent vertical layer was programmed, it showed about 150 mV V_{th} shift due to *Z*-directional interference. When adjacent vertical channels were programmed, the stored electrons changed the electrostatic potential of the selected channel. This "back-gate bias effect" was strongly affected by the *Z*-directional pitch. ONO, polysilicon, and buried-oxide thicknesses had to be optimized. These sources of interference came from electrostatic potential shift caused by the stored charge of an adjacent cell rather than from real charge injection in the selected cell. When there is real charge injection into the selected cell, it is called program disturb.

The primary program disturb mechanism in 3D vertical gate NAND was the *Z*-disturb. The *Z*-disturb occurred, for example, when the top poly layer was being programmed, the bottom poly layer was self-boosted, and the channel potential was raised. When both layers were boosted, there was a 600 MV smaller program disturb. *Z*-disturb-enhanced programming occurs when the second layer is also boosted, causing programming to be faster in the first layer. *Z*-disturb may introduce irregular programming speed when switching the programming code during PV. V_{pass} disturb requires a larger V_{pass} in order to increase the boosted channel potential. This in turn increases the V_{pass} disturb. In a 3D NAND flash, the many unselected pages normally introduce more V_{pass} disturb stress.

Programming algorithms were developed to reflect the understanding gained of the interference and disturb effects in the 3D vertical gate NAND device. A single ISPP sequence with reduced programming shots was developed that reflected the well-behaved ISPP V_{th} distribution, which had been found without tail bits. A layer-programming method was developed in which many layers were separated into three groups during programming. By doing this, each inhibited layer would only see one adjacent layer programmed so that *Z*-disturb was limited by one side only. In addition, each programmed layer always saw two adjacent layers boosted so the *Z*-disturb-enhanced programming was fixed during PV. Finally, the WL (WL) WL(n) was programmed to pre-PV, which was slightly smaller than the final PV. Programming WL($n+1$) would then introduce WL interference and raise WL(n)'s V_{th} slightly. Then WL(n) was programmed to the final PV. Iterating the pre-PV sequence would automatically compensate the WL interference. The optimized programming algorithms helped tighten the PV distribution. The wide memory window required less than 10 bits of ECC per page. The average number of page PV cycles was less than 40. It was believed that this made the device suitable for practical applications.

4.4.13 3D Vertical Gate NAND Flash Program and Read and Fail-Bit Detection

A 3D vertical gate NAND flash was discussed in June of 2013 by Macronix and National Tsing Hua University [47]. This device used circuit techniques to overcome degradations in speed, yield, and reliability that result from cross-layer process variations. These techniques included: layer-aware program, verify, and read (LAPV&R), layer-aware bitline precharge (LABP), and wave-propagation fail-bit-detection (WPFBD) schemes.

For LAPV&R, the profile of cells in a 3D vertical gate NAND array differ across layers because the etching processes do not produce perfectly vertical features. This affects program behavior, such as speed and V_{th}, and also creates disturb effects across layers. LAPV&R requires layer-aware V_{th} sensing to permit different V_{th} between layers and corresponding verify and read operations. The WPFBD scheme proposed takes only one cycle to detect a failed bit. For a design using j-bit error correction code (ECC), the detection cycle is repeated *j* times. For a 16Kb page with eight layers, WPFBD has a nine-times-faster fail-bit-detection (FBD) time than binary search FBD. A two-layer 3D vertical gate NAND test chip with 256Mb capacity was used to confirm that these various circuit techniques attain various target cell program V_{th} in each layer. They also need to attain a 40% reduction in sensing margin loss due to background pattern dependency with less than 0.1% area penalty for a gigabit-scale 3D vertical gate NAND device. The page size of the device was 512 bytes per page.

4.4.14 3D p-Channel Stackable NAND Flash with Band-to-Band Tunnel Programming

A p-channel 3D stackable NAND flash that uses new programming and erasing methods was discussed by Macronix in June of 2013 [48]. By using a p-channel 3D NAND, the disadvantage was avoided of GIDL-induced hole erase of the floating body n-channel NAND. This resulted in a very efficient FN hole erase and negligible disturb of the SSL and GSL devices. The p-channel NAND array permitted a new FN erase selection method. It also provided a new bit-alterable erase that allowed small-unit random code overwrite without block erase. In addition, the band-to-band tunneling–induced hot electron programming method provided lower operating voltage and enhanced scaling of the periphery. The device was implemented in a 37.5 nm half-pitch 3D vertical gate junction-free NAND architecture. A 3D schematic of the p-channel NAND array is shown in Figure 4.79 [48]. The memory device was junction free, while a P+ diffusion junction was formed at only one side of the long-channel SSL and GSL. The WL half-pitch was 37.5 nm, and the WL channel length was 25 nm.

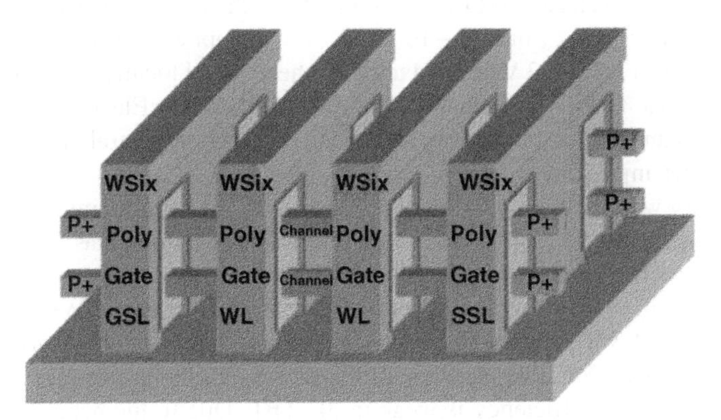

Figure 4.79 3D schematic of p-channel vertical gate stacked NAND array. (Based on H.T. Lue *et al.*, (Macronix), VLSI Technology Symposium, June 2013 [48].)

Most 3D NAND flash use n-channel devices in the array. A drawback of the use of an n-channel device was the need to use high +20 V FN programming bias so that the large CMOS WL drivers consume significant die area. For a negative FN block erase, GIDL-induced erase was needed. During a GIDL erase, a moderate negative voltage was applied to SSL and GSL, and a high positive voltage was applied to the BL and CSL to generate GIDL current at the SSL–GSL junction edge. Electrons and holes were generated by GIDL, and the holes were swept into the NAND string. The holes collected in the channel raised the channel potential, which led to negative FN hole-tunneling injection to the device. Drawbacks of GIDL erase include the following: the erase transient had time lag due to the long minority carrier (hole) generation time, and erase speed was dependent on GIDL current and SSL–GSL junction profile. Increasing GIDL current meant that higher negative voltages had to be applied to the SSL–GSL devices, which disturbed them during block erase. As a solution, this study proposed using a p-channel NAND with the lower-voltage hot electron programming. This permitted use of smaller CMOS design rules. GIDL erase was avoided in the p-channel NAND because channel hole erase could be easily induced without GIDL. The bit-alterable erase enabled a simple overwrite of any small unit without needing to use block erase.

In the p-channel 3D NAND architecture used, the TFT device had an undoped polysilicon channel and no junctions inside the NAND array. The P+ diffusion junction was formed only outside the long channel SSL and GSL devices. The junction-free NAND gave excellent short-channel device performance. The WL half-pitch was 37.5 nm with typical channel length of 25 nm. The NAND array used a 64-WL BL string. The device was a double-gate TFT BE CT device. The p-channel TFT NAND had similar read current to the n-channel device and had good subthreshold behavior. It was speculated that this result might mean that the hole mobility in a polysilicon TFT device was comparable to the electron mobility rather than slower, as found in a bulk device. Block erase was achieved using a high negative voltage applied to all WLs. Holes were easily induced in the p-channel NAND. While n-channel NAND often had an initial time lag for erase, there was no initial erase lag in p-channel 3D NAND, and the memory window was larger. The SSL–GSL had negligible disturb during erase.

For band-to-band hot electron programming, rather than using a large BL bias, local self-boosting was used to create a high virtual drain bias by channel boosting. A positive +9 V bias was applied to the selected WL(n) to cut off the p-channel NAND string, as shown in Figure 4.80 [48]. A large V_{pass} bias of −10 to −13 V was applied to WL($n+1$) and WL(63). The selected BL had $V_{cc} = -3.3$ V applied to boost the channel locally. A high band-to-band generation rate was found between WL(n) and WL($n+1$). Electron-hole pairs were generated by the band to band tunneling and accelerated by the lateral electric field, which induced hot carrier injection.

A large memory window with good inhibit was found. The programming was found to be more efficient with larger V_{pass} due to the larger boosted virtual drain bias. The V_{pass} stepping method was more efficient for programming and provided linear programming, with V_{th} linearly proportional to V_{pass}. When several pages shared the same WL, V_{pass} could be applied on unselected SSL. Because unselected SSL could not be turned off, self-boosting did occur. There was no hot hole injection observed for WLs with negative gate bias. This might mean that hot hole efficiency is weak in 3D TFT. Due to the need for a large V_{pass}, programming in sequence from G1 to G62 avoided having the negative V_{pass} disturb the already programmed cells.

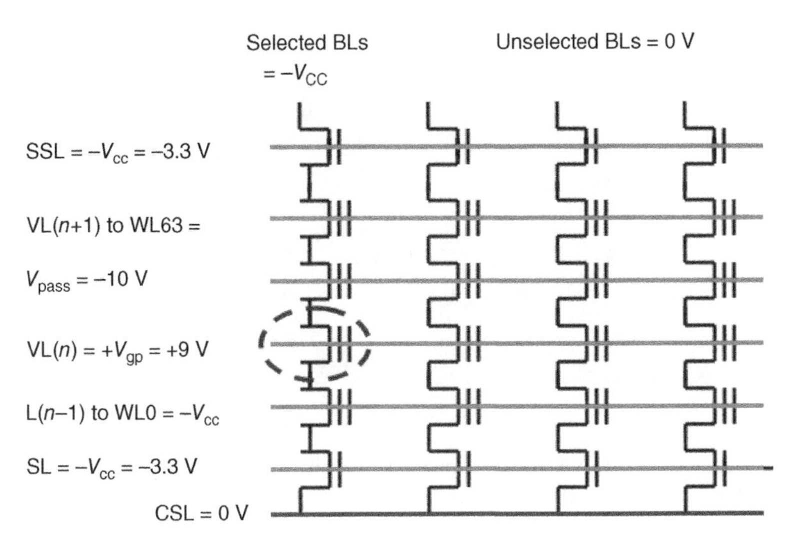

Figure 4.80 Band-to-band hot-electron programming method for p-channel 3D NAND. (Based on H.T. Lue *et al.*, (Macronix), VLSI Technology Symposium, June 2013 [48].)

For bit-alterable erase, the global self-boosting of the p-channel NAND could be done for negative FN erase select by inverting the bias polarity of the conventional n-channel NAND. This self-boosting erase technique showed good erase inhibit. The bit-alterable erase could also be performed in page mode to increase throughput.

An overwrite feature was available. When the device was already programmed by band-to-band hot electron injection, a bit erase on any small unit of the array could be performed without block erase. The operation bias in band-to-band hot electron injection and bit erase could be shifted simply by $+V_{cc}$ so that negative voltage on the BL was avoided. It was possible to limit the maximum array operation voltage below an absolute 15 V, which permitted smaller CMOS design rules that could save area in the periphery.

4.4.15 A Bit-Alterable 3D NAND Flash with n-Channel and p-Channel NAND

A new bit-alterable dual-channel 3D NAND flash memory that had both n-channel and p-channel NAND characteristics was discussed in December of 2013 by Macronix [49]. The NAND channel was junction free with no dopant used inside the array. The drain side of the channel near the SSL was an N+ doped junction, while the source side of the channel near GSL was a P+ doped junction. A 3D schematic of an array with these characteristics is illustrated in Figure 4.81 [49].

An advantage of this operation was that the carrier source for both positive FN programming and negative FN erase could be provided by either the N+ drain or the P+ source without waiting for the GIDL-generated minority carrier for the floating body 3D NAND. This resulted in faster FN speed than is possible with a conventional 3D NAND.

Both positive and negative FN operations had a suitable inhibit method, which enabled a bit-alterable flash memory. A positive pass-gate read voltage induced an n-type virtual S/D for the

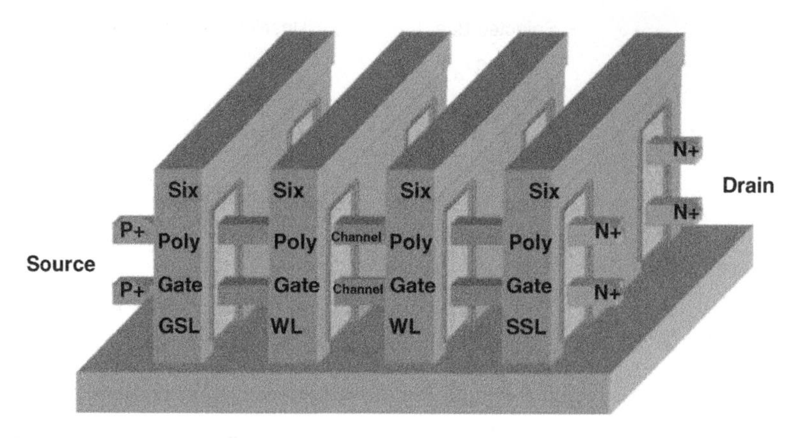

Figure 4.81 Dual-channel 3D NAND with drain-side N+ bit-line and source-side P+ bit-lines. (Based on H.T. Lue, *et al.*, (Macronix), IEDM, December 2013 [49].)

center WLs, which gave n-channel behavior. A negative pass-gate read voltage induced a p-type virtual S/D, which provided p-channel behavior. I_d-V_g characteristics with very small leakage current resulted from both n-channel and p-channel reads.

For the array decoding method, a drain-side (N+) BL was vertically shared and connected to BLs in the metal one (M1) layer. On the source side (P+), the source line was horizontally shared and separately decoded for the various layers. The staircase source line contact was made every 128 ML1 BLs to reduce the source line loading. The source lines were then connected to the ML2 layer near the source line decoder. This array decoding method is illustrated in Figure 4.82 [49].

The read method for the array consisted of applying +2 V to the selected source line while the unselected source line was set at 0 V. The selected BL was set at 0 V, while the unselected

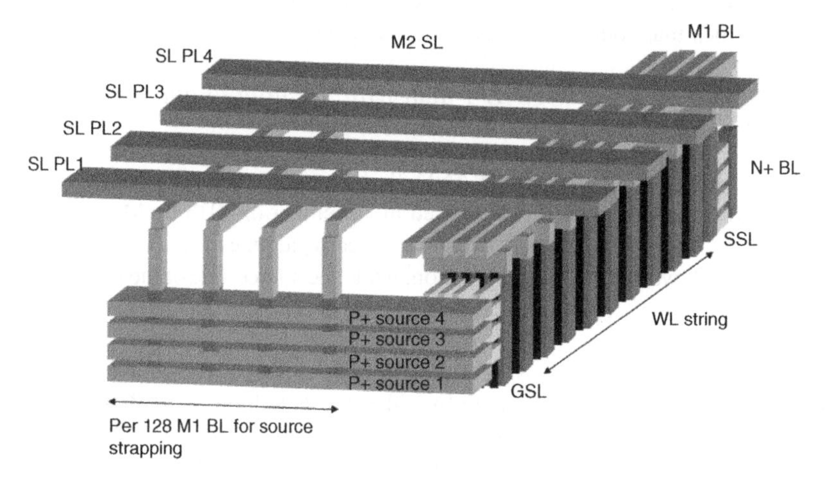

Figure 4.82 Array decoding method for the dual-channel vertical gate NAND flash. (Based on H.T. Lue *et al.*, (Macronix), IEDM, December 2013 [49].)

BL was set to +2 V. The source-side PN diode was necessary to prohibit the formation of a sneak path during read. The selected BL could be set at either 0 V or set slightly positive to obtain smaller background leakage. The typical polysilicon PN diode in the NAND array showed good on and off characteristics with small leakage and junction breakdown greater than 9 V.

The two sensing methods, both n-channel and p-channel read, were compared, and the trapped charge in the space between the WLs was identified. This both provided characterization of the lateral charge profile and also offered the potential for creating another storage node in the WL space inside the array.

References

1. Choi, J. and Seol, K.S. (June 2011) 3D Approaches for non-volatile memory, (Samsung). VLSI Technology Symposium.
2. Hsiao, Y.H., Lue, H.T., Hsu, T.H. *et al.* (16 May 2010) A critical examination of 3D stackable NAND flash memory architectures by simulation study of the scaling capability, (Macronix). IMW.
3. Tanaka, H. *et al.* (June 2007) Bitcost scalable technology with punch and plug process for ultra high density flash memory, (Toshiba). VLSI Technology Symposium.
4. Goda, A. and Parat, K. (December 2012) Scaling directions for 2Dand 3D NAND cells, (Micron). IEDM.
5. Clarke, P. (2013) Toshiba to Build Fab for 3D NAND Flash, EE Times (July 2).
6. Fukuzumi, Y. *et al.* (December 2007) Optimal integration and characteristics of vertical array devices for ultra-high density, bit-cost scalable flash memory, (Toshiba). IEDM.
7. Komori, Y. *et al.* (December 2008) Disturbless flash memory due to high boost efficiency on BiCS structure and optimal memory film stack for ultra high density storage device, (Toshiba). IEDM.
8. Maeda, T. *et al.* (June 2009) Multi-stacked 1G cell/layer pipe-shaped BiCS flash memory, (Toshiba). VLSI Circuits Symposium.
9. Katsumata, R. *et al.* (June 2009) Pipe-shaped BiCS flash memory with 16 stacked layers and multi-level-cell operation for ultra high density storage devices, (Toshiba). VLSI Technology Symposium.
10. Ishiduki, M. *et al.* (December 2009) Optimal device structure for pipe-shaped BiCS flash memory for ultra high density storage device with excellent performance and reliability, (Toshiba). IEDM.
11. Nitiyama, A. and Aochi, H. (June 2010) Bit cost scalable (BiCS) flash technology for future ultra high density storage devices, (Toshiba). VLSI Technology Symposium.
12. Ohshima, J. (16 April 2012) State-of-the-art flash memory technology, looking into the future, (Toshiba), GSA/ SEMATECH Memory Conference.
13. Yanagihara, Y., Miyaji, K., and Takeuchi, K. (May 20, 2013) Control gate length, spacing and stacked layer number design for 3D-stackable NAND flash memory, (University of Tokyo). IMW.
14. Nitayama, A. and Aochi, H. (June 2013) Bit cost scalable (BiCS) technology for future ultra high density storage memories, (Toshiba). VLSI Technology Symposium.
15. Samsung (2013) Samsung Starts Mass Producing Industry's First 3D Vertical NAND Flash, Press release, August 6.
16. Kim, J. *et al.* (June 2008) Novel 3-D structure for ultra high density flash memory with VRAT (vertical-recess-array-transistor) and PIPE (planarized integration on the same plane), (University of California Los Angeles, Samsung). VLSI Technology Symposium.
17. Kim, J. *et al.* (June 2009) Novel vertical-stacked-array-transistor (VSAT) for ultra-high-density and cost-effective NAND flash memory devices and SSD (solid state drive), (Samsung). VLSI Technology Symposium.
18. Jang, J. *et al.* (June 2009) Vertical cell array using TCAT (terabit cell array transistor) technology for ultra high density NAND flash memory, (Samsung). VLSI Technology Symposium.
19. Cho, W.S. *et al.* (June 2010) Highly reliable vertical NAND technology with biconcave shaped storage layer and leakage controllable offset structure, (Samsung). VLSI Technology Symposium.
20. Nowak, E. *et al.* (12 June 2012) Intrinsic fluctuations in vertical NAND flash memories, (Samsung). VLSI Technology Symposium.
21. Jeong, M.K., Joe, S.M., Shin, H. *et al.* (2011) High-density three-dimensional stacked NAND flash with common gate structure and shield layer, (Seoul National University). *IEEE Transactions on Electron Devices*, **58** (12), 4212.

22. Ji, J., Park, B.G., Lee, J.H., and Shin, H. (16 May 2010) A comparative study of the program efficiency of gate all around SONOS and TANOS flash memory (Seoul National University), IMW.
23. Seo, M.S., Park, S.K., and Endoh, T. (May 16, 2010) The 3-dimensional vertical FG NAND flash memory cell arrays with the novel electrical S/D technique using the extended sidewall control gate (ESCG), (Tohoku University). IMW.
24. Seo, M.S. and Endoh, T. (June 2011) New design method of the 3-dimensional vertical stacked FG type NAND cell arrays without the interference effect, (Tohoku University, JST-CREST). VLSI Technology Symposium.
25. Seo, M.S., Park, S.K., and Endoh, T. (2011) 3-D vertical FG NAND flash memory with a novel electrical S/D technique using the extended sidewall control gate, (Tohoku University, Hynix). *IEEE Transactions on Electron Devices*, **58** (9), 2966.
26. Seo, M.S., Lee, B.H., Park, S.K., and Endoh, T. (22 May 2011) A novel 3-D vertical FG NAND flash memory cell arrays using the separated sidewall control gate (S-SCG) for highly reliable MLC operation, (Tohoku University, Hynix). IMW.
27. Seo, M.S., Lee, B.H., Park, S.K., and Endoh, T. (2012) Novel concept of the three-dimensional vertical FG NAND flash memory using the separated-sidewall control gate, (Tohoku University, Hynix). *IEEE Transactions on Electron Devices*, **59** (8), 2018.
28. Whang, S.J. *et al.* (December 2010) Novel 3-dimensional dual control-gate with surrounding floating-gate (DC-SF) NAND flash cell for 1 Tb file storage application, (Hynix). IEDM.
29. Yoo, H.S. *et al.* (22 May 2011) New read scheme of variable Vpass-read for dual control gate with surrounding floating gate (DC-SR) NAND flash cell, (Hynix). IMW.
30. Noh, Y. *et al.* (June 2012) A new metal control gate last process (MCGL process) for high performance DC-SF (dual control gate with surrounding floating gate) 3D NAND flash memory, (Hynix). VLSI Technology Symposium.
31. Aritome, S. *et al.* (2013) Advanced DC-SF cell technology for 3-D NAND flash, (Hiroshima University, Hynix). *IEEE Transactions on Electron Devices*, **60** (4), 1327.
32. Seo, M.S., Choi, J.M., Park, S.-K., and Endoh, T. (20 May 2012) Highly scalable 3-D vertical FG NAND cell arrays using the sidewall control pillar (SCP), (Tohoku University, Hynix). IMW.
33. Jeong, M.K. *et al.* (December 2012) Characterization of traps in 3-D stacked NAND flash memory devices with tube-type poly-Si channel structure, (Seoul National University, Hynix). IEDM.
34. Shim, K.S. *et al.* (20 May 2013) Inherent issues and challenges of program disturbance of 3D NAND flash cell, (Hynix). IMW.
35. Lee, J.W., Jeong, M.K., Park, B.G. *et al.* (10 June 2012) 3-D stacked NAND flash memory having lateral bit-line layers and vertical gate, (Seoul National University), Silicon Nanoelectronics Workshop.
36. Seo, J.Y. *et al.* (10 June 2012) Investigation into the effect of the variation of gate dimensions on program characteristics in3D NAND flash array, (Seoul National University), Silicon Nanoelectronics Workshop.
37. Kim, W.J. *et al.* (June 2009) Multi-layered vertical gate NAND flash overcoming stacking limit for terabit density storage, (Samsung). VLSI Technology Symposium.
38. Lue, H.T. *et al.* (June 2010) A highly scalable 8-layer 3D vertical-gate (VG) TFT NAND flash using junction-free buried channel BE-SONOS device, (Macronix). VLSI Technology Symposium.
39. Hung, C.H. *et al.* (June 2011) A highly scalable vertical gate (VG) 3D NAND flash with robust program disturb immunity using a novel PN diode decoding structure, (Macronix). VLSI Technology Symposium.
40. Chang, K.P. *et al.* (20 May 2012) Memory architecture of 3D vertical gate (3DVG) NAND flash using plural island-gate SSL decoding method and study of its program inhibit characteristics, (Macronix). IMW.
41. Chen, C.P. *et al.* (June 2012) A highly pitch scalable 3D vertical gate (VG) NAND flash decoded by a novel self-aligned independently controlled double gate (IDG) string select transistor (SSL), (Macronix). VLSI Technology Symposium.
42. Chen, S.H. *et al.* (December 2012) A highly scalable 8-layer vertical gate 3D NAND with split-page bit line layout and efficient binary-sum MiLC (minimal incremental layer cost) staircase contacts, (Macronix). IEDM.
43. Hung, C.H. *et al.* (December 2012) Design innovations to optimize the 3D stackable vertical gate (VG) NAND flash, (Macronix). IEDM.
44. Hsiao, Y.H. *et al.* (December 2012) Modeling the variability caused by random grain boundary and trap-location induced asymmetrical read behavior for a tight-pitch vertical gate 3D NAND flash memory using double-gate thin-film transistor (TFT) device, (Macronix). IEDM.
45. Yang, Z. *et al.* (14 May 2013) Investigation of shape etching on multi-layer SiO2/Poly-Si for 3D NAND architecture, (Macronix). ASMC.

46. Hsieh, C.C. *et al.* (June 2013) Study of the interference and disturb mechanisms of split-page 3D vertical gate (VG) NAND flash and optimized programming algorithms for multi-level cell (MLC) storage, (Macronix). VLSI Technology Symposium.
47. Hung, C.H. *et al.* (June 2013) 3D stackable vertical-gate BE-SONOS NAND flash with layer-aware program-and-read schemes and wave-propagation fail-bit-detection against cross-layer process variations, (Macronix, National Tsing Hua University), VLSI Circuits Symposium.
48. Lue, H.T. *et al.* (June 2013) A novel bit alterable 3D NAND flash using junction-free p-channel device with band-to-band tunneling induced hot-electron programming, (Macronix). VLSI Technology Symposium.
49. Lue, H.T. *et al.* (December 2013) A novel dual-channel 3D NAND flash featuring both n-channel and p-channel NAND characteristics for bit-alterable flash memory and a new opportunity in sensing the stored charge in the word-line space, (Macronix), IEDM.

5

3D Cross-Point Array Memory

5.1 Overview of Cross-Point Array Memory

Cross-point array architecture has a potential for achieving a fast, random access, nonvolatile memory with the highest density and smallest cell size possible. The recent exploration of nonvolatile resistance memory elements that can be placed at the cross-point of word-lines and bit-lines has helped realize this potential. If this memory element is no larger than the pitch of these lines (2F), the cell is $2F \times 2F$ ($4F^2$) in area. If this cross-point array can be stacked with another cross-point array, then the density of memory elements is doubled. If stacked with 10 arrays, then the density of memory elements is multiplied by 10. An example of a cross-point array stack with memory elements is shown in Figure 5.1. The cross-point array architecture is considered a potential future competitor for large, high-density NAND flash memory applications. An optimum cross-point switch memory would consist of minimum-size memory elements at the cross-points of a stacked array of perpendicular wires.

Theoretically, this stack can be very high and the wire grid very small, allowing the maximum density of data storage to be obtained. An individual cell would need to be externally selectable without incurring current leakage from neighboring unselected cells or from partially selected cells on the same word-lines or bit-lines. This could be achieved with a selector device in series with the memory elements that permits current to flow only in the selected devices and turns off any unselected or half-selected device. This selector would need to be associated with the memory element and be independent of the substrate.

Without a selector, both the on and off resistance of the selected memory element would need to be smaller than the resistance of an unselected or half-selected memory element to avoid parasitic leakage current in these devices.

The parasitic resistance of the wires would need to be close to zero, or at least be much smaller than the resistance of the low-resistance state (LRS) of the memory element so that the same voltage could be applied across every memory element at every wire intersection in the array. This is a difficult criterion for a scaled array because the resistance of a conductive wire increases as the diameter is reduced. If the resistivity of the wire could be reduced, the parasitic resistance would also be reduced. Otherwise the size of a single cross-point array block will be limited.

Vertical 3D Memory Technologies, First Edition. Betty Prince.

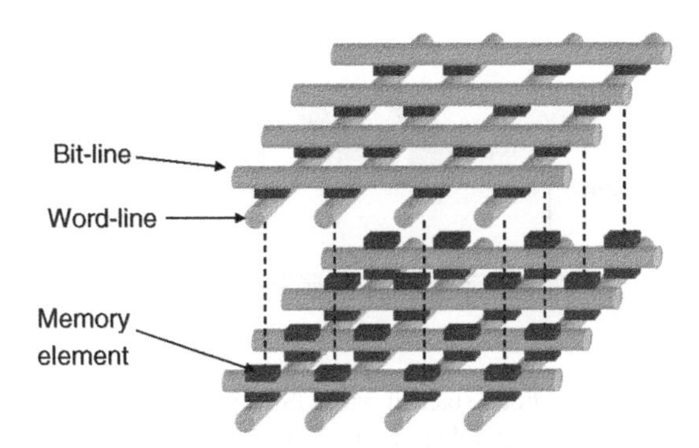

Figure 5.1 Typical cross-point array stack with memory elements at the cross-points.

Each memory element would need to be nonvolatile with a fast access time and low programming current while occupying no more area than what is taken up by the half-pitch of the wire used in the cross-point array. Peripheral support circuitry would need to occupy minimal silicon area to ensure that the array efficiency is high.

This chapter explores the many approaches that have been investigated for solving these challenges and developing optimal cross-point array memories.

5.2 A Brief Background of Cross-Point Array Memories

5.2.1 Construction of a Basic Cross-Point Array

Historically, the requirement has been that each memory element has an access device, either a diode or transistor, in the silicon substrate. This limits the number of memory devices that can be stacked. The access or select device serves to reduce leakage paths between neighboring elements. Figure 5.2(a) shows the leakage paths present in a cross-point array without an associated pass device for each memory element [1,2]. Neighboring devices to the selected cell, on both the selected word-line and selected bit-line, can potentially pass leakage current. These are called half-selected devices because they have only the bit-line or the word-line selected. A

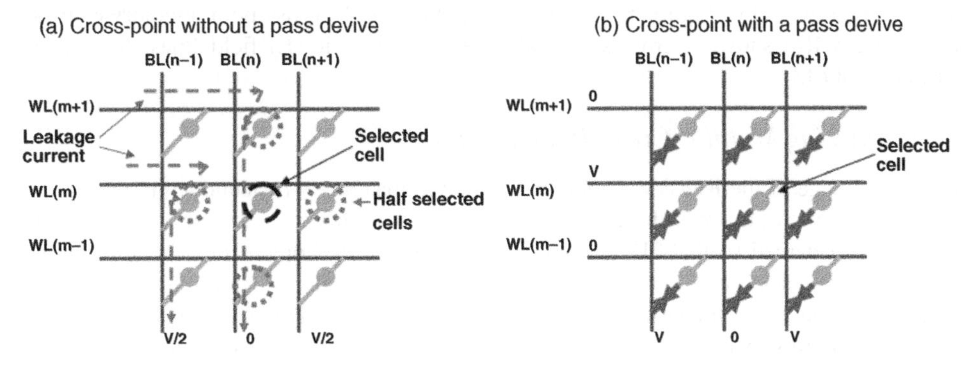

Figure 5.2 Cross-point memory array showing leakage currents (a) without; and (b) with a pass device. (Based on Y. Zhang *et al.*, (Stanford University), VLSI Technology Symposium, June 2007 [2].)

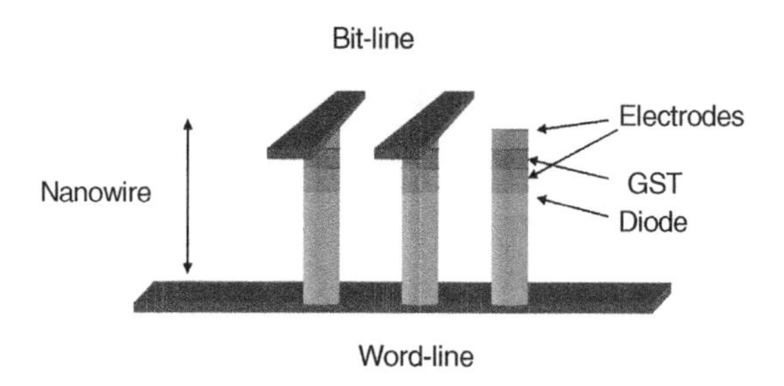

Figure 5.3 Vertical nanowire Ge diode stacked with GST phase-change RAM. (Based on Y. Zhang *et al.*, (Stanford University), VLSI Technology Symposium, June 2007 [2].)

cross-point array using a diode + memory element, as shown in Figure 5.2(b), can eliminate these "sneak" leakage paths [1,2].

The requirement that memory cells in an array with both a bit-line and word-line contact need an access device in the substrate to reduce leakage has historically limited the number of layers of memory cells that could be placed vertically to two [1,2].

A vertical diode nanowire in series with a vertical memory element can reduce the area required for an access device and permit the device to be placed in the array above or below the memory element rather than in the substrate. This eliminates the limitation on number of layers in a cross-point memory array. An example is an early cross-point array proposed by Stanford University in June of 2007 [2]. The memory cell in this array was made of a vertical germanium (Ge) nanowire diode stacked with a $Ge_2Sb_2Te_5$ (GST) phase-change memory (PCM) element, as shown in Figure 5.3 [2]. The resulting cell area was $4F^2$, and the limitation on number of layers in the cross-point array could be relaxed. Both minimizing lateral dimensions and increasing the number of layers are considerations in increasing the density and reducing the cost of the process.

Several different types of resistance RAM (ReRAM) memories have been considered for use in cross-point arrays. Cross-point array ReRAMs, in their simplest form, are formed by a layer of nonconducting material sandwiched between two conductors. These two conductors can be wires perpendicular in the plane so that the memory device is at the cross-point of a memory array with a metal–oxide memory element, NiO in this case, as shown in Figure 5.4 [3]. Figure 5.4(a) shows a top view of the cross-point memory element, and Figure 5.4(b) shows a cross-sectional view with electric (E) field lines through the indicated NiO [3].

5.2.2 Stacking Multibit Cross-Point Arrays

A key advantage of the ReRAM is that the switching material is nonconducting except in a small area between the crossed metal lines. This means, in this case, that only the top metal needs to be patterned so that 3D stacked arrays can be easily implemented in conventional complementary metal oxide semiconductor (CMOS) technology with multiple memory elements on one select transistor, as shown in Figure 5.5 [3]. This structure, discussed by Stanford University in June of 2009 [3], illustrates a multibit, multilayer cross-point ReRAM array using one access transistor along the bit-line direction. Word-lines are out of the page. This was an early alternative to having a select device with every memory element. In the

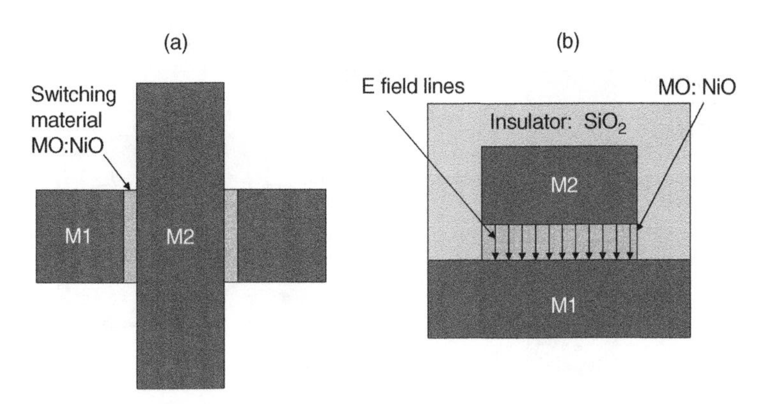

Figure 5.4 Cross-point memory element with NiO metal–oxide deposited between crossing metal lines: (a) top view; (b) cross-sectional view showing electric (E) field lines. (Based on B. Lee and H.S. P. Wong, (Stanford University), VLSI Technology Symposium, June 2009 [3].)

conventional multibit per select device array, such as shown in Figure 5.5, dielectric 1 is an insulator, such as SiO_2, and dielectric 2 is the metal–oxide switching material, NiO in this case.

A special case occurred when the two dielectrics were exchanged so that dielectric 2, the dielectric between the crossed metal lines, was the insulating SiO_2, and dielectric 1 was the NiO metal–oxide. This case is shown in Figure 5.6 [3].

This structure had small localized conduction paths through the metal–oxide: NiO where filaments were formed that connect M1 and M2. The rest of the NiO remained in the high-resistance state (HRS). The concentrated electric field near the bottom corners of the M2 effectively confined the conduction path, resulting in improved uniformity of resistance values. Due to the filamentary conduction property of the metal–oxide, it was important to confine the conduction path of the current in a transition metal–oxide (TMO) ReRAM cell because it affected the uniformity of the operating voltage and current as well as the distribution of the LRS. The confined memory cells showed lower RESET current (about 90 μA), larger ON/OFF current ratio, and improved LRS uniformity compared with a conventional stacked, non-confined structure.

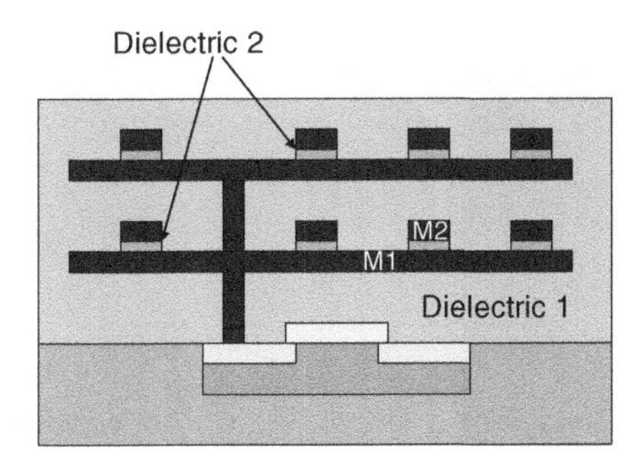

Figure 5.5 Multibit, multilayer cross-point ReRAM structure using one transistor. (Based on B. Lee and H.S.P. Wong, (Stanford University), VLSI Technology Symposium, June 2009 [3].)

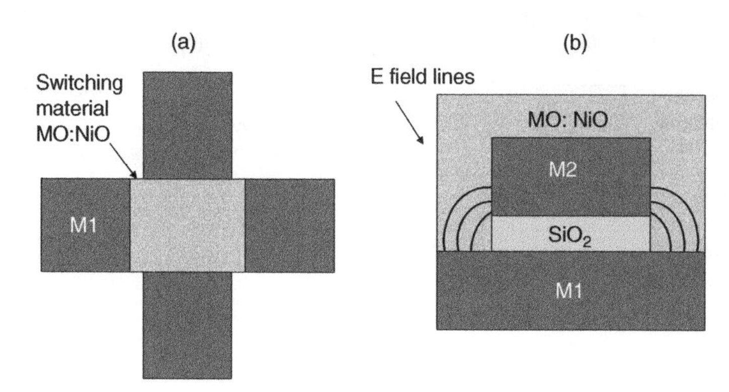

Figure 5.6 Cross-point element with insulator and switching material exchanged so that the SiO_2 insulator is between the metal lines, and the metal–oxide switching material is covering the M2: (a) top view; and (b) cross-sectional view showing the E field lines. (Based on B. Lee and H.S. P. Wong, (Stanford University), VLSI Technology Symposium, June 2009 [3].)

5.2.3 Methods of Stacking Cross-Point Arrays

Cross-point architectures can be vertical or horizontal. A cross-point array using an ReRAM with a NiO switching element was discussed by Samsung in June of 2009 [4]. The ReRAM array used a vertical cross-point architecture. The vertical cross-point architecture could increase memory density in the lateral chip area. It could also reduce costs by processing multilayer cell stacks simultaneously. Examples are shown in Figure 5.7 [4] of (a) horizontal cross-point architecture and (a) vertical cross-point architecture.

The ReRAM cell was formed in the vertical cross-point architecture, as shown in Figure 5.8, where two ReRAM cells were formed one on each side of a vertical electrode [4]. These cells had a ReRAM element and a diode. Both unipolar and bipolar switching mode were demonstrated in the same cell.

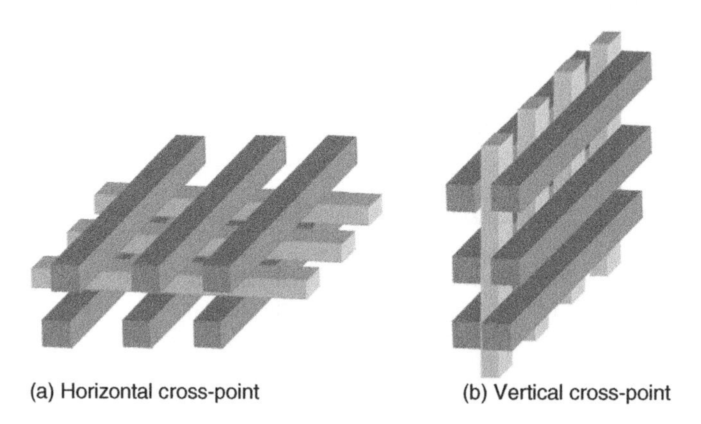

Figure 5.7 Examples of cross-point array architecture: (a) horizontal; and (b) vertical. (Based on H.S. Yoon, (Samsung), VLSI Technology Symposium, June 2009 [4].)

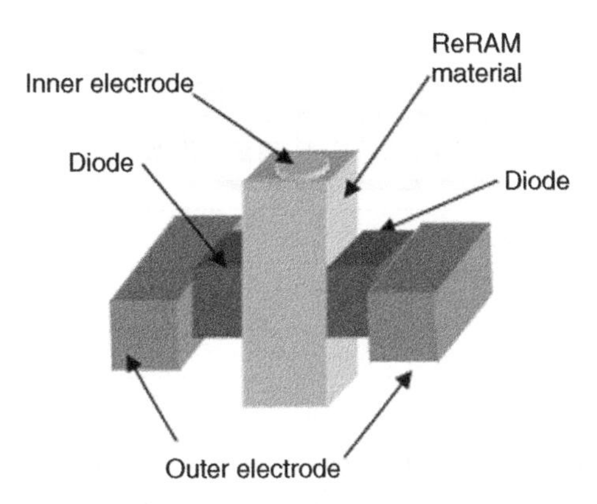

Figure 5.8 Illustration of two ReRAMs in the vertical cross-point architecture. (Based on H.S. Yoon, (Samsung), VLSI Technology Symposium, June 2009 [4].)

The NiO ReRAM cell was modeled as a filamentary conduction cell that was ON or SET in the LRS when the filament was formed across the NiO dielectric between the two conductive plates and OFF or RESET in the HRS when the filament was broken [4]. These memory cells showed low RESET currents of about 90 μA, large ON/OFF ratios, and improved uniformity compared with a conventional structure. The cell size in this early technology was $48 \times 48\,nm^2$ [4].

The size of a ReRAM cell has been shown to be scalable to $10 \times 10\,nm^2$ or less in cross-section depending on the cell-switching mechanism. In December of 2011, IMEC and KU Leuven reported on a $10 \times 10\,nm^2$ HfO_2-based ReRAM cell with an Hf–HfO_x resistive element stack [5]. It was operated in bipolar mode. Switching energy per bit was <0.1 pJ, and endurance was more than 5×10^7 cycles.

5.2.4 Stacking Cross-Point Layers for High Density

A 3D cross-point cell ReRAM array that stacked eight layers of metal cross-point cells was reported in July of 2009 by Fudan University [6]. The density was over 260% higher than a single-layer one transistor–one resistor (1T1R) cell. An operational algorithm was used to inhibit miswrite and misread caused by sneak current when no access device was present. This was expected to reduce power consumption. The cost was lowered by eliminating the select device with every cell.

Another early cross-point memory cell with multiple layer stacking using ReRAM technology was discussed in February of 2010 by Unity Semiconductor, now Rambus Semiconductor [7]. A 64Mb test chip was built in 130 nm technology with a $0.17\,\mu m^2$ cell size. Sensing current was less than 100 nA, and programming current was in the μA range. Cross-point array selection was demonstrated. Unity's "CMOx" process cross-point memory layers were formed on top of a completed CMOS wafer. A schematic cross-section of the array is shown in Figure 5.9 [7]. Five metal layers are used to create four memory layers. The metal word-lines are out of the page. If a memory layer was being read, the adjacent memory layer above or below it would be affected by the READ. A word-line was shared by two memory layers so it would load bit-lines on both layers.

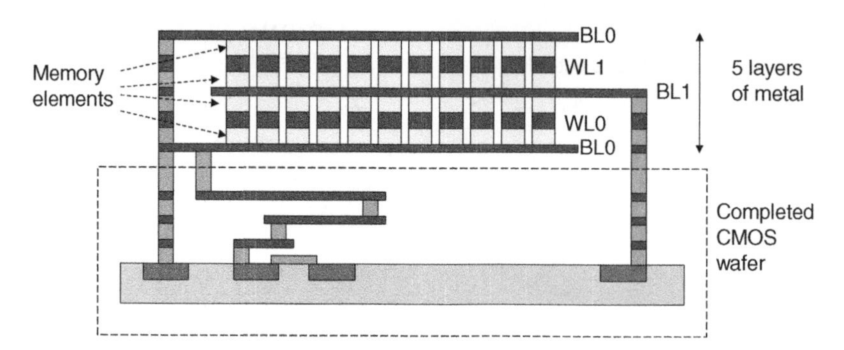

Figure 5.9 Schematic cross-section of the memory array using five added metal layers above the completed CMOS wafer. (Based on C.J. Chevallier *et al.*, (Unity Semiconductor), ISSCC, February 2010 [7].)

Half-voltage selection of the memory cell was used for program or erase. For programming, a positive voltage was applied on the top electrode of the cells along a word-line, while a negative voltage was applied to the bottom electrode of the cells along one or more bit-lines. The cells at the intersection of a word-line and a bit-line received the full programming voltage, while other cells on the selected line saw half the programming voltage. Other cells were grounded. For erase, the voltage polarity was reversed. Writing used two cycles for erase and program. It was not necessary to erase an entire block prior to programming pages, which resulted in improved write throughput. Individual pages could be erased. Program data was stored in one latch, while indication of whether a particular cell should be modified was stored in another. A different selection technique was used for a read. Leakage during read was addressed but only partially resolved for a large array. The ReRAM cross-bar array was integrated on top of the CMOS logic layers to reduce the area overhead of the ReRAM peripheral control circuits and the CMOS logic.

5.2.5 An Example of Unipolar ReRAM

ReRAMs exist as both bipolar and unipolar memories. Both bipolar and unipolar ReRAMs can be used in cross-point arrays. A Bipolar ReRAM switches using both positive and negative voltage where one polarity is the SET or ON state, and the other polarity is the RESET or OFF state. The unipolar ReRAM switches to both the high- and low-resistance states using only one polarity—positive or negative.

A typical unipolar NiO ReRAM switching current–voltage (I–V) curve is shown in Figure 5.10 [4]. The switching bias for a unipolar ReRAM for the SET–ON state and for the RESET–OFF state uses, in this case, only positive applied voltage. The current is constrained as the voltage is increased to produce the SET state, or the LRS, and the current is unconstrained as voltage is increased to produce the RESET state, or the HRS. For reading, all cells on the selected vertical word-line can be read at one time using the same number of current-sensing circuits as horizontal bit-lines [4].

A vertical cross-bar ReRAM was demonstrated exhibiting unipolar switching with a NiO dielectric. The cell architecture was modeled without diodes in 35 nm design rules with a $4F^2$

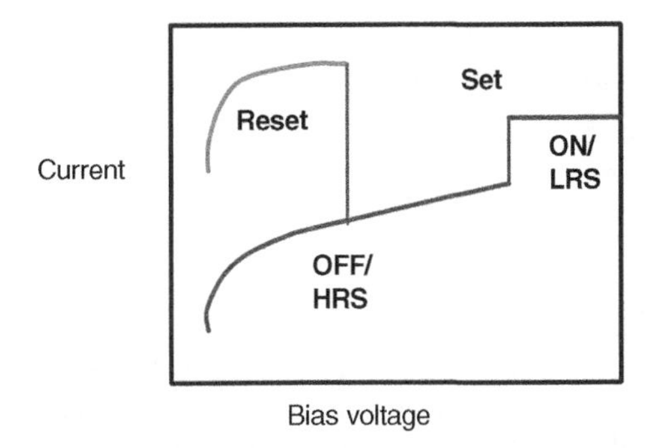

Figure 5.10 Typical unipolar ReRAM switching I–V curve showing both SET and RESET with positive bias voltage. The ON/LRS and OFF/HRS states are indicated. (Based on H.S. Yoon, (Samsung), VLSI Technology Symposium, June 2009 [4].)

cell that had 16 stacked layers with 32 horizontal lines and 512 vertical lines, resulting in a memory density of over 500Gb.

5.2.6 An Example of a Bipolar ReRAM

A ReRAM element can be bipolar; that is, it can switch in both positive and negative directions. An example of bipolar metal–oxide ReRAM device operation characteristics was described in May 2012 by SEMATECH, Stanford, CNSE, and the University of Albany and are shown in Figure 5.11 [8]. Initially, a conductive filament was formed with a limited applied current, called the compliance current, which determined the diameter of the filament. After the initial "forming" operation, the device cycled between a low-resistance ON(SET) state and a high-resistance OFF(RESET) state of the opposite voltage polarity. Filament forming is shown in

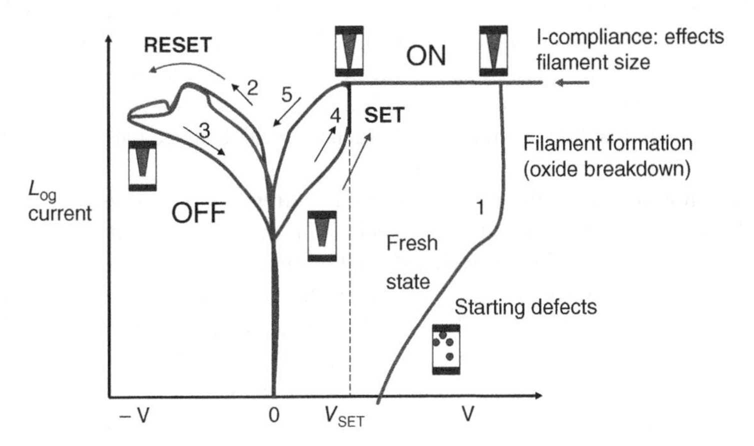

Figure 5.11 DC bipolar metal–oxide ReRAM device operation I–V curve showing SET at a positive bias and RESET at a negative bias. Indications of the state of the conductive filament at each position are also shown. (Based on D.C. Gilmer *et al.*, (SEMATECH, Stanford University, CNSE University of Albany), IMW, 20 May 2012 [8].)

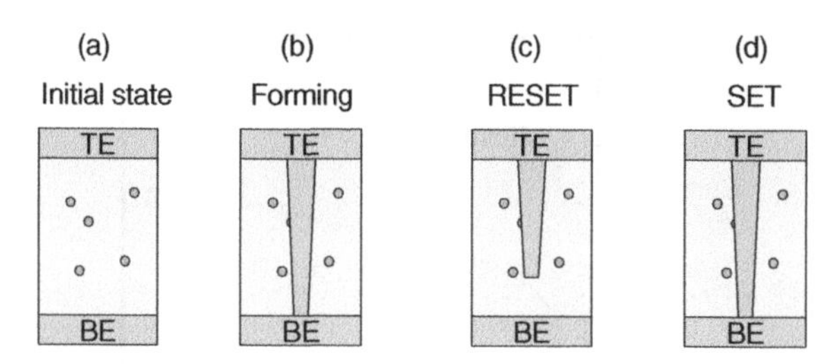

Figure 5.12 Illustration of programmed states in a bipolar ReRAM element: (a) starting defects in the dielectric enable filament formation; (b) during forming, a voltage is applied across the element and a conductive filament forms that shorts the top and bottom electrodes of the ReRAM, with diameter determined by the compliance current; (c) during a RESET operation done by reversing the direction of the applied voltage, a gap forms in the filament, which increases the resistance because the current must now cross the dielectric in the gap, and a high-resistance OFF state results; (d) during SET operation, a forward voltage across the element reforms the filament across the gap, resulting in a low-resistance ON state. (Based on D.C. Gilmer *et al.*, (SEMATECH, Stanford University, CNSE University of Albany), IMW, 20 May 2012 [8].)

the I–V curve in Figure 5.11 [8] along with subsequent cycling between ON(LRS) and OFF (HRS) states. This figure shows an ohmic (linear) LRS(ON) and a non-ohmic (nonlinear) HRS (OFF). These are related to a formed (SET) conductive filament and its rupture (RESET). The filament was formed in this case due to dielectric breakdown.

A schematic illustration of the programmed states of a bipolar ReRAM element is shown in Figure 5.12 [8], where a conductive filament is formed in the dielectric between the top electrode (TE) and the bottom electrode (BE) of a ReRAM element.

5.2.7 Basic Cross-Point Array Operation with a Diode Selector

An illustration of the operation of a cross-point array with a ReRAM and a diode selector was given in February of 2013 by Peking University [9]. This ReRAM was unipolar, as it operated with only positive voltage. This positive voltage permitted a conventional diode to be used in series with the ReRAM to avoid sneak current leakage. The analysis was intended to address issues caused by circuit and device interaction, such as sneak leakage paths in the cross-point array, that could degrade the array performance.

Simulations showed that while a large ON/OFF current ratio of states for the ReRAM resulted in large readout margin and was essential for an array without selector devices, a selector connected in series with the ReRAM could eliminate the requirement for high ON-state resistance (R_{on}). The readout margin was found to be sensitive to the nonlinearity of the I–V characteristics of R_{on}, which was the LRS. Nonlinearity in the I–V characteristics of the selector device helped provide a larger readout margin. The fitted I–V curves from a ReRAM with a series diode selector based on the device model are shown in Figure 5.13(a) for the ReRAM and Figure 5.13(b) for the diode selector. The simulated I–V curves of the combined one diode–one resistor (1D1R) memory cell are shown in Figure 5.13(c) [9]. It is clear that the current of the LRS was strongly constrained by the selector device in the one selector–one resistor (1S1R) cell compared to the case of the ReRAM I–V curve without the diode selector. In Figure 5.13(c), it can be seen that both the LRS and HRS of the selected memory element at a

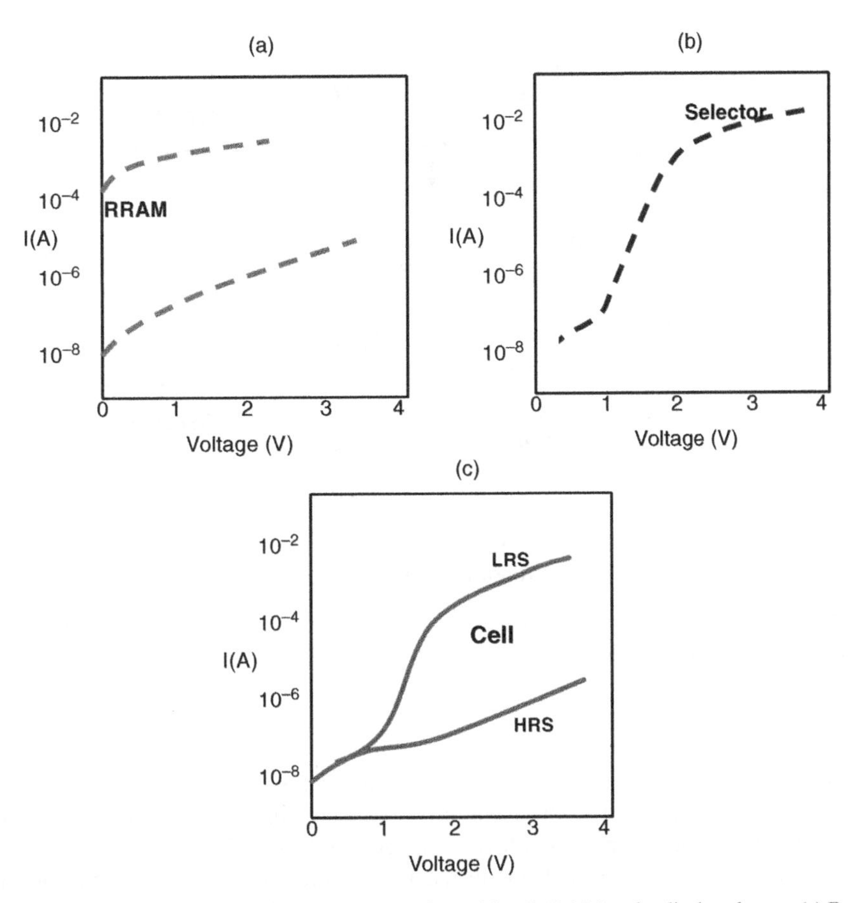

Figure 5.13 Illustration of cross-point array operation with a ReRAM and a diode selector: (a) ReRAM I–V curves, (b) diode selector I–V curve; (c) combined I–V curve of 1D+1R cell. (Based on Y. Deng, *et al.* (2013) (Peking University), *IEEE Transactions on Electron Devices*, 60(2), 719 [9].)

given read voltage have lower resistance than the OFF state of unselected devices with lower applied voltage. This rectification effect permits control of sneak leakage paths in unselected and half-selected cells in large cross-point array stacks with unipolar ReRAMs and select diodes.

5.2.8 Early Test Chip Using a ReRAM Cross-Point Array with Diode Selector

A full test chip of a metal–oxide 4GB ReRAM in 24 nm technology was shown in February of 2013 by Toshiba and Sandisk [10]. The two-terminal cell used a diode selector in a memory array that was stacked above CMOS peripheral circuitry. The memory array was divided into 16 bays, each with 128 blocks divided into 4 stripes. Small blocks were used to reduce the effect of parasitic resistance in the word-lines and bit-lines. The word-line drivers were placed under the array. The block size was 2 K bit-lines by 4 K word-lines for each

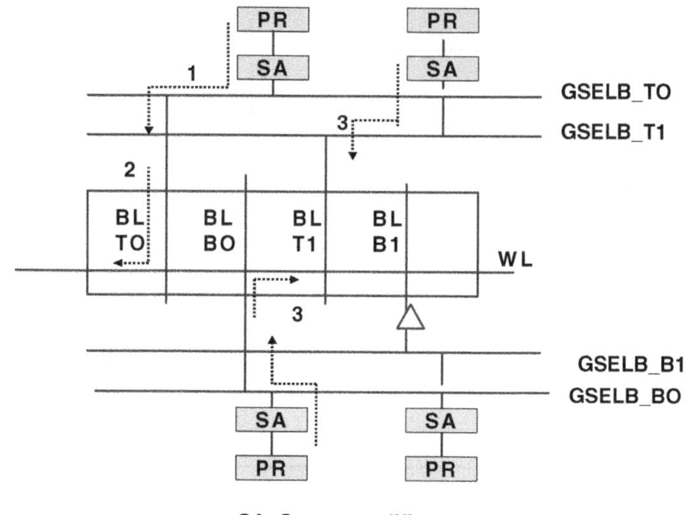

Figure 5.14 Pipelined array control during the write operation. (Based on T.Y. Liu *et al.*, (Sandisk, Toshiba), ISSCC, February 2013 [10].)

layer. The area efficiency was high because much of the peripheral circuitry was under the memory array including array control circuitry, sense amplifier, page buffer, and voltage regulator drivers. When a pair of blocks was selected, all 64 sense amplifiers in the associated bay were activated and connected to the selected blocks. The bit-lines were multiplexed to the bus and then connected to the sense amplifiers in other blocks according to the decoding [10].

Array control was pipelined during the write operation. A pipelined word-line–bit-line control reduced address switching overhead caused by the bit-line charge and discharge times. As the selected cells driven from one side of the block were being written, the data for the next write cycle was read from the page buffer. When the word-line pulse ended, the bit-lines driven from the other side were charged quickly for the next write pulse. This hid the time to fetch the next set of data and to charge the select lines. The pipelined array control is shown in the following figures. A schematic of the array control circuit diagram is shown in Figure 5.14 [10], and the corresponding timing diagram is shown in Figure 5.15 [10].

During the read operation, the selected word-line was biased at one voltage while the unselected word-lines and bit-lines were biased so that the unselected memory cells had no leakage current along the selected path. Read latency was 40 μs with a sensing cycle of 1.5 μs. Write latency was 230 μs.

A 32Gb cross-point array was made in 24 nm technology. The cell size was $24 \times 24\ \text{nm}^2$, and the chip size was $130.7\ \text{mm}^2$. A NAND-compatible interface was used with a page size of 2KB.

The chip achieved high memory density by putting shared and distributed control circuitry under the memory array. The impact of the large block size was reduced by the pipelined array control, the bias compensation, and charge pump efficiency optimization [10].

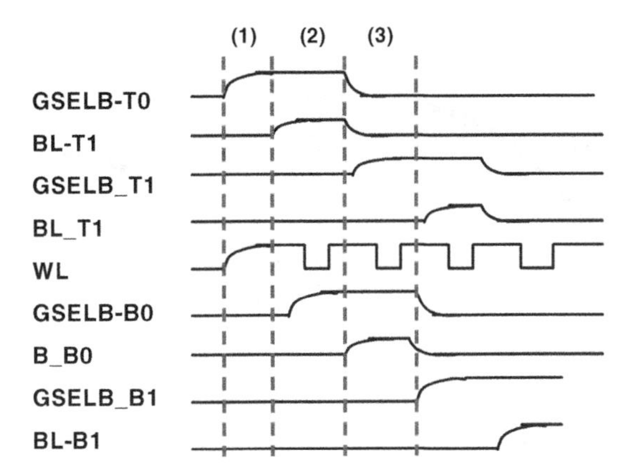

Figure 5.15 Timing diagram for pipelined array control during the write operation. (Based on T.Y. Liu *et al.*, (Sandisk, Toshiba), ISSCC, February 2013 [10].)

5.3 Low-Resistance Interconnects for Cross-Point Arrays

5.3.1 Model of Low Resistance Interconnects for Cross-Point Arrays

In order for cross-point array memories to reach high densities, the interconnects must be scalable to minimal dimensions and the parasitic resistance of the interconnects must be very small. In May of 2012, Stanford University discussed cross-point resistive memory array scaling [11]. The impact of Cu interconnect scaling on write–read margin, energy dissipation, speed, and reliability were checked for wire sizes down to the sub–10 nm node.

The resistivity increase due to wire scaling resulted in degraded write and read windows, increased interconnect energy, and increased wire latency. The current density increase required for programming caused increases in Cu electromigration. Performance degradations were dependent on memory device parameters and memory array size. Results indicated that if the on-resistance (R_{on}) were less than 100 KΩ and array size greater than 1Mb, then write margin was less than 55%, read margin less than 5%, and wire energy greater than 1 pJ for wire size smaller than 20 nm.

A trade-off was found between the size of R_{on} and the speed. A large R_{on} value could tolerate a small R_{off}–R_{on} ratio, but a too-high R_{on} could result in slow speed. A careful device and interconnect co-optimization was found to be required to meet performance specifications for cross-point memory arrays at the sub–10 nm nodes.

A quantitative analysis was done of the effect of the size of the Cu wire on write and read margin, energy dissipation, speed, and reliability of the cross-point memory array. This analysis was done for wire sizes below 10 nm. Significant degradation of array performance with wire scaling and dependencies on memory device parameters and array sizes, which change with wire size, indicated that a new design methodology of cross-point resistive memory was needed to reduce the wire-scaling effects. Possible solutions include using wire materials with better conductivity and scalability, such as graphene and carbon nanotubes (CNTs), using memory arrays with smaller partition sizes, strapping the bit-lines and word-lines with wider wires, and using memory elements with larger resistance and resistance ratios. The cross-point memory array was modeled as a passive mxn resistor network with bistable

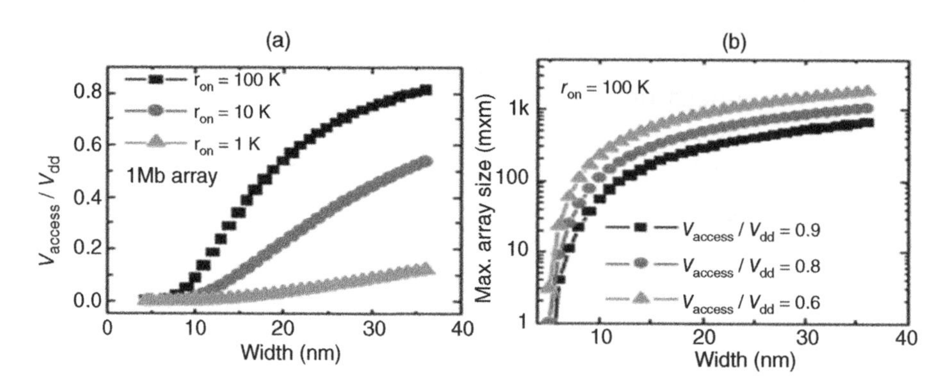

Figure 5.16 (a) V_{access}/V_{dd} degradation with wire scaling for various R_{on} values in an $m \times m$ 1Mb array; (b) maximum array size vs. width for various V_{access}/V_{dd}. (J. Liang *et al.*, (Stanford University), IMW, May 2012 [11], with permission of IEEE.)

switching material at either R_{on} or R_{off}. The memory array was assumed driven from one end with the cell that suffers the largest parasitic wire resistance located at the farthest corner of the array. For the write operation, a $V_{dd}/2$ write method was used. The worst-case array pattern was found when all cells were at R_{on}, which resulted in the smallest accessed voltage. A larger array suffered a larger degradation than a small array.

The degradation of V_{access}/V_{dd} with wire scaling for various R_{on} values in an mxm 1Mb array is shown in Figure 5.16 [11]. Figure 5.16(a) shows that the write margin depended strongly on the value of R_{on}. The smaller the R_{on}, the worse the V_{access}/V_{dd} signal level due to the increased voltage division between the wire and the memory cell on-resistance as R_{on} decreased. This meant that to write successfully in a scaled array, the array either needed to be operated at higher V_{dd} and be divided into smaller memory partitions, or the bit-line/word-line must be strapped by a wider metal layer that would decrease array efficiency and increase chip size. Figure 5.16(b) shows the maximum allowable memory partition size for an mxm partition or the longest distance between vias for a metal strap as the wire shrinks for different V_{access}/V_{dd}. The conclusion was that a memory partition smaller than 230×230 or a via distance less than 230 cells long was needed to ensure V_{access}/V_{dd} greater than 0.6 at the 10 nm node.

For the read operation, V_{th} was supplied on the selected word-line, and all other word-lines and bit-lines were grounded. This enabled parallel readout of all bits on the same row. Current difference was sensed by sense amps connected in series with each bit-line. A strong reduction in read margin ($\Delta V/V_r$) with wire size was observed. The two sources of degradation were found: one was the increase of copper wire resistance with scaling, and the other was the increased voltage loss at a larger read current leading to read voltage being smaller when reading R_{on} than reading R_{off}.

Two main sources of energy dissipation in wires were considered. Dynamic energy is dissipated when the word-lines and bit-lines are charged and discharged. Static energy is consumed when the memory cell is read or written. Typically, the write operation consumes more energy. The energy of a 1Mb square cross-point memory array was calculated and is shown in Figure 5.17 [11].

The static write energy depends strongly on the memory R_{on} values and on the programming time. Static wire energy can become significant for a smaller R_{on} and for longer times. The

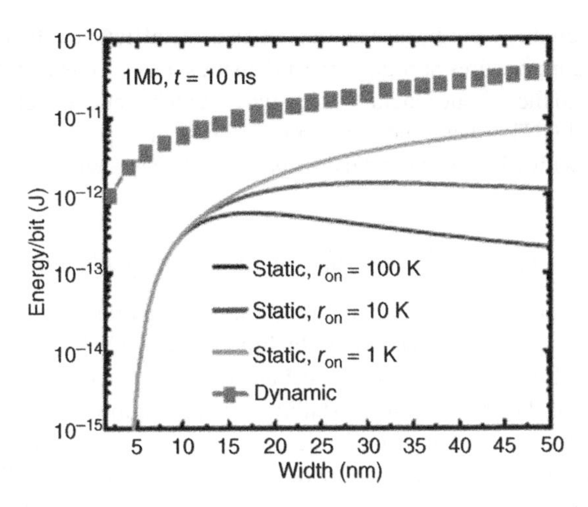

Figure 5.17 Energy/bit of a 1Mb square cross-point array vs. array width. (J. Liang *et al.*, (Stanford University), IMW, May 2012 [11], with permission of IEEE.)

sharp decrease in static energy at very low widths is a result of significant degradation in write margin and so cannot be used.

The latency of copper wire was also considered, as it increases significantly with scaling. The larger the memory array and the smaller the wire width, the longer the wire delay. Wire latency was found to increase significantly as the wire width scaled down, as shown in Figure 5.18 [11] approaching 100 ns at <5 nm width for a 1Mb cross-point memory array. This latency is greater than that of various memory elements being considered for cross-point arrays. For example, the latency of PCM was reported at 30 ns, conductive bridge (CB) random access memory (RAM) at 5 ns, spin-transfer torque (STT) RAM at 2 ns, and ReRAM at 0.3 ns. It was suggested that strapping the bit-line and word-lines with wider wires might

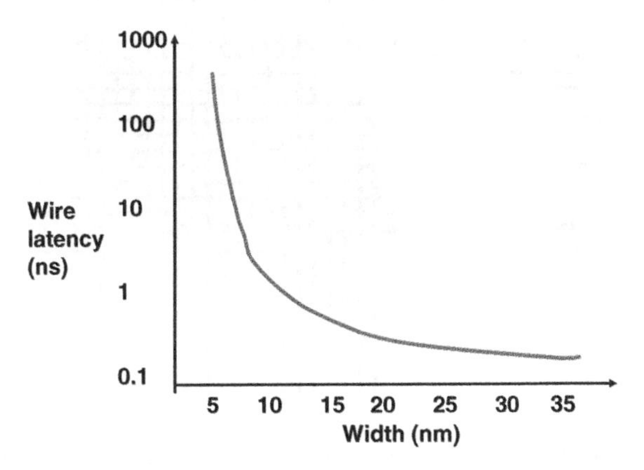

Figure 5.18 Wire latency vs. wire width for copper wire in a 1Mb cross-point array. (Based on J. Liang *et al.*, (Stanford University), IMW, June 2012 [11].)

be a solution, although this would cause an impact on array efficiency. In addition to latency, electromigration in copper wires at high current densities needs to be considered as an issue.

Possible solutions to these issues include using wires with better conductivity and scalability such as graphene and CNTs, using memory arrays with smaller partition sizes, strapping bit-lines and word-lines with wider wires and using memory elements with larger resistance values and ratios.

5.3.2 A Cross-Point Array Grid with Low-Resistivity Nanowires

An issue for cross-point arrays is achieving a tight nanoscale cross-point grid with sublithographic pitch with low-resistivity nanowires. In July of 2011, CEA-LETI, the University of Milano-Bicocca, and Ecole Polytech. Fed. de Lausanne discussed a multispacer patterning technique for making layers of polycrystalline silicon nanowires with sublithographic pitch [12]. The process exclusively used micron resolution and CMOS processing steps. Single spacers operated as polysilicon nanowire field effect transistors (FETs). The capability was also shown to lay a spacer perpendicular to a set of parallel spacers in a cross-point organization. The cross-point density was extrapolated from a 4×1 array to $10^{10}\,cm^{-2}$. The efforts were intended as a framework for a nanowire cross-point array where data could be stored at the cross-points and decoders were provided for on the rows outside the array. A baseline architecture for the array with decoders is shown in Figure 5.19 [12].

5.3.3 A Cross-Point Array Using Two Nickel Core Nanowires

A cross-point ReRAM memory array made of two perpendicular Ni core nanowires with a NiO shell was discussed by the Politecnico di Milano and Lawrence Berkeley Labs in September of

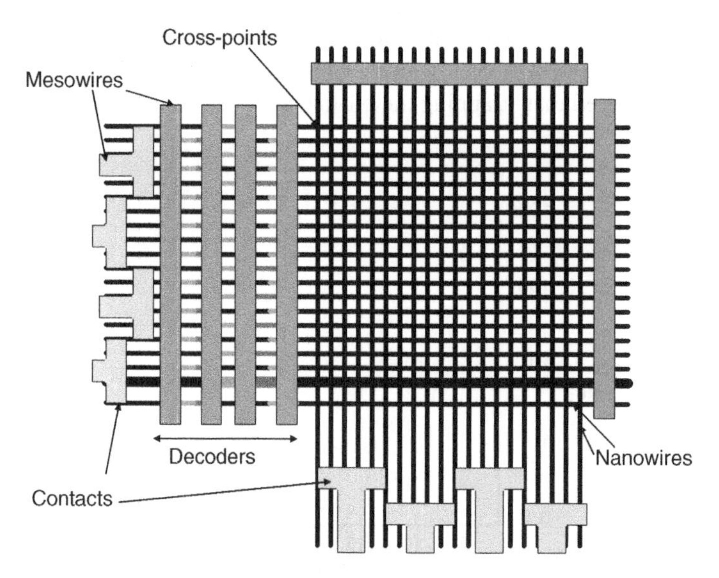

Figure 5.19 Baseline architecture for cross-point array with decoders. (Based on M.H.B. Jamaa *et al.* (2011) (CEA-LETI, University of Milano-Bicocca, Ecole Polytech. Fed. de Lausanne EPFL), *IEEE Transactions on Nanotech*, 10(4), 891 [12].)

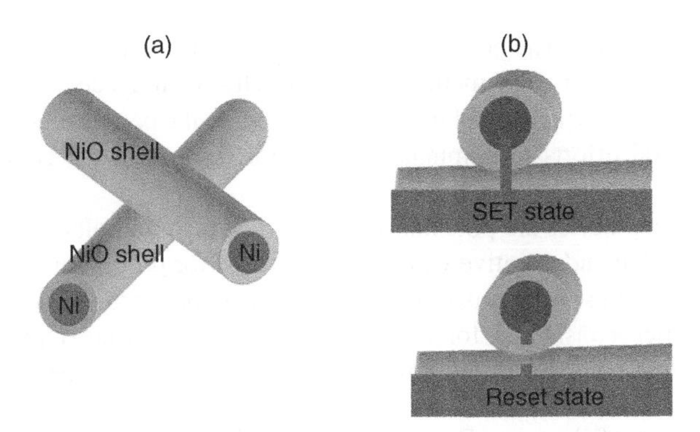

Figure 5.20 Unipolar switching at cross-point junction between nanowires: (a) crossed NiO–Ni nanowires; and (b) showing the low-resistance SET state when the nanowires are connected and the high-resistance RESET state when the nanowires are disconnected. (Based on C. Cagli *et al.*, (Politecnico di Milano, Lawrence Berkeley National Laboratory), ESSDERC, 12 September 2011 [13].)

2011 [13]. Reversible formation and disruption of a conductive filament a few nanometers in size occurred within an insulating metal–oxide, such as NiO, TiO, or HfOx. The density of the array could approach a terrabit/cm^2 (Tb/cm^2). The array formation was by self-assembling nanostructures. Unipolar resistance switching occurred in the NiO shell at the cross-point junction between nanowires, as shown in Figure 5.20 [13]. In this core–shell Ni–NiO nanowire, the metallic core is the top–bottom contact and the interconnect, and the NiO-oxide shell is the active switching layer. Electrical characterization showed unipolar switching. If a half-pitch F is equal to the nanowire size and is assumed to be 10 nm, the cell area for a 4F^2 crossbar array would be 400 nm^2. This corresponds to a density of 0.25 Tb/cm^2, if a single memory layer is assumed with one bit per cell. A single nanowire crossbar ReRAM based on Ni–NiO was characterized, and unipolar SET–RESET switching at the crossing was demonstrated [13].

5.3.4 Resistive Memory Using Single-Wall Carbon Nanotubes

An AlOx-based resistive switching memory using single-wall carbon nanotubes (SWCNTs) as contact electrodes was discussed in June of 2012 by Stanford University and Hong Kong University of Science and Technology. The metallic CNT electrode permitted the conductive filament of the ReRAM to be confined to a 1–2 nm scale, which provided better uniformity and lower programming current. The CNT also provided significantly lower bit-line and word-line resistance for the memory array compared with copper for <10 nm dimensions. In addition, the CNT had high electromigration immunity. The Al–AlOx–CNT device switched over 10^4 cycles with less than 5 µA programming current. The device could be scaled to 6×6 nm^2 using a CNT–AlOx–CNT cross-point structure. Retention was 10^5 s, and read immunity was 10^5 cycles. The device switched successfully up to 10^4 cycles with less than 10 µA switching current [14].

5.4 Cross-Point Array Memories Without Cell Selectors

If cross-point array memories are to be successful as the future of NAND flash, then they need to be as dense as, have as high memory capacity as, and follow the cost curve currently occupied by the NAND flash. The process complexity and cost could potentially be reduced by not using cell selectors with each memory element. This would need to be accomplished without reducing the array efficiency of the chip or degrading the parameters or reliability. This section reviews several efforts to eliminate the selector devices in the cross-point memory array.

Bipolar resistance memories are potentially more robust than unipolar resistance memories because both the positive and negative voltages can be used for programming and erase. They are also more difficult to use with a two-terminal unidirectional diode switch. For this reason many of the memories considered for use without cell selectors are bipolar switching. A resistance memory that switches through the origin is also called a "memristor."

5.4.1 Early Model of Bipolar Resistive Switch in Selectorless Cross-Point Array

The advantages and limitations of passive cross-point arrays at the nanoscale level were reviewed in 2007 by the University of Aachen [15]. These structures are regular and dense but due to their passive behavior require active devices to restore signal levels. Advantages of cross-point array structures include their inherent potential for $4F^2$ cell size, their scalability to nanometer dimensions, and their potential for stackability. A drawback of passive cross-point arrays is parasitic conducting paths, which can disturb read operation and can require large sensing circuits that reduce the density advantage of the small cell size. Another drawback is the reduction in area of the cross-point at nanoscale geometries, which reduces the maximum cell size that determines the reliable distinction between two states [15].

High-performance CMOS read and write circuits with small area and low power consumption are essential for large cross-point arrays. Write of the resistive memory at the cross-point occurs with switching between a high- and low-resistance state. A model of the I–V characteristics for writing to a bipolar resistive cross-point memory is shown in Figure 5.21 [15]. When a positive voltage is applied across the resistive material, the material switches to a low-resistive state, R_{on}, and remains in that state when the voltage is removed. This is defined as a write 0. When a

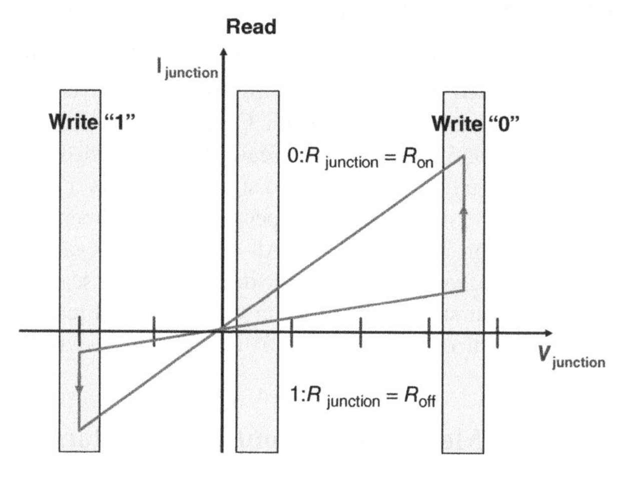

Figure 5.21 Model of I–V characteristics of bipolar resistive switch. (Based on A. Flocke, T.G. Noll, (Aachen University), ESSCIRC, 2007 [15].)

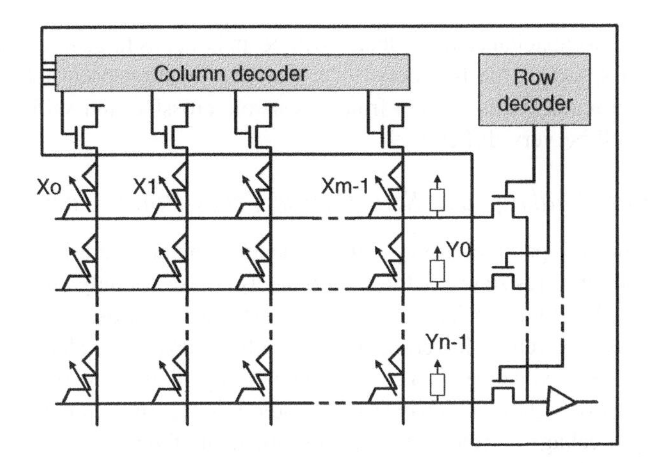

Figure 5.22 Basic configuration of cross-point array read circuitry with CMOS periphery. (Based on A. Flocke, T.G. Noll, (Aachen University), ESSCIRC, 2007 [15].)

negative voltage is applied across the resistive material, it switches to a nonvolatile high-resistive state, R_{off}, and is referred to as write 1. To read, a positive voltage that is smaller than the write voltage is applied to the cross-point, and the current flowing through the junction is determined. It is assumed in this model that the I–V curve is linear and goes through the origin [15].

The basic configuration of read circuitry for a cross-point array with CMOS periphery is shown in Figure 5.22 [15]. This includes a column and a row decoder for an mxn array.

For an ideal read operation with no parasitic sneak current paths, the read operation can be represented as

$$\frac{\Delta V}{V_{pu}} = \frac{V_{out,hi}}{V_{pu}} - \frac{V_{out,lo}}{V_{pu}} = \frac{1}{1+\sqrt{\frac{R_{on}}{R_{off}}}} - \frac{1}{1+\sqrt{\frac{R_{off}}{R_{on}}}}$$

where $\Delta V =$ is the maximum voltage swing, and V_{pu} is the pull-up voltage for

$$R_{pu} = \sqrt{R_{on} - R_{off}}$$

If parasitic sneak current paths are present, then the worst cases for reading the crossbar are:

1. reading an R_{off} junction with all non-accessed junctions set to R_{on}, and
2. reading an R_{on} junction with all non-accessed junctions set to R_{off}.

There were several conclusions regarding this analysis. Small voltage-level differences and resulting small sense margins were expected to require the use of sense amplifiers even for small cross-point arrays such as 8 × 8. For reasonably large cross-point arrays, it was found that an R_{off}/R_{on} ratio of several orders of magnitude was necessary. Defective junctions were expected to require fault-tolerance techniques that would add additional logic in the CMOS periphery. Resistance–capacitance (RC) delays will be an issue, as large R_{on} or R_{off} values will reduce the readout speed. If the I–V curve of a resistive junction does not show nonlinear characteristics, then reading information from a resistive crossbar and shifting it to CMOS-compliant levels will be very difficult.

5.4.2 Sneak Path Leakage in a Selectorless Cross-Point Array

Intrinsic drawbacks of a passive cross-point architecture include parasitic leakage paths in unselected cells and series resistance of the interconnects. These not only limit the signal swing but also eliminate the high-density advantage of the stacked cross-point architecture by requiring large sensing circuits, lowering the array efficiency, and reducing the maximum allowable array size. The advantage of a cross-point memory architecture without cell selectors is that it will permit word-lines and bit-lines to be laid out in the minimum metal pitch allowed by the technology. Leakage currents must be controlled, and the peripheral circuitry must be capable of accessing the memory elements in the dense pitch. The cross-point memory architecture tends to suffer from large sneak path leakages that result in high power dissipation and small sensing margin.

Sneak path leakage for the read operation for a cross-point array with and without an access device was discussed by Samsung in December of 2007 [16]. Figure 5.23(a) [16] illustrates sneak path leakage during a read through the unselected cells in a 3D 2 × 2 array with no access device. The selected word-line is at $V = V_{read}$, and the selected bit-line is at $V = 0$. The selected cell is in the HRS. Leakage is through cells in the LRS. Figure 5.23(b) shows the rectified read operation with an access device in the 2 × 2 array. The cells, which are switched off by the diode in line with the memory element, eliminate the leakage.

The parasitic series resistance of the interconnects can also degrade the output signal and imposes an additional limitation on the maximum allowable single-layer array density. As the process scales, the diameter of the interconnects is reduced. Because $R = \rho\, l/d$, where d is the diameter of the interconnect, ρ is the resistivity, and l is the length of the interconnect, as d is

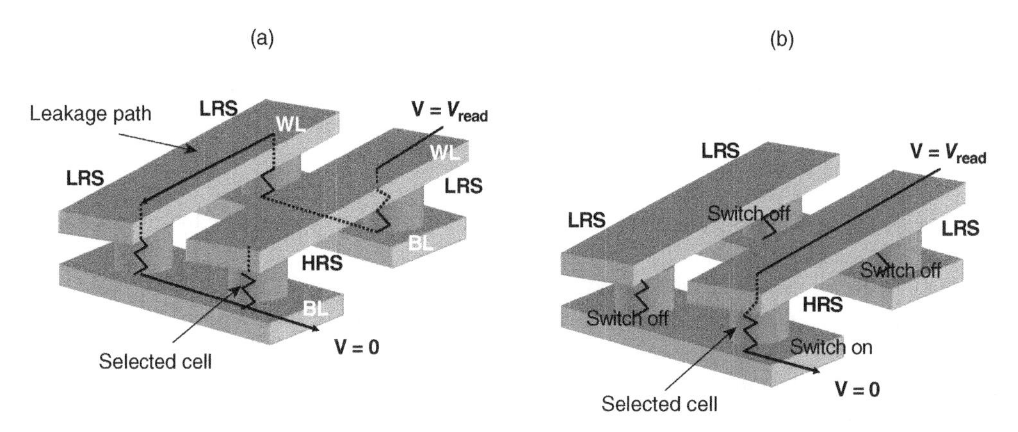

Figure 5.23 Read operation for (a) 2 × 2 array with no access device showing current leakage through the nonselected cells; and (b) with an access device showing the current leakage eliminated. (Based on M.J. Lee *et al.*, (Samsung), IEDM, December 2007 [16].)

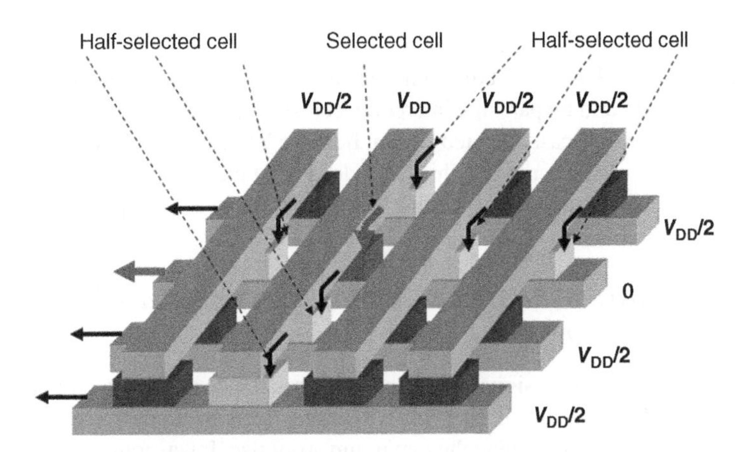

Figure 5.24 Cross-point memory architecture using bistable memory element. (Based on J. Liang and H.S.P. Wong (2010) (Stanford University), *IEEE Transactions on Electron Devices*, 57(10) [17].)

reduced, the resistance increases for constant length and resistivity. Because the usual rationale in scaling the array technology is to increase its capacity, it can be assumed that the length of the interconnect will not necessarily be decreased as the array is scaled.

Characteristics of cross-point memory arrays without cell selectors were discussed by Stanford University in June of 2010 and in October of 2010 [17,18]. The device requirements of a resistive cross-point array were studied under worst-case write and read conditions with focus on the data pattern dependence of the array. The cross-point memory architecture simulation was made with a bistable memory element. V_{dd} was applied to the word-line of the selected cell and 0 V to the bit-line. The unselected cells on the selected word-line and bit-line were all half-selected. All other unselected cells had $V_{dd}/2$ applied to both the word-line and the bit-line. Misprogramming and misreading can happen due to the sneak path leakage in the half-selected and unselected cells in a 2D cross-point array, as shown in Figure 5.24 [17].

During a write operation, an additional voltage drop in the interconnects caused by leakage current can result in the selected cell not receiving the voltage required for a successful write. During the read operation, parasitic conducting paths in unselected cells can degrade the output signal so it is difficult to distinguish the two memory states.

For a successful write operation to both the high- (V_{hrs}) and low- (V_{lrs}) resistive states, the switching threshold of the selected cell ($V_{t\text{-select}}$) must be larger than either the maximum of V_{hrs} or V_{lrs}. In addition, the voltage drop on the unselected cells ($V_{t\text{-unselect}}$) must be smaller than the minimum values of V_{hrs} or V_{lrs} so that the unselected cells are not misprogrammed. In the case illustrated in Figure 5.24 of a $V_{dd}/2$ writing method, the selected word-line is biased to V_{dd}, the selected bit-line is grounded, and all unselected word-lines and bit-lines are biased to $V_{dd}/2$. For the simulation, $V_{dd} = 1.8$ V and $V_{dd}/2 = 0.9$ V were used. Because $V_{t\text{-unselect}} = 1\text{ V} > 0.9$, misprogramming is excluded as long as the thresholds remain stable. $V_{hrs} = 1.5$ and $V_{lrs} = 1.0$, which are less than $V_{dd} = 1.8$ V, so write occurs to both states.

For the read operation, the selected word-line $= V_r$ and all other word-lines and bit-lines are grounded. This maximizes throughput by reading multiple outputs at one time. It has the disadvantage of increasing the power dissipation. For a nondestructive read, $V_r < V_{t\text{-unselect}} = 1.0$ V.

The current difference between READ "1" and READ "0" is sensed by connecting all bit-lines to sense amplifiers. The current difference is degraded by multiple parasitic current paths, which distort the signal, and by parasitic interconnect resistance. As a result, the signal-to-noise ratio is degraded. The worst-case selected cell for both the READ and WRITE operations is the one with the smallest voltage across it. This cell is located at the farthest corner from the word-line and bit-line voltage sources because the loss of voltage is caused primarily by parasitic resistance of the interconnects.

5.4.3 Effect of Parasitic Resistance on Maximum Size of a Selectorless Cross-Point Array

The large degradation of output signal due to parasitic resistance of the interconnects was also analyzed by Stanford University in 2010 [17,18]. The effects of memory cell resistance and resistance ratio were used to determine the maximum array size. It was found that the number of cells in the array could reach 10^6 with a signal swing greater than 50% of the read voltage when R_{on} is $>3\,M\Omega$ and $R_{off}/R_{on} > 2$. A large memory cell resistance value can reduce the power consumption and eliminate the requirement for a large resistance ratio which, in turn, can reduce the need for using cell-selection devices.

The parameter requirements of a resistive cross-point memory array were defined under worst-case write and read operations. The parasitic leakage paths in unselected cells and the series resistance of the interconnects decreases the signal swing. This reduced the advantage of the small array area by adding large sensing circuits and reducing the maximum array size. It was determined that memory cells with high on and off resistance values maximize array size and minimize power consumption. The high resistance values eliminated the requirement for a large R_{off}–R_{on} resistive ratio and avoided including select devices that would add to process complexity and cost. The bias scheme for a read operation is shown in Figure 5.25 [17,18].

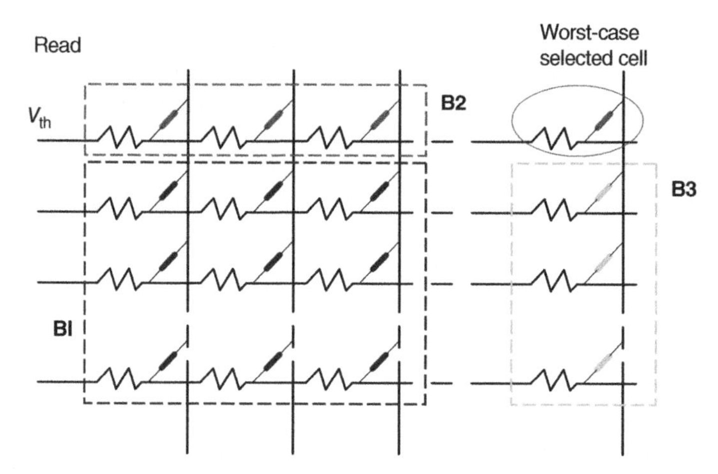

Figure 5.25 Bias for a read operation in a cross-point memory array without cell selection devices with the worst-case selected cell indicated. The array is divided into groups in which all elements share the same memory states. (Based on J. Liang and H.S.P. Wong (2010) (Stanford University), *IEEE Transactions on Electron Devices*, 57(10) [17].)

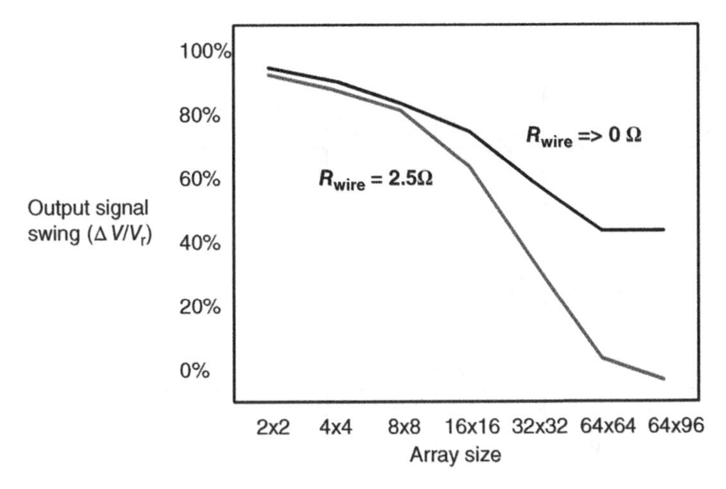

Figure 5.26 Output signal swing during a read operation as a function of array size and resistance of the wire. (Based on J. Liang and H.S.P. Wong (2010) (Stanford University), *IEEE Transactions on Electron Devices*, 57(10) [17].)

In Figure 5.25, the worst-case selected cell is indicated as the one farthest from the row/column voltage sources. Devices on the selected cell word-line are shown in group B2 and on the selected cell bit-line in group B3. These devices are half-selected. Unselected devices are in group B1. All devices in a group share the same memory state.

To read, the selected row is at V_r and all other rows and columns are grounded. The current difference is sensed by connecting the columns to a sense amplifier. The worst-case signal swing $\Delta V/V_r$ degrades with increase of array size. The worst-case array pattern is different for READ "0" and READ "1." It also differs for various reading methods and device parameters. In this instance, the worst-case pattern for read "1" is when the cells on the selected word-line (B1) are in HRS (R_{off}) and all other cells are at LRS (R_{on}). The worst-case pattern for read "0" is when all the unselected cells—B1, B2, B3—are at LRS (R_{on}).

The worst-case output signal swing, $\Delta V/V_r$, can be calculated, and it degrades with an increase in array size, as shown in Figure 5.26 [17]. Decreasing the resistance of the wire (R_{wire}) improves $\Delta V/V_r$ for large array size, as shown in this figure. If the resistance of the wire approaches zero, the $\Delta V/V_r$ becomes independent of the number of bit-lines in the array. This analysis shows the dependence of the signal during a read operation on the resistance of the wire and the array size.

Decreasing the resistance of the wire significantly improves output swing ($\Delta V/V_t$) for large array sizes. It was determined that when R_{on} was >3 MΩ and $R_{off}/R_{on} > 2$, the number of cells in the array could be as large as 10^6 with a signal swing of >50% of the read voltage. A small ratio of resistance states was found acceptable as long as the resistance values of both states were large [18].

For writing, the selected row is biased to V_{dd}, the selected column is grounded, and unselected rows and columns are biased at $V_{dd}/2$, as shown in Figure 5.27 [18]. An examination of pattern dependence of allowable array size in a cross-point structure without cell selectors was done. For the write operation, the cells were divided into three groups where B1 is the selected row, B2 is unselected rows and columns, and B3 is the selected column. All cell states

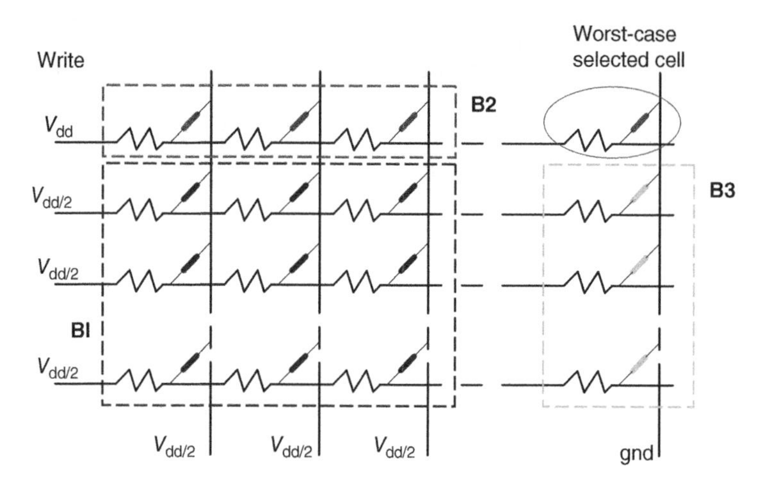

Figure 5.27 Bias for a write operation. (Based on J. Liang, H.S.P. Wong, (Stanford University), IITC, p. 1, 6 June 2010 [18].)

in one group are assumed to be the same but can be different from other groups. Figure 5.27 illustrates three worst-case bias conditions for a write.

A simulation was used to show the results of the voltage across the access cell of a 32 × 32 array. The worst-case pattern during the write operation was found to be when all cells were in the R_{on} LRS. This was because the voltage drop along the interconnects degraded most when all cells were at R_{on} due to the large current flow. The allowable array size for a successful write operation was calculated. The maximum array size for a successful write for the worst-case array pattern was determined as a function of the number of columns and rows. For a square array this was about 25 rows × 25 columns.

The previous analysis showed the array size limited due to sneak current leakage and parasitic resistance. A solution to this issue is to increase the resistance values of the memory elements because large resistance values reduce the impact of the parasitic resistance. The ratio of the two resistance states of the memory element is also important in determining maximum array size.

For the write operation, the maximum array size is only dependent on R_{on} in the worst case. The minimum requirement for R_{on} increases linearly with the number of memory elements. Increasing R_{on} from 1 kΩ to 3 MΩ increases the maximum number of cells in the array to 10^6. In this case it eliminates the requirement for a memory cell–selection device to cut off leakage paths. High resistance values reduce the power dissipation in the array. Power consumption as high as 0.1 W can occur as the number of cells in the array increases up to 10^6 with $R_{on} = 5$ kΩ. To reduce power consumption, the resistance value of the memory cells must be scaled up to make a large array feasible. High R_{on} can, however, degrade the write and read speeds of the circuit.

For the read operation, there are two competing mechanisms that arise from a change in R_{on}: increasing R_{on} increases the resistance value, which helps reduce the sneak path leakage and lowers the signal degradation caused by interconnect resistances. However, increasing R_{on} at a given R_{off} reduces the R_{off}–R_{on} resistance ratio, which reduces the output-signal swing. The resistance ratio therefore has a dominant effect on maximum array size for large resistance values. At large R_{off} the ratio between R_{off} and R_{on} can be degraded to 2× and still achieve a large sense margin. This reduces the requirements for a selection device but does not eliminate them due to the effects of signal-to-noise ratio (SNR) and sense amplifier offset.

5.4.4 Effect of Nonlinearity on I–V Characteristics of Selectorless Memory Element

Other characteristics studied included nonlinearity of the I–V characteristics of the memory cells. Nonlinearity requires a tradeoff between memory cell resistance values and the resistance ratio, which must be considered during device design. The effect of nonlinearity of the memory element was modeled using a sinh function [17,18].

The function is of the form

$$I = k \cdot \sinh(\alpha \cdot V)$$

where k is a fitting constant and α is the nonlinear coefficient. As α increases, the nonlinearity increases. This function is illustrated in Figure 5.28, where it is shown to fit well with the I–V characteristics of a Ti-doped NiO ReRAM [17,18]. The inset in Figure 5.28 illustrates the generic I–V characteristics of a single memory element with different nonlinear coefficients.

The three possibilities for the nonlinear behavior of the memory elements are (i) $\alpha(\text{off}) = \alpha(\text{on})$, (ii) $\alpha(\text{off}) > \alpha(\text{on})$ and (iii) $\alpha(\text{off}) < \alpha(\text{on})$, where $\alpha(\text{off})$ and $\alpha(\text{on})$ are nonlinear coefficients for the memory element in the high- and low-resistance states, respectively. If $\alpha(\text{off})$ is equal to or greater than $\alpha(\text{on})$, the nonlinearity increases the value of the resistance

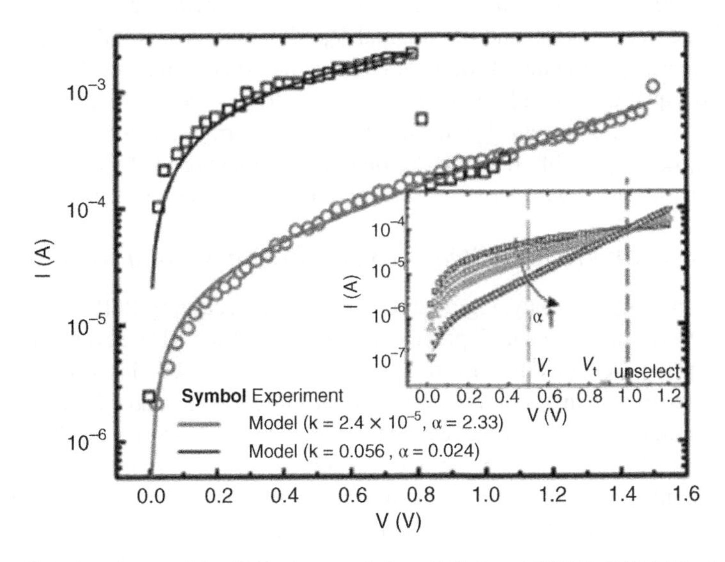

Figure 5.28 Sinh function model of I–V characteristics of Ti-doped NiO ReRAM. The inset illustrates the generic I–V characteristics of a single memory element with different nonlinear coefficients. (J. Liang and H.S.P. Wong (2010) (Stanford University), *IEEE Transactions on Electron Devices*, 57(10) [17], with permission of IEEE.)

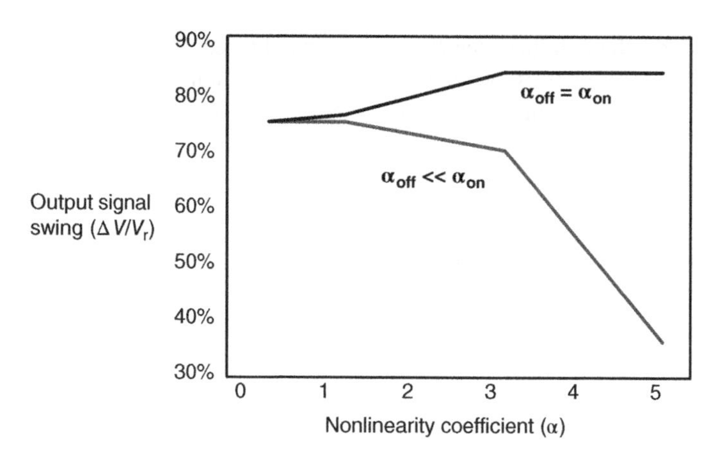

Figure 5.29 Illustration of output signal swing as a function of the nonlinearity coefficient α for the cases α(off) = α(on) and α(off) < α(on). (Based on J. Liang and H.S.P. Wong (2010) (Stanford University), *IEEE Transactions on Electron Devices*, 57(10) [17].)

without reducing the R_{off}–R_{on} ratio. This results in a positive effect on the output swing. In the case where α(off) is less than α(on), however, the increase in nonlinearity reduces the resistance ratio and could possibly degrade the maximum signal margin. This is illustrated in Figure 5.29 [17], which shows the output signal swing as a function of α for the cases when α(off) is less than or equal to α(on) [17,18].

When the resistance value is small (R_{off} = 1 MΩ) and the original R_{off}–R_{on} ratio is large, then the increase in α results in an optimal value of output signal swing ($\Delta V/V_r$). If, however, the resistance is large (R_{off} = 1 GΩ) and the original R_{off}–R_{on} ratio is small, then an increase in α decreases the output signal swing. It was noted that, in the experimental results reported to date, α(on) is usually equal to or smaller than α(off), so the nonlinearity normally has the positive effect of increasing the output signal swing and the memory array size. Stronger nonlinearity in R_{on} than in R_{off} could possibly degrade the output signal swing of the cross-point memory array, particularly in cases with large resistance values when the resistance ratio dominates.

5.4.5 Self-Rectifying ReRAM Requirements in Cross-Point Arrays

A numerical framework for analyzing requirements of self-rectifying ReRAM cells for use in cross-point arrays was discussed in September of 2012 by IMEC, KU Leuven, and ESAT [19]. The relationship between maximum array size and cell characteristics was analyzed. This included characteristics such as nonlinearity, absolute current level, and ON/OFF current ratio. Optimal bias conditions were determined, and the advantage of these bias conditions compared with a conventional half-voltage bias method was discussed. Cross-point memory arrays were analyzed, taking into consideration both the read and write operations. It was found that large nonlinearity and low current level were important because they permitted a larger array size. The LRS/HRS current ratio did not appear to be critical when a small read window was required, such as with single-level cells. It was, however, important when a larger read window was required, as it is with multi-level cells. If the unselected bit-lines and word-lines are biased to optimally distribute the voltage over the bit-line half-selected and word-line half-selected and nonselected cells, it allows for a much larger array size than does a fixed half-voltage bias scheme [19].

5.4.6 A Cross-Point Array Model for Line Resistance and Nonlinear Devices

A crosspoint array model for line resistance and nonlinear device characteristics was discussed by Global Foundries in April of 2013 [20]. The model can be solved using matrix algebra and can be used for statistical analysis. The model initially used linear bistable switching devices that could be switched between an HRS, or OFF state, and an LRS, or ON state. Devices were characterized by R_{on} and R_{off} with the ON/OFF ratio defined as R_{off}/R_{on}. The model was expanded to include various device characteristics, where HRS and LRS were described by nonlinear I–V behaviors.

It was shown that nonlinearity improved sensing margin by making sneak paths more resistive. Resistance is voltage dependent with higher resistance at lower voltage. Because unselected devices tend to have lower voltage than selected devices, they are more resistive. Nonlinearity was shown to help the write operation by improving voltage delivery, as voltage was shown to degrade less for nonlinear devices than for linear devices due to more resistive sneak paths.

There is a nonlinear device solution that permits crossbar arrays with diodes or nonlinear select devices to be analyzed. Calculations using this model showed that voltage and current degradation due to line resistance were not negligible, even for small crossbar arrays. This presented a constraint on array feature size. The analysis showed that diode and nonlinear select devices significantly improved the sensing margin of the read operation and the voltage window of the write operation. Sneak paths and line resistance limited feasible cross-point array sizes, while nonlinearity and select diodes improved sensing margin, write voltage window, and power efficiency by suppressing leakage through sneak paths.

In December of 2013, GlobalFoundries further discussed the technology and design tradeoffs important to optimize a crossbar memory array [21]. A model was developed to assess array functionality with different nonlinear, asymmetric selector parameters, bias methods, and array designs. Among the results were the following: array V_{DD} should be maximized within power and efficiency constraints, and partial bias methods can be chosen based on selector types and design targets. In addition, the impact of line resistance, contact resistance, and memory variability was analyzed. It was found that parallel access and self-selecting design may improve array performance.

5.5 Examples of Selectorless Cross-Point Arrays

5.5.1 Example of Nonlinearity in a Selectorless Cross-Point Array

A 54 nm technology $4F^2$ selectorless crossbar array 2Mb ReRAM test chip was discussed in June of 2012 by Hynix and Hewlett-Packard [22]. A TiO_x–Ta_2O_5-based ReRAM was used with a W–TiN–TiO_x–Ta_2O_5–TiN–W stack. The proposed switching mechanism involved the migration of oxygen vacancies for SET and RESET after the initial forming operation. Memory cell characteristics included nonlinearity, $I_{op} < 10\,\mu A$, and $V_{op} < 3$ V. An illustration of the nonlinear bipolar I–V curve is shown in Figure 5.30 [22] for a TiO–Ta_2O_5 ReRAM stack with different TiO ratios where $TiO_y < TiO_x < 2$ [22].

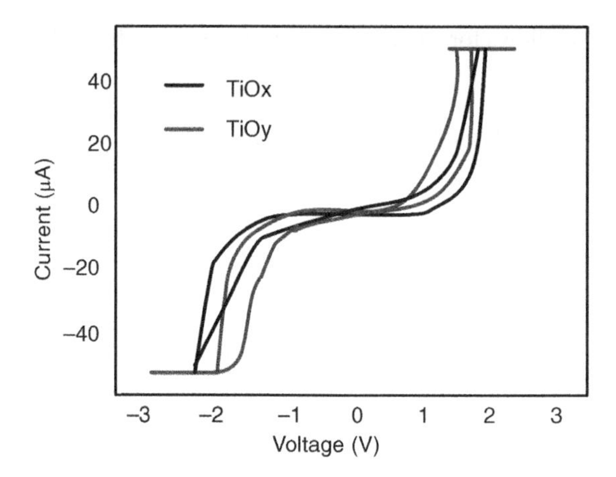

Figure 5.30 Illustration of nonlinear bipolar I–V curve for TiO–Ta_2O_5 stack with different TiO ratios. (Based on H.D. Lee *et al.*, (Hynix, Hewlett-Packard), VLSI Technology Symposium, June 2012 [22].)

The feasibility of read and write operations at <500 ns without a selector was investigated. A schematic of the four-plane array structure and row and column contact scheme are shown in Figure 5.31 [22], where the word-lines are out of the page [22].

The $4F^2$ selectorless crossbar array 2Mb ReRAM macro was integrated with core and peripheral circuits. Final decoders are under the cell cross-point array region, as shown in the figure, to increase the array efficiency of the macro. Cell characteristics can be modified by optimizing materials. The ON/OFF current ratio was improved by using different write conditions. Read and write specifications for memory operation were implemented so as to minimize sneak current through unselected cells.

5.5.2 *Example of High-Resistive Memory Element in Selectorless Cross-Point Array*

In September of 2011, Stanford University discussed another example of a selectorless cross-point array using an ReRAM memory element in a test chip in 180 nm CMOS technology with a four-layer 3D cross-point array [23]. This ReRAM cell did not use an individual selection

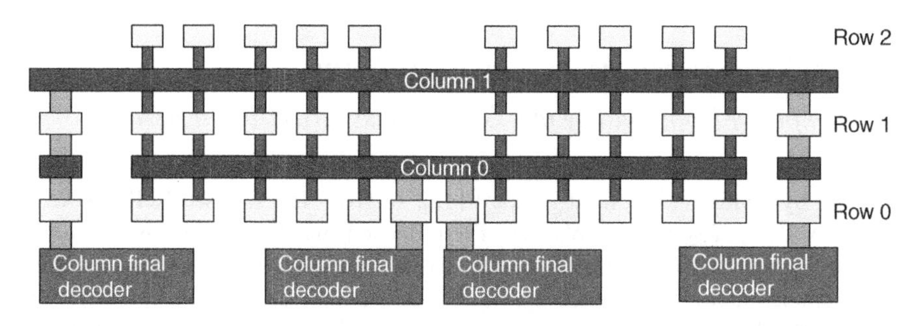

Figure 5.31 Schematic of four-plane array structure with column contact scheme. (Based on H.D. Lee *et al.*, (Hynix, Hewlett-Packard), VLSI Technology Symposium, June 2012 [22].)

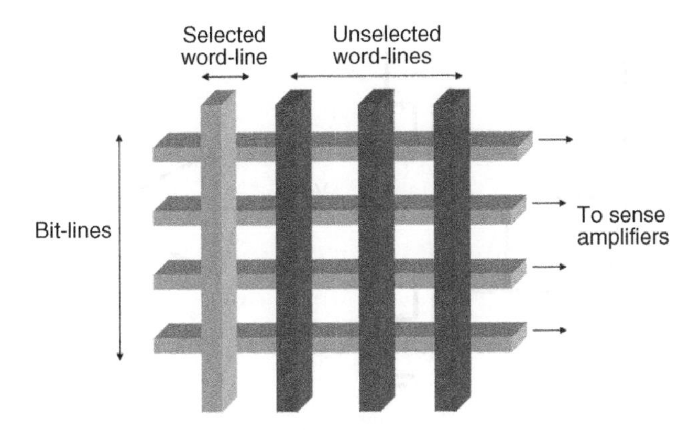

Figure 5.32 Memory array during a read operation. (Based on E. Ou and S.S. Wong (2011) (Stanford University), *IEEE Journal of Solid-State Circuits,* 46(9), 2158 [23].)

device, which required that the resistance of the material used in the metal–oxide ReRAM be high enough to regulate leakage current in both the low-resistance and high-resistance states. In addition, the metal–oxide needed to have low write energy requirements to avoid using extra peripheral circuitry and avoid limiting scalability. In this case, the ReRAM used an HfO_2-based resistance change material with the LRS (R_{on}) = 2 kΩ and the HRS (R_{off}) = 1 MΩ. Novel architecture and circuit techniques were used that minimized leakage current while maintaining a high bit density. The performance of a 65 nm 8Gb memory chip was simulated. Access time was 104 ns with power dissipation of 61.2 mW and area efficiency of 91.3%, which made the 3D cross-point memory competitive with NOR flash in read time and NAND flash in area efficiency.

The design challenges for controlling sneak current during the read operation without individual cell selectors were considered for a 2D cross-point array. The memory array was accessed word-line-by word-line. During a read, the selected word-line was raised to V_{read}, and read current was driven in parallel through the bit-lines, as shown in Figure 5.32 [23]. Unselected word-lines were terminated with high impedance, and each bit-line was connected to an individual sense amplifier at one end. This meant that the sense amplifiers provided the only current path to ground. If the voltage difference between bit-lines was minimized, then the leakage current was also. The sense amps needed to maintain the bit-lines at nearly constant voltage regardless of the memory cell state.

Current-sensing amplifiers reduce bit-line voltage swing more than voltage-sensing amplifiers. To fit in the bit-line pitch, a simple diode-connected n-type metal–oxide–silicon (NMOS) transistor with a current mirror was used to detect the bit-line read current. This single-gain-stage current-sensing amplifier is shown in Figure 5.33 [23]. The mirrored current was compared to a bias current set by the gate voltage V_{bias} on the PMOS transistor in the reference branch. This controlled the voltage signal V_{out}, which was buffered and latched.

An array with 32 word-lines was determined to provide the most area-efficient solution for a cross-point array with the given memory cell characteristics. The number of bit-lines was limited by the amount of current that a word-line could drive during write where cells would be programmed in parallel. With a programming current of 40 μA per cell, 32 bit-lines required a total programming current of about 1.2 mA, which was determined to be manageable. The

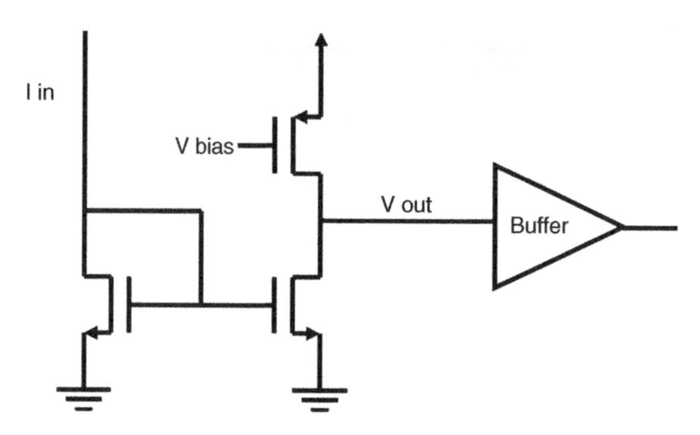

Figure 5.33 Single gain-stage current-sensing amplifier for cross-point array read. (Based on E. Ou and S.S. Wong (2011) (Stanford University), *IEEE Journal of Solid-State Circuits*, 46(9), 2158 [23].)

chosen memory block size was therefore 32 bit-lines by 32 word-lines, and there were no select transistors within the 32 × 32 cell block. Block bit-lines were connected to global bit-lines through bit-line select transistors. During a page access, an entire column of blocks was read in parallel by selecting a single word-line. Horizontal routing was used to ensure that control signal lines did not interfere with word-lines when multiple layers of memory were used.

Because the cross-point array did not require individual access devices, it could be integrated into a 3D stacked structure with multiple array layers. The resulting array was expected to have a bit density greater than that of single and multi-level NAND flash made in the same technology node. Device requirements became more stringent with multiple memory layers because an increasing number of vertical interconnects added additional leakage paths. The sense amplifier design was verified to be capable of tolerating increased leakage current from multiple layers when the memory cell variations were within target device specifications. The array efficiency could be improved by folding the bit-line selection transistors under the cross-point memory array, as shown in Figure 5.34 [23].

With multiple stacked layers, word-lines were shared between two layers of bit-lines, but only one layer of bit-lines was active at a time. This meant that the word-line drivers would drive the same current in a read as they would in the single-layer design. The additional bit-lines in the unselected layer had no effect on the leakage current during read or write because they were floating when unselected with no path to ground. The only added area overhead is at the array edge where an additional set of bit-line contacts must be inserted. This gave an effective array efficiency of 88.5% for two layers of memory cells in a 32 × 32 block.

With some added decoder complexity, four memory layers could be built up by adding two sets of word-lines. The top and bottom word-line layers could share a set of select transistors because they delivered current to different bit-line layers. All word-line driver transistors were connected to a global bus with the contacts at the array edge. This decreased the array efficiency to 71.7%. All four edges of the memory array were now used by vias. As a result, the array efficiency would be reduced if more than four layers were added. In a 65 nm process, a 32 × 32 bit cross-point array had an area of 17.3 μm^2. Simulations showed that an 8Gb memory architecture could be accessed with power and speed competitive with those of NOR flash. A four-layer cross-point memory had a significantly greater bit density than a NAND flash made in the same technology.

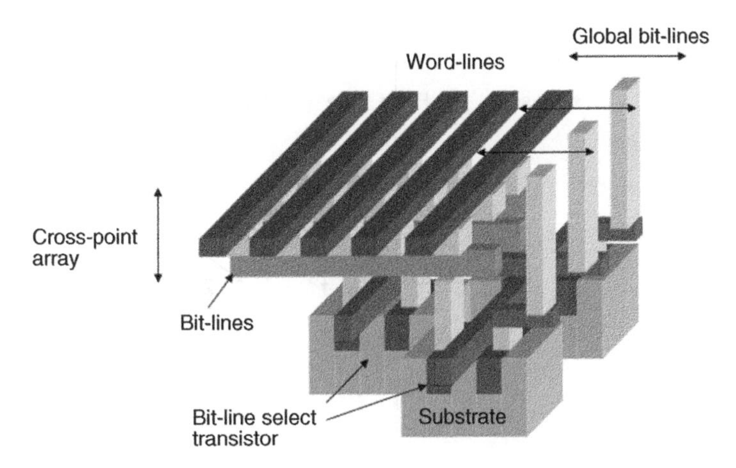

Figure 5.34 Illustration of bit-line select transistors folded under the 2D cross-point memory array. (Based on E. Ou and S.S. Wong, (Stanford University), *IEEE Journal of Solid-State Circuits,* 46(9), 2158 [23].)

5.5.3 Design Techniques for Nonlinear Selectorless Cross-Point Arrays Using ReRAMs

Some design techniques for building cross-point resistive memory arrays without using a select device for each memory element were described by Unity Semiconductor in November of 2011 [24]. Bipolar resistive elements were used. A memory cell selection technique for allowing the use of low-voltage transistors was described. This technique also helped solve current resistance (IR) drop, electromigration, and disturb issues in the selected array lines. An illustration of the nonlinear hysteresis I–V curve of the Unity Semiconductor memory element is shown in Figure 5.35 [24].

Challenges for designing resistive element cross-point array memories include memory cell selection, constraints limiting cross-point array size, and achieving high array efficiency (ratio of array area to chip size). Array efficiency was achieved by putting the peripheral CMOS circuitry under the cross-point array.

Historically, two selection methods have been used. In the first, a positive half-select voltage was applied on the selected word-line, and a negative half-select voltage was applied on the selected bit-line. The selected cell received the full select voltage. Generation of the negative voltage can be a drawback. In the second, an all-positive voltage was used, with the reference level moving from ground to an intermediate voltage so that the half-select voltage was equal on both sides of the reference voltage. The challenge of this biasing method was that the unselected memory cells on both the selected word-line and bit-line received the half-select voltage so that leakage current flowed during memory operation, which limited the size of a purely resistive cross-point array.

This study proposed a selection method for read that removed the half-select condition on the bit-lines during sensing. The selected word-line was biased at full voltage, and the unselected word-lines were initially at a reference level. Bit-lines were precharged to the same reference level and left floating. When the word-line was selected, a current flowed in all the bit-lines,

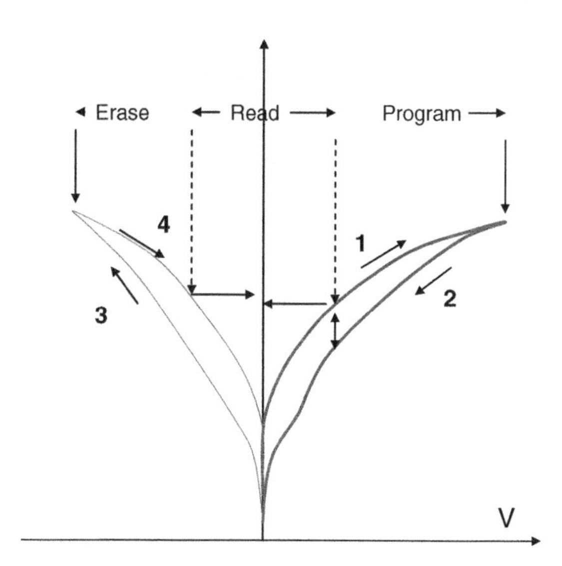

Figure 5.35 Illustration of bipolar nonlinear hysteresis I–V curve of memory element. (Based on B. Bateman, (Unity Semiconductor), ASCC, 14 November 2011 [24].)

which charged them to a saturation voltage. Sensing needed to be done before the bit-lines reached saturation because a selected memory cell in the low-resistance state would conduct more current than one in the high-resistance state. This means that there was an interval where the bit-line voltage could have been differentiated based on the state of the memory cell. The window for sensing depended on the nonlinearity of the memory cell I–V curve and the number of unselected memory cells on the bit-line. The resistive memory element used had a nonlinear exponential I–V curve. To reduce bit-line length, a local bit-line architecture was used, where the local bit-lines were connected to the global bit-line through low-voltage NMOS pass gates. Constraints limiting the word-line length included electro-migration and voltage drop from the driver to the last cell on the word-line due to the metal line resistance. A word-line voltage regulation technique was proposed together with an unselected word-line biasing scheme. The word-lines were segmented into sections of 256-cell pairs including cells above and below the selected word-line. A tracking signal indicated the voltage level at the end of the word-line.

5.5.4 Film Thickness and Scaling Effects in Cross-Point Selectorless ReRAM

In order to develop stackable cross-point ReRAMs without select transistors to replace current NAND flash technology, performance and reliability must be improved by reducing memory device area and by scaling film thickness. In May of 2011 Gwangju IST discussed scaling of ZrOx–HfOx and Ir–TiOx stacks [25]. Both of these ReRAMs used a filament formation–rupture mechanism by oxygen ion movement. It is important in scaling to continue to meet device criteria such as high-speed operation, low-power switching, switching uniformity, endurance, and retention time for the cross-point array.

For both ReRAM stacks, the SET voltage (V_{set}) and resistance of the HRS (R_{hrs}) were found to be a function of the active device area. With scaling, $V_{forming}$, V_{set}, and R_{hrs} increased. For

HfOx, the increase in V_{set} with scaling was attributed to a reduced number of defects, where the formation of a conducting filament could be due to connection of defects in the active memory area. In a large device, a small voltage can form a filament, but with reduced defects, a higher voltage, such as 1.1 V, is required to form a filament. For the LRS the same resistances were observed regardless of device area because the R_{lrs} was due to the resistance of the conductive filament and not of the entire dielectric stack. For the HRS a dependence of the resistance on area was observed because R_{hrs} was that of the volume of dielectric in the stack [25].

To confirm a temperature effect on resistive switching, the speed vs. temperature was measured between 300 and 480 K. The response speed was found to increase significantly with temperature, so if additional heating could be provided during program and erase, speed could be increased without changing retention. The effect of area scaling on the local heating of the filament was studied, and it was found that small area devices switch faster due to a local heat effect. The reduced area when scaled caused current density to increase significantly. This induced the high temperature during bias, which caused switching to run faster. It was estimated that nanodevices ran about 100 °C hotter than large-area devices. Nanodevices showed more uniform and sharp distribution of V_{set}/V_{reset}, and the distribution of resistance in LRS/HRS was improved. These improvements were attributed to the increase in defects in the active memory areas of large devices. Oxygen vacancies are known to cause intrinsic defects in HfO_2 film, which might be responsible for the formation of oxygen vacancy filaments during switching. It was speculated that if the defects could be minimized, then switching could be controlled. Various thicknesses of TiN bottom electrode were used to show that decreasing the active area thickness caused values of V_{set}/V_{reset} and R_{lrs}/R_{hrs} to decrease [25].

5.5.5 Vertical HfOx ReRAM 3D Cross-Point Array Without Cell Selector

Fabrication of a double-layer stacked HfOx vertical ReRAM was discussed in December of 2012 by Stanford University and Peking University [26]. The use of electrode–oxide interface engineering on TiON permitted a nonlinear I–V characteristic, which was suitable for a selectorless array. The ReRAM showed good performance, with RESET current <50 μA, switching speed of 50 ns, switching endurance $>10^8$ cycles, half-selected read disturbance immunity $>10^9$ cycles, and retention $>10^5$ s at 125 °C.

The 3D vertical ReRAM was formed on a bottom SiO_2 layer, as shown in Figure 5.36 [26]. First, multiple Pt–SiO_2 layers were deposited by evaporation and low-pressure chemical vapor

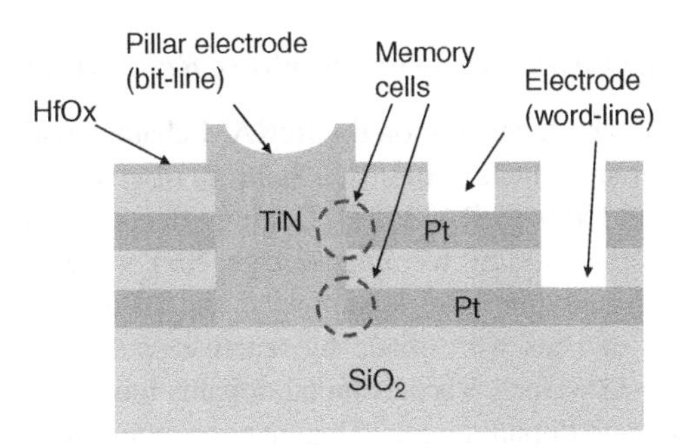

Figure 5.36 Schematic cross-section of stacked TiN–HfOx–Pt ReRAM cells. (Based on H.Y. Chen *et al.*, (Stanford University, Peking University), IEDM, December 2012 [26].)

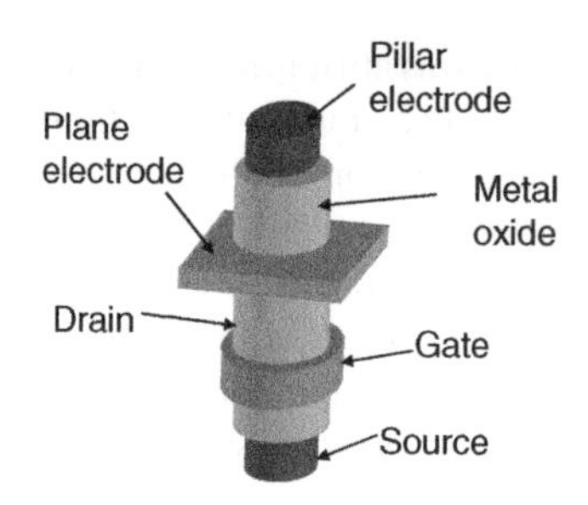

Figure 5.37 Schematic of vertical MOSFET transistor. (Based on H.Y. Chen *et al.*, (Stanford University, Peking University), IEDM, December 2012 [26].)

deposition. A trench was dry etched down to the bottom SiO_2 layer, and TiN was deposited by sputtering to fill the trench as the pillar electrode. The Pt electrode for word-line contact to each plane was then exposed by dry etching. This formed several cell stacks of TiN–HfOxPt vertically, each with a separate plane electrode contact. The TiN plug was the bit-line.

A vertical metal–oxide–semiconductor field-effect transistor (MOSFET) was stacked with the pillar electrode to serve as the bit-line selector. This enabled random access capability for individual cells in the array. A schematic of this transistor is shown in Figure 5.37 [26].

A new write method was proposed for the 3D cross-point architecture. With this write method, V_w was applied on the selected cell's word-line, and $V_w/2$ was applied on all the unselected cells' word-lines to avoid unintentional writing. The source line of the selected cell's pillar was turned on, and its bit-line grounded. A new read method was also proposed. For read, V_r was applied on the selected cell's word-line. The source line that controlled the selected cells was turned on, and the data of a row of cells was read out by the sense amplifier [26].

Analysis showed that for such a 3D selectorless array, a large R_{on} of about 100 kΩ from the nonlinear I–V helped to reduce the sneak path current and that a low interconnect resistance using metal planes as word-lines reduced the unwanted voltage drop on the interconnect. Simulation showed that megabit-scale arrays without cell selectors could be achieved. A vertical transistor was used as the bit-line selector. Simulations suggested that a large R_{on} and a low word-line plane resistance could enable a megabit-scale array without a cell selector. Further increasing the nonlinearity in the ReRAM I–V characteristics can be expected to permit an even larger memory array [26].

5.5.6 Dopant Selection Rules for Tuning HfOx ReRAM Characteristics

Because selectorless arrays are dependent on the ReRAM characteristics, it is important to be able to tune these characteristics. In June of 2013, Stanford discussed a systematic approach to tuning device parameters in HfOx ReRAM [27]. Doping effects were used to demonstrate the tunability of key parameters including forming voltage, SET voltage, and ON/OFF ratio. In order to provide universal guidelines for dopant selection, 12 dopants were analyzed.

TiN–HfO_x–Pt ReRAM devices were made by reactive sputtering of HfO_x and TiN top electrodes on Pt–Ti–SiO_2–Si wafers. Various metal dopants including Al, Zr, Ta, W, and Ni, were co-sputtered with Hf and dopant targets. Dopant concentrations were controlled. Switching characteristics of the undoped and doped devices were measured. The results showed significant tunability of the ReRAM characteristics by doping.

Guidelines for using these dopants to affect device characteristics were given as follows: (i) Hf-like dopants such as Zr, Ti, and Si, can be used to reduce formation energy without major changes in switching parameters. (ii) Weak p- and n- type dopants such as Al, La, Ta, and W can be used for a larger reduction of formation energy and for more stable filaments. The devices showed lower forming and switching voltages and better uniformity but worse ON/OFF ratio. (iii) Strong p- and n- dopants such as Sr, Ni, and Cu may dramatically change the switching parameters and shrink the ON/OFF ratio. (iv) Higher doping concentration will enhance the dopant–filament interaction and produce a larger change in device characteristics. An example of a ReRAM using these rules was demonstrated.

5.5.7 High-Resistance CB-ReRAM Memory Element to Avoid Sneak Current

Using the dual V_{th} characteristics of a multilayer HfO_2–SiO_2–Cu-GST conducting bridge resistive RAM to make a zero transistor–one resistor (0T1R) cell for a crossbar array was discussed in June of 2012 by Macronix [28]. Voltage thresholds (V_{th}) were used to store the logic state, which left all devices in an HRS, so a separate isolation device was not needed to eliminate sneak current leakage.

The stack of the CB RAM device is shown in Figure 5.38(a) [28] as Cu-GST–SiO_2–HfO_2–W. The HfO_2 layer was the storage node or memory layer. The SiO_2 is normally free of a conductive bridge. A conducting filament bridge (CB) in the HfO_2 denoted the low-threshold $V_{th(L)}$ state, as shown in Figure 5.38(b). The absence of the filament was the high-threshold state $V_{th(H)}$, as shown in Figure 5.38(c). The SiO_2 served to reduce the SET–RESET current. During read of a $V_{th(L)}$ state, a conductive filament formed briefly in the SiO_2 layer, as shown in Figure 5.38(d), and then spontaneously ruptured. During read of a $V_{th(H)}$ state, no filament formed in either layer.

In a CB ReRAM array that used this cell, no select device was required because all cells were in a HRS, resulting in low array leakage. By operating between the two thresholds, a cross-point array could be constructed using the CB RAM alone because the cell was always in a HRS except briefly during read. The spontaneous rupture of the filament in the SiO_2 layer after

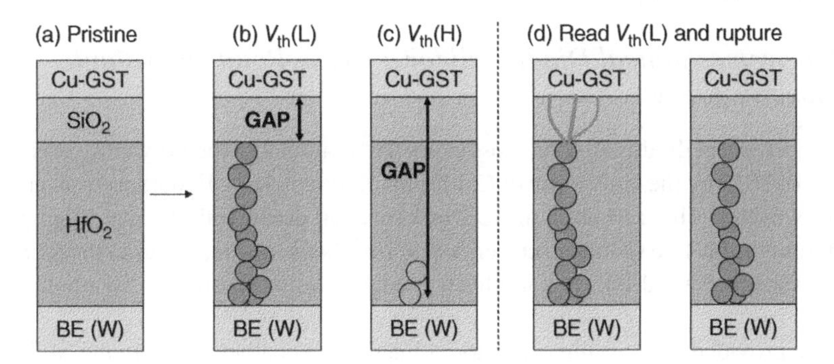

Figure 5.38 Schematic of operation of a double-layer conductive bridge ReRAM showing (a) the unprogrammed state; (b) the $V_{th(L)}$ state; (c) the $V_{th(H)}$ state; and (d) read of the $V_{th(L)}$ state with a temporary resistance bridge through the SiO_2 layer that ruptures immediately after read. (Based on F.M. Lee *et al.*, (Macronix), VLSI Technology Symposium, June 2012 [28].)

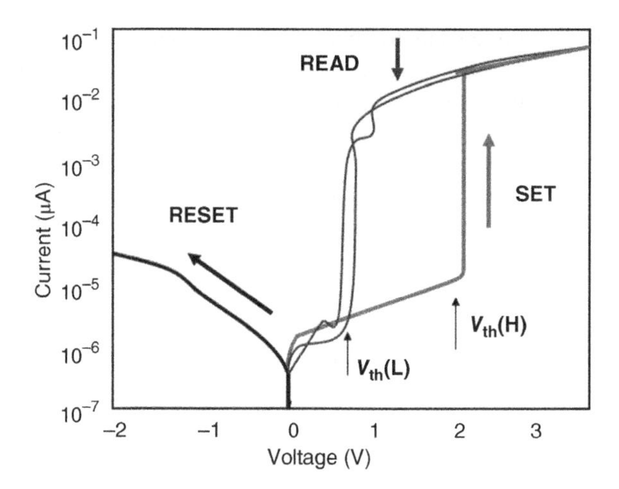

Figure 5.39 Illustration of I–V curve showing DC operation of 0T1R CBRAM double cell storing the memory state in a voltage threshold. (Based on F.M. Lee *et al.*, (Macronix), VLSI Technology Symposium, June 2012 [28].)

read of the $V_{th(L)}$ state reduced the switching current significantly and also reduced the read disturb. The DC operation of the 0T1R CB ReRAM cell is shown in Figure 5.39 [28]. The SET and RESET operations are indicated along with the temporary read curve for $V_{th(L)}$.

By using the dual V_{th} characteristics of a multilayer HfO_2–SiO_2–Cu-GST CB-RAM, a one-resistor (1R) cell with no access device (0T1R) could be made. This is similar to the function of the NAND flash memory cell. The V_{th} of the cell was determined by the CB in the HfO_2 layer only. The CB in the SiO_2 was only present during read and spontaneously dissolved after read. The spontaneous rupture of the filament in the SiO_2 layer reduced the switching current and also reduced the read disturb. The device operated between two V_{th}, which permitted the memory to be used without a select transistor since the cell was always at HRS except during read. The memory state was stored as the V_{th} of this CB ReRAM, similar to the operation of a NAND flash memory [28].

5.5.8 *Electromechanical Diode Cell for a Cross-Point Nonvolatile Memory Array*

An electromechanical diode cell for a cross-point $4F^2$ nonvolatile memory array was discussed in February of 2012 by the University of California, Berkeley [29]. Endurance was greater than 10^4 set–reset cycles. This cell used an electromechanical diode with the word-line physically separated from the bit-line so that no current could flow between these lines in the RESET state. In the SET state, the word-line was in physical contact with the bit-line, so a p–n diode was formed.

The diode provided a rectifying capacitor voltage (CV) characteristic. In the SET state, the attractive electrostatic force exerted by the diode's built-in electric field and the surface adhesion force were greater than the spring restoring force of the word-line beam that maintained the state. The (a) RESET and (b) SET state configurations of this cell are shown Figure 5.40 [29].

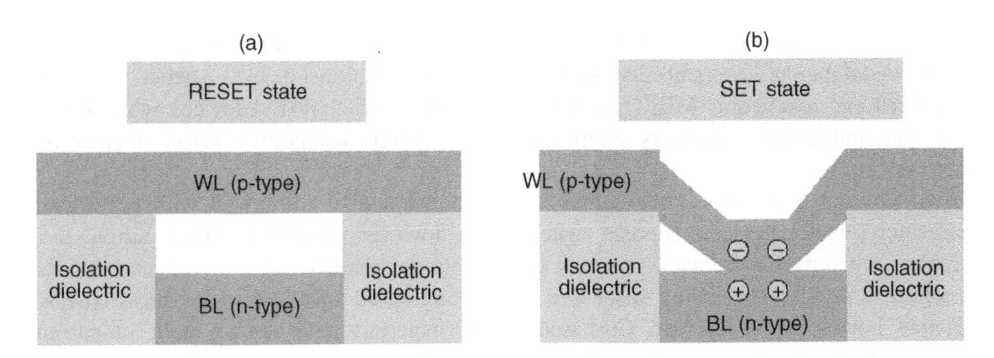

Figure 5.40 Schematic of (a) RESET; and (b) SET states of electromechanical diode cell. (Based on W. Kwon *et al.* (2012) (University of California Berkeley), *IEEE Electron Device Letters*, 33 (2) [29].)

The voltage in the selected word-line was 0 V during programming of the cell. For the SET operation, the voltage on the selected bit-line was large enough to bend the top part of the cell downward into contact with the lower part of the cell, where a diode is formed. The surface adhesion force and the diode's built-in electric field maintained the configuration when the voltage on the bit-line was removed. For the RESET operation, the voltage on the selected bit-line was reduced to less than the release voltage, which released the upper part of the cell to return to its original state. During both SET and RESET, the unselected word-lines were floating and unselected bit-lines were at 0 V. For a read, the selected bit-line was set to 0 V and the selected word-line was set to a voltage greater than zero. During a read, the unselected word-lines were at 0 V and the unselected bit-lines were floating.

5.6 Unipolar Resistance RAMs with Diode Selectors in Cross-Point Arrays

5.6.1 Overview of Unipolar ReRAMS with Diode Selectors in Cross-Point Arrays

Unipolar ReRAMs have both the SET and RESET states in a single quadrant. They can be used with a two-terminal diode selector. These unipolar RAMs include both metal–oxide unipolar ReRAMS and phase-change ReRAMs. The two terminal diode selectors can be either polysilicon diodes or metal–oxide diodes.

A typical resistance change memory cross-point array requires a selection device for each memory element. A lateral MOSFET or bipolar junction transistor normally requires a cell size of 10–15 F^2, where F is the minimum feature size for the process technology. A diode selection device can result in a cell size of $4F^2$ if stacked with the memory element in a 3D cross-point memory array. Considerations in selecting a diode type include the required processing temperature of the memory element and the diode selection element. Another consideration is whether the memory element needs bipolar programming, which requires a bidirectional diode or elimination of the selection device entirely. If the feature size can be maintained at $4F^2$, then with n memory layers, the effective cell size becomes $4F^2/n$ [23].

There are several general types of select devices suggested for use as drivers for memory elements in crossbar arrays. These include the unidirectional diode devices, such as polysilicon junction devices, oxide junction devices, and oxide rectifiers, which must be used with unidirectional memory elements. In addition, there are bidirectional devices, such as the mixed ionic electronic conduction (MIEC) and Ovonyx threshold switch (OTS) devices, which can be used with bidirectional memory elements. These various technology select devices are discussed in this section.

Two-terminal unipolar switches have been demonstrated to work well with polysilicon diodes for selector devices. Bipolar switching is, however, faster and lower current than unipolar switching for the same memory device, but an adequate selector for large bipolar switching arrays is still being discussed. Tunneling-based metal–insulator–metal (MIM) selectors have poor on-current. Cu-based MIEC materials have shown poor on-current densities in bipolar mode even though they have good on-current densities in unipolar mode. Polysilicon n–p–n devices show somewhat improved on-currents. A vanadium oxide (VO_2) selector has high on-current density but poor ON/OFF current ratio. A drain-induced-barrier-lowering (DIBL) – based polysilicon selector has been proposed along with an epitaxial punch-through-diode-based selector for bipolar switching [30].

5.6.2 A Unipolar ReRAM with Silicon Diode for Cross-Point Array

The unipolar ReRAM based on a Ni electrode–HfOx dielectric–n+ silicon substrate structure can be integrated with a silicon diode as a select device for application in a crossbar architecture. This stack was discussed in March of 2011 by Nanyang Technical University, National University of Singapore, Soitec, and Fudan University [31]. To integrate a ReRAM device in a crossbar architecture, interference caused by leakage current through neighboring cells with LRS must be minimized to avoid readout errors. A unipolar ReRAM can be integrated in a 1D1R crossbar architecture without leakage currents in the half-selected devices using a low-cost silicon diode.

This unipolar operating ReRAM was based on a Ni electrode–HfOx dielectric–n+ silicon substrate structure. HfOx was deposited by atomic layer deposition (ALD) on oxidized n-type silicon. A Ni top electrode was used. HfOx was considered a promising ReRAM dielectric because it was compatible with scaled CMOS processing. Analysis showed that oxygen ions from the HfOx layer diffused into the Ni top electrode, forming a NiOx interfacial layer between the HfOx and Ni electrode. This NiOx layer was important for the unipolar switching in this device, as the formed and ruptured conductive filament during resistive switching occurred in this interface layer, as shown in Figure 5.41 [31].

The device had an R_{off}/R_{on} ratio $> 10^3$, retention $>10^5$ s at 150 °C, pulse switching endurance $>10^5$ cycles, and programming speed about 50 ns. It had almost 100% device yield on a 6-inch wafer. While the basic structure was Ni–HfOx–n+Si, a SiO_2 interfacial layer formed between the HfOx and the N+ Si substrate during the ALD process. The use of Pt as the top electrode was tried, and no switching characteristics could be measured.

It was concluded that the unipolar resistive switching could be explained by the rupture and formation of conductive filaments in association with a local Joule heating–induced reduction–oxidation (redox) inside the NiO_x interfacial layer, similar to switching in a NiOx-based ReRAM. It was found that despite the fluctuation in HRS values, resistance ratios >3 orders of magnitude were maintained. Switching endurance with a pulse width of 50 ns showed no degradation after 10^5 cycles. The RESET current (I_{reset}) was about 800 μA, which was a

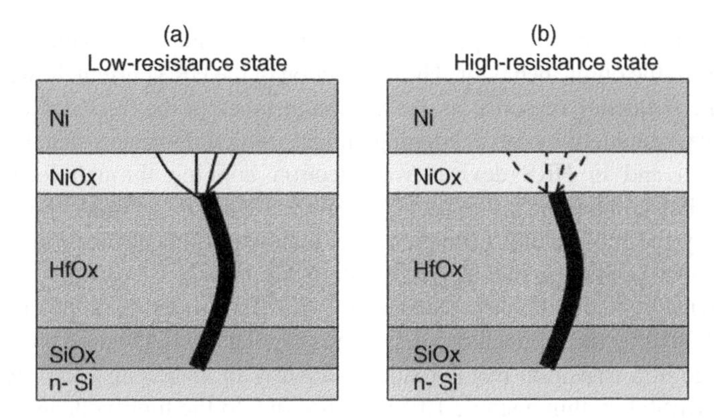

Figure 5.41 Diagram of filament in Ni–HfOx–n+Si (a) formation (SET); and (b) rupture (RESET). (Based on X.A. Tran *et al.* (2011) (Nanyang Technological University, NUS, Soitec, Fudan University), *IEEE Electron Device Letters*, 32(3) [31].)

drawback because it limited the dimension of the select diode. A typical I–V curve of the positive DC sweep is shown in Figure 5.42 [31].

To activate the device initially, a forming process was used with a forming voltage of about 3 V. The current was limited (compliance current) to 10^{-4} A. During the SET process, a positive voltage sweep to $V_{set} = 2.1$ V with compliance current of 100 μA triggered abrupt conduction, and the resistance switched from the HRS to the LRS. Another positive voltage sweep to 1.1 V without compliance current caused the abrupt decrease of current, and the resistance switched back to the HRS. The yield of the device was nearly 100%, and endurance was more than 100 switching cycles.

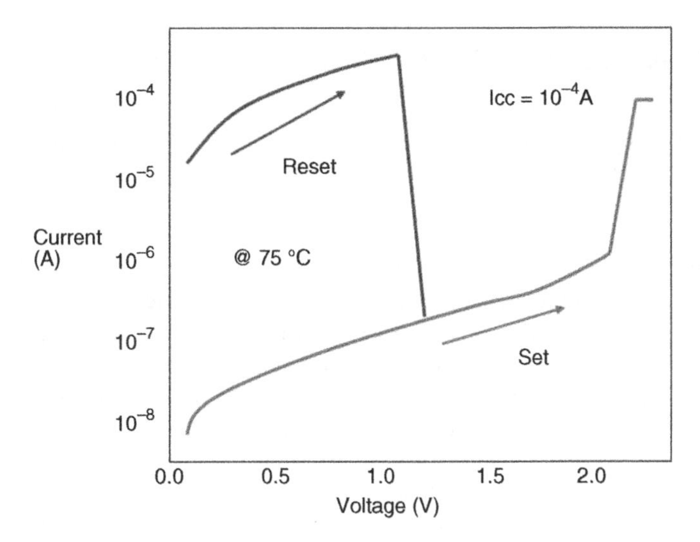

Figure 5.42 Typical I–V curve of positive DC sweep in Ni–HfOx ReRAM cell. (Based on X.A. Tran *et al.* (2011) (Nanyang Technological University, NUS, Soitec, Fudan University), *IEEE Electron Device Letters*, 32(3) [31].)

The LRS current was found independent of temperature and linear with applied voltage, which implied metallic ohmic behavior. The conduction of the HRS current, however, relied on temperature. The frequency response of the AC conductance of the fresh device fit well with the AC electrical conductivity of a hopping model, which indicated that the conduction transport in fresh and in HRS devices was electron hopping through localized oxygen vacancies in the HfOx under the influence of an electric field. The unipolar resistive switching was explained by the rupture and formation of conductive filaments with local Joule heat–induced reduction–oxidation inside the NiOx interfacial layer.

An attempt to identify conditions for a stable SET–RESET process in unipolar ReRAMs was made in June of 2010 by the Politechnico di Milano [32]. The ReRAM used a voltage-induced resistance change in a transition metal–oxide. Instability in switching could be due to SET occurring after RESET within a single RESET pulse due to the high voltage across the cell. This RESET–SET instability could result in long program-verify loops to achieve the required cell state. The SET process consisted of the formation of a conductive filament through the oxide. The RESET transition consisted of the conductive filament being dissolved by diffusion and oxidation. The crossover of SET and RESET times as a function of voltage was found to permit a stable SET process at high voltage and a RESET process at low voltage. That is, RESET was shown slower than SET for voltage below a crossover value and faster than SET for voltages above a crossover value. The unipolar switching was compatible with diode-selected high-density crossbar arrays. A stochastic SET–RESET model was created to describe a circuit using NiO as the active switching layer and to indicate conditions to avoid instability. These conditions were that RESET time must be above and SET time below a specified value and that load resistance must be low enough to prevent excessive voltage after RESET. The impact of RESET–SET instability can be reduced for small-series resistance in the array and large-cell resistance in the SET state.

5.6.3 CuOx–InZnOx Heterojunction Thin-Film Diode with NiO ReRAM

A two-layer stacked 8 × 8 bit cross-point array with one diode and one resistor was discussed in December of 2007 by Samsung [16]. The $0.5 \times 0.5\ \mu m^2$ cells used stacked ReRAM with a p-CuO_x–n-$InZnO_x$ heterojunction thin-film oxide diode that showed increased current density over a previous p-NiO_5–n-TiO_x oxide diode. The storage element was a Ti-doped NiO ReRAM.

Results of the study showed no limitation to the number of stacks allowed. All fabrication was done at room temperature. Bistable switching was shown for the 1D1R memory with good behavior for both the diode and storage nodes. Forward current density for the p-CuO_x–n-IZO_x diodes was over 10^4 A/cm^2, and operation voltage for the storage node with diode attached was around 3 V. For high-density compatibility the study used an oxide diode to rectify unwanted leakage current paths. Oxide diodes have more flexibility in processing technology than silicon based diodes because they can be made over any substrate, even at room temperature. An issue was the forward current density of oxide diodes, which is usually limited by material issues. The oxide-based NiO ReRAM had process compatibility with the oxide-based p–n diode. The two-layer stack was assembled with a shared bit-line, as shown in Figure 5.43 [16].

The I–V curves of the upper layer (a) NiO ReRAM, (b) diode, and (c) the combined 1D1R stack are shown in Figure 5.44 [16]. The increase in program and erase voltages for the 1D1R cell was due to the resistance of the diode element. The I–V curves of the lower layer (a) NiO

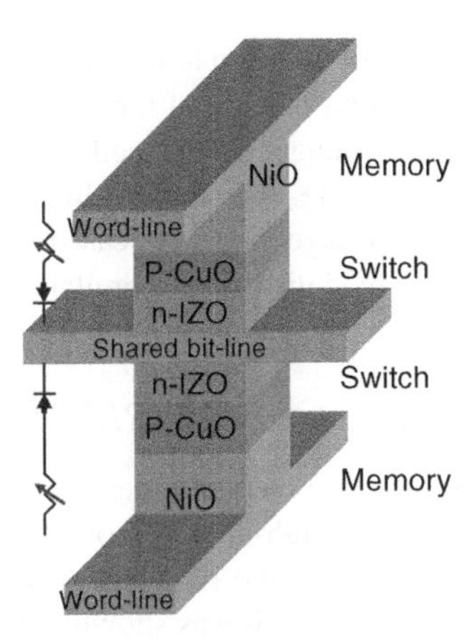

Figure 5.43 Two-layer NiO ReRAM and p-CuO_x–n-IZO_x diode stack. (Based on M.J. Lee, *et al.*, (Samsung), IEDM, December 2007 [16].)

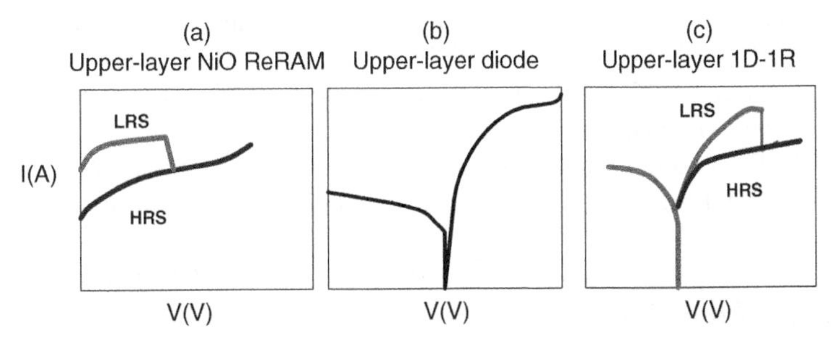

Figure 5.44 I–V curves of (a) upper layer NiO ReRAM; and (b) upper oxide diode and (c) combined upper layer of ReRAM and diode. (Based on M.J. Lee *et al.*, (Samsung), IEEE IEDM, December 2007 [16].)

ReRAM, (b) diode, and (c) combined 1D1R stack are shown in Figure 5.45 [16]. The increase in program and erase voltages for the 1D1R cell was due to the resistance of the diode element.

Layer-to-layer interference was tested, and no significant interference was detected. The 8×8 bit array was tested, and the individual cells were programmed. Key factors in choosing

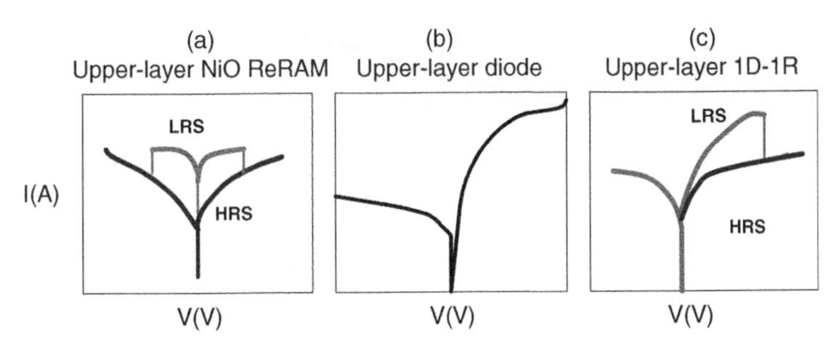

Figure 5.45 I–V curves of lower layer NiO ReRAM and lower oxide diode. (Based on M.J. Lee *et al.*, (Samsung), IEEE IEDM, December 2007 [16].)

oxide diodes were the ease of fabrication at room temperature and the potential for fabricating over metal substrates. The higher current density of the p-CuO_x–n-IZO_x diodes meant it was possible to reduce the operational voltage of the cells from 4.5 to 3 V. No significant degradation between the two-stack diode layers was observed. The program voltage was increased from 0.6 to 2 V, and erase voltages were increased from 1.6 to 3 V. Leakage current paths were rectified, so there was no read disturb. No limitation was seen in stacking multiple layers to increase memory density.

5.6.4 Unipolar NiO ReRAM I_{reset} and SET–RESET Instability

Unipolar ReRAM based on NiO transition metal–oxide was discussed in May of 2011 by the Politecnico di Milano and IMM-CNR for high-density crossbar architecture solutions [33]. The issue was that the RESET (I_{reset}) current in the NiO ReRAM needed to be reduced to allow scaling of the diode-selected crossbar arrays. It was shown that I_{reset} reduction was limited by the competition between SET and RESET transitions in high-resistive conductive filaments. The current overshoot effect can cause uncontrolled SET. To minimize this overshoot effect, V_{set} needed to be kept low. SET–RESET instability is a SET transition to low resistance followed by a RESET transition to a RESET state with too high a resistance. SET–RESET instability and SET transition can cause reliability issues. SET–RESET instability in a crossbar array can lead to "over-SET failure," where the excessive I_{reset} cannot be supplied by the scaled select diode. It was proposed that minimum I_{reset} for pulsed RESET of the cell use an incremental RESET algorithm, as shown in Figure 5.46 [33]. Minimum I_{reset} was found to require controlled SET starting from $R < 10^7\ \Omega$ to avoid overshoot effects.

5.6.5 HfOx–AlOy Unipolar ReRAM with Silicon Diode Selector in Cross-Point Array

A fast unipolar ReRAM with a Ni-electrode–HfOx–AlOy–p+Si structure was discussed in June of 2011 by Nanyang TU, Peking University, A*STAR, NUS, and Globalfoundries [34]. This structure was compatible with a low-cost silicon diode selector for 3D cross-point implementation. The ReRAM had an $R_{off} - R_{on}$ resistance ratio = 10^5, 100% device yield on a 6-inch wafer, good cycle-to-cycle and device-to-device uniformity of switching parameters, pulse switching endurance $>10^6$ cycles, temperature retention $>10^5$ s @ 120 °C, and operating stability at >100 °C without threshold resistive switching, SET–RESET speed of 10/30 ns, full

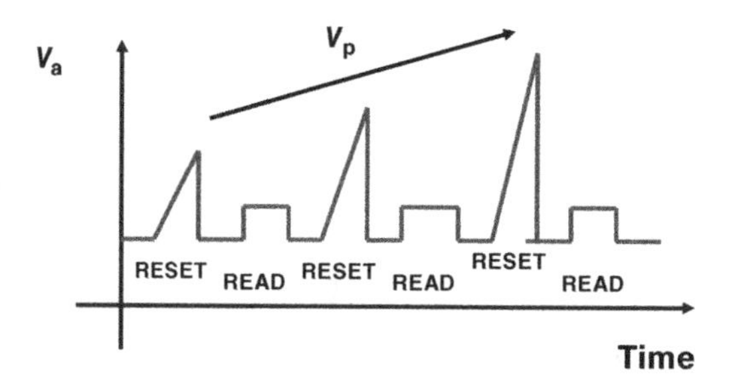

Figure 5.46 Example of incremental RESET algorithm showing possible behavior. (Based on F. Nardi *et al.*, (Politecnico di Milano, IMM-CNR), IMW, May 2011 [33].)

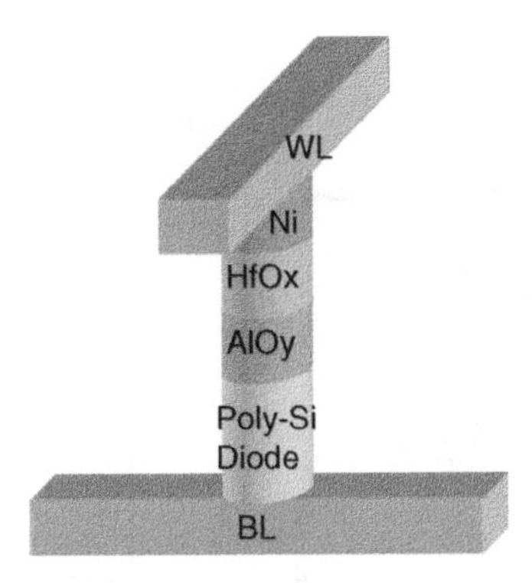

Figure 5.47 Schematic of Ni–HfO_x–AlO_y–p+Si ReRAM + diode selector in 3D crossbar configuration. (Based on X.A. Tran *et al.*, (Nanyang Technological University, Peking University, A*STAR, NUS, GlobalFoundries), VLSI Technology Symposium, June 2011 [34].)

CMOS compatibility, and p+Si bottom electrode that avoids the use of Pt. A schematic of the ReRAM stack along with the silicon diode selector in the 3D crossbar configuration is shown in Figure 5.47 [34].

Comparison of the characteristics of the AlOy–HfOx- and HfOx-based ReRAM stacks were done. AlO–HfO was found to have endurance $>10^5$, ON speed of 10 ns, and OFF speed of 30 ns, whereas HfOx had endurance $>10^3$ and ON and OFF speeds of 50 ns. Retention time was $>10^5$ s at 120 °C for both configurations.

A typical I–V curve of the positive DC sweep of Ni–HfOx–p+ silicon and Ni–HfOx–AlOy–p+ silicon is shown in Figure 5.48 [34]. The lower SET and RESET voltages can be seen along with the endurance stability of the AlOx–HfOx after 10^6 pulses.

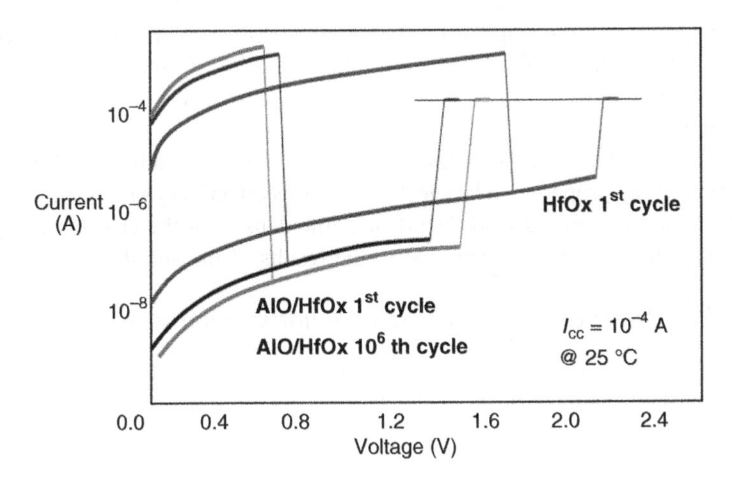

Figure 5.48 Typical I–V curve of Ni–HfOx–p+Si after 1 cycle and Ni–HfOx–AlOy–p+Si after first cycle and after 10^6 cycles. (Based on X.A. Tran *et al.*, (Nanyang Technological University, Peking University, A*STAR, NUS, GlobalFoundries), VLSI Technology Symposium, June 2011 [34].)

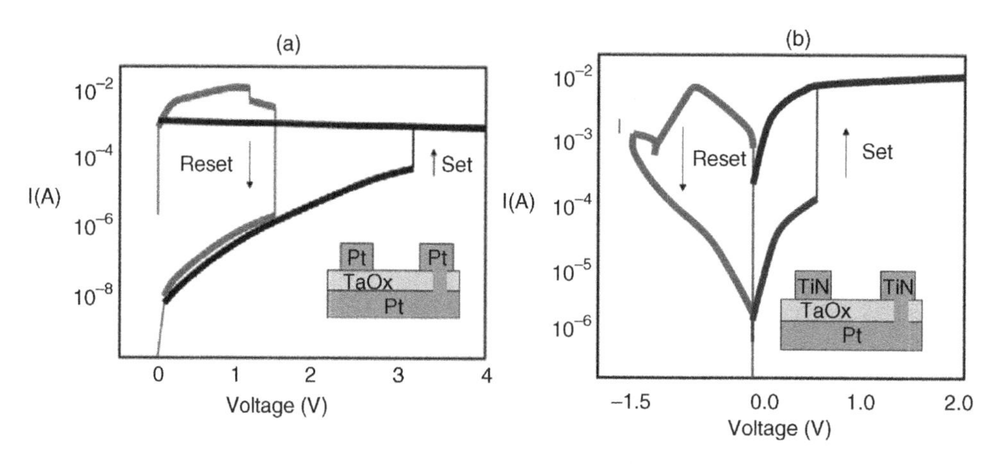

Figure 5.49 Schematic illustration of (a) unipolar I–V curve of Pt–TiOx–Pt ReRAM memory device; and (b) bipolar I–V curve of TiN–TaOx–Pt selector device. (Based on Y. Huang *et al.* (2012) (IME Peking University), *IEEE Transactions on Electron Devices* [35].)

5.6.6 TiN–TaOx–Pt MIM Selector for Pt–TaO$_x$–Pt Unipolar ReRAM Cross-Point Array

A bipolar TiN–TaO$_x$–Pt ReRAM device used as a dynamic selector for a unipolar Pt–TaO$_x$–Pt ReRAM crossbar array was discussed by IME Peking University in August of 2012 [35]. The selector was used to suppress sneak current in the crossbar array and permit a larger array. The bipolar ReRAM was found to act as a good selector, with the sneak current reduced by two orders of magnitude as estimated by a $V_{read}/2$ voltage scheme. Simulations showed that, with the sneak current suppressed to this level, the maximum size of the crossbar array with the bipolar ReRAM selector could be increased to >1Mb. The interaction of the two devices reduced the sneak current. The unipolar I–V curve of the Pt–TaO$_x$–Pt ReRAM device is shown in Figure 5.49(a) [35]. The Bipolar I–V curve of the TiN–TaO$_x$–Pt selector device is shown in Figure 5.49(b) [35]. The overlay of the I–V curves of the two devices when they are stacked is shown in Figure 5.50 [35].

Both devices were fabricated. The unipolar ReRAM device with a bipolar ReRAM selector showed good switching properties. The sneak current was effectively suppressed by two orders of magnitude. With the reduced read disturbance, the density of the crossbar array could be increased to more than $10^3 \times 10^3$ according to the results of the simulation.

5.6.7 Self-Rectifying Unipolar Ni–HfOx Schottky Barrier ReRAM

A self-rectifying HfO$_x$-based ReRAM, using a Schottky barrier rather than a select device, was made using materials available in the wafer fab. It was discussed in December of 2011 by Nanyang TU, Peking University, A*STAR, NUS, Globalfoundries, Soitec, and Fudan University [36]. The self-rectifying behavior in the LRS was >10^3 at 1 V. The unipolar resistive switching was forming-free, and it had a wide read-out margin in a high-density cross-point memory array. The number of word-lines was >10^6 for worst-case conditions. The Ni–HfO$_x$–n+ silicon device showed a pronounced rectifying effect for the LRS. When the memory

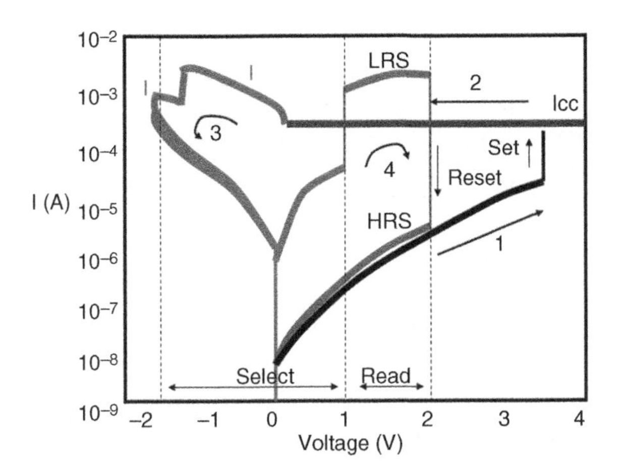

Figure 5.50 I–V curve of TiN–TaOx–Pt selector and Pt–TiOx–Pt ReRAM device. (Based on Y. Huang *et al.* (2012) (IME Peking University), *IEEE Transactions on Electron Devices* [35].

was set to LRS, the current in the positive direction of LRS was $>10^3$ larger than the negative current value for an 0.5 V reading voltage. With this forward–reverse current ratio for LRS, crosstalk could be suppressed without connecting a diode. HRS current does not change when the polarity of the applied voltage is changed. The self-rectifying effect in LRS was studied, and the effective Schottky barrier height at 0.5 V applied voltage was calculated.

A model was used to represent the self-rectifying effect in the LRS of the Ni–HfO_x–n+Si silicon device. During programming, Oxygen vacancy traps could be introduced in the HfO_x–SiO_x dielectrics whose energy levels might align with the mid-gap of the silicon substrate. During reverse bias, injection of electrons from the trap state in the oxide into the silicon electrode occurred. Due to the Schottky barrier at the n+ silicon and SiO_x–HfO_x junction, the current transport was suppressed. The Schottky barrier would not affect the HRS current transport [37]. In April of 2012, this group plus SUS&T Shenzhen discussed further the self-rectifying HfO_x-based unipolar ReRAM with NiSi electrode. A typical I–V curve of DC sweep of the NiSi–HfO_x–TiN is shown in Figure 5.51 [38].

The ReRAM has a NiSi–HfO_x–TiN structure, which is compatible with the NiSi source/drain in advanced CMOS technology. The ReRAM is CMOS process compatible, has self-rectifying behavior in the LRS, shows well-behaved memory performance including an R_{off}–R_{on} resistive ratio of $>10^2$, retention $>10^5$ s at 125 °C, and wide readout margin for dense cross-point arrays that is the number of word-lines $>10^6$ as a worst case. A self-rectifying ReRAM has rectification in the LRS, so crosstalk can be reduced without connecting a serial diode. The self-rectifying effect can improve misreading in crossbar memory without adding separate diodes. This self-rectifying unipolar HfO_x-based ReRAM is made with a TiN top electrode and a NiSi bottom electrode. The self-rectifying effect in the LRS can be seen in the difference in conduction ($>10^3$), with a positive and negative applied bias while in the SET state. In the SET process, a positive voltage sweep of 0–2.8 V abruptly triggers conduction and resistance, switching from the HRS to the LRS. Current compliance of 100 μA is applied to protect against permanent breakdown. A voltage sweep of 0–1.6 V causes abrupt decrease in current as the device switches back to HRS.

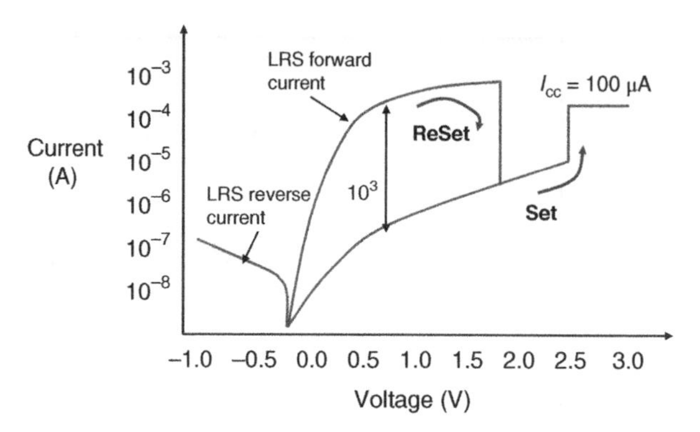

Figure 5.51 Typical I–V curve of self-rectifying unipolar Schottky barrier ReRAM. (Based on X.A. Tran, *et al.* (2012) (Nanyang Technological University, Peking University, NUS, Soitec, Fudan University, SUS&T Shenzhen), *IEEE Electron Device Letters*, 33(4) [38].)

Using a highly doped n+/p+ silicon as the bottom electrode in a unipolar ReRAM with a Ni-electrode–HfO_x structure was discussed in January of 2013 by Nanyang TU, NUS, Soitec, and SUST China [39]. With heavily doped p+ silicon as the bottom electrode, the devices showed a coexistence of bipolar and unipolar resistive switching. By substituting heavily doped n+ silicon, the switching behavior changed to that of a self-rectifying unipolar device. The asymmetry and reproducible rectifying behavior in the n+ $SiHfO_x$–Ni device resulted from the Schottky barrier of defect states in the SiOx–HfOx junction and N+ silicon substrate. Both rectifying characteristics and a high forward current density were observed in the Ni–HfO_x–n+ silicon device. The sneak current path in the conventional cross-point architecture was significantly suppressed.

5.6.8 Schottky Barriers for Self-Rectifying Unidirectional Cross-Point Array

A cross-point resistive switching memory with a self-formed Schottky barrier that worked without a selection device was discussed in June of 2012 by the Gwangju Institute [40]. The Schottky barrier was formed by a reduction–oxidation reaction at the Al–$Pr_{0.7}Ca_{0.3}MnO_3$ (Al–PCMO) interface and worked as a selection device. The polycrystalline PCMO film was sputtered at 650 °C on a Pt–Ti–SiO_2–Si substrate followed by an Al layer and Pt capping layer, as shown in Figure 5.52 [40]. Also shown is the structure with an electrode diameter of 50 nm.

In the HRS, an AlO_x layer occurred by oxygen migration. The I–V curve of the Al–PCMO device is displayed in Figure 5.53 [40], which shows the reduction and oxidation of AlO_x at the Al–PCMO interface. This I–V curve can be considered as three regions. In the LRS, the curve had a Schottky diode type of behavior (metal–semiconductor [MS]). In the HRS, AlO_x is formed in region II and the device is a MOS structure. Region III is a transition region, changing between MS and MOS. Cross-point cell operation was confirmed using a 50 nm electrode. Data read at 10^5 cycles with a read time of 10 μs was confirmed along with SET–RESET operation under worst-case conditions for 10^4 times [40].

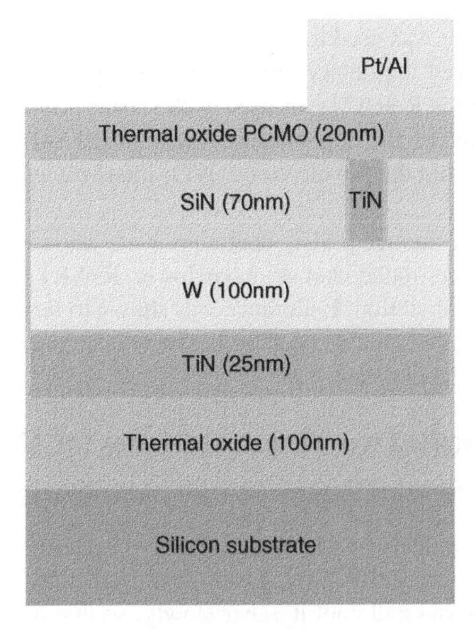

Figure 5.52 Cross-point ReRAM with self-formed Schottky barrier. (Based on M. Jo *et al.*, (Gwangju Institute), VLSI Technology Symposium, June 2010 [40].)

5.6.9 Thermally Induced Set Operation for Unipolar ReRAM with Diode Selector

The use of a thermally induced SET operation for unipolar WO_x ReRAM was explored by Macronix in June of 2013 [41]. The SET operation of a unipolar metal–oxide ReRAM is difficult because the electric field is in the direction for RESET rather than SET. As a result, unipolar operation causes strong device degradation from the high current required for the SET

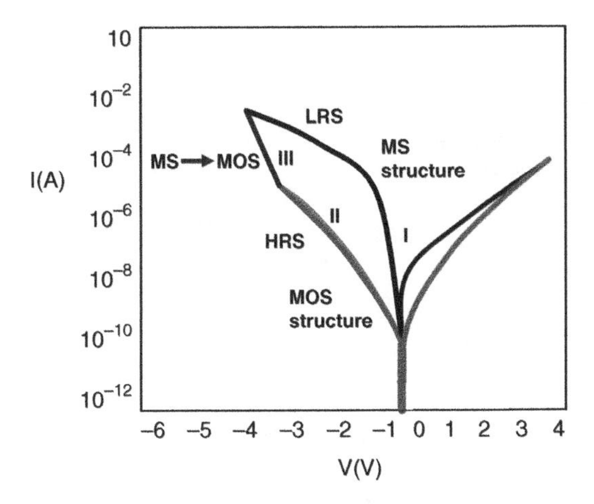

Figure 5.53 Illustration of I–V curve of the Al–PCMO device. (Based on M. Jo *et al.*, (Gwangju Institute), VLSI Technology Symposium, June 2010 [40].)

operation. An external heater was used to provide a thermally induced SET operation without high current, which improved the device performance and permitted a higher density 1D1R array. The top electrode of the ReRAM was used as the heater. By forcing current through the metal line, the temperature of the top electrode was raised sufficiently to cause thermal diffusion of oxygen away from the top electrode–WO_x interface without requiring the electric field across the WOx layer.

Cells with the thermally induced SET operation were shown to have similar RESET behavior to those in bipolar mode and to have lower RESET voltage than when using conventional unipolar SET operation. Endurance was shown to be greater than 70 000 cycles. The operation is expected to be scalable because heater voltage decreases with bit-line scaling for the device number stayed constant.

5.7 Unipolar PCM with Two-Terminal Diodes for Cross-Point Array

5.7.1 Background of Phase-Change Memory in a Cross-Point Array

PCMs used as switching elements in a cross-point array have used both diode and transistor switches in the array. PCMs are ReRAMs that change resistance by using a heating element to heat the phase-change material and cool it either slowly, so that a crystalline structure forms that has low resistance, or rapidly, so that the material remains in an high-resistance amorphous state. The two resistances represent the two states of a memory device.

A typical I–V curve of a PCM is shown in Figure 5.54. Rapid cooling yields the amorphous state and slow cooling the crystalline state. The resistance of the two states is different. If a threshold voltage below the switching voltage is used to read, a two-state resistive memory results. The SET, RESET, and READ levels are shown. The PCM also has a threshold shift completely in the amorphous state. This is indicated in Figure 5.54.

For a write operation, an electrical current pulse leads to internal temperature change that either melts and rapidly quenches a volume of amorphous chalcogenide materials for the HRS or a pulse that holds the volume at a lower temperature for long enough to recrystallize for the LRS. Figure 5.55 illustrates the PCM write temperature vs. time for both SET and RESET.

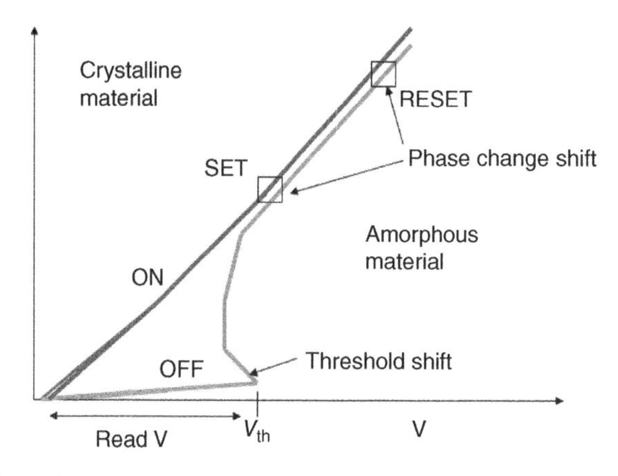

Figure 5.54 Phase-change memory I–V curve showing SET, RESET, and READ operations.

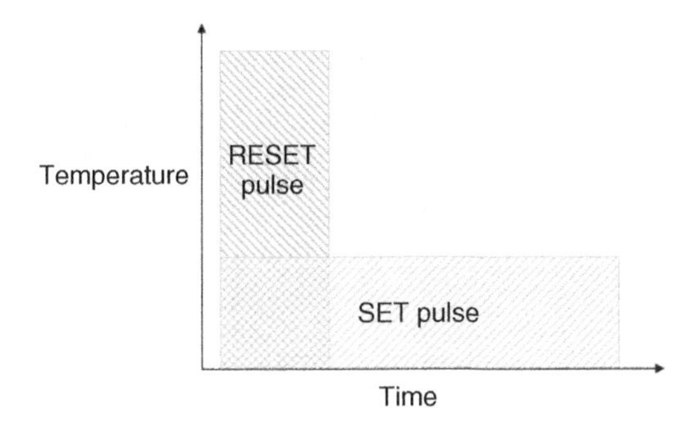

Figure 5.55 Phase-change resistance memory write temperature vs. time for SET and RESET.

5.7.2 PCMs in Cross-Point Arrays with Polysilicon Diodes

An early cross-point array that used a PCM element and a two-terminal polysilicon selection diode was described in June of 2009 by Hitachi [42]. The polysilicon diode was made using a low-thermal-budget process with 80 nm design rules. It had low contact resistivity, high 8 mA/ cm^2 on-current density, and low 100 μA/cm^2 off-current density. Reset current was 160 μA. The properties of the polysilicon diode made the set–reset operation of the $4F^2$ cell possible. It permitted a smaller cell size and had half the reset current of an earlier epitaxial silicon diode.

PCM devices using a polysilicon access device and PCM technology have the potential to make a multilayered cross-point memory with very high density that is suitable for solid-state disk applications. The PCM required a low operating voltage, and the $4F^2$ cross-point array had a large reset current of about 8 mA/cm^2. The polysilicon diode permitted fabrication of a memory array in an upper layer above the silicon substrate so that part of the peripheral circuitry could be under the memory array and the total chip size be reduced. The polysilicon diode was made on a low-resistance tungsten electrode word-line so that the resistance would be low enough for array operation. The contact resistance between the polysilicon and the top and bottom electrodes was necessary to increase the on-current of the poly diode. The TiN top contact bit-line had the resistance lowered by inserting Ti between the TiN and the p-doped polysilicon. A cross-sectional view of the GST PCM and polysilicon diode is shown in Figure 5.56 [42].

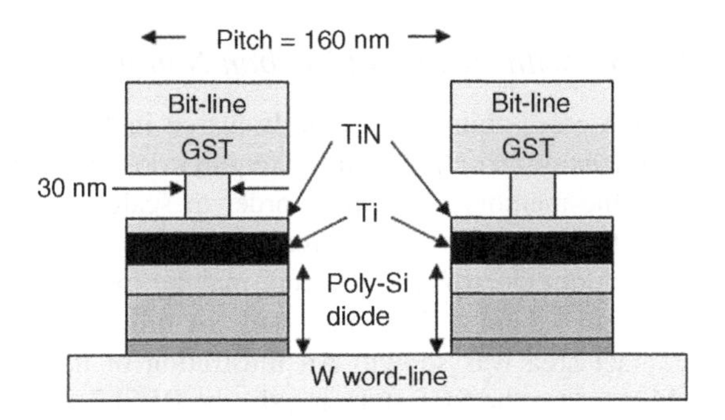

Figure 5.56 Schematic cross-section of GST PCM and polysilicon diode cross-point array. (Based on Y. Sasago *et al.*, (Hitachi), VLSI Technology Symposium, June 2009 [42].)

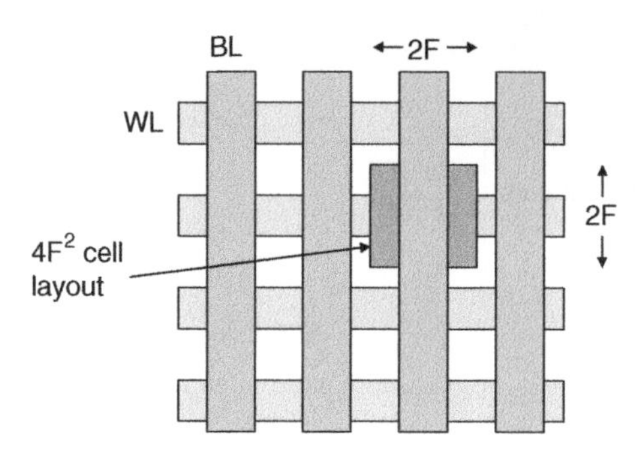

Figure 5.57 $4F^2$ cell layout with no strapping contacts in cell. (Based on S.H. Lee *et al.*, (Hynix), IEDM, December 2011 [43].)

The process flow for the memory array used a self-aligned process and started by depositing the diode materials on the word-line then patterning the stack, followed by depositing the interlayer dielectric and planarizing. The electrode and diode were then patterned so that a $4F^2$ pillar-shaped polysilicon diode resulted. Interlayer dielectric was deposited and planarized, then the GST chalcogenide and tungsten were deposited and patterned. A $4F^2$ cell in 80 nm design rule was made, along with a single 30 nm cell. I_{reset} at 30 ns was 160 μA. Simulations showed this could be reduced to 135 μA using a pillar-type PCM and to 10 μA using a recessed pillar PCM with an air gap [42].

A $4F^2$ 1Gb single-layer cross-point memory array where $F = 42$ nm was described by Hynix in December of 2011 [43]. The array used a GST PCM element and a polysilicon access device on a metal word-line. The cell had an 84 nm pitch with a unit cell size of 84×84 nm^2 (0.007 μm^2 cell size). The 33.21 mm^2 chip had 16 partitions of 64Mb each. Two-level Cu, W, and Al interconnects were used in the memory cell array and core/peripheral circuit. A fully integrated chip had functionality and reliability demonstrated.

The fully integrated chip contained a $4F^2$ cell structure with the access device in series on the metal word-line with no strapping contact within the cell array. The $4F^2$ layout in Figure 5.57 [43] shows the bit-line and word-line orientation. The memory element device was put under the cell block, and a self-aligned process was used. The technology was considered a platform for further multilayer cross-point array stacking.

5.7.3 Cross-Point Array with PCM and Carbon Nanotube Electrode

A scaled PCM for use in a cross-point array was discussed in June of 2011 by Stanford University [44]. A high-density cross-point array requires low wire resistance and low programming current for the memory element in order to scale the selector device. The PCM device used in this study had a RESET current of 1.4 μA and SET current of 0.5 μA. CNTs were used for the bottom electrode of the cell, making the lithography independent of critical dimension down to 1.2 nm at the 2.5 nm node. A fully functional PCM cell with sub–5 nm^2 effective contact area was shown. An illustration of the cell cross-section is shown in Figure 5.58 [44] for (a) the SET state and (b) the RESET state. An illustration of the switching behavior is shown in Figure 5.59 [44] for the first and for the hundredth cycle [44].

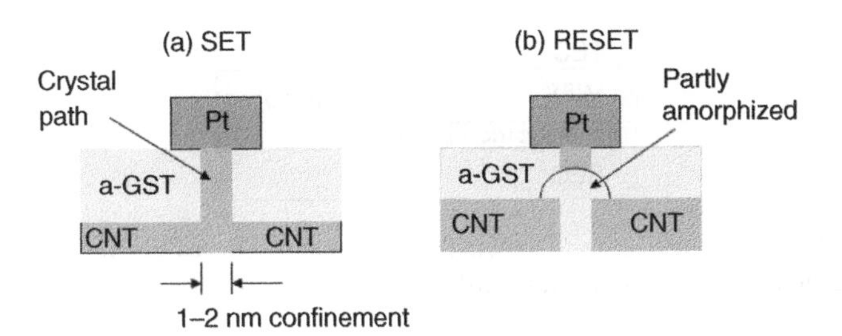

Figure 5.58 Illustration of cross-section of the cell (a) SET state; and (b) RESET state. (Based on J. Liang, (Stanford University), VLSI Technology Symposium, June 2011 [44].)

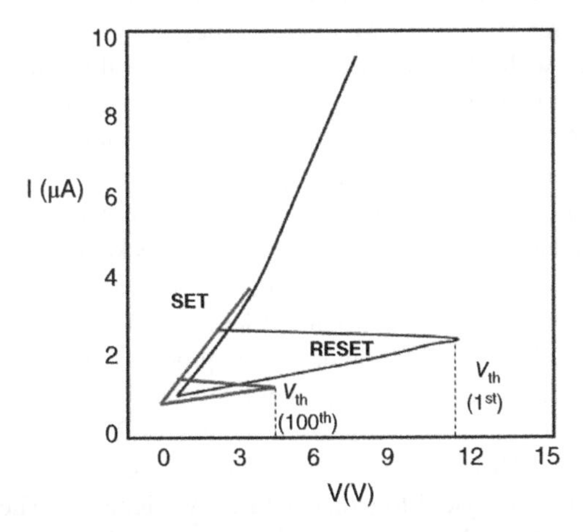

Figure 5.59 I–V curve of the CNT cell switching behavior, showing V_{th} for the 1st and the 100th cycle. (Based on J. Liang, (Stanford University), VLSI Technology Symposium, June 2011 [44].)

This cell had the effective confinement of the conducting chalcogenide path at 1–2 nm due to the dimension of the CNT electrode. It had I_{reset} of 1.4 μA. For comparison, I_{reset} is 200 μA for a 45 nm PCM technology. The small dimensions and reset current make this memory potentially suitable for a high-density cross-point array.

5.7.4 Cross-Point Array with MIEC Access Devices and PCM Elements

Access devices for cross-point arrays based on Cu–ion motion in novel Cu containing MIEC materials were discussed by IBM in June of 2010 [45]. These MIEC access devices were shown to provide the very high-current densities needed for PCM operation and also have high ON/OFF current ratios. Back-end-of-line (BEOL) fabrication of MIEC devices is at a temperature below 400 °C. The MIEC material was used to fill a 250 nm via etched into a 40 nm thick silicon nitride that stops at the inert bottom electrical contact. The stack consisted of the negative top electrode contact, the MIEC material, a silicon nitride dielectric that contains the via, and the back electrode contact, as shown in Figure 5.60 [45].

Operationally, the negative bias on the top electrode contact pulls Cu+ ions away from the bottom electrode contact, which leaves vacancies that act as acceptors. A gradient is induced in

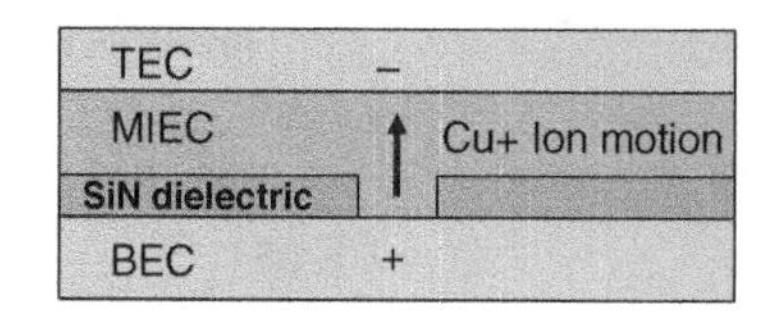

Figure 5.60 MIEC access device stack showing Cu ion motion. (Based on K. Gopalakrishnan, *et al.*, (IBM), VLSI Technology Symposium, June 2010 [45].)

the hole density, which is exponentially dependent on bias and drives a steady-state hole current. The large current densities are possible because the fraction of displaceable copper atoms is high. These MIEC access devices are scalable and require no breakdown, which permits nondestructive reads of a PCM device. They have the potential for combining multilevel cells together with 3D stacking for very high-density nonvolatile memory [45].

Compact arrays of BEOL-friendly access devices based on Cu-containing MIEC materials were discussed by IBM in June of 2011 [46]. MIEC selection devices could be used as an access device for PCMs because they had the required high current densities and the large ON/OFF ratios. These access devices could also be used with a bipolar memory switch. The scaled access devices showed large voltage margin (V_m), low leakage near 0 V of <10 pA, and endurance $>10^8$ cycles at high current densities. Figure 5.61 illustrates the I–V curve of a single MIEC access device showing leakage currents of less than 10 pA near 0 V [46]. The device used a simplified CMOS-compatible diode-in-via process. A simple one-mask BEOL-compatible chemical-mechanical polishing process was used.

BEOL-friendly access devices were discussed in June of 2012 by IBM [47]. These were based on Cu-containing MIEC materials integrated in series with PCM elements into 512 × 1024 bit arrays at 100% yield together with PCM elements. The current provided by the MIEC device was >200 μA. Leakage was <10 pA, and high voltage margin was 1.5 V, as

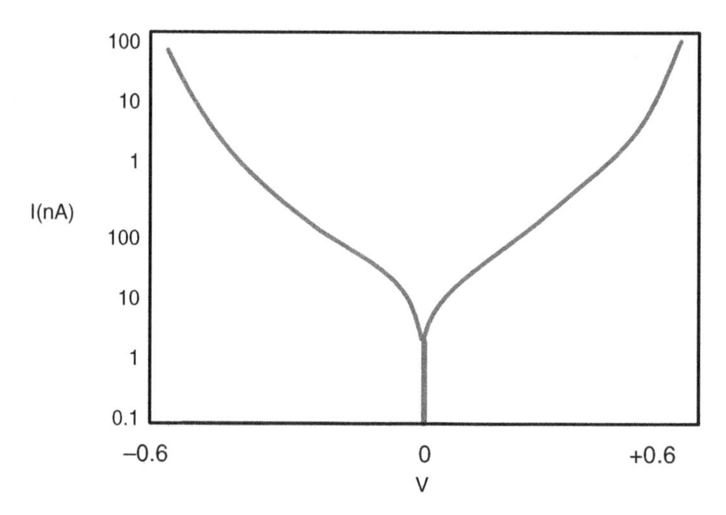

Figure 5.61 Illustration of I–V curve of single MIEC access device. (R.S. Shenoy, *et al.*, (IBM), VLSI Technology Symposium, June 2011 [46].)

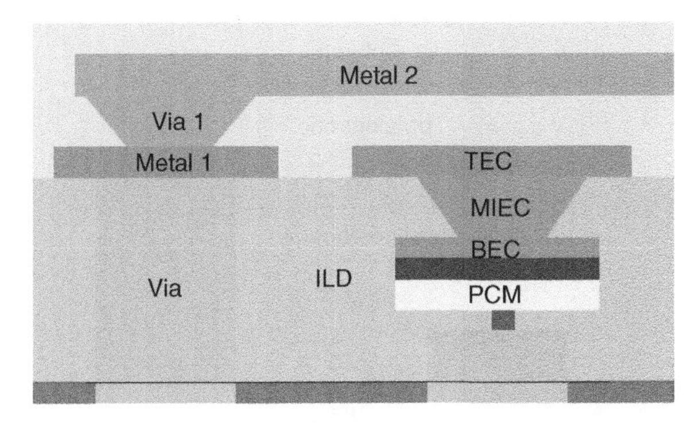

Figure 5.62 Schematic cross-section of MIEC devices in series with PCM. (Based on G.W. Burr *et al.*, (IBM), VLSI Technology Symposium, June 2012 [47].)

required for large cross-point arrays. The target application was to compete with the NAND flash.

The MIEC devices had large ON/OFF ratios in large cross-point arrays with high voltage margin and leakage below 10 nA. The MIEC devices had bipolar diode-like characteristics, so bipolar operation was possible with the MIEC device. MIEC access devices were shown to support processing temperatures up to 500 °C. The 512Kb arrays with integrated MIEC access devices were shown at 100% yield. The MIEC devices were demonstrated with mushroom-cell PCM memory elements with 35 nm heater electrodes that were integrated using the keyhole transfer method followed by MIEC, as shown in Figure 5.62 [47].

5.7.5 Threshold Switching Access Devices for ReRAM Cross-Point Arrays

A threshold switching device can be used as an access device with a bipolar memory switch. They are compatible with a PCM resistance switch because they can be made of the same material and have the same response to electrical, thermal, and stress conditions.

A 64Mb test chip array with a cell consisting of a PCM element with a threshold switching access device that used the amorphous state threshold switching curve of the PCM was discussed in December of 2009 by Intel and Numonyx [48]. The threshold switching device was called an Ovonyx threshold switch (OTS). The devices could be built above the substrate with the chip organized in a true cross-point array. The companies indicated the breakthrough was an early research milestone and that a product based on the stacked PCM might still be years away. One layer of the array was fully integrated with CMOS technology. The memory cell stack included rows and columns sandwiched between metals M2 and M3. A schematic cross-section of the PCM stacked with the threshold switching device as one memory layer in a cross-point array is shown in Figure 5.63 [48]. This was expected to be a $4F^2$ cell in a cross-point array.

The arrays were stacked on top of CMOS circuits that were used for decoding, sensing, and logic functions. RESET speed was 9 ns, and endurance was 10^6 cycles. There was 1 V of dynamic range achieved between SET and RESET. In this element, a thin-film two-terminal threshold switching device was used as the selector of the memory cell. The technology was stackable over back-end CMOS technology. The thin-film PCM cell used low-temperature

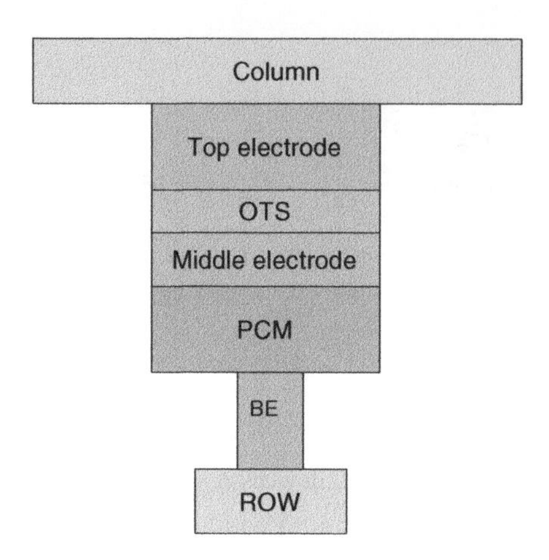

Figure 5.63 Schematic cross-section of stacked PCM and threshold switching device in cross-point array. Based on D.C. Kau *et al.*, (Intel, Numonyx), IEDM, December 2009 [48].)

processing, which made it compatible with the BEOL metallization of the current CMOS technology [48].

The SET state was shown to have lower threshold voltage than the RESET state. The higher threshold voltage of the RESET state was due to the serial connection of the amorphous OTS and amorphous PCM and was the sum of these two RESET states. To access a bit, V_{access} was applied to the selected column and 0 V to the selected row. Deselected row and column biases needed to be chosen so that the voltage across each unselected cell was less than the minimum V_{th} in the block. A 64Mb PCM plus OTS cross-point test chip in 90 nm CMOS technology was shown. Cell sizes were from 40 to 230 nm. One memory layer was placed between the second and third levels of Cu interconnect [48].

5.7.6 p–n Diode Selection Devices for PCM

The relative advantages of different driving devices for PCM using 3D numerical device simulation were discussed in March of 2011 by Hong Kong University of Science & Technology, IBM, Qimonda, and Macronix. The candidates considered included p–n diodes and vertical gate-all-around (GAA) MOSFETs. Performance parameters considered included cell size, current drive, disturb immunity, power dissipation and scalability. In large device technologies, p–n diodes were found superior, but scaling to 65 nm and beyond appeared to improve the performance of GAA MOSFETs. As a result, GAA MOSFETs outperformed p–n diodes in very scaled technologies [49].

Both driver types were found to achieve satisfactory reset current in device size and supply voltage tradeoffs. The p–n diodes had high current drive and low disturb current in larger technology nodes; however, the diode voltage could not be scaled adequately, and the disturb current was worse due to the increasing gain of the lateral p–n–p transistors. For this reason, the performance of GAA MOSFET arrays was found superior for scaling to 65 nm and beyond [40].

5.7.7 Epitaxial Diode Selector for PCM in Cross-Point Arrays

In December of 2011, Samsung discussed a 20 nm node PCM cell technology for use in a crossbar array architecture with an epitaxial diode access device for the PCM cell [50]. The diode integration process and improved implantation technology were used to obtain the required diode on-current with low I_{off}. A confined cell structure was used to reduce power. New bottom electrode materials were developed to reduce RESET current to below 100 μA. A 20 nm PCM device was developed, and a $4F^2$ PCM cell was used. The unit cell consisted of a silicon diode and a confined cell. The diodes were made by selective epitaxial growth. On top of the diode, a bottom electrode contact was made of a dash-type highly resistive material followed by a confined phase-change material cell structure. In the resulting cell, I_{off} was <1 pA and I_{reset} was <100 μA. Program disturb was checked, and it was found that the programming did not affect the adjacent cell during 10^7 cycles. The 20 nm phase-change RAM was found still viable with potential for scaling.

5.7.8 Dual-Trench Epitaxial Diode Array for High-Density PCM

The design and fabrication of a dual deep-trench epitaxial p–n junction diode array in 130 nm CMOS technology was discussed in August of 2011 by SMIC, Microchip, and the Chinese Academy of Science [51]. A 16 × 16 diode array with a $5F^2$ 0.196 μm^2 cell size was made with good electrical properties. Current drivability was more than 12.5 mA/μm^2. The ON/OFF current ratio was greater than ninefold, and cross-talk immunity was good. A schematic of the deep-trench epitaxial diode array is shown in Figure 5.64.

This dual-trench epitaxial diode could be used as the access device for a high-density PCM. The disturbance current during program was investigated. The disturbance current was the reverse-biased current of the neighbor diode and the bipolar current from the parasitic p–n–p bipolar junction transistor formed from the neighbor diodes. When the forward voltage was at 2 V, the ratio of the disturbance current of the nearest neighbor diode to programming current was 1.34%. The disturbance current of the nearest neighbor diode along the same word-line was found to be the worst case.

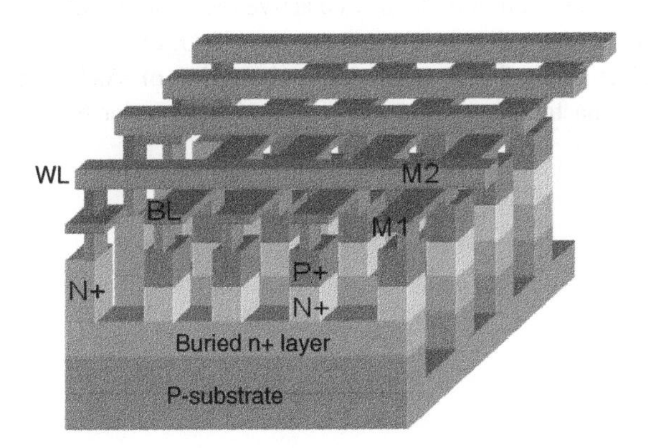

Figure 5.64 3D schematic of the deep trench epitaxial diode array. (Based on C. Zhang *et al.*, (SMIC, Chinese Academy of Science, Microchip), *IEEE Electron Device Letters*, 32(8), 1014 [51].)

5.8 Bipolar Resistance RAMS With Selector Devices in Cross-Point Arrays

5.8.1 VO_2 Select Device for Bipolar ReRAM in Cross-Point Array

A VO_2 select device for a bipolar 1T1R ReRAM in a cross-point array was reported by the Gwangju Institute of Science and Technology in November of 2011 [52]. VO_2 had a first-order metal-insulator transition property along with ON/OFF ratio >50. Switching speed was <20 ns, and current density (J) > 10^6 A/cm^2. Large-area VO_2 devices were found to be prone to defects, but small VO_2 devices, those with an active area <5×10^4 nm^2, had good switching uniformity and were essentially defect free. This was attributed to reduced numbers of defects in small samples and to the metal-insulator transition of the VO_2. The VO_2 select device was thought to have potential for use in cross-point array bipolar ReRAM applications. The VO_2 device used had a Pt–VO_2–Pt stack. A schematic cross-section of the VO_2 selection device is shown in Figure 5.65 [52].

A ZrO_x–HfO_x bipolar ReRAM was integrated with theVO_2 select device to obtain an improved readout margin. An I–V curve of the bipolar ReRAM and that of the VO_2 select device separately is illustrated in Figure 5.66. An LRS of the selected cell, an HRS of a selected cell, and an LRS of an unselected cell are indicated.

The VO_2 selection device was connected in series with the ZrO_x–HfO_x bipolar ReRAM. The rectifying behavior was confirmed, as shown in Figure 5.67, where an LRS of the selected cell, an HRS of a selected cell, and an LRS of an unselected cell are again indicated. Memory size was improved based on the readout margin. The VO_2 selection device offered good selection properties that could suppress crosstalk issues and improve readout margin [52].

5.8.2 Threshold Select Devices for Bipolar Memory Elements in Cross-Point Arrays

A scalable, threshold select device for bipolar memory elements in cross-point arrays was described by Samsung in December of 2012 [53]. It was made of nitridized AsTeGeSi chalcogenide glass, which showed threshold switching using electronic charge injection. These AsTeGeSiN switches were able to reduce the endurance limitations experienced previously due to changes in Te concentration in the device active region and scaled to 30 nm with an on-current of 100 μA. A memory cell was fabricated using a TaO_x resistance memory element with the AsTeGeSiN select device. The degradation behavior of AsTeGeSiN switching was analyzed, and N_2 plasma treatment was used to provide a barrier to Te loss so that a high

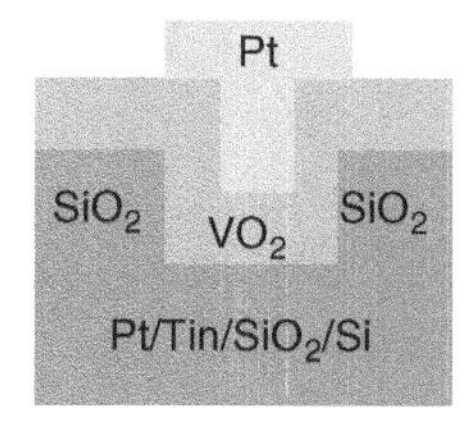

Figure 5.65 Schematic cross-section of the VO_2 selection device. (Based on M. Son *et al.* (2011) (Gwangju Institute of Science and Technology), *IEEE Electron Device Letters*, 32(11), 1579 [52].)

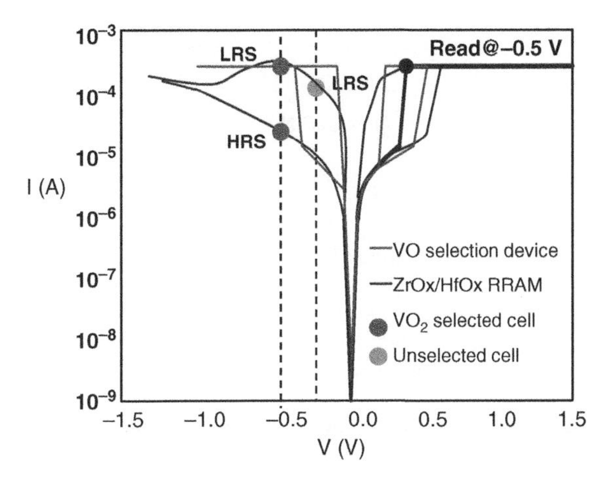

Figure 5.66 Illustration of I–V curves of (a) a VO_2 select device; and (b) a ZrO_x–HfO_x bipolar ReRAM. (Based on M. Son *et al.* (2011) (Gwangju Institute of Science and Technology), *IEEE Electron Device Letters*, 32(11), 1579 [52].)

Te concentration was maintained even after annealing. The N_2 plasma treatment created a thin SiN layer on the surface. Cycling performance was $>10^8$ at a pulse width of 1 μs. Devices of various sizes were made, and all showed threshold switching. The current density increased for smaller-sized devices, indicating a filamentary switching mechanism. For a $30 \times 30\,nm^2$ cell, the current density was $1.1 \times 10^7\,A/cm^2$, which is close to the values of current density for a silicon diode. High temperature measurements showed that the amorphous phase was maintained past 600 °C [53].

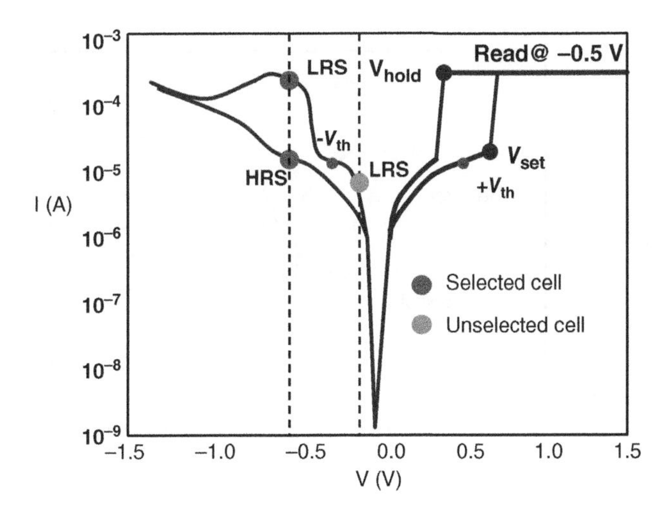

Figure 5.67 Illustration of rectifying behavior of VO_2 select device with bipolar ReRAM. (Based on M. Son *et al.* (2011) (Gwangju Institute of Science and Technology), *IEEE Electron Device Letters*, 32(11), 1579 [52].)

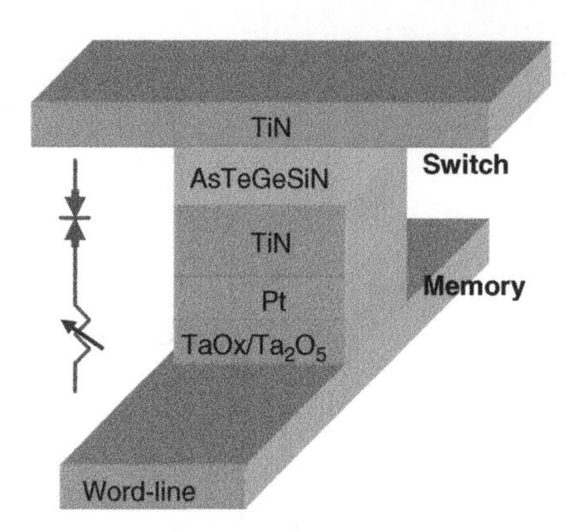

Figure 5.68 Schematic of AsTeGeSiN threshold switching device. (Based on M.J. Lee *et al.*, (Samsung), IEDM, December 2012 [53].)

A one switch–one resistor memory cell was made with a W bottom electrode and a thin AlO_x layer for passivation of the W. The TaO_x resistive memory was made by reactive sputtering and then oxidized to form a Ta_2O_5 layer. The middle electrode is a Pt–TiN double layer. Next, a switch layer of AsGeTeSiN was deposited and capped with a TiN electrode. This stack is shown in Figure 5.68. The individual switching behavior is shown in Figure 5.69 for the diode and memory element separately. The combined I–V measurement of the switch and memory cell are shown in Figure 5.70. The operation window for unselected cells in the ON and OFF states is indicated [53].

Select device characteristics included maximum current density of 1.1×10^7 A/cm^2 @ 30 nm node as well as selectivity at ΔI at I_{set}, I_{read} (1.2 V_{set}) of 100 at 30 µm and 1000 at 30 nm. Endurance was 10^3 cycles for DC and 10^8 cycles for pulse. The maximum processing temperature was 200 °C. The device was stable to 500 °C [53].

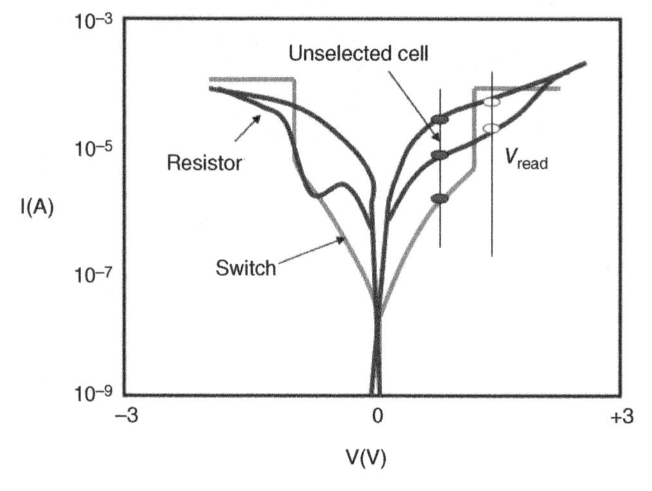

Figure 5.69 Individual switching behavior for diode and memory element. (Based on M.J. Lee *et al.*, (Samsung), IEDM, December 2012) [53].)

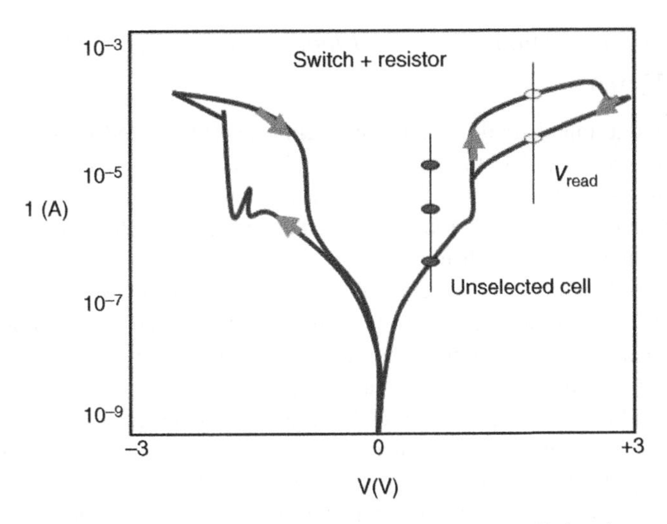

Figure 5.70 Illustration of combined I–V for the switch and memory cell showing rectification effect. (Based on M.J. Lee *et al.*, (Samsung), IEDM, December 2012 [53].)

5.8.3 Vertical Bipolar Switching Polysilicon n–p–n Diode for Cross-Point Array

A vertical bipolar switching polysilicon n–p–n diode with ambipolar operation for a cross-point memory was discussed in June of 2011 by National Taiwan Normal University and ITRI [54]. A p–n diode can be used with unipolar operation for PCM, but for the bipolar programming found in many ReRAMs, a bidirectional turn-on behavior is required for the switching driving device. The structure of the n–p–n diode is shown in Figure 5.71.

The ambipolar switching diode with polysilicon n–p–n had high $J_{on} = 0.1\ mA/cm^2$ for an area $= 2.25 \times 10^{-8}\ cm^2$. Both negative and positive biases made the positive shift of the J–V curve with DC stress of 100 s. Reliability was sufficient for $>10^6$ cycles of 100 ns operation with 0.1 s total stress at 4 V. $V_{bd} = 2.25 \times 10^8\ cm^2$ was estimated >1 V for 10 years [54].

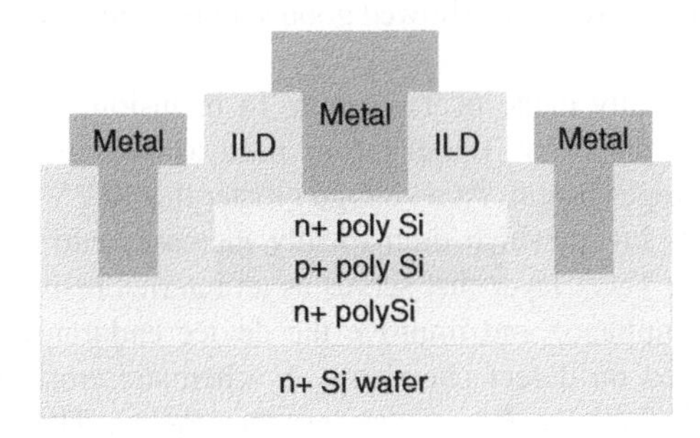

Figure 5.71 Structure of vertical n–p–n diode with ambipolar operation. (M.H. Lee *et al.*, (National Taiwan Normal University, ITRI), DRC, June 2011 [54].)

5.8.4 *Two-Terminal Diode Steering Element for 3D Cross-Point ReRAM Array*

The status of two-terminal diode steering elements for 3D stacked crossbar memory was reviewed by Applied Materials in May of 2012 [55]. The process, integration, and device scaling requirements were considered for the steering elements for metal–oxide ReRAM cells. Ion implantation and the activation required to achieve the 50 nm tall diode pillar required for the 3D sub–2 nm memory were disucssed.

The two-terminal diode steering element for use in a 3D PCM crossbar switch was discussed further by Applied Materials in June of 2012 [56]. Process, integration, and device scaling requirements of the steering element were noted along with the required ion implantation and activation for making a 50 nm tall diode pillar for a sub–20 nm 3D crossbar memory. For the phase-change RAM, high-density integration was required. Other issues were programming current density, thermal instability during integration, and material reliability during operation. The solution for high current density was to reduce the volume of the material and use more isolation material. The electrode needed to be designed to introduce heat during operation so that the material could operate at lower current density. The deposition and integration processes could be improved to address the reliability issue. Due to heat transfer considerations, the phase-change material could not be exposed to temperatures above 500 °C. For this reason, high-temperature rapid thermal annealing activation was determined to be unsuitable, and laser anneal was considered a potential alternative.

5.8.5 *Varistor-Type Bidirectional Switch for 3D Bipolar ReRAM Array*

Because a bipolar ReRAM is more uniform and has higher endurance than a unipolar device, it is considered better suited for large-scale array applications. However, this means a bidirectional selection device would be useful to suppress sneak currents. A varistor-type bidirectional switch solves issues faced by earlier selection devices that had insufficient current density at SET and RESET operations along with low selectivity and poor endurance. A varistor-type bidirectional (VBS) switch with good selection properties was discussed by Gwangju IST and Hynix in June of 2012 for use as a select device in a ReRAM array [57]. The highly nonlinear VBS had current density $>3\times10^7$ A/cm^2 and high selectivity at 10^4. The nonlinear I–V characteristics were explained as multilayer tunnel barriers formed by the Ta included in thin TiO_2. The 1S1R device showed good suppression of leakage current with $>10^4$ reduction at $V_{read}/2$.

Varistors are conventionally made of TiO_2 with Ta inclusion, which increases tunneling current. At voltages less than 0.7 V, the electrons need to overcome the TaO_x–TiO_2–TaO_x barriers that suppress leakage current. At a voltage greater than 0.7 V, the electrons can tunnel directly through the TiO_2 layer, which dramatically increases current. At voltages >1.2 V, current is saturated by the TaOx layer that serves as an internal resistor. The internal resistor was found to control compliance and improve the device endurance. The Pt–TaO_x–TiO_2–TaO_x–Pt device was based on defect chemistry. A schematic cross-section of the stack is shown in Figure 5.72 [57]. Initially, $R_{(TaOx)}$ was less than $R_{(TiO2)}$. The TaOx(1) supplied Ta to the TiO_2 during device fabrication. The TaO_2 buffer oxide acted as an internal resistor,

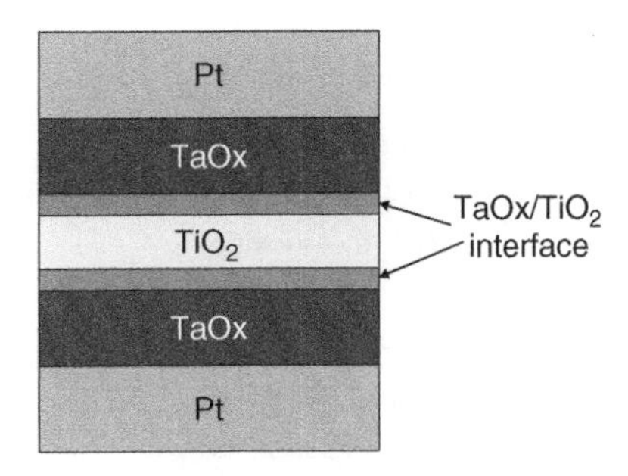

Figure 5.72 Pt–TaO_x–TiO_2–TaO_x–Pt varistor-like bidirectional switch for ReRAM device. W. Lee *et al.*, (Gwangju Institute of Science and Technology, Hynix), VLSI Technology Symposium, June 2012 [57].)

providing self-compliance at high voltage. The defective TiO_2 was formed by Ta incorporation at the TaO_x–TiO_2 interface. The Ta-doped TiO_2 enabled electron creation [57].

5.8.6 Bidirectional Threshold Vacuum Switch for 3D $4F^2$ Cross-Point Array

A 3D stackable, bidirectional threshold vacuum switching (TVS) selector was described by NDL, NARL, NCTU, Mesoscope Technology, Fu-Jen University, CYCU, and the University of California, Berkeley, in December of 2012 [58]. This selector used the same WOx material as the ReRAM element. It provided current density of $>10^8$ A/cm^2 and selectivity of $>10^5$. A stress test at high current density showed a $>10^8$ cycle capability for RESET and SET. A mechanism based on recombination of oxygen ions and vacancies was suggested for the observed volatile switching of the TVS selector. Using the threshold characteristics of the TVS selector, a two-step reading waveform was developed that could be used for 3D stackable $4F^2$ cross-point ReRAM applications.

A vacuum layer was used in the film stack of the 3D ReRAM structure with TVS selector and TMO memory. An SiN sacrificial layer was used to create the vacuum gap. The memory element and selector were made using a thermal-budget-free wafer-scale electrochemical oxidation. A schematic diagram of the stack is shown in Figure 5.73. The vacuum gap is defined by a SiN sacrificial layer. The TVS selector's linear switching characteristics are illustrated in Figure 5.74 [58].

The 3D stackable TVS selector was capable of current density of $>10^8$ A/cm^2, selectivity $>10^3$, and 10^8 endurance cycles. The operating mechanism was believed to depend on the oxygen ion and vacancy dynamics at the interface. A two-step reading method enlarged the sensing window to make the device suitable for a 4Mb/block cross-point memory array. The two-step reading method could turn on the TVS selector of the selected cell while simultaneously turning off the other selectors in the memory array. Differential input voltages in the second step of the reading scheme between selected and unselected cells assisted in the reduction of sneak current in the memory array. Because the sneak current reduction by the high resistance of the TSV-off is much higher than the high resistance of the TMO memory, there was sufficient sensing window for a 2048×2048 bit (4Mb/block) cross-point memory array.

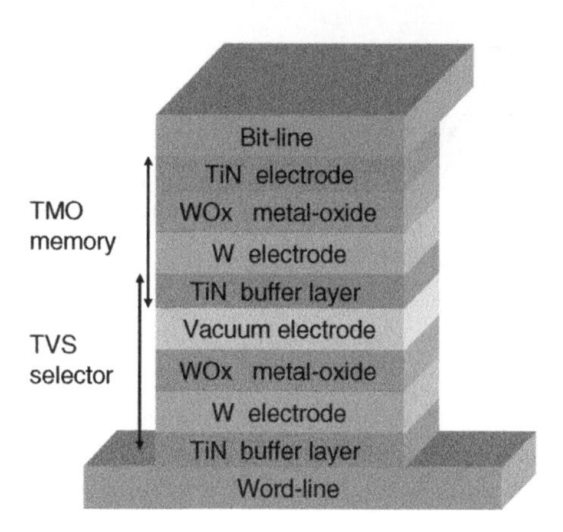

Figure 5.73 Schematic diagram of TVS vacuum selector and TMO memory. (Based on C.Y. Ho *et al.*, (NDL, NARL, NCTU, Mesoscope Tech., Fu-Jen University, CYCU, University of California Berkeley), IEDM, December 2012 [58].)

5.8.7 Bidirectional Schottky Diode Selector

An 8Mb two-layer cross-point 180 nm ReRAM macro with a bidirectional Schottky diode selector was discussed by Panasonic in February of 2012 and again in January of 2013 [37,59]. The macro had 443MB/s write throughput, consisting of a 64b parallel write bus in 17.2 ns cycle time. A bipolar TaO_x ReRAM was used. Cell stack for the memory was Ir–Ta_2O_5–TaO_x–TaN and for the bidirectional diode was TaN–SiN_x–TaN, which comprised two back-to-back metal–semiconductor–metal (MSM) Schottky diodes. The bottom TaN of the resistive element and the top TaN of the diode was shared. A schematic cross-section of the array structure of the ReRAM macro is shown in Figure 5.75 [37,58].

Sneak current in the bipolar cross-point cell array was reduced by several methods: a bidirectional diode was used as a select element, a hierarchical bit-line structure was used for

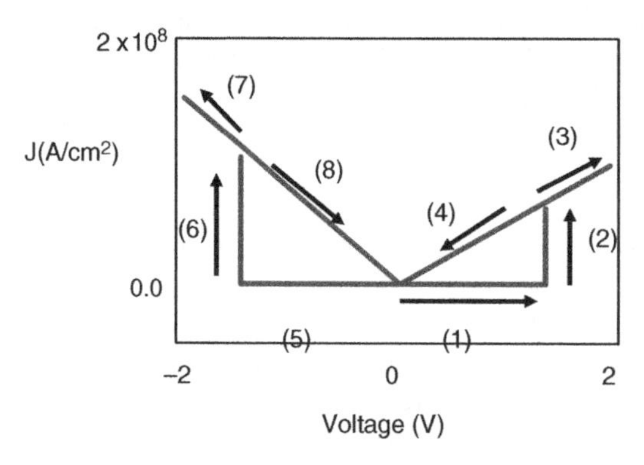

Figure 5.74 TVS selector linear switching characteristics. (Based on C.Y. Ho *et al.*, (NDL, NARL, NCTU, Mesoscope Tech., Fu-Jen University, CYCU, University of California Berkeley), IEDM, December 2012 [58].)

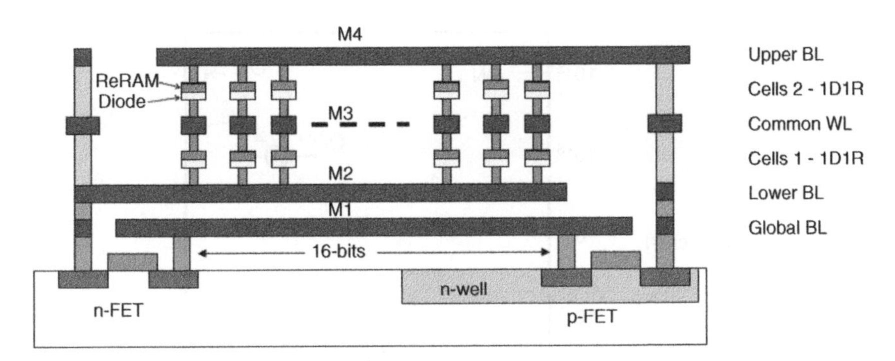

Figure 5.75 Schematic cross-section of two-layer ReRAM macro. (Based on A. Kawahara *et al.*, (Panasonic), ISSCC, 19 February 2012 [37,59].)

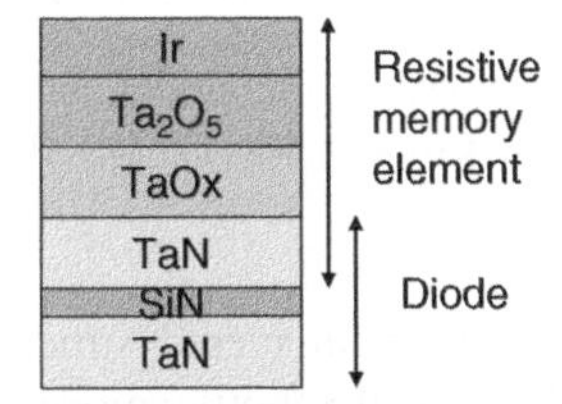

Figure 5.76 MSM diode and bipolar resistive switching element. (Based on A. Kawahara *et al.* (2013) (Panasonic), *IEEE Journal of Solid-State Circuits*, 48(1), 178 [59].)

multilayered cross-point memory with fast and stable current control, a multibit write architecture was implemented that has fast write and suppresses sneak current. A schematic cross-section of the MSM diode and TaN–TaO_x–Ti_2O_5–Ir bipolar resistive switching element is shown in Figure 5.76 [59].

The Schottky barrier height in the diode is determined by the difference between the TaN electrode's work function and the SiN_x semiconductor material's bandgap. The OFF current is controlled by the SiN_x thickness. The ON current is increased by reducing the nitrogen concentration, because the SiN_x bandgap decreased and the Schottky barrier height is lowered. An energy band diagram in Figure 5.77 illustrates the OFF state and ON state of the bidirectional diode [59].

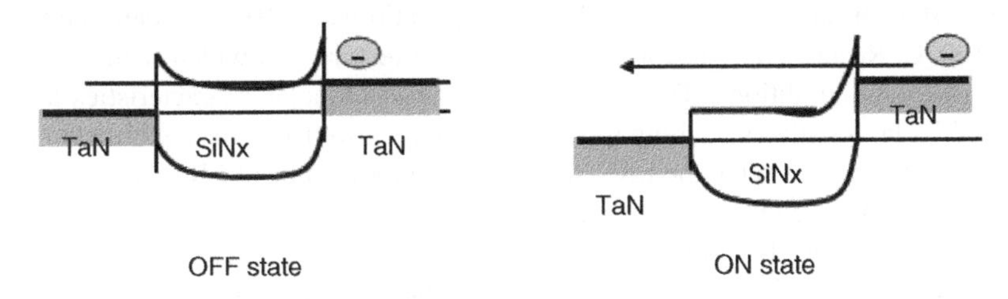

Figure 5.77 Schematic of the MSM diode energy band diagram. (Based on A. Kawahara *et al.* (2013) (Panasonic), *IEEE Journal of Solid-State Circuits*, 48(1), 178 [59].)

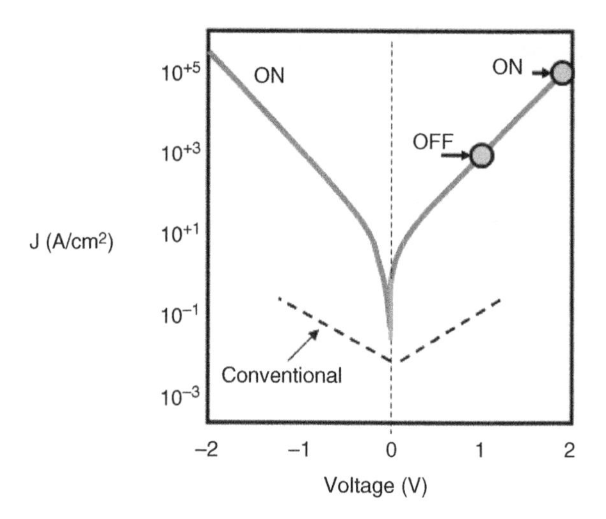

Figure 5.78 J-V curve of the bidirectional MSM diode. (Based on A. Kawahara *et al.* (2013) (Panasonic), *IEEE Journal of Solid-State Circuits*, 48(1), 178 [59].)

The MSM diode could be used for multilayered structures at fabrication temperatures <200 °C. The diode is composed of identical Schottky diodes connected back to back as TaN–SiN_x–TaN. The J–V curve of the MSM diode is shown in Figure 5.78 [59]. The curve is symmetrical in both bias directions. The measured ON current density is 1×10^5 A/cm^2, which is comparable to the required switching current of 140 μA, given the dimensions of the diode. The measured off-current density is 700 A/cm^2 with an ON/OFF ratio of 140. The J–V curve was easily controlled using the nitrogen concentration and the thickness of the SiN_x film.

The organization of the 8Mb macro was 64 K pages × 64 bits × 2 layers, and the macro size was 3.0 × 3.5 mm. The internal bus width was 64-bit with 16-bit I/O. Error correction code (ECC) was used with 7 parity bits per 64 data bits. Write throughput was 443 MB/s with a read latency of 25 ns.

5.8.8 *Bipolar ReRAM with Schottky Self-Rectifying Behavior in the LRS*

A resistive switching RAM using Na–AlO_y–n+ silicon was described by NUS, Soitec, and Southern University of Science and Technology of China in October of 2012 [60]. This device showed well-behaved bipolar memory characteristics with self-rectifying behavior in the LRS. RESET and SET currents were 50 μA. The R_{off}–R_{on} ratio was $>10^3$, and data retention was $>10^4$ s at 100 °C. Readout margin for the crossbar architecture was widen with $N > 25$ word-lines for worst-case conditions. The device also showed unipolar characteristics at positive voltages. In bipolar resistive switching mode, the device showed a rectifying effect in the LRS. After the ReRAM cell was set to the LRS, the positive current value of LRS was more than 700 times larger than the negative current value in the 0.2 V range.

It was found that control samples with top electrodes of TiN, Cu, and Al also showed self-rectifying behavior. For this reason it was postulated that the rectifying property resulted from the AlOx and n+ silicon interface, which forms an effective Schottky barrier with a height of

0.31 eV. During memory programming, defects can be introduced into the AlO_y dielectric whose energy levels form a Schottky barrier with a conduction band of n+ silicon in the LRS. This could suppress current transport, resulting in a self-rectifying effect. The Schottky barrier height does not change with device area, so it is expected to scale. The result is a ReRAM with self-rectifying behavior in the LRS, which makes it a potential memory element in a high-density cross-point memory array.

5.8.9 Self-Rectifying Bipolar ReRAM Using Schottky Barrier at Ta–TaO_x Interface

A self-rectifying Schottky Barrier bipolar Ta–TaO_x–TiO_2–TiO_x–Ti ReRAM with Ta top electrode (TE) and Ti bottom electrode (BE) was discussed in June of 2013 by NCTU and Winbond [61]. The device, which was intended for high-density 3D storage class memory, had over 10^{12} cycles endurance; was forming free, self-compliant, self-rectifying (>10^5), and MLC capable; and used a room temperature process and fab-friendly materials. Fabrication consisted of room temperature sequential deposition of 70 nm of TiO_x and 20 nm TaO_x followed by deposition and patterning of 100 nm thick Ta top electrode. The result was an amorphous TiO_x interfacial layer between polycrystalline TiO_2 and Ti with amorphous TaO_x. The interfacial TiO_x and TaO_x layers were oxygen deficient.

The self-rectifying cell showed stable bipolar resistive switching using opposite polarities of SET and RESET voltages. MLC operation showed negligible degradation after 1000 cycles. Retention was >10^4 s, and read disturb immunity was shown at −2 V for >10^9 cycles. Switching was thought to originate from oxygen migration in the TaO_x with electron injection from the Ta electrode at negative bias controlled by the Schottky barrier at the Ta–TaO_x interface.

5.8.10 Diode Effect of Pt–$In_2Ga_2ZnO_7$ Layer in TiO_2-type ReRAM

The resistance switching behavior in a ReRAM with stack of Pt–$In_2Ga_2ZnO_7$ (IGZO) – TiO_2–Pt was discussed in April of 2012 by Seoul National University [62]. This ReRAM was investigated for potential use in diode-free memory integration. The IGZO layer functioned as the semiconductor layer because it showed accumulation or depletion of carriers depending on the polarity of the bias. Electroforming occurred only with the IGZO depleted due to the limited background leakage current flow. Repeated SET–RESET operation was observed under depletion. While RESET was possible, SET was impeded by the high background current flow in the IGZO layer under the accumulation condition. The IGZO thin film had an amorphous structure and an easy fabrication process at low temperatures, so it could be made in a multilayer crossbar memory array. The electron accumulated configuration of the IGZO layer made electroforming impossible with a negative bias applied to the IGZO layer. On the other hand, the carrier depletion of the IGZO layer, with limited electron injection at the Pt–TiO_2 interface when positive bias was applied to the IGZO layer, made electroforming possible. After the forming operation, which created the conductive filament, SET–RESET operation was possible when the IGZO layer was positively biased. When the IGZO layer was negatively biased, SET was interfered with, thereby creating a diode-like effect after forming [62].

5.8.11 Confined NbO_2 as a Selector in Bipolar ReRAMs

The use of NbO_2 with threshold switching characteristics as a selector device with ReRAMs was discussed in December of 2013 by POSTECH, University of Tsukuba, Nagoya University, and Stanford [63]. Both 1S1R NbO_2–TaO_x and hybrid NbO_2–Nb_2O_5 devices were evaluated, and the feasibility of using an NbO_2 selector device was confirmed. Bidirectional selector devices with high selectivity are required to suppress sneak path leakage in bipolar ReRAMs. An NbO_x selector had good performance but relatively high operation current. As a result, the team investigated the threshold characteristics of an NbO_2 selector with reduced leakage current due to a heat-confined 3D architecture.

A 1S1R structure with NbO_2 selector and TaOx-based ReRAM were made with a shared TiN layer for a middle electrode. In addition, an NbO_2–Nb_2–Nb_2O_5 hybrid device was made with a sputter-deposited NbO_2 layer. The compliance current condition was established, and then deposition times of the NbO_2 film were investigated. The dependence of threshold voltage and ON/OFF ratio on NbO_2 thickness was shown. V_{th} was found proportional to film thickness, and a higher ON/OFF current ratio was found for a thicker sample. There was good uniformity in ON and OFF states. Threshold characteristics were not degraded at high ON current density up to 10^6 cycles. Stable V_{th} characteristics were found up to 125 °C. Compared with a typical ReRAM without a selector, the 1S1R device showed a sufficient current reduction in the 1/2 V_{read} region and a good ON/OFF ratio of >8 for V_{read}. The hybrid device also showed promise.

5.9 Complementary Switching Devices and Arrays

5.9.1 Complementary Resistive Switching for Dense Crossbar Arrays

Sneak leakage paths in passive crossbar arrays can be overcome by two bipolar antiserial memristive elements that act as complementary resistive switches (CRSs) for high-density crossbar arrays. These were discussed in February of 2011 by the Institute of Solid State Research Julich and Aachen University [64]. Vertical integration of the CRS was shown using $CuSiO_2$–Pt bipolar resistive switches. A schematic cross-section of a CRS switch is shown in Figure 5.79 [64].

In a CRS, a memory cell is formed by two bipolar resistively switching elements, vertically stacked in an antiserial configuration on top of each other. Such CRS cells were made and characterized. The CRS device stack was 30 nm Pt, 20 nm Cu, 20 nm Pt, 20 nm Cu, 20 nm SiO_2, 5 nm Ti, and 40 nm Pt. The LRS and HRS were defined by electrochemical formation and dissolution of nanometer-sized metallic copper filaments. Information was stored in the CRS cell by the combination of the resistance states of the two memristive elements.

A schematic overview of signals during CRS operation is shown in Figure 5.80(a). The two memristive elements formed a voltage divider with three possible combinations. The signal in this case was physically applied to the electrode on the top, and the bottom electrode was grounded. Figure 5.80(b) illustrates the measured I–V characteristic of a CRS cell.

The resistance ratio, R_{off}/R_{on}, was greater than 1500, which was expected to permit array sizes of more than 3000×3000 elements. Switching speed was less than 120 µs. The conclusion was that gigabit passive crossbar arrays were possible [64]. The CRS concept was also discussed in 2010 by the same institution [65].

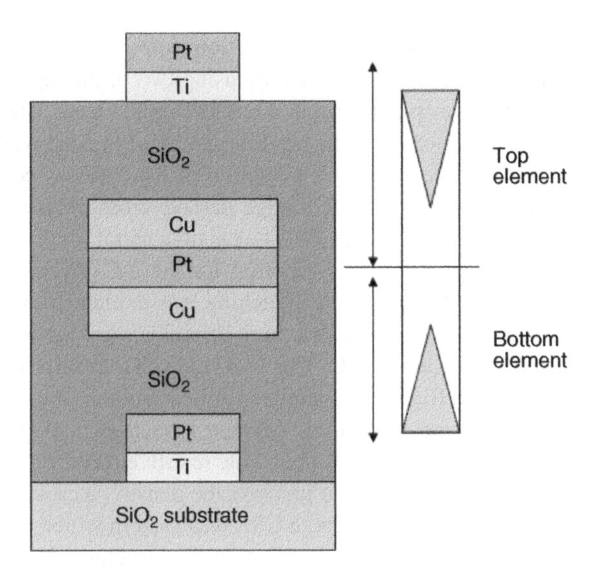

Figure 5.79 Schematic cross-section of a CRS device. (R. Rosezin *et al.* (2011) (Institute of SS Research Jülich, Aachen University), *IEEE Electron Device Letters*, 30(2) [64].)

5.9.2 CRS Memory Using Amorphous Carbon and CNTs

A CNT-based electrode for a bipolar ReRAM device was studied by Stanford University, the University of California, Berkeley, and Hong Kong Polytechnical University in November of 2011 [66]. The use of the CNT significantly reduced the size of the active device area to an average diameter of 1.2 nm. An amorphous carbon (a-C) layer was used between a metal top

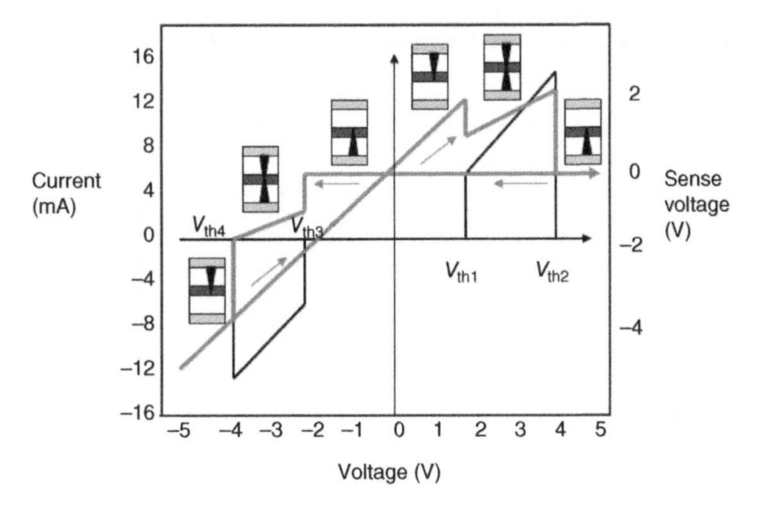

Figure 5.80 CRS cell operation showing element switching, I–V characteristics, and sense voltage. (R. Rosezin *et al.* (2011) (Institute of SS Research Jülich, Aachen University), *IEEE Electron Device Letters*, 30(2) [64].)

electrode and a CNT bottom electrode. This configuration showed bipolar switching behavior. A carbon-based CRS memory element was made by using two of the bipolar devices back to back. The stack configuration was planar metal–a-C–CNT–a-C–metal. The CRS reduced the sneak current in the cross-point memory. The bit information of the cell was stored in a high-resistance state, which reduced the power consumption of the memory cell.

This CRS device could potentially enable a large passive cross-point array with the CRS serving an integrated array cell selection device. The capping metal on the a-C determined the type of switching. Unipolar switching was reported for the a-C layer with an inert metal electrode such as W and Cr. In this case, the switching was explained as the formation and rupture of a carbon chain. Bipolar resistive switching behavior was reported for fast diffusing metal, such as Cu, Ag, and Au, on top of a-C. This was believed to result from electrochemical metallization of the conductive filaments—similar to the conductive bridge memory. This study used Ag or Au as the top metal electrode because bipolar switching was thought to be more controllable than unipolar switching. A nanoscale resistive memory based on the metal–a-C–CNT–metal stack was demonstrated. To increase the density, a cross-point memory was attempted. To eliminate sneak leakage current, a CRS was used in which two bipolar resistive memory cells were connected back to back. In the previous structure, the two-series memory cells were made simultaneously. The carbon-based lateral CRS structure was made with two bipolar metal–a-C–CNT–metal cells connected back to back by a common CNT as a metal–a-C–CNT–a-C–metal device. The I–V curves of the CRS memory cell are shown in Figure 5.81 [66].

The difference lies in the pattern. For the metal–a-C–CNT–a-C–metal device, both ends of the CNT were in contact with a-C–metal. Initially, both memory cells were in the HRS. When a positive voltage sweep was applied, one cell was changed to the LRS. The other remained at the HRS, acting as a voltage divider. If the voltage was swept further, both cells would be in the LRS. With a negative voltage sweep, one cell is RESET to the HRS, and then both are in the LRS. These four distinct threshold voltages enable the cell to have four different states: ON, OFF, 0, and 1.

The endurance was low, as the cell failed after 11 cycles [66].

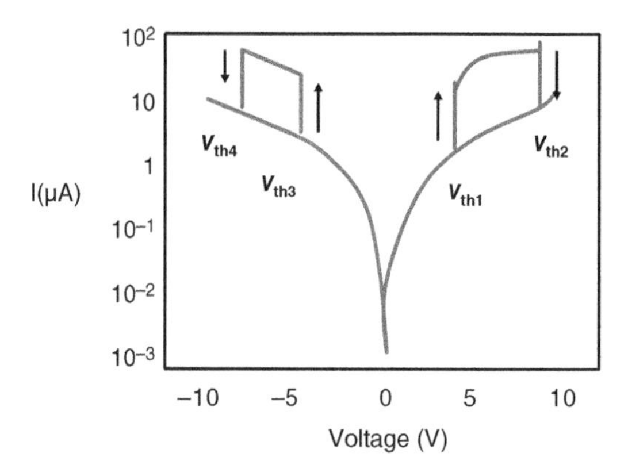

Figure 5.81 I–V curves of complementary resistance switching memory cell made with CNT. (Based on Y. Chai *et al.* (2011) (Stanford University, University of California Berkeley, Hong Kong Polytech University), *IEEE Transactions on Electron Devices*, 58(11), 3933 [66].)

5.9.3 Complementary Switching in Metal–Oxide ReRAM for Crossbar Arrays

In December of 2011, the Politecnico di Milano discussed using CRS in metal–oxides as selection devices with high-density crossbar ReRAMs [67]. The challenge has been finding a select device with suitable supply current and ON/OFF swing. The CRS concept is based on a two-ReRAM stack, which can potentially solve the sneak path issue in crossbar arrays. The team discussed complementary switching in single-stack nonpolar ReRAM devices. First, complementary switching characteristics were described by use of simulations. The operation under DC-pulsed voltages were shown, and methods were discussed to improve the resistive window. A unified model for unipolar, bipolar, and complementary switching in oxide ReRAM was proposed. It was concluded that a CRS is a possible method for solving the sneak-path problem for future ReRAM crossbar arrays.

The CRS was originally described as a stack of two CB-RAMs that are antiserially connected, resulting in a memory device with an inert top electrode and bottom electrode separated by a common active electrode, which in this case was Cu. The two binary logic states necessary to store a bit correspond to the conductive filament shunting, either the top switching layer or the bottom switching layer. At the end of a program operation, the overall device was always in a high-resistance state, which helped to avoid leaky paths during read operation. To switch from the HRS of CB-RAM(1) to the HRS of CB-RAM(2), a bias was applied in one direction so that a SET process occurred, forming a conductive filament in the CB-RAM(1) and dissolving the conductive filament in CB-RAM(2). The application of the opposite bias reversed the operation. The conductive filament was alternately moved from one layer to the other, avoiding the presence of leaky paths during the read operation. The read operation was performed by applying positive top electrode voltage above the set voltage V_{set} [67]. Figure 5.82 shows I–V characteristics from biasing a CB-RAM CRS device with a Pt–solid electrolyte–Cu–solid electrolyte–Pt stack with positive and negative voltage sweeps [67].

In this study, a complementary switching device was made from metal–oxide ReRAMS using Pt–ZrO_x–HfO_x–metal–HfO_x–ZrO_x–Pt and Pt–ZrO_x–HfO_x–ZrO_x–Pt stacks. This demonstrated that both CB-RAMs and M–O ReRAMS could be used for making such a device. A complementary switching device was simulated using single-stack nonpolar metal–oxide ReRAM devices. The complementary switching device relied on the natural asymmetry of the RESET state, where a depleted gap could be selectively created close to the bottom electrode

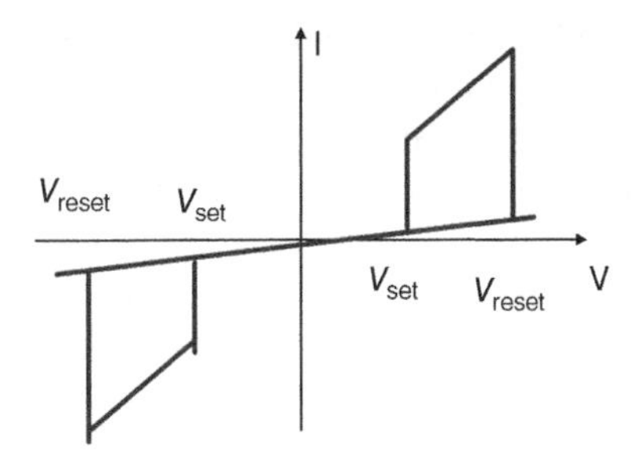

Figure 5.82 I–V characteristics of voltage sweeps of antiserial connected CBRAMs using (Pt–SE–Cu–SE–Pt). (Based on F. Nardi *et al.*, (Politecnico di Milano), IEDM, December 2011 [67].)

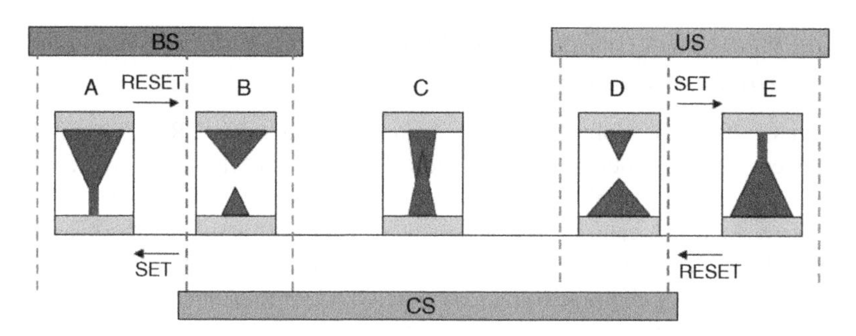

Figure 5.83 Mechanism while changing polarity of applied bias in CRS. (Based on F. Nardi *et al.*, (Politecnico di Milano), IEDM, December 2011 [67].)

or the top electrode by application of a positive or a negative voltage. The initial shape was made by the application of a negative RESET, which induced a depletion of conductive defects close to the bottom electrode. An applied positive voltage caused defects to migrate toward the bottom electrode, resulting first of all in conductive filament reconnection followed by depletion of the conductive filament at the top electrode side. Each program sequence resulted in a RESET operation, which left the memory element in a high-resistive state—therefore also avoiding the presence of leaky paths during read operations. The experimental metal–oxide ReRAM structure used was a simple TiN–HfO_x–TiN stack with HfO_x as the active layer material. A uniform Hf concentration profile permitted a fully symmetric nonpolar ReRAM device. The voltage levels $V_{set} = 0.5$ V and $V_{reset} = 0.7$ V showed only a slight asymmetry while the polarity of the applied bias was changed, as shown in Figure 5.83 [67].

Several issues were found. The study found the I_{on}/I_{off} ratio needed to be improved, perhaps by choice of materials, and the resistance window degraded after about 15 cycles due to the lack of current compliance during complementary switching. It was believed that both architectural and material alternatives should be tried. In this study, CRS was shown in single-stacked nonpolar ReRAM devices, both in simulations and in experiments with symmetric HfOx ReRAM. It was also used to explain the coexistence of bipolar and unipolar switching in the same device [67].

5.9.4 CRSs Using a Heterodevice

ReRAM devices with ZrOx–HfO_x bilayer films were discussed in December of 2010 by the Gwangju Institute of Science and Technology [68]. They used back-to-back connection of two ReRAM devices to confirm the feasibility of a diode-free cross-point array. Precise control of the oxygen vacancy concentration in the HfO_2 layer was achieved by depositing a thin 2–14 nm layer of Zr metal. The resulting diode-free cross-point array had wide readout margin and stable data read.

Operating voltage control in CRSs was discussed further in April of 2012 by the Gwangju Institute of Science and Technology [69]. An HfOx-based ReRAM with a TiO_x-based ReRAM was used for the switch. To control the operating voltages of the CRS, ReRAMs with asymmetric SET and RESET voltages were used. This provided a wider voltage window for the READ process, high switching speed, high reliability, and a greater than 10× readout margin from the heterodevice switch.

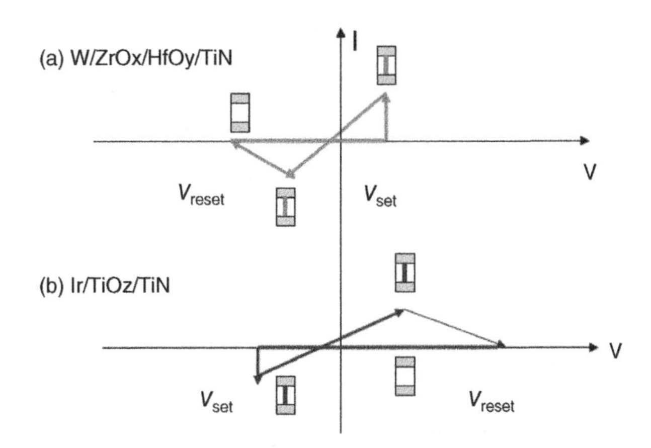

Figure 5.84 I–V curves of HfO_x and TiO_x ReRAMs: (a) HfO_x-based ReRAM with positive SET voltage and negative RESET voltage; and (b) a TiO_x-based ReRAM with negative SET voltage and positive RESET voltage. (Based on D. Lee (2012) (Gwangju Institute of Science and Technology), *IEEE Electron Device Letters*, 33(4), 600 [69].)

A heterodevice CRS was proposed that uses one of the following methods to control the operation voltages and CRS behavior: (i) the reset voltage of each ReRAM was equal to or greater than its set voltage; (ii) both ReRAMS must show self-compliance current behavior that can suppress the overflow of current to prevent breakdown during the read in the heterodevice complementary resistance switch; and (iii) in one-side biasing region, either positive or negative, the ReRAMs should have opposite switching behaviors. The latter can be achieved by the antiserial connection of similar ReRAMs or serial connection of different ReRAMs. In this case, the operating voltages of the CRSs could be controlled by the heterodevice structure. Figure 5.84(a) illustrates an HfO_x-based ReRAM with positive SET voltage and negative RESET voltage, and 5.84(b) shows a TiO_x-based ReRAM having negative SET voltage and positive RESET voltage [69]. An illustration of the heterodevice complementary resistive device using these two ReRAMs is shown in Figure 5.85 [69].

The HfO_x-based ReRAM has a small V_{set}, while the TiO_x-based ReRAM has a large V_{reset} in the positive bias region. Due to the clear difference between the SET voltage of the HfO_x-based ReRAM and the RESET voltage of the TiOx-based ReRAM, there was a broad ΔV. Due to the advantageous properties of each ReRAM, such as fast switching speed and self-current compliance, the heterodevice CRS showed reliable and fast pulse switching in the write and erase processes. The low current level at half V_{read} enabled the heterodevice complementary switch to have a significantly higher number of words in the cross-point array applications [69].

5.9.5 *Self-Selective W–VO_2–Pt ReRAM to Reduce Sneak Current in ReRAM Arrays*

A W–VO_2–Pt ReRAM that exhibited combined threshold switching and bipolar resistive switching was studied in May of 2012 by Gwangju IST [70]. The device simultaneously showed self-selective performance and memory switching by electroforming. This phenomenon avoided sneak current, which meant the device had potential for future high-density cross-point arrays.

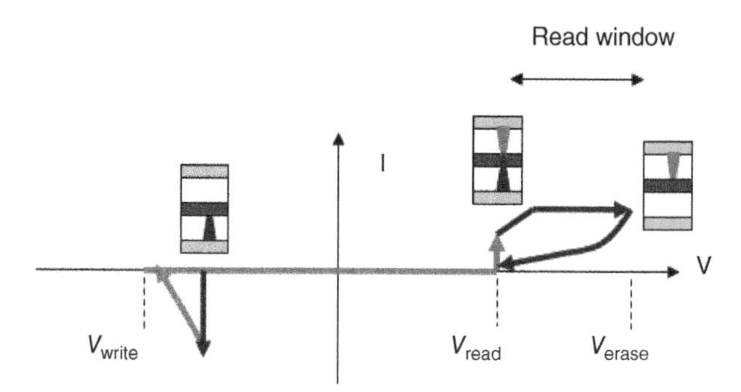

Figure 5.85 Heterodevice CRS using HfOx-based and TiOx-based ReRAM. (Based on D. Lee (2012) (Gwangju Institute of Science and Technology), *IEEE Electron Device Letters*, 33(4), 600 [69].)

It was found that using W instead of Pt for the top electrode improved memory performance for cycling, pulse endurance, and retention. This was attributed to a self-formed WO_x–VO_x interface, which had the potential to avoid sneak current in cross-point arrays. It is known that electroformed vanadium oxides show both threshold switching and bipolar memory switching, depending on the oxygen content. A device was experimentally made with both threshold and bipolar switching capabilities. The I–V curve for the W–VO_x–Pt hybrid device in this cross-point array is shown in Figure 5.86 [70]. An illustration of the cross-point array with selected and unselected devices used in the I–V curve is shown Figure 5.87 [70].

The I–V characteristics of the hybrid type device showed a HRS in the low-voltage region due to the self-selective behavior of the VO_2 switching element and two counterclockwise

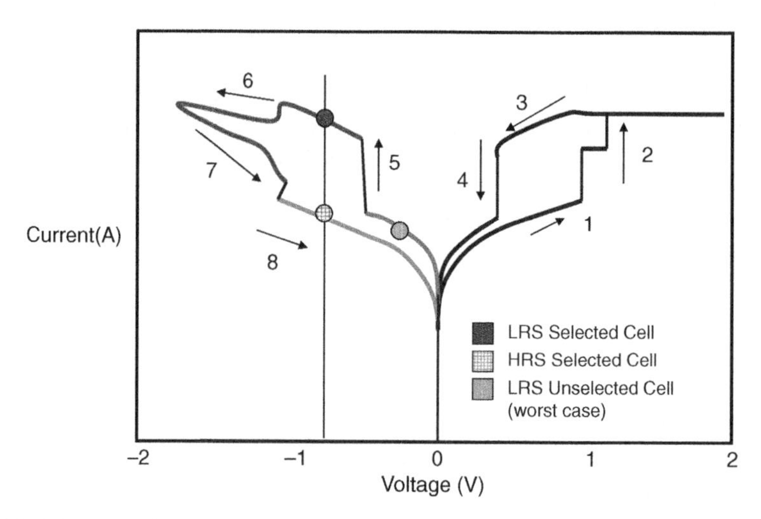

Figure 5.86 I–V curve for W–VOx–Pt hybrid ReRAM device. (Based on M. Son *et al.* (2012) (Gwangju Institute of Science and Technology), *IEEE Electron Device Letters*, 33(5), 718 [70].)

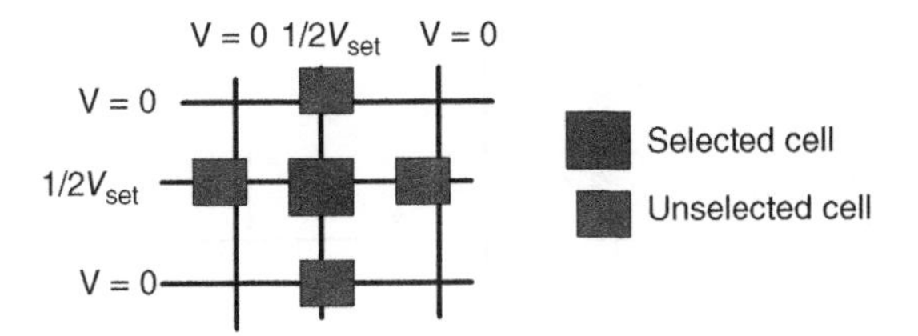

Figure 5.87 Cross-point array with selected and unselected devices. (Based on M. Son *et al.* (2012) (Gwangju Institute of Science and Technology), *IEEE Electron Device Letters*, 33(5), 718 [70].)

loops in the higher-voltage region at both bias polarities. In the cross-point array, the self-selectivity of the VO_2 switch could reduce the sneak path through the ReRAM because the leakage current through the ReRAM was determined by the VO_2 switch—even in the worst case, where all the unselected cells were in the LRS. By applying a voltage lower than +/− V_{th} to the other cells, they can be kept in the OFF state, preventing them from interfering with the selected cell during the read operation [70].

5.9.6 Hybrid Nb_2O_5–NbO_2 ReRAM with Combined Memory and Selector

A W–NbO_x–Pt ReRAM with both threshold switching and memory switching characteristics was discussed in June of 2012 by the Gwangju IST and Hynix [71]. The threshold switching characteristics of NbO_2 include stability at 160 °C, 22 ns switching speed, good switching uniformity, and scalability to 10 nm. By oxidizing the NbO_2, it was possible to make a very thin Nb_2O_5–NbO_2 stacked layer for a hybrid memory with both threshold switching and memory switching. Without adding a selector device, a 1Kb cross-point hybrid memory device without SET–RESET disturbance up to 10^6 cycles was shown. Bipolar memory switching for W–NbO_x–Pt is illustrated in Figure 5.88(a), and threshold switching is illustrated in Figure 5.88(b) [71].

A hybrid memory device with both selector and memory characteristics was made. The hybrid memory structure was attained by alternatively stacking 10 nm of NbO_2-x film for threshold switching and 10 nm of $Nb_2O_{5\text{-}x}$ film for memory switching. These were formed by controlling the oxygen concentration on a Pt–Ti–SiO_2–Si substrate. A Pt or W top electrode was sputtered.

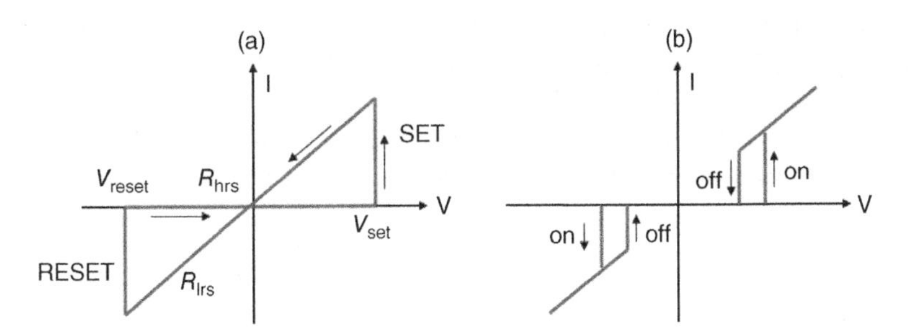

Figure 5.88 Curves for W–NbOx–Pt bipolar ReRAM with (a) bipolar memory switching characteristics; and (b) threshold switching characteristics. (Based on S. Kim, *et al.*, (Gwangju Institute of Science and Technology, Hynix), VLSI Technology Symposium, June 2012 [71].)

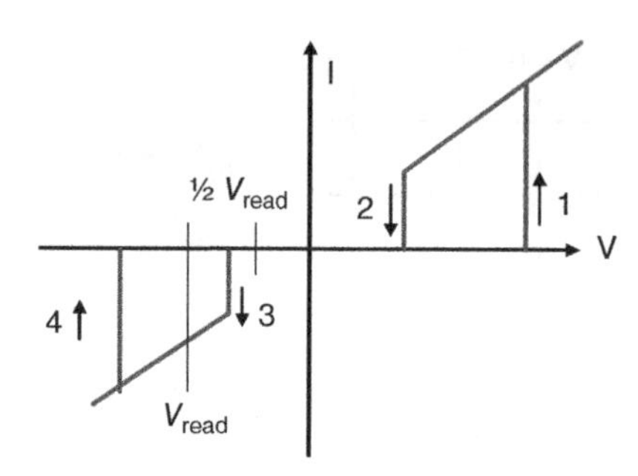

Figure 5.89 Simulated switching properties of the hybrid device. (Based on S. Kim, *et al.*, (Gwangju Institute of Science and Technology, Hynix), VLSI Technology Symposium, June 2012 [71].)

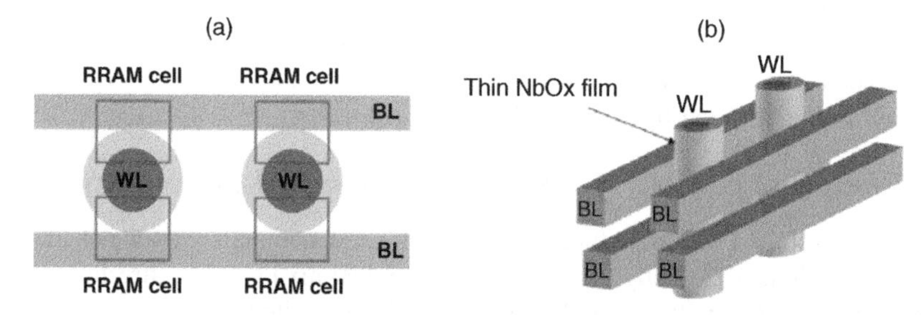

Figure 5.90 3D stacked hybrid Nb_2O_5–NbO_2 ReRAM with combined memory and selector: (a) top-down view; (b) 3D schematic. (Based on S. Kim, *et al.*, (Gwangju Institute of Science and Technology, Hynix), VLSI Technology Symposium, June 2012 [71].)

An illustration of the switching properties of the hybrid device as simulated is shown in Figure 5.89 [71]. A top view and 3D view of the individual ReRAM cell is shown in Figure 5.90(a) and (b). A dense 3D array of this cell can be constructed vertically. A 1Kb cross-point hybrid memory array was made that had no SET or RESET disturbances up to 10^6 cycles without a selector [71].

5.9.7 Analysis of Complementary ReRAM Switching

Complementary ReRAM switching using HfO_2-based bipolar ReRAM elements was discussed in August of 2012 by IMEC and ESAT-KUL [72]. This analysis was extended to include complementary resistive switching stacks that have been shown in other studies made from other materials. In this analysis, the I–V switching curves of the complmentary switching ReRAM cells are decomposed into the intrinsic switching characteristics of the individual constituent elements.

Analysis of different types of CRS cells showed that similar intrinsic switching behaviors occurred in very different types of bipolar switching ReRAMs. There was a strong material dependence of the characteristic switching voltage. The CRS cell structure used consisted of two back-to-back stacked TiN–HfO_2–Hf elements, as shown in Figure 5.91 [72].

For this HfO_2 CRS cell, SET–RESET of the bottom–top element happened at positive voltages with respect to the top electrode, while RESET–SET of the bottom–top element

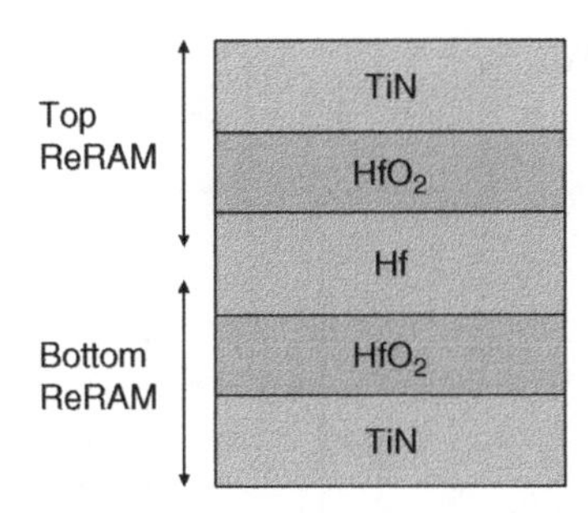

Figure 5.91 CRS cell structure consisting of back-to-back stacked HiN–HfO_2–Hf elements. (Based on D.J. Wouters *et al.*, (IMEC, ESAT-KUL), *IEEE Electron Device Letters*, August 2012 [72].)

occurred at negative voltages. This meant that the total I–V triangular sweep could be divided into two regions. Between the end of positive SET and the start of negative RESET, the bottom element is in the LRS. Between the end of negative SET and the start of positive RESET, the top element is in the LRS. At each point, because the total resistance was known ($R_{total} = V/I$), the other resistance could be calculated. The experiment was done with the stacked TiN–HfO_2–Hf–HfO_2–TiN stack. The resulting I–V curve of switching characteristics of the CRS is as shown in Figure 5.92 [72].

A derivation analysis using characteristics of resistive elements with serial load resistors was done. This analysis was then extended to include previous studies with similar results. The stacks used in the previous studies included TiN–HfO_2–Hf–TiN, Pt–SiO_2–GeSe–Cu, Pt–$Ta_2O_{5\text{-}x}$–$TaO_{2\text{-}s}$–Pt, Au–CNT–a-C–Au. While the physical operation differed among these stacks, the results showed similar intrinsic switching behavior with a constant SET voltage parameter that was dependent on the cell material [72].

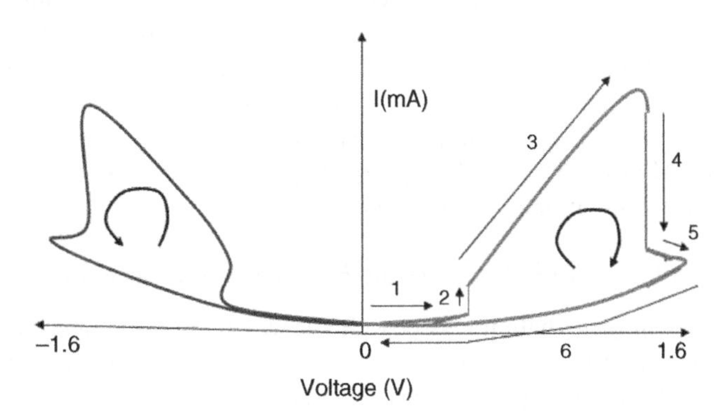

Figure 5.92 Calculated I–V curve of switching characteristics of the HfO_2-based CRS: (1) increase of V_{re} until (2) first snapback part of SET of bottom ReRAM, (3) second part of SET of bottom ReRAM at constant voltage V_{trans}, (4) start of RESET of top ReRAM snapforward, (5) second part of RESET of top ReRAM, and (6) sweep back of applied voltage. The I–V loop direction is opposite for bottom and top ReRAMs. (Based on D.J. Wouters *et al.*, (IMEC, ESAT-KUL), *IEEE Electron Device Letters*, August 2012 [72].)

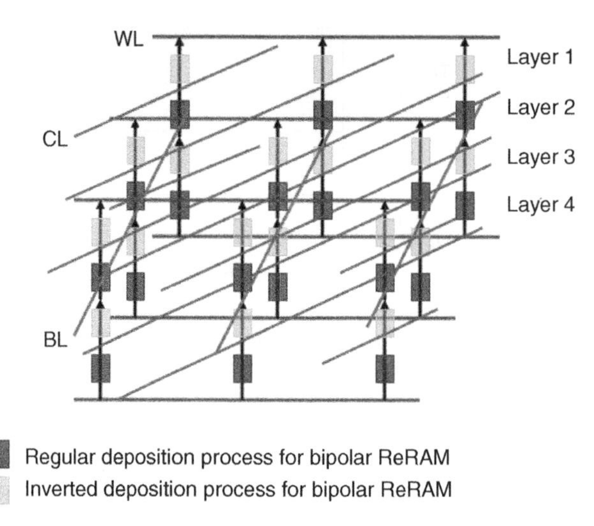

Figure 5.93 Proposed 3D stacking array for bipolar ReRAM using alternately inverted layers. Based on Y.C. Chen *et al.* (2012) (New York University, Nanyang Technological University, ICF International), *IEEE Transactions on Nanotechnology*, II(5), 948 [73].)

5.9.8 Complementary Stacked Bipolar ReRAM Cross-Point Arrays

Two stacked bipolar ReRAM cross-point array structures were discussed in September of 2013 by ICF International, New York University, and Nanyang Technical University [73]. The arrays of these two bipolar ReRAMs were formed by alternating the deposition of ReRAM material in forward and reverse sequences. The interleaved structure helped maintain signal margin and proper programming voltage while suppressing the impact of sneak paths and current leakage. Isolation layers are eliminated. Bipolar ReRAMs were built using material with high R_{on} capability. The capacity of a memory island was increased to 8Kb organized as eight layers of 32 × 32 bit cross-point arrays. The worst-case configuration sense margin was more than 20% of the maximum sensing voltage. One of the two proposed stacking configurations is shown in Figure 5.93 [73]. Four memory layers are shown, each of which consists of a ReRAM crossbar array. Complementary stacking was used; a regular memory stack was followed with one in a reversed deposition order.

5.9.9 Complementary Switching Oxide-Based Bipolar ReRAM

Complementary switching in a hafnium oxide ReRAM was discussed in January of 2012 by the Politecnico di Milano and University of Texas [74]. These were HfO_x ReRAMS, where the logic bit can be encoded in two high-resistance levels, making it immune from leakage currents and other related sneak-through effects in the crossbar array. The complementary switching physical mechanism was described through simulation results using an ion migration model for polar switching. Results from pulsed characterization were shown, indicating that complementary switching can be operated for at least 10 ns. Complementary switching in HfO_x was demonstrated and explained using an ion-migration model. The complementary switching mechanism was supported by the different switching behaviors in the HfO_x layers with a symmetric or asymmetric composition profile. SET and RESET voltages were studied as a function of the pulse width and indicated an Arrhenius dependence on local temperature. The impact of forming and current compliance on complementary switching was discussed. The conclusion was that complementary switching in ultrathin HfO_x is a feasible solution for passive 3D crossbar arrays with no selector device [74].

5.10 Toward Manufacturable ReRAM Cells and Cross-point Arrays

5.10.1 28 nm ReRAM and Diode Cross-Point Array in CMOS-Compatible Process

A 3D ReRAM made in a 28 nm high-κ metal gate CMOS-compatible process was discussed in December of 2013 by TSMC and ITRI [75]. The cross-point 3D ReRAM array was formed by a stacked 30 nm Cu via and Cu metal line single-damascene process. The ReRAM cell used a TaON resistive film with a Cu via as the top electrode and Cu metal as the bottom electrode. The TaON film was a composite of a backend metal glue layer of Ta and TaN available in the 28 nm Cu damascene process. The cross-point 3D ReRAM used no extra TMO film or additional process steps to reduce cost.

The 28 nm node TaN-based unipolar ReRAM cell had a cell size of $30 \times 30\ nm^2$. It used a Cu via as a top electrode, which was the word-line, and an orthogonal Cu metal line for the bottom electrode, which was the bit-line. The stacked storage node was Ta–TaN–TaON. The unit structure could be reproduced vertically by stacking more Cu layers to form a 3D ReRAM cell string. The unit area of a single stacked cell string was reduced to only four times the via size using the 28 nm CMOS design rules. All materials were compatible with the CMOS logic backend process. An illustration of the unipolar SET and RESET characteristics of the 28 nm TaN-based ReRAM are shown in Figure 5.94 [75].

To eliminate the sneak current for unselected cells of the 3D cross-point structure, a TaOx diode was made and stacked on top of the 28 nm TaN-based ReRAM. An illustration of a cross-point array using both ReRAM and diode in M1 to M7 of the BEOL of a 28 nm conventional high-κ metal gate Cu CMOS process is shown in Figure 5.95 [75].

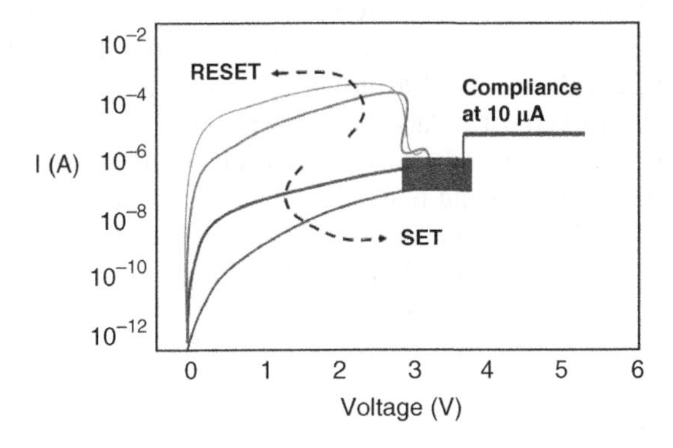

Figure 5.94 Unipolar I–V curve of 28 nm TaN-based ReRAM in conventional 28 nm high-κ metal gate CMOS process. (Based on M.C. Hsieh *et al.*, (TSMC, ITRI), IEDM, December 2013 [75].)

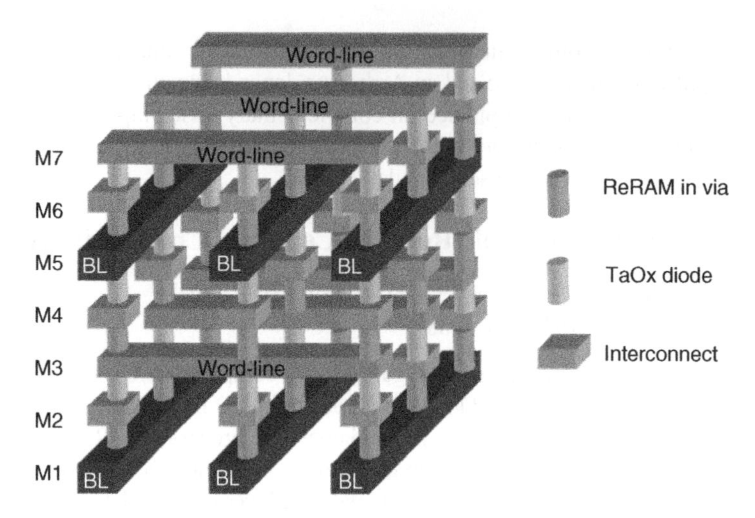

Figure 5.95 Schematic illustration of a stacked 3D cross-point array in 28 nm conventional high-κ metal gate CMOS process with TaOx diode stacked on top of a TaN-based ReRAM cell. (Based on M.C. Hsieh *et al.*, (TSMC, ITRI), IEDM, December 2013 [75].)

A TiON-based bipolar ReRAM cell was also made in a conventional Al-based 180 nm CMOS backend process. The stack was W–Ti–TiN–TiON–Ti–TiN–AlCu, where W was the top electrode and AlCu was the bottom electrode. Cell size was $0.22 \times 0.22\,\mu m^2$.

The 28 nm TaON-based Cu ReRAM cell had a larger LRS–HRS window than the TiON-based Al ReRAM. However, the I–V characteristics showed that SET and RESET were faster for the bipolar operation than for the unipolar operation.

5.10.2 Double-Layer 3D Vertical ReRAM for High-Density Arrays

In December of 2013, National Chiao Tung University, the Taiwan National Nanodevice Lab, and Winbond discussed the self-rectifying Ta–TaO_x–TiO_2–Ti ReRAM from Section 5.8.9 configured as a double-layer stack of two vertical ReRAM cells intended for high-density data storage [76]. The interfacial vertical bipolar ReRAM cell was found to overcome the intrinsic tradeoff between operating current and variability in filamentary ReRAMs.

The cell was formed with two Ti (100 nm)–SiO_2 (100 nm) layers deposited alternately, patterned, and etched into vertical pillars. Next, 40 nm of TiO_2 and 20 nm of TaO_x were deposited sequentially on the pillar sidewalls, as shown in Figure 5.96(a). The contacts to the horizontal Ti electrodes of the top and bottom cells were masked and patterned, and the Ta vertical electrode was then deposited and patterned. A conceptual diagram of a high-density 3D array using this cell is shown in Figure 5.96(b) [76].

The effective device area at the vertical sidewall was $0.2\,\mu m^2$. Device characteristics resembled those of the previous planar device, which had a larger area. The ON/OFF ratio was greater than 10 and could only be read at negative bias. The rectifying ratio at +/− 2 V was greater than 3. The maximum RESET current was in the sub-μA range. Retention was greater than 10^4 s, and read disturb immunity was greater than 10^9 cycles. Pulse endurance using a 1 μs write pulse was greater than 10^{10} cycles.

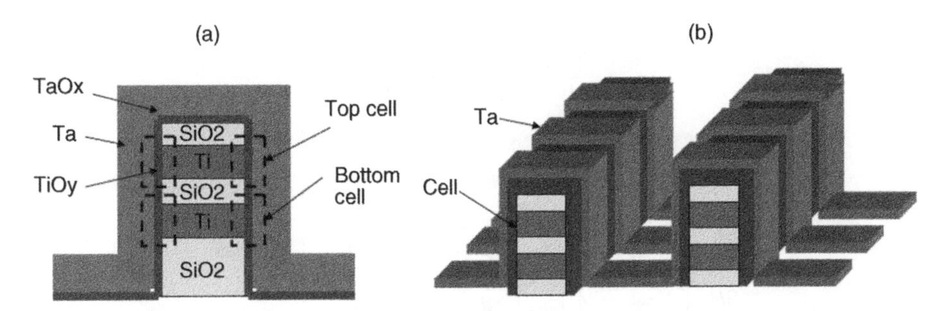

Figure 5.96 Double-layer vertical ReRAM: (a) schematic cross-section of two stacked cells with Ti bit-line and Ta word-line; and (b) 3D schematic of potential configuration of high-density array. (Based on C. W. Hsu *et al.*, (NCTU, NNDL, Winbond Electronics), IEDM, December 2013 [76].)

5.10.3 Study of Cell Performance for Different Stacked ReRAM Geometries

A study of 3D ReRAM arrays was done by Peking and Stanford Universities in December of 2013 [77]. This study compared the cell performance for a horizontal cross-point array to a vertical pillar-around geometry and showed the performance to be comparable. Four configurations were considered. In the first, the early 3D stacked ReRAM arrays had the memory cell between the cross-point of the word-line and the bit-line. The second had the memory cell between the horizontal word-line and the vertical pillar. The third had the memory cell between the plane electrode and the vertical pillar, and the fourth had the metal core shell pillar structure.

The array performance for read–write, energy, and speed for the different 3D architectures were investigated by SPICE simulation, which showed that the horizontal stacked ReRAM was superior but has higher bit cost. It was shown that increasing the number of stacks while keeping the total bits the same and while scaling the feature size was critical for reducing RC delay and energy consumption.

5.11 STT Magnetic Tunnel Junction Resistance Switches in Cross-Point Array Architecture

5.11.1 High-Density Cross-Point STT Magnetic Tunnel Junction Architecture

The design of a cross-point array architecture using an STT magnetic tunnel junction (MTJ) memory element was discussed in May and September of 2012 by the University of Paris Sud, STMicroelectronics, Imperial College London, and CNRS [78,79]. In this design the mean area per word corresponded to only two transistors, which were shared by 64 bits. This structure led to a significant improvement in data density to 1.75 F^2/bit. The proposed cross-point architecture for the STT MTJ elements included a cross-point array of MTJs for data storage, a cross-point array of reference MTJ, write circuits on the right side, and read circuits on the left side, as shown in Figure 5.97 [79].

The word-lines and bit-lines addressed the two terminals of the MTJ at the cross-point. The storage density was determined by the minimum distance between two MTJs. There were two select transistors per N bit word (in this case $N = 4$). Two lines of MTJs are for data reference

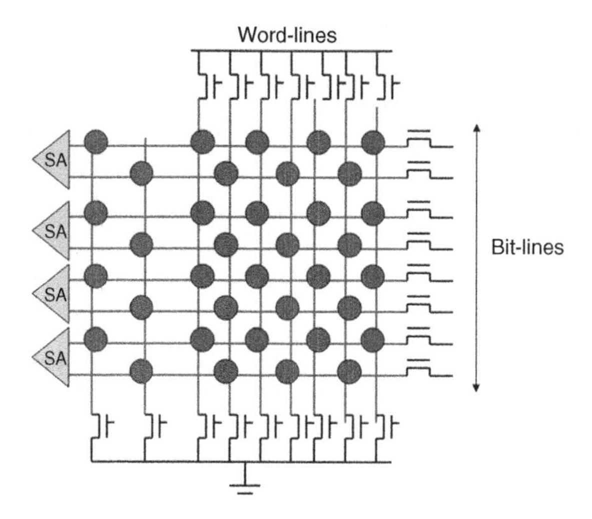

Figure 5.97 Cross-point architecture for the STT MTJ elements. (Based on W. Zhao *et al.*, (Université Paris-Sud), *IEEE Transactions on Nanotechnology*, 11(5), 907, Figure 4 [79].)

and read of the odd and even words by the sense amplifier. There is one write–read circuit per bit. Contacts between MTJs and CMOS circuits are only on the edge of the MTJ array. Special techniques are discussed to address the sneak current and low-speed issues of conventional cross-point architecture. To avoid sneak currents, two design considerations were developed. One was balancing the sensing structure so that there were the same number of MTJs in both branches of the sense amplifier. There was one reference MTJ and M storage MTJ in each side of the sense amplifier, where M was the number of words in the cross-point STT MTJ array. The odd and even words were read with the different reference MTJ, which were associated with two branches of the sense amplifier. This permitted the disturbance of the sneak currents from the MTJs to be reduced during data sensing. The second design consideration was parallel data reading. N sense amplifiers were used to detect the data in parallel. For each TMJ associated with the same word, the parallel reading allowed them to avoid the sneak currents from the other word because there weren't any floating nodes on the bit-lines. The bit-lines were all precharged to V_{read} before data sampling, and the selected word-line was grounded during sensing.

An STT MTJ SPICE model was used, which included precise experimental parameters such as inputs from a 65 nm MTJ technology from STMicroelectronics. This model included precise experimental parameters, including cell area, data access speed, and power, which were either calculated or simulated to show the expected performance level of this architecture.

The switching mechanism of the MTJ in the cross-point array is shown in the schematic circuit diagram in Figure 5.98 [78]. In (a), for the selected word, the PMOS is active and programs the selected bit to "0." In (b), for the selected word, the NMOS is active to program the selected bit to "1." For unselected words, both the NMOS and PMOS are always inactive [78].

Two special techniques were discussed to deal with sneak currents and speed issues. The first was a balanced sensing structure where there were the same number of MTJs in both branches

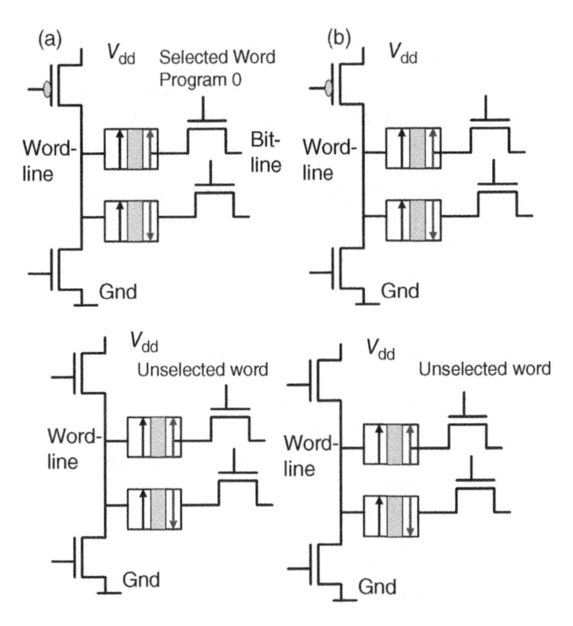

Figure 5.98 Schematic circuit diagram of STT MTJ switching mechanism in cross-point array. (W. Zhao *et al.*, (IEF, Université Paris-Sud, UMR, CNRS, STMicroelectronics), IMW, 20 May 2012 [78].)

of the sense amplifier. There were one reference MTJ and *M* storage MTJs in each side of the sense amplifier. The odd and even words were read with different reference MTJs associated with the two branches of the sense amplifier. This let the disturbance of the sneak currents from the same bit address or from different word-lines be reduced during data sensing. The second design technique to limit sneak currents involved implementing parallel data reading. *N* sense amplifiers were used to detect data in parallel. For each MTJ associated in the same word, parallel reading permitted them to avoid the impact of sneak currents from the other word because there were no other floating nodes in the bit-lines. This cross-point architecture could potentially overcome the limitations of the cross-point architecture and permit the STT-MTJ to be used as a universal memory in a digital computing system [80].

References

1. Prince, B. (2007) Vertical NV memories as an alternative to scaling, (Memory Strategies International). NVMTS.
2. Zhang, Y. *et al.* (June 2007) An Integrated phase change memory cell with Ge nanowire diode for cross-point memory, (Stanford University). VLSI Technology Symposium.
3. Lee, B. and Wong, H.S.P. (June 2009) NiO resistance change memory with a novel structure for 3D integration and improved confinement of conduction path, (Stanford University). VLSI Technology Symposium.
4. Yoon, H.S. (June 2009) Vertical cross-point resistance change memory for ultra-high density non-volatile memory applications, (Samsung) VLSI Technology Symposium.
5. Govoreanu, B. *et al.* (December 2011) 10×10 nm^2 Hf/HfO_2 crossbar resistive RAM with excellent performance, reliability and low-energy operation, (IMEC, KU Leuven). IEDM.
6. Zhang, J. *et al.* (23 July 2009) A 3D RRAM using stackable 1TXR Memory cell for high density application, (Fudan University). IEEE ICCCAS.
7. Chevallier, C.J. *et al.* (February 2010) A 0.13 μm 64 Mb multi-layered conductive metal-oxide memory, (Unity Semiconductor). ISSCC.

8. Gilmer, D.C. *et al.* (May 20 2012) Asymmetry, vacancy engineering and mechanism for bipolar RRAM, (SEMATECH, Stanford University, College of Nanoscale Science and Engineering Albany). IMW.
9. Deng, Y. *et al.* (2013) RRAM crossbar array with cell selection device: A device and circuit interaction study, (Peking University). *IEEE Transactions on Electron Devices*, **60**(2), 719.
10. Liu, T.Y. *et al.* (February 2013) A 130.7 mm^2 2-layer 32Gb ReRAM memory device in 24 nm technology, (Sandisk, Toshiba). ISSCC.
11. Liang, J., Yeh, S., Wong, S.S., and Wong, H.S. Philip (May 2012) Scaling challenges for the cross-point resistive memory array to sub-10nm node—an interconnect perspective, (Stanford University). IMW.
12. Jamaa, M.H.B., Micheli, G.C.G.De., and Leblebici, Y. (2011) Polysilicon nanowire transistors and arrays fabricated with the multispacer technique, (CEA-LETI, University of Milano-Bicocca, École Polytechnique Fédérale de Lausanne). *IEEE Transactions on Nanotechnology*, **10**(4), 891.
13. Cagli, C., Nardi, F., Ielmini, D. *et al.* (12 September 2011) Nanowire-Bbsed RRAM crossbar memory with metallic core-oxide shell nanostructure, (Politecnico di Milano, Lawrence Berkeley National Lab). ESSDERC.
14. Wu, Y., Chai, Y., Chen, H.Y. *et al.* (June 2012) Resistive switching AlOx-based memory with CNT electrode for ultra-low switching current and high density memory application, (Stanford University, Hong Kong University of Science and Technology). VLSI Technology Symposium.
15. Flocke, A. and Noll, T.G. (2007) Fundamental analysis of resistive nano-crossbars for the use in hybrid nano/CMOS-memory (Aachen University). ESSCIRC.
16. Lee, M.J. *et al.* (December 2007) 2-stack 1D-1R cross-point structure with oxide diodes as switch elements for high density resistance RAM applications, (Samsung). IEDM.
17. Liang, J. and Wong, H.S.P. (2010) Cross-point memory array without cell selectors—device characteristics and data storage pattern dependencies, (Stanford University). *IEEE Transactions on Electron Devices*, **57**(10), 5231.
18. Liang, J. and Wong, H.S.P. (6 June 2010) Size limitation of cross-point memory array and its dependence on data storage pattern and device parameters, (Stanford University). IITC, p. 1.
19. Zhang, L., Cosemans, S., Wouters, D.J. *et al.* (September 2012) Analysis of the effect of cell parameters on the maximum RRAM array size considering both read and write, (IMEC, KU Leuven, ESAT). ESSDERC.
20. Chen, A. (2013) A comprehensive crossbar array model with solutions for line resistance and nonlinear device characteristics, (GlobalFoundries). *IEEE Transactions on Electron Devices*, **60**(4), 1318.
21. Chen, A. (December 2013) Comprehensive methodology for the design and assessment of crossbar memory array with nonlinear and asymmetric selector devices, (GlobalFoundries). IEDM.
22. Lee, H.D. *et al.* (June 2012) Integration of $4F^2$ selector-less crossbar Array 2 Mb ReRAM based on transition metal oxides for high density memory applications, (Hynix, Hewlett-Packard). VLSI Technology Symposium.
23. Ou, E. and Wong, S.S. (2011) Array architecture for a nonvolatile 3-dimensional cross-point resistance-change memory, (Stanford University). *IEEE Journal of Solid-State Circuits*, **46**(9), 2158.
24. Bateman, B., Siau, C., and Chevallier, C. (14 November 2011) Low power cross-point memory architecture, (Unity Semiconductor). ASSCC.
25. Lee, J., Park, J., Jung, S., and Hwang, H. (8 May 2011) Scaling effect of device area and film thickness on electrical and reliability characteristics of RRAM, (Gwangju Institute of Science and Technology). IITC, pp. 1.
26. Chen, H.Y., Yu, S., Gao, B. *et al.* (December 2012) HfOx based vertical resistive random access memory for cost-effective 3d cross-point architecture without cell selector, (Stanford University, Peking University). IEDM.
27. Zhao, L. *et al.* (June 2013) Dopant selection rules for extrinsic tunability of HfOx RRAM characteristics: A systematic study, (Stanford University). VLSI Technology Symposium.
28. Lee, F.M., Lin, Y.Y., Lee, M.H. *et al.* (June 2012) A novel cross point one-resistor (0T1R) conductive bridge random access memory (CBRAM) with ultra low SET/RESET operation current, (Macronix). VLSI Technology Symposium.
29. Kwon, W., Jeon, J., Hutin, L., and Liu, T.J.K. (2012) Electromechanical diode cell for cross-point nonvolatile memory arrays, (University of California, Berkeley). *IEEE Electron Device Letters*, **33**(2), 131.
30. Srinivasan, V.S.S. *et al.* (2012) Punchthrough-diode-based bipolar RRAM selector by Si epitaxy, (IIT, Bombay). *IEEE Electron Device Letters*, **33**(10), 1396.
31. Tran, X.A. *et al.* (2011) A high-yield hfox-based unipolar resistive RAM employing Ni electrode compatible with Si-diode selector for crossbar integration, (Nanyang Technological University, National University of Singapore, Soitec, Fudan University). *IEEE Electron Device Letters*, **32**(3), 396.
32. Ielmini, D. (2010) Reset-set instability in unipolar resistive-switching memory, (Politechnico di Milano). *IEEE Electron Device Letters*, **31**(6), 552.

33. Nardi, F., Cagli, C., Ielmini, D., and Spiga, S. (May 2011) Reset current reduction and set-reset instabilities in unipolar NiO RRAM, (Politecnico di Milano, IMM-CNR). IMW.
34. Tran, X.A. *et al.* (June 2011) High performance unipolar AlOy/HfOx/Ni based RRAM compatible with Si diodes for 3D application, (Nanyang Technical University, Peking University, A*STAR, NUS, GlobalFoundries). VLSI Technology Symposium
35. Huang, Y. *et al.* (2012) A new dynamic selector based on the bipolar RRAM for the crossbar array application, (IME Peking University). *IEEE Transactions on Electron Devices*, **59**(8), 2277.
36. Tran, X.A. *et al.* (December 2011) Self-rectifying and forming-free unipolar HfOx based high performance RRAM built by fab-available materials, (Nanyang Technological University, IMEPeking, A*STAR, NUS, Global-Foundries, IME CAS Beijing, Soitec, Fudan University). IEDM.
37. Kawahara, A. *et al.* (19 February 2012) An 8 Mb multi-layered cross-point ReRAM macro with 443 MB/s write throughput, (Panasonic). ISSCC.
38. Tran, X.A. *et al.* (2012) A self-rectifying HfOx-based unipolar RRAM with NiSi electrode, (Nanyang Technical University, Peking University, NUS, Soitec, Fudan University, SUS&T Shenzhen). *IEEE Electron Device Letters*, **33**(4), 585.
39. Tran, X.A., Zhu, W., Liu, W.J. *et al.* (2013) Self-selection unipolar HfOx-based RRAM, (Nanyang Technological University, NUS, Soitec, SUST China). *IEEE Transactions on Electron Devices*, **60**(1), 391.
40. Jo, M. *et al.* (June 2010) Novel cross-point resistive switching memory with self-formed Schottky barrier, (Gwangju Institute of Science and Technology). VLSI Technology Symposium.
41. Chien, W.C. *et al.* (June 2013) A novel high performance WOx ReRAM based on thermally-induced SET operation, (Macronix). VLSI Technology Symposium.
42. Sasago, Y. *et al.* (June 2009) Cross-point phase change memory with 4F^2 cell size driven by low contact-resistivity poly-Si diode, (Hitachi). VLSI Technology Symposium.
43. Lee, S.H. *et al.* (December 2011) Highly productive PCRAM technology platform and full chip operation: Based on 4F^2 (84 nm Pitch) cell scheme for 1 Gb and beyond, (Hynix). IEDM.
44. Liang, J., Jeyasingh, R.G.D., Chen, H.Y., and Wong, H.S. Philip (June 2011) A 1.4 μA reset current phase change memory cell with integrated carbon nanotube electrodes for cross-point memory application, (Stanford University). VLSI Technology Symposium.
45. Gopalakrishnan, K. *et al.* (June 2010) Highly-scalable novel access device based on mixed ionic electronic conduction (MIEC) materials for high density phase change memory (PCM) arrays, (IBM). VLSI Technology Symposium.
46. Shenoy, R.S. *et al.* (June 2011) Endurance and scaling trends of novel access-devices for multi-layer crosspoint-memory based on mixed-ionic-electronic-conduction (MIEC) materials, (IBM). VLSI Technology Symposium.
47. Burr, G.W. *et al.* (June 2012) Large-scale (512 Kbit) integration of multilayer-ready access devices based on mixed-ionic-electronic-conduction (MIEC) at 100% yield, (IBM). VLSI Technology Symposium.
48. Kau, D.C. *et al.* (December 2009) A stackable cross point phase change memory, (Intel, Numonyx). IEDM.
49. Li, L., Lu, K., Rajendran, B. *et al.* (2011) Driving device comparison for phase-change memory, (Hong Kong University of Science and Technology, IBM, Qimonda, Macronix). *IEEE Transactions on Electron Devices*, **58**(3), 664.
50. Kang, M.J. *et al.* (10 December 2011) PRAM cell technology and characterization in 20 nm node size, (Samsung), IEDM.
51. Zhang, C. *et al.* (2011) Design and fabrication of dual-trench epitaxial diode array for high-density phase-change memory, (SMIC, Chinese Academy of Science, Microchip). *IEEE Electron Device Letters*, **32**(8), 1014.
52. Son, M. *et al.* (2011) Excellent selector characteristics of nanoscale VO_2 for high-density bipolar ReRAM applications, (Gwangju Institute of Science and Technology). *IEEE Electron Device Letters*, **32**(11), 1579.
53. Lee, M.J. *et al.* (December 2012) Highly-scalable threshold switching select device based on chalcogenide glasses for 3D nanoscaled memory arrays, (Samsung). IEDM.
54. Lee, M.H., Kao, C.Y., Yang, C.L. *et al.* (June 2011) Reliability of ambipolar switching poly-Si diodes for cross-point memory applications, (National Taiwan Normal University, Industrial Technology Research Institute). DRC.
55. Ping, E.X., Erokhin, Y., Gossmann, H.J., and Khaja, F.A. (14 May 2012) Two-terminal diode steering element for 3D x-bar memory, (Applied Materials). International Workshop on Junction Technology.
56. Ping, E.X., Erokhin, Yl., Gossmann, H.J., and Khaja, F.A. (June 2012) Two-terminal diode steering element for 3D X-Bar memory, (Applied Materials). VLSI Technology Symposium.
57. Lee, W. *et al.* (June 2012) Varistor-type bidirectional switch (J_{max}>107 A/cm^2, selectivity 104) for 3D bipolar resistive memory arrays, (Gwangju Institute of Science and Technology, Hynix), VLSI Technology Symposium.

58. Ho, C.Y. *et al.* (December 2012) Threshold vacuum switch (TVS) on 3D-stackable and $4F^2$ cross-point bipolar and unipolar resistive random access memory, (NDL, NARL, National Chiao Tung University, Mesoscope Technology, Fu-Jen University, CYCU, University of California Berkeley). IEDM.
59. Kawahara, A. *et al.* (2013) An 8 Mb multi-layered cross-point ReRAM macro with 443 MB/s write throughput, (Panasonic). *IEEE Journal of Solid-State Circuits*, **48**(1), 178.
60. Tran, X.A., Zhu, W., Liu, W.J. *et al.* (2012) A self-rectifying AlOy bipolar RRAM with Sub-50-uA set/reset current for cross-bar architecture, (NUS, Soitec, SUST China). *IEEE Electron Device Letters*, **33**(10), 1402.
61. Hsu, C.W. *et al.* (June 2013) Self-Rectifying bipolar TaOx/TiOx/TiO_2 RRAM with superior endurance over 1012 cycles for 3D high-density storage-class memory, (National Chiao Tung University, Winbond). VLSI Technology Symposium.
62. Seok, J.Y. *et al.* (2012) Resistive switching in TiO_2 thin films using the semiconducting In-Ga- Zn-O electrode, (Seoul National University). *IEEE Electron Device Letters*, **33**(4), 582.
63. Cha, E. *et al.* (December 2013) Nanoscale (10nm) 3D vertical ReRAM and NbO_2 threshold selector with TiN electrode, (Pohang University of Science and Technology, University of Tsukuba, Nagoya University, Stanford University). IEDM.
64. Rosezin, R., Linn, E., Nielen, L. *et al.* (2011) Integrated complementary resistive switches for passive high-density nanocrossbar arrays, (Institute of SS Research Jülich, Aachen University). *IEEE Electron Device Letters*, **30**(2), 191.
65. Linn, E. *et al.* (2010) Complementary resistive switches for passive nanocrossbar memories, (University of Aachen, Forschungszentrum Jülich). *Nature Materials*, **9**, 403.
66. Chai, Y. *et al.* (2011) Nanoscale bipolar and complementary resistive switching memory based on amorphous carbon, (Stanford University, University of California, Berkeley, Hong Kong Polytechnic University). *IEEE Transactions on Electron Devices*, **58**(11), 3933.
67. Nardi, F., Balatti, S., Larentis, S., and Ielmini, D. (December 2011) Complementary switching in metal oxides: Toward diode-less crossbar RRAMs, (Politecnico di Milano). IEDM.
68. Lee, J. *et al.* (December 2010) Diode-less nano-scale ZrOx/HfOx RRAM device with excellent switching uniformity and reliability for high-density cross-point memory applications, (Gwangju Institute of Science and Technology). IEDM.
69. Lee, D. (2012) Operation voltage control in complementary resistive switches using heterodevice, (Gwangju Institute of Science and Technology). *IEEE Electron Device Letters*, **33**(4), 600.
70. Son, M. *et al.* (2012) Self-selective characteristics of nanoscale VOx devices for high-density ReRAM applications, (Gwangju Institute of Science and Technology). *IEEE Electron Device Letters*, **33**(5), 718.
71. Kim, S. *et al.* (June 2012) Ultrathin (<10 nm) Nb_2O_5/NbO_2 hybrid memory with both memory and selector characteristics for high density 3D vertically Stackable RRAM applications, (Gwangju Institute of Science and Technology, Hynix). VLSI Technology Symposium.
72. Wouters, D.J. *et al.* (2012) Analysis of complementary RRAM switching, (IMEC, ESAT-KUL). *IEEE Electron Device Letters*, **33**(8), 1186.
73. Chen, Y.C., Li, H., Zhang, W., and Pino, R.E. (2012) The 3-D stacking bipolar RRAM for high density, (New York University, Nanyang Technological University, ICF International). *IEEE Transactions on Nanotechnology*, **II**(5), 948.
74. Nardi, F., Balatti, S., Larentis, S. *et al.* (2013) Complementary switching in oxide-based bipolar resistive-switching random memory, (Politecnico di Milano, University of Texas). *IEEE Transactions on Electron Devices*, **60**(1), 70.
75. Hsieh, M.C. *et al.* (December 2013) Ultra high density 3D via RRAM in pure 28 nm CMOS process, (TSMC, ITRI). IEDM.
76. Hsu, C.W. *et al.* (December 2013) 3D vertical TaOx/TiO_2 RRAM with over 10^3 self-rectifying ratio and sub-μA operating current, (National Chiao Tung University, National Nano Device Lab, Winbond). IEDM.
77. Deng, Y. *et al.* (December 2013) Design and optimization methodology for 3D RRAM arrays, (Peking University, Stanford University, National Chiao Tung University). IEDM.
78. Zhao, W. *et al.* (20 May 2012) High density spin-transfer torque (STT)-MRAM based on cross-point architecture, (IEF Paris-Sud, UMR, CNRS, STMicroelectronics). IMW.
79. Zhao, W., Chaudhuri, S., Accoto, C. *et al.* (2012) Cross-point architecture for spin-transfer torque magnetic random access memory, (Université Paris-Sud). *IEEE Transactions on Nanotechnology*, **11**(5), 907.
80. Zhao, W., Chaudhuri, S., Accoto, C. *et al.* (2012) Cross-point architecture for spin-transfer torque magnetic random access Memory, (Université Paris-Sud, STMicroelectronics, Imperial College London). *IEEE Transactions on Nanotechnology*, **11**(5), 907.

6

3D Stacking of RAM–Processor Chips Using TSV

6.1 Overview of 3D Stacking of RAM–Processor Chips with TSV

Three-dimensional (3D) chips stacked and connected with vertical through-silicon-vias (TSV) entered the market in about 2013 led by stacked chips of homogeneous memory and followed in 2014 by early stacked heterogeneous memory and logic chip systems. The 3D configuration can replace shrinking the planar chip, which is nearing its limit for being a cost-effective technology. It can also replace two-dimensional (2D) chips of memory integrated with logic, which have developed some system issues such as noise due to long interconnects. Three-dimensional chip stacking with TSV provides smaller footprints, higher performance, lower power, and higher reliability through less system noise. Early use of configurable passive interposers helped connect standard chips in three dimensions. Standards for chips redesigned for better placement of 3D connections followed. Early standards for chips reconfigured for 3D TSV packaging have been posted, and at least one industry consortium is discussing new standards for 3D stacked chips using TSV.

Design issues include both new computer-aided design (CAD) tools for 3D optimization and methodologies for repartitioning systems in three dimensions. Both systems of homogenous die, such as memory systems and field-programmable gate arrays (FPGA), and systems of heterogeneous die can benefit from vertical 3D TSV integration. Early examples of heterogeneous systems benefitting from TSV include multimedia systems and conventional memory hierarchies that gain performance and reduce power by being redesigned for the short TSV interconnects. As an illustration of this technology, a 3D schematic diagram of a solid-state drive (SSD) subsystem integrated using vertical TSV is shown in Figure 6.1 [1]. This stack includes an SSD controller, DRAM, ReRAM, and NAND flash memory, all connected using TSV.

The TSVs consist of vertical holes that are drilled or etched in the silicon, oxidized, and then filled with a conducting metal, which is usually copper. They provide very short interconnects between points on different chips, resulting in a small, high-performance system. Initial applications are expected to be those where performance rather than cost is the main concern, such as high-end computers. Challenges of 3D stacking using TSV include instability, heat

Vertical 3D Memory Technologies, First Edition. Betty Prince.

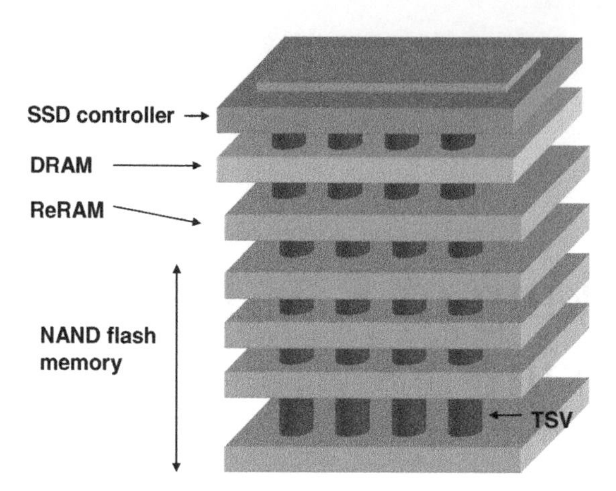

Figure 6.1 3D schematic diagram of TSV-integrated SSD. (Based on C. Sun *et al.*, (University of Tokyo, ASP-DAC), 22 January 2013 [1].)

dissipation, and thermal stress due to coefficient of expansion issues between the copper vias and the insulating material and silicon. Ensuring interconnection between chips from different manufacturers is also an issue.

Shrinking the size of planar chips cannot continue because the tightened space between paths means more interference between wires, and the tighter circuitry produces more thermal energy because the thinner wires and smaller vias have higher resistance. Vertical interconnections offer the promise of shorter wires which reduce signal delay, more interconnections between functions which can increase bandwidth, and connection of heterogeneous chips in a single, small-footprint package without extreme lateral scaling which means complete systems on a chip.

Configurable passive interposers with TSV were used initially to permit interconnection of different conventional chips. Early 2.5D interposer technology placed bumped flip chips side by side on an interposer that redistributed the system connection of the chips and passed the connections down through copper vias to the package. An example of two bumped flip chips integrated side by side on a passive interposer with TSV that redistributes the wiring is shown in Figure 6.2. An early pioneer of interposer technology was Xilinx, which used TSVs to integrate four FPGAs into one larger FPGA containing almost 7 billion transistors [2]. Power consumption for the single chip was less than a fifth of that required to operate the individually packaged FPGAs with traditional interconnection. Another early pioneer of interposer

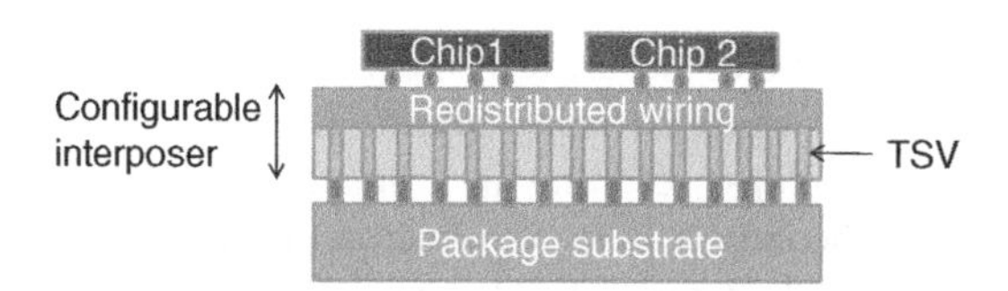

Figure 6.2 Illustration of 2.5D interposer technology showing two bumped flip chips integrated on a configurable passive interposer with TSV, which redistributes the wiring and connects to the package substrate.

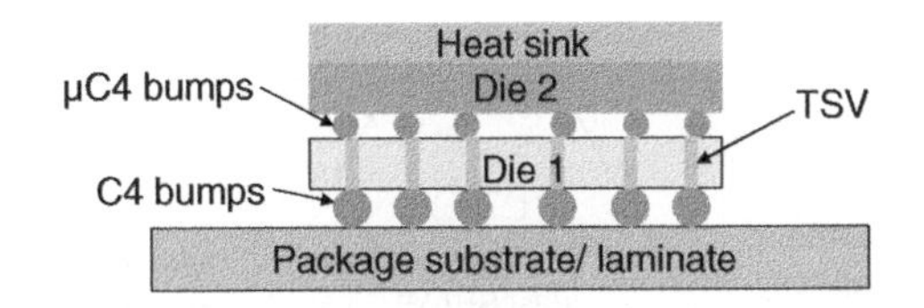

Figure 6.3 High-speed die face down on a second die that redistributes wiring using TSV (Based on S.S. Iyer *et al.*, (IBM) VLSI Technology Symposium, June 2009 [7].)

technology was IBM, which used interposers to integrate high-speed chips in different technologies, such as silicon–germanium (SiGe) and complementary metal–oxide silicon (CMOS), using deep trench decoupling capacitors on the interposer [3]. The redistributed wiring on the interposer could be connected to a package using copper TSV technology.

The JEDEC standards group introduced standards for memory chips redesigned for efficient 3D interconnections. The specification for a double data rate fourth generation (DDR4) synchronous dynamic random access memory (SDRAM) (JESD79-4) that supported 3D stacking was posted in September of 2012 [4]. This specification defined an interface with 3.2 giga-transfers per second (GT/s) and signaling at 1.6 GT/s. The DDR4 SDRAM interface supported stacks of up to eight memory devices presenting only a single signal load. The standard for wide-input/output (I/O) mobile DRAM, (JESD229) wide-I/O single data rate (SDR), was released by the JEDEC JC42 Standard Committee in January of 2012 [5]. This standard also supported the use of TSV to connect DRAM to logic in a 3D integrated circuit (IC). The standard specified a 512 bit data interface, which doubled the bandwidth of the low-power DDR2 specification without increasing power consumption. The Hybrid Memory Cube Consortium, which was led by Samsung and Micron, was intended to accelerate industry adoption of stacked DRAM chip technology by defining a new specification for 3D memory chips with new architectures supporting the 3D approach [6].

Both systems of heterogeneous dice and systems of homogeneous dice can benefit from vertical 3D TSV integration. An illustration of an early 3D heterogeneous integration of two dice that was described by IBM in June of 2009 is shown in Figure 6.3 [7].

Homogeneous die systems with TSV connections between individual dice are already available from companies such as Xilinx with an FPGA and Samsung with a 32GB DRAM registered dual-inline-memory module. An early 3D memory chip using TSV was discussed by Samsung in January of 2012 [8]. This 1.2 V 1Gb SDRAM was composed of two identical 50 nm technology dice connected with 40 μm pitch TSVs. The overall package yield was 76%, and there was no difference in performance between the top and bottom die. The integrated chip had four channels with 512 input–output (DQ) pins. The read operating power was 330.6 mW during four-channel operation at 50 MHz clock, which provided a 12.8 GB/s data bandwidth.

Wide I/O is also being used for TSV connections between heterogeneous dice to improve performance over that of chips with different properties all on the same chip. An example is the wide-I/O DRAM with a 256-bit bus width using four channels with 64 bits/channel that was described by Samsung in June of 2013 [9]. A TSV system in package (SIP) was used with short TSV connections between the bus of the application's processor chip and that of the wide-I/O DRAM to obtain a bandwidth of 25.6 GB/s. The cost of a chip that integrates heterogeneous dice using TSV may initially be more expensive than planar technology with embedded RAM,

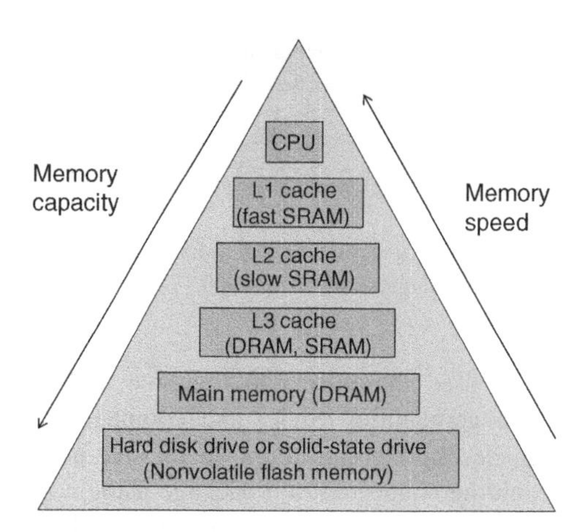

Figure 6.4 Classical memory hierarchy including CPU, cache, main memory, and hard drive, showing memory capacity and speed trends.

so the bandwidth, latency, and power improvement must compensate for the additional cost [5].

Several applications are positioned to make use of heterogeneous stacked die chips with short TSV connections, including multimedia and memory hierarchy systems. Multimedia applications, such as smart phones, can benefit from the high performance, low power consumption, and compact size of multicore processors and DRAMs stacked using TSV.

The performance of the classical memory hierarchy can also be improved using 3D processor cores on one chip stacked using TSV with cache and DRAM on another chip. The memory hierarchy consists of a processor with several levels of fast static random access memory (SRAM) cache, where level 1 (L1) is the fastest and level 3 (L3) is the slowest, together with an even slower but larger DRAM main memory and a hard drive that may be NAND flash. Figure 6.4 shows a memory hierarchy.

A more efficient memory hierarchy may be obtained by repartitioning the chips to use the wide data bus and short interconnects made available by the TSV stacking technology. While the L1 cache is usually embedded in the processor chip, the SRAM L2 cache can be 3D stacked using TSV with the DRAM main memory to improve the speed of the connection. There can be several processor cores with each one stacked using short TSV connections with a separate DRAM core to provide high bandwidth. A storage-class memory, which is a new category of nonvolatile random access memory (NV-RAM), may be included in the TSV stack. The storage-class memory may be implemented in one of the new NV-RAM memories currently in development, such as the phase-change memory (PC-RAM) or the resistance RAM (ReRAM). The storage drive can be either a hard disk drive (HDD) or an SSD, which is normally a NAND flash memory. A NAND flash hard drive could also be 3D stacked.

For example, Kyushu University, in August of 2011, discussed a high-performance 3D processor in which several dice were stacked using TSV technology to implement microprocessor unit (MPU) cores, cache, and DRAM in a single-chip footprint [10]. The effectiveness of 3D stacking depends on the characteristics of the target application. In this case, the cache architecture consisted of a small, fast SRAM and a large, stacked DRAM. The cache was

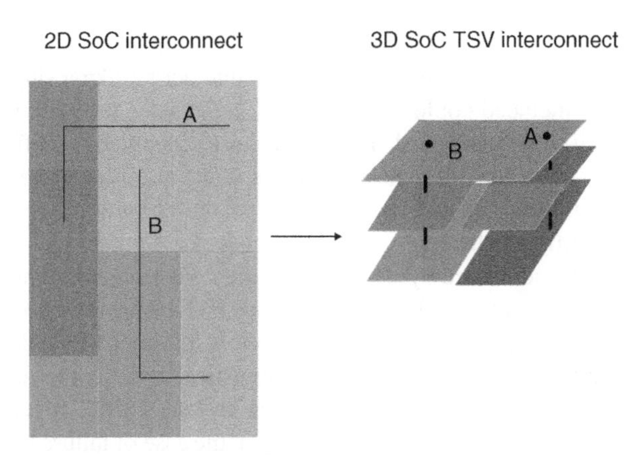

Figure 6.5 Illustration of shortening interconnects and reducing footprint between logic and memory chips using 2D and 3D design tools. (Based on C.H. Lin *et al.*, CODES+ISSS11 (ITRI), 9 October 2011 [11].)

implemented in SRAM on the same die used for the processor core. The DRAM die was stacked on the processor core to store cache lines, and the SRAM was used for tags.

Design issues for TSV stacking include methods for repartitioning processor chips and their memory in 3D internally to improve performance and reduce power. Figure 6.5 illustrates the potential for shortening interconnects between logic and memory elements in a 3D TSV design with adequate 3D design tools [11]. In this figure, the length of an interconnect between logic and memory elements in a 2D design and a 3D design are compared. The interconnects in the 3D design are shortened significantly. The chip footprint is also reduced.

An early paper from Intel in 2004 described repartitioning 3D die for TSV stacking [12]. The team described 3D die stacking as the bonding of two dice either face to face (FTF) or face to back (FTB) to construct a 3D structure. FTF was used in this case because it offers a higher density of die-to-die interconnects than FTB. It was suggested that with sufficiently dense interconnects a processor as complex as an iA32 MPU could be repartitioned between the two dice to improve performance and power using the short TSV connections between the chips. This study showed that a 3D implementation could potentially improve performance and power by 15%.

New design tools for optimizing 3D repartitioning are also needed. For example, there is a benefit to using 3D TSV to shorten the interconnects between macros that have been repartitioned from a planar 2D chip to several stacked chips.

The benefits to be gained from designing specifically for 3D rather than recasting a 2D-optimized design into the third dimension were discussed in December of 2013 by North Carolina State University and Synopsys [13]. A number of approaches were explored for creating 3D-specific designs including memory on logic and logic partitioning. A high-bandwidth, low-latency bus was designed to move data in a 3D stack, and a corresponding 3D processor was discussed. The design of the 3D-specific communications layers improved the power efficiency of computing.

An early stacked-die cube with chips redesigned to optimize key processing functions in the vertical stack was discussed by the University of Rochester in 2008 [14]. This system had 3D synchronization circuitry that ran at 1.4 GHz. The stacked chips were redesigned to optimize the key processing functions vertically with all layers interacting like a single system. It was projected that the chips included could be compacted in the same technology to 10% of their current size and run at 10 times the speed. The architecture accounted for different impedances occurring from chip to chip, for different operating speeds, and for different power requirements. The chip had millions of holes drilled to join the chips with vertical connections between transistors in different layers.

Process issues are being addressed by several companies. At least one company, Tezzaron, is offering a generic technology that stacks two logic chips FTF using a thinned top die [15]. The diameter of the TSVs are 1.2 μm, and the pitch is 5 μm. A minimum TSV density is required, and dummy TSVs are used to satisfy this rule. A logistical issue is the question of who is responsible if the stack fails or if the thinned wafers are broken. TSMC, for example, indicated that it has struggled with the question of responsibility in the case of failure [2]. For this reason, TSMC made the decision to do much of its TSV wafer manufacturing and packaging in house. There are other issues, including liability, industry standards, and many manufacturing challenges, that need to be solved.

Beyond TSV connections, there is effort being put into developing wireless connections using CMOS-compatible inductor coils to pass signals between stacked chips and additional effort being put into the development of optical connections between chips.

6.2 Architecture and Design of TSV RAM–Processor Chips

6.2.1 Overview of Architecture and Design of Vertical TSV Connected Chips

Several system architectures are expected to benefit from the 3D stacking of memory and processors using TSV connections. These include systems requiring high-performance memory with many independent high-capacity banks, high-performance systems with wide buses between the processor and memory, massive parallel processing systems with many cores, systems that repartition the memory hierarchy, and systems for mobile applications with low power requirements. The benefits of vertical TSV packaging for 3D ICs include increased functionality in a smaller form factor, improved circuit delay due to shorter interconnects, reduced parasitic capacitance and inductance, and integration of different components with incompatible manufacturing processes. Challenges for this new technology include 3D design, system validation, limitation in the number of TSVs, and energy and noise management. Reliability issues include stress from differences in thermal expansion between the copper in the silicon and coupling effects.

6.2.2 Repartitioning for Performance by Increasing the Number of Memory Banks

Repartitioning the memory in 3D can improve memory performance significantly by increasing the number of high-capacity memory banks. A 3D DRAM hybrid memory cube architecture repartitioned in 3D was described by Micron Technology in June of 2012 [16]. This architecture stacked four DRAM dice and a logic die into a single chip with low latency, low power consumption, high bandwidth, and high density. Three-dimensional chip stacking using TSVs underlay the system architecture approach. A system schematic is shown in Figure 6.6 [16].

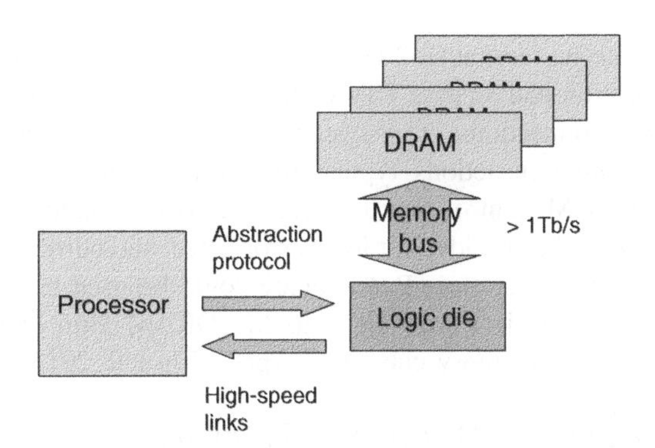

Figure 6.6 System schematic of 3D TSV hybrid memory cube. (Based on J. Jeddeloh and B. Keeth, (Micron), VLSI Technology Symposium, June 2012 [16].)

This hybrid memory cube is a stack of heterogeneous dice consisting of a homogeneous stack of 1Gb DRAM chips and a logic chip. Each 1Gb DRAM layer is optimized for high bandwidth and concurrency. TSV and fine pitch copper pillar interconnects are used. DRAM logic is off-loaded to a high-performance logic chip that handles DRAM sequencing, refresh, data routing, error correction, and high-speed interconnect to the host.

There are thousands of TSV connections in the Z direction. These short connections result in improved power dissipation. The hybrid memory cube was made with 1866 TSVs on a 60 µm pitch. Energy per bit was 3.7 pj/bit for the DRAM layers and 6.78 pj/bit for the logic layer, resulting in 10.48 pj/bit total compared to 65 pj/bit for existing DDR3 modules with similar functionality.

The DRAM chips were 68 mm^2 1Gb dice in 50 nm technology segmented into 64 autonomous partitions. Each partition had two independent memory banks totaling 128 banks per hybrid memory cube. Each partition supported a closed page and full cache line transfers of 32 to 256 bytes. Memory vaults were vertical stacks of DRAM partitions with 16 banks, as illustrated in Figure 6.7, which permitted more than 16 concurrent operations per stack [16].

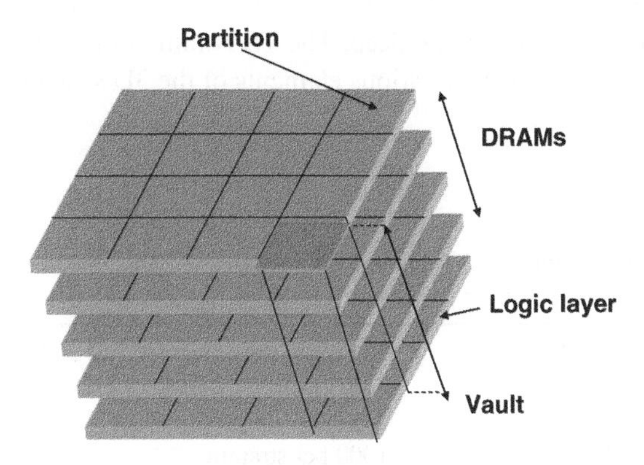

Figure 6.7 Hybrid memory cube with DRAM stack and logic layer. (Based on J. Jeddeloh and B. Keeth, (Micron), VLSI Technology Symposium, June 2012 [16].)

The TSV signaling rate was less than 2 Gb/s, requiring fewer high-speed transistors in the DRAM logic. Each partition had 32 data TSV connections plus command, address, and error correction code (ECC) connections. The system latency was low, and the DRAM was optimized for the random transactions typical of multicore processors. Memory control was at the memory. DRAM control was by read and write commands, which hid bank conflicts within the cube. ECC provided for local error detect and correct. Soft errors could be dynamically repaired during operation. Redundancy could be used to map out failing TSV connections. Self-test and repair were built into the DRAM chip along with diagnostic capability. Production of this memory cube was expected in late 2013.

6.2.3 Using a Global Clock Distribution Technique to Improve Performance

Another architecture for improved memory performance involved using a global clock distribution technique for 3D stacked chips with the clock tree and grids shorted between strata. This architecture was described by IBM in June of 2012 [17]. The technique was implemented in a two-layer 45 nm silicon-on-insulator (SOI) eDRAM test chip using TSV technology. The embedded DRAM (eDRAM) test chip was $5.6 \times 10.9\,\text{mm}^2$ in area and used TSV and microbumps for vertical connections. The two-layer 3D chip had a target clock frequency of 2 GHz. In each layer, a 2D clock tree was designed that considered the added loads due to the 3D interconnects. At points within the clock trees and grids, TSVs were used to short the two layers. Most of the TSVs were used for power and ground, with about 1% being used for shorting the clock. Multistrata global clock simulations were done to ensure that performance met specifications. The existing IBM "2D tree driven grid global clock" design tools were extended to support the 3D shorting-based method.

6.2.4 Stacking eDRAM Cache and Processor for Improved Performance

The IBM "Power7" processor L3 cache and a "processor proxy" layer formed a 3D system operating at 2 GHz that stacked a 45 nm eDRAM layer and logic blocks with data bandwidth of 450 Gb/s. IBM discussed this system in February of 2012 [18]. This 3D stack was constructed with 50 µm pitch µC4 bumps that joined the front side of one thick chip to TSV connections on the back side of a thinned chip. The TSVs were Cu-filled oxide-isolated vias that were about 20 µm in diameter and less than 100 µm deep. The TSV minimum pitch was 50 µm. The chip size was $5.6 \times 10.9\,\text{mm}^2$. Numbers of various elements of the 3D system prototype are shown in Table 6.1 [18].

Table 6.1 Numbers of Various Elements of 3D TSV DRAM–Processor Stack Prototype [18].

eDRAM Stack	176 Mb 4-chip stack/80 Mb 2-chip stack
Wafer Test Pads	256 power, 64 signal
Off-Stack C4s	2481
Signal TSV	1000 per stratum
Power TSV	5600 per stratum

The two-die stack showed operation with interstratum memory accesses equivalent to 2D at 2.7 GHz. A process-dependent stage delay difference of about 15% was detected between the two strata.

6.2.5 Using Decoupling Scheduling of the Memory Controller to Improve Performance

Another reconfiguration of the memory architecture for improved memory performance, the universal memory architecture (UniMA), enabled interoperability by decoupling the scheduling of device operations from the memory controller. The UniMA architecture was described in October of 2011 by the University of Illinois [19]. The decoupling was done by using a bridge chip at each memory module to perform local scheduling. The performance impact of local scheduling of device operations was also studied. This new memory architecture framework was intended to replace the older DDR SDRAM architecture, which required the memory controller to track the internal status of memories and schedule the timing of device operations. This tracking and scheduling function introduced a rigidity that some have found an obstacle in new memory systems.

A prototype implementation of UniMA on top of a DDR memory bus was presented and its efficiency evaluated with different workloads. The simulation results showed that UniMA improved memory system efficiency for memory-intensive workloads due to increased parallelism among memory modules. Performance improvement over conventional DDR memory architecture was about 3.1%.

Limitations of 3D die stacking of wide-I/O DRAM on logic using TSV were discussed by Qualcomm in May of 2012 [20]. The wide-I/O DRAM was found to offer good power efficiency in terms of bandwidth per unit of power due to the low I/O loads enabled by 3D stacking with TSV technology. However, some of the degrees of freedom normally associated with 3D integration schemes were limited.

6.2.6 Repartitioning Multicore Processors and Stacked RAM for Improved Performance

Many options for multicore architectures have been considered, with the main benefit being improved memory bandwidth from stacked memories. For example, Intel has shown a teraflop MPU chip with 80 core design using memory-on-logic architecture. Each MPU core connects to a 256KB SRAM with 12 GB/s bandwidth [21].

A massive parallel-processing multicore processor system architecture can benefit from 3D TSV stacking. A TSV-based fully functioning massive parallel-processing, multicore processor logic chip with stacked SRAMs was described by Georgia Technical Institute, KAIST, and Amkor in February of 2012 [15]. The 130 nm CMOS processor stack, called the 3D-MAPs, had 33 million transistors, 50 000 TSV, and 50 000 FTF connections. It ran at 1.5 V and consumed 4 W of power. The logic chip consisted of 64 general-purpose processor cores running at 277 MHz. The memory used 256KB SRAMs. Fabrication was in 130 nm GlobalFoundries six-metal CMOS technology using Tezzaron TSV and bonding technology. Packaging was done by Amkor. The Tezzaron technology stacked two logic chips FTF using a thinned top die [15]. The diameter of the TSVs were 1.2 μm, and the pitch was 5 μm. A minimum TSV density was required, and dummy TSV were used to satisfy this rule.

Table 6.2 Footprint and Bonding Style of 3D-MAPS V2 Logic and DRAM Technology.

Logic Footprint	10 × 10 mm
DRAM Footprint	20 × 12 mm
Bonding Style	F2F and F2B

While the 3D-MAPs version one (V1) technology used SRAMs stacked with logic, there was ongoing development of the 3D-MAPs V2, which stacked two logic chips with three DRAM chips. The 3D-MAPS V2 logic and DRAM technology had two tiers of logic and three of DRAM. The logic consisted of 128 cores. Memory capacity was 256MB of 2 Kb/cycle DDR SDRAM and 512KB of 4 Kb/cycle SRAM. 150 K TSVs were used along with 185 K FTF connections. Active power dissipation was 175 MHz at a frequency of 175 MHz. Footprint and bonding style of this technology are shown in Table 6.2 [15].

The GlobalFoundries process was modified to include Tezzeron TSVs, which are etched into the silicon and filled with tungsten (W). Tungsten was used rather than copper because it has a lower coefficient of thermal expansion mismatch with the surrounding silicon. The devices and metal layers were patterned, and the wafers were flipped and bonded. One wafer was then thinned until the TSVs were exposed on the backside. The two-layer FTF bonded stack used TSVs for I/O. Because the wafers were bonded before thinning, wafer breakage was not an issue. An illustration of the cross-section of the 3D stack of the 3D-MAPS V2 is shown in Figure 6.8, where TSVs are used for I/O and DRAM access [15].

Another 3D parallel architecture core is the ITRI 3D-PAC (Parallel Architecture Core) technology, which was described in October of 2011 [11]. In this architecture the SRAM was stacked directly on top of the logic chip to form a heterogeneous multicore computing platform for multimedia applications. The logic chip was a general purpose MPU, a dual programmable digital signal processing (DSP) core, and various subsystems and peripherals. A virtual platform was used for early architecture exploration, power estimation, and hardware–software (HW–SW) co-simulation.

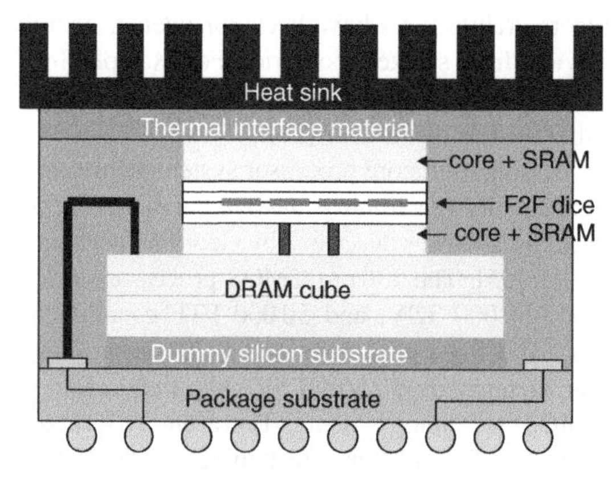

Figure 6.8 Schematic cross-section of 3D stack of 3D-MAPS V2 chip. (Based on D.H. Kim *et al.*, (Georgia IT, KAIST, Amkor), ISSCC, February 2012 [15].)

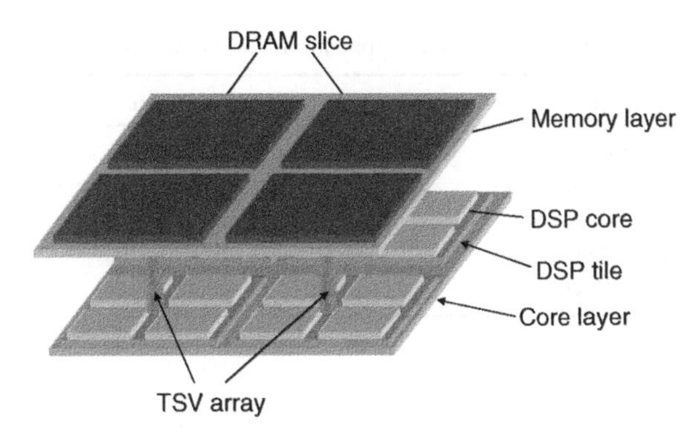

Figure 6.9 Schematic diagram of the 3D-iSPA architecture. (Based on T. Zhang *et al.*, (Penn State University) (NTHU, ITRI, MWSCAS), 7 August 2011 [22].)

Even though the stacked approach with TSV could increase capacity and performance, the placement and organization was important to achieve this design goal. The electronic design automation (EDA) requirements for system-level concurrent design for this 3D IC had a 3D timing and power analyzer for stacked chips with TSV. It also included thermal- and timing-driven 3D floor planner, 3D placer, and 3D router as well as 3D layout management tools that took stacked chips with TSV into consideration. The number of TSVs used became an important additional design constraint.

The 3D intelligent signal processor array (iSPA) was a multicore processor using 3D TSV stacking. This 3D embedded multicore DPS processor used TSV between the cores and the memory serving that core. It was described in August of 2011 by Pennsylvania State University, National Tsing Hua University, and ITRI. A schematic diagram of the 3D-iSPA architecture is shown in Figure 6.9 [22].

This processor was targeted for multimedia applications and had two unique design features to support this application. It leveraged 3D DRAM stacking to improve performance and had a wider memory data bus to better utilize the vertical TSV. It was shown that duplicating the memory controller could provide a 10-fold increase in speed. The wider memory bus had an impact on performance due to limitations of on-chip interconnects.

Multicore processors generate more memory access requests than do single-core processors. To avoid memory contention, the embedded multicore processor emphasized the acceleration of multimedia applications that introduce demand on high-volume regional memory access. This local memory access led to consideration of distributed local memory systems with high bandwidth so that no core suffers from data starvation. The 3D IC permitted designers to provide the individual cores with local memory with high bandwidth. This system attempted to apply 3D DRAM stacking to an embedded processor to accommodate memory access locality and high memory bandwidth. The primary multimedia applications supported by the 3D-iSPA are listed in Table 6.3 [22].

The workload analysis of these applications demonstrated strong memory access locality with high-memory bandwidth requirements. Two properties of interest were that the memory access had high locality so each processor element was only responsible for the assigned macro-blocks and that the applications had high memory bandwidth for streaming data.

Table 6.3 Multimedia Applications for the 3D-iSPA.

Function	Application
JPEG Encoding	Cellphone
H.264 Encoding	Cellphone, vehicle
Object Tracking	Camera, monitor
Face Tracking	Cellphone
Line Recognition	Vehicle

The 3D-iSPA was designed to have 16 commercial DSP MPUs to accelerate image processing. The roadmap showed the number of processing elements increasing to 128 in the future. Two or more layers could be required for this many cores. To use the memory access locality, every four processing elements were classified as a tile. A hierarchical advanced extensible interface (AXI) crossbar was used as the backbone interconnect. Every tile had a local AXI crossbar, and an AXI bridge was used for global access. A tile had an associated memory controller to handle the local memory access. Multiple independent memory slices were placed on the memory layer so that every tile had a private memory slice. TSVs were used as signal carriers between layers. Many memory controllers and a wider memory bus were evaluated to optimize the advantages of 3D TSV technology. It was concluded that the characteristics of the multimedia applications were particularly favorable to the 3D TSV chip integration technique.

Memory bottlenecks that occur in a 1TB/s 1Gb DRAM can be resolved by using a multicore configuration and 3D TSV interconnects. In April of 2011, Hitachi described a 1TB/s data rate 1Gb 3D TSV DRAM stacked on a 16-core CPU with 512-bit I/Os using TSVs [23]. A DRAM core with five-stage pipelined architecture was developed to reduce the data-bus operation cycle to 2 ns. The proposed DRAM architecture improved power efficiency. The TSV method reduced parasitic capacitance of the interconnects between the DRAM and the CPU. The multicore architecture reduced the length of the data bus on the DRAM by having a core-to-core connection between the DRAM and processor cores using the TSV. A 45 nm 1Gb DRAM was designed with a chip size of 51.6 mm^2 using $4F^2$ cells. The density was about five times higher than that of an embedded DRAM. Circuit simulations were used to verify the 2 ns data bus operation, 8 ns memory array operation, and 16 Gb/s I/O operation. Power consumption was 19.5 W with power efficiency of 51.3 GB/s/W. This was claimed to be an order of magnitude higher than that of conventional DRAMs.

Another multicore MPU chip with 80 cores used memory-on-logic stacked TSV architecture to produce a teraflop datarate MPU chip. The 80-core design using stacked memory-on-logic architecture was discussed by Intel in December of 2011 [21]. Each MPU core in this stack connected to a 256KB SRAM with 12 GB/s bandwidth. The heterogeneous DRAM stacked chip was designed to improve performance and energy efficiency by tightly integrating an SRAM row cache in a DRAM chip. A new floor plan and various architectural techniques were developed to exploit the benefits of 3D stacking technology. Multicore simulation results with memory-intensive applications indicated that by tightly integrating a small SRAM row cache with its corresponding DRAM array, performance could be improved by 30% while reducing dynamic energy by 31%.

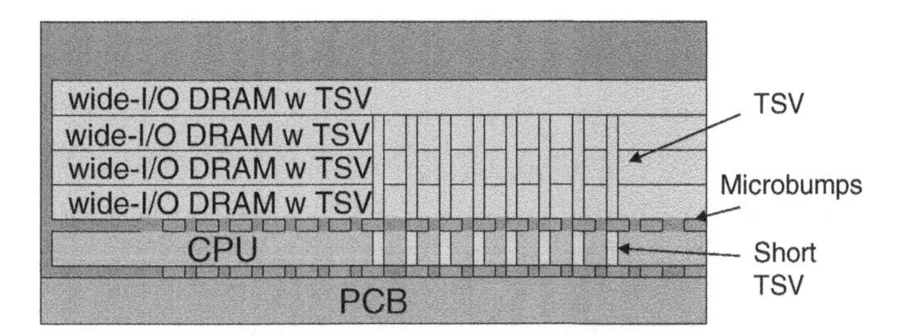

Figure 6.10 Four stacked wide-I/O DRAM chips using short TSV connections to the CPU data bus. (Based on J.S. Kim *et al.* (2012) (Samsung), *IEEE Journal of Solid-State Circuits*, 47(1), 107 [8].)

6.2.7 Increasing Performance and Lowering Power in Low-Power Mobile Systems

Mobile DRAM is used in portable electronics due to its low power consumption. The mobile system also requires high capacity and high performance. High capacity can be obtained by assembling multiple chips in a package, but then maintaining low power and high bandwidth become more difficult. Dynamic power consumption can be reduced by scaling the operating voltage and also by using a very wide I/O combined with a lower operating frequency to obtain a high data bandwidth. Stacking multiple wide-I/O chips using TSV technology can permit a wide bus in a small footprint with short connections between the memory and the processor, as shown in Figure 6.10. Power consumed in standby and self-refresh still must be reduced [8].

This 1.2 V 1Gb density mobile SDRAM was described by Samsung in January of 2012 [8]. The chip was composed of two stacked 50 nm technology dice connected with 40 μm pitch TSVs. The overall package yield was 76%, and there was no difference in performance between the top and bottom die. The chip had four channels with 512 DQ pins so that during four-channel operation at 50 MHz clock rate the data bandwidth was 12.8 GB/s. The read operating power was 330.6 mW. A block-based dual period refresh scheme was used to reduce self-refresh current while maintaining minimum chip size. The architecture had 16 segmented 64Mb arrays. Bandwidth was 12.8 GB/s using microbump pads with 1 pF loading. The memory density was attained using both microbumps and TSV stacking. All four channels were independent, and each channel could be composed of two or four banks with 128 DQ. The chip had four partitions symmetric to the chip center, and each partition had 4×64 Mb arrays, peripheral circuits, and microbumps. Each channel had its own input pins, but external power pins and internal voltage generators were shared. Each bank had a 2KB page depth.

6.2.8 Increasing Performance of Memory Hierarchies with 3D TSV Integration

The classical memory hierarchy, shown in Figure 6.11, is used in high-performance server systems to provide good performance at a reasonable cost by allocating memory between very fast, expensive memory, such as SRAM, used for cache next to the processor and slower but lower-cost DRAM used for lower-level cache and main memory.

3D integration can add value to high-performance server systems. Benefits include higher computational density, lower power consumption, and increased modularity over 2D versions.

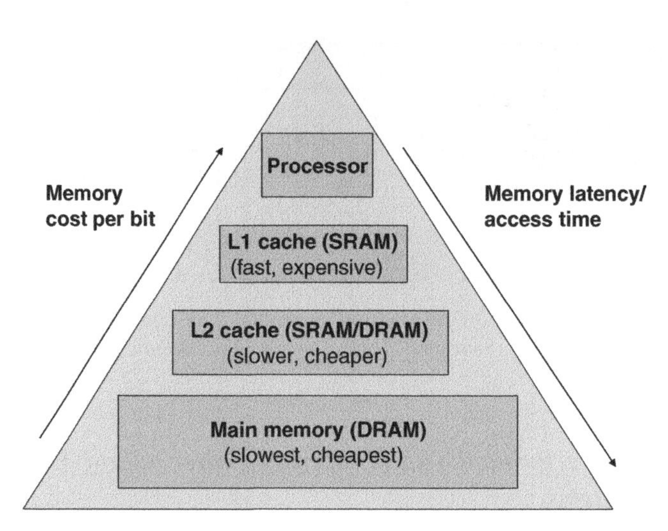

Figure 6.11 Classical memory hierarchy for server systems, showing decreasing cost per bit with increasing memory latency/access time when moving down the pyramid.

In June of 2011, IBM discussed integration of eDRAM in high-end server systems where the focus was on 3D TSV stacking, which affected chip design in fundamental ways [24]. A multidisciplinary, co-design approach was used to attain the benefits of 3D integration of high-density memory and high-performance server chips. Considerations included the degree of stacking, the number of layers, and the degree of activity in each layer. The chip infrastructure included power and ground distribution, global clock distribution, and I/Os. Three-dimensional integration affects all of these. Effective exploitation of the 3D technology requires incorporation of the 3D effects into the design of the functional 2D elements. An illustration of a stacked 3D multichip module package is shown in Figure 6.12 [24].

Process variations in ICs can impact performance, leakage, and stability. This is particularly an issue in high-density chips, such as DRAMs, when the DRAMs are stacked as last-level cache on multicore processors in a 3D configuration. Variations in bank speed create nonuniform cache accesses in the 3D structure. In June of 2012, the University of Pittsburgh discussed development of a model for the process variation in a four-layer DRAM used to characterize the latency and retention time variations among different banks to the core [25]. Cache migration schemes were developed to use fast banks while limiting the cost of migration. A performance benefit was shown in exploiting fast memory banks through migration. It was

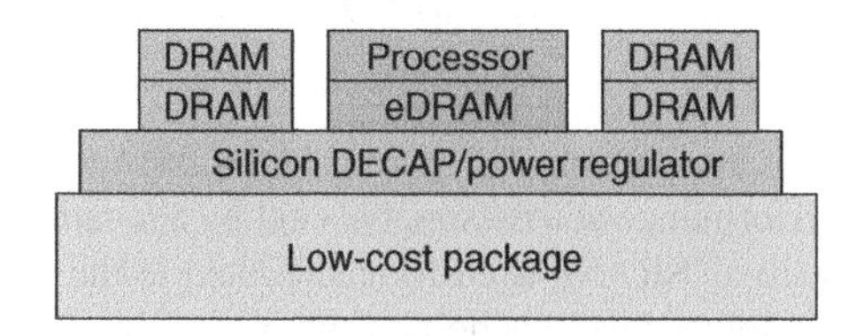

Figure 6.12 Stacked 3D MCM package with DRAM, processor, and eDRAM (J.L. Burns, (IBM), VLSI Technology Symposium, June 2011 [24].)

found that on average a variation-aware management can improve the performance over baseline by about 16.5%. It also put the system only 0.8% away in performance from an ideal memory with no process variation.

Separate magnetoresistive RAM (MRAM) and CMOS logic chips can be integrated in a stacked configuration to make a processor with MRAM cache. The impact of a stacked MRAM cache on a microprocessor core was evaluated by Pennsylvania State University in May of 2011 [26]. It was modeled and compared to a simulation that showed that MRAM stacking can provide competitive instruction-per-cycle performance along with a reduction in power consumption compared with a similarly configured SRAM cache.

The performance of an SSD hierarchy can be improved by integration with 3D TSV stacking. In January of 2013, the University of Tokyo discussed a 3D TSV-integrated stacked hybrid DRAM–ReRAM multilevel cell (MLC) NAND SSD architecture with a NAND-like interface [1]. The device had a sector-access overwrite policy for the ReRAM. Intelligent data-management algorithms were proposed to suppress data fragmentation and excess usage of the MLC NAND. The resulting device exhibited an 11-fold performance increase, a 6.9-fold improvement in endurance, and a 93% write energy reduction. A 68% energy reduction was achieved by using 3D TSV interconnects from the conventional MLC NAND SSD. The ReRAM write and read latency would need to be less than 3 μs to achieve these improvements, and the required endurance for the ReRAM was 10^5. A NAND-type ReRAM interface sector access overwrite policy was found necessary. A 3D schematic diagram of the SSD is shown in Figure 6.1 and included SSD controller, DRAM, ReRAM, and NAND flash memory, all connected using TSV. A block diagram of the 3D TSV integrated hybrid DRAM–ReRAM–MLC NAND SSD is shown in Figure 6.13 [1]. The ReRAM used a NAND-like interface.

6.2.9 Adding Storage-Class Memory to the Memory Hierarchy

The performance of the traditional memory hierarchy can be improved by adding a new type of memory, the storage-class memory (SCM), to the classical memory hierarchy system, as illustrated in Figure 6.14. The SCM is both random access and nonvolatile, and it has random access time or latency between that of the volatile DRAM main memory and the slow nonvolatile HDD or SSD.

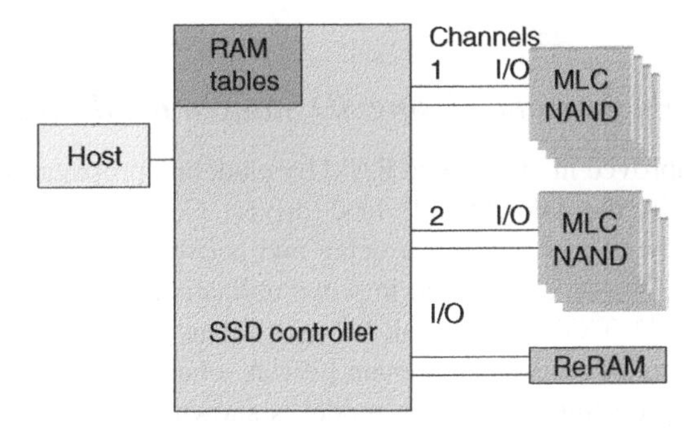

Figure 6.13 Block diagram of 3D TSV hybrid DRAM–ReRAM–MLC NAND SSD. (Based on C. Sun *et al.*, (University of Tokyo), ASP-DAC, 22 January 2013 [1].)

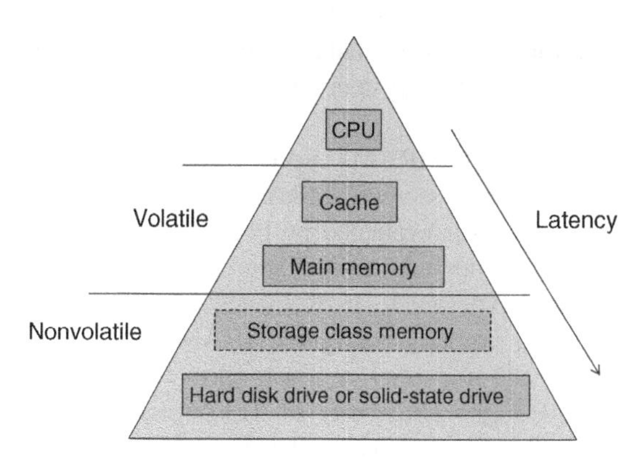

Figure 6.14 Inclusion of nonvolatile random access storage-class memory in a conventional memory hierarchy.

Several types of newly emerging memories fall under the SCM specification type. These include MRAM, PC-RAM, and ReRAM. These memories are all nonvolatile and random access.

SCMs can act as a cache for the HDD or can be used with SSD as a buffer for the DRAM. It is also possible to use an SCM as a write buffer in NAND flash with a hybrid SSD memory system. Such a hybrid memory system has potential to show significant reduction in power consumption over a conventional system.

In the memory hierarchy, 3D stacked integration can be used to improve the performance between the storage class memory and the other memories in the hierarchy. A 3D TSV integrated hybrid ReRAM and MLC NAND SSD was discussed in June of 2012 by the University of Tokyo for PC, server, and smart phone applications [27]. The ReRAM had a NAND-like interface and sector access overwrite policy. Intelligent data management algorithms were used to suppress data fragmentation and also suppress excess usage of the MLC NAND by storing hot data in the ReRAM. The system had an 11-fold performance increase, 6.9-fold endurance enhancement and 93% write energy reduction compared with a conventional MLC NAND SSD. To achieve this, ReRAM write–read latency was less than 3 µs. Endurance for ReRAM was 10^5 cycles. The TSV interconnects reduced energy by 68%. A circuit schematic of a hybrid SSD system with SCM and NAND flash memory is shown in Figure 6.15 [28].

6.2.10 Improving Performance Using 3D Stacked RAM Modeling

Performance can be improved in 3D stacked RAM by modeling propagation delay in the TSVs and capacitive delay in the DRAM cell load driver. Constraints on 3D DRAM peak performance include row-to-row activation delay and active bank window. Row activation can be balanced across DRAM channels to improve utilization of the DRAM. An analysis of propagation delay in 3D TSV stacked DRAM chips was done in May of 2012 by the Universities of Alabama, Ulsan, and Yeungnam [29]. A schematic illustration of the 3D TSV stacked DRAM and logic setup used in the model is shown in Figure 6.16 [29].

The physical-level design of 3D stacked TSV DRAM ICs was analyzed by performing design and electrical circuit model extraction. The traditional three-transistor model was used

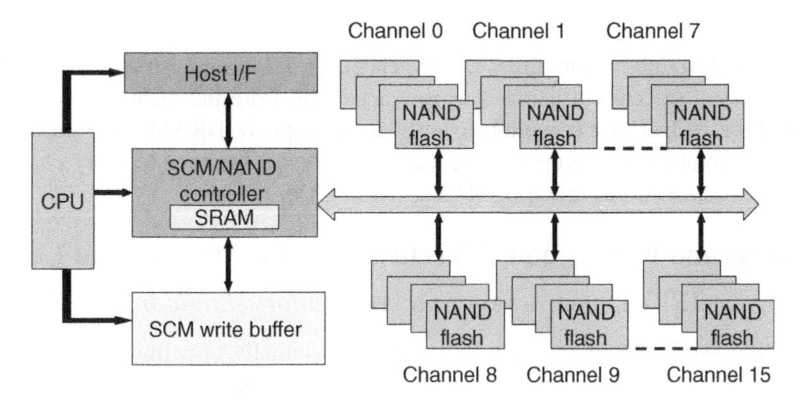

Figure 6.15 Circuit schematic of the hybrid SSD and NAND flash memory system. (Based on K. Takeuchi, (University of Tokyo), ISLPED, August 2011 [28].)

in the simulation. Propagation delay in the TSVs was represented by the RC time constant and by capacitive delay in the DRAM cell load driver during pull up and pull down in CMOS. Time-domain reflectometetry (TDR) and eye diagram analysis were used to validate the models. The conclusion was that the proposed propagation delay model can be used for various fast, high-density memories [29].

A design strategy using an electronic system level (ESL) virtual platform to develop 3D memory architecture for a heterogeneous multicore system was described by ITRI in April of 2013 [30]. Based on this virtual platform, designers were expected to be able to swiftly obtain the 3D stacking interface for better system performance, improved energy efficiency, and better utilization of the TSV. A possible stacking architecture and memory interface meeting the design and performance requirements was evaluated for the target system. Real multimedia H.264 decoding experiments indicated the stacking system could attain about 30% improvement on performance and 20% improvement on energy savings compared to a 2D system.

An analysis was done of the impact of constraints on 3D stacked DRAM peak memory performance in March of 2011 by Postech [31]. These constraints included row-to-row

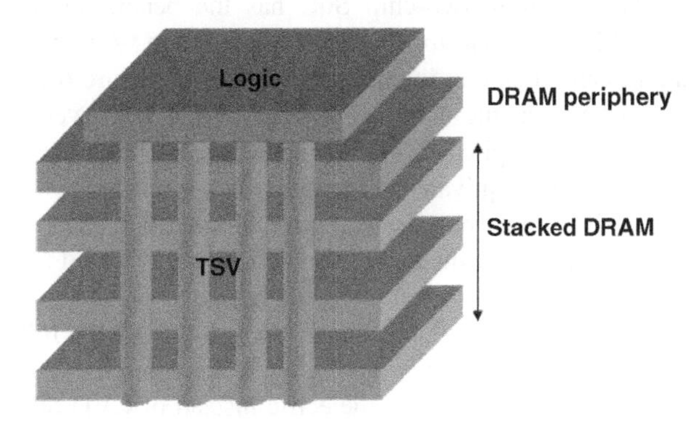

Figure 6.16 Schematic cross-section of 3D TSV stacked DRAM–logic in model. (Based on S. Kannan *et al.*, (University of Alabama, University of Ulsan, Yeungnam University), ISCAS, 20 May 2012 [29].)

activation delay (T_{rrd}) and four active bank windows (t_{FAW}). It was proposed to balance the budget of DRAM row activation across DRAM channels. To accomplish this, an intermemory controller received the current demand of row activation from the memory controllers and redistributed the budget to the memory controllers to improve DRAM performance. It was shown experimentally that sharing the row activation budget between memory channels can give an improvement in utilization of 3D stacked DRAM.

6.3 Process and Fabrication of Vertical TSV for Memory and Logic

6.3.1 Passive TSV Interposers for Stacked Memory–Logic Integration

Configurable passive interposers with TSV were used initially to permit interconnection of different conventional chips. The 2.5D interposer technology placed bumped flip chips side by side on an interposer, which redistributed the system connection of the chips and passed the connections down through copper vias to the package. The interposer was considered "passive" because it contained no active circuit elements. Many of the TSV fabrication processes announced have been of the 2.5D passive interposer technology type.

In early 2012, production 3D IC integration entailed die stacking with wire bonds and package-on-package (PoP) stacking. The 3D IC chip stacking integration was in development for memory and logic integration using active interposers, TSV, and microbumps. A passive interposer layer with TSVs could provide redistribution of interconnects between two stacked conventional heterogeneous chips. It helped provide the small footprint of a vertical chip stack and some of the short-wire improvements of the TSV without requiring a full redesign of the planar chips.

Early stacking with TSV interposers used side-by-side chip stacking. This type of interposer use was called 2.5D TSV chip integration. Silicon interposers with TSVs could be used for integrating several heterogeneous chips in a single-chip footprint. Shinko discussed the development process for a silicon interposer as early as May of 2008. The chips could be side by side, as in a conventional wirebonded chip package, but the use of the silicon interposer permitted a tighter redistribution of the wiring between the chips into a single system-on-chip (SoC) -type chip. This 2.5D integration provides some of the speed and power benefits of 3D chip integration without the need for new design tools for chip redesign. The silicon interposer has copper TSV and copper wiring on both sides. It provides a flat surface to form fine wiring and has high reliability because it shares a coefficient of thermal expansion with the silicon chips attached to it. The resulting two-chip SoC has the benefit of heterogeneous chip integration with a reliability similar to a single silicon chip. An example of two-chip side-by-side integration using a silicon interposer is shown in Figure 6.17 [32]. To achieve high-speed transmission between the chips, fine 4 μm wiring lines were integrated on the silicon interposer along with fine-pitched 40 μm solder bumps. This configuration of silicon module is expected to be nearly equivalent in performance to an SoC.

The fabrication of the silicon interposer began with thinning the silicon wafer to a 200 μm thickness, which could be handled without a carrier. TSVs were formed by deep reactive ion etching (DRIE). Thermal oxidation to form an insulating layer of SiO_2 was followed by Cu electroplating to fill the vias. Chemical mechanical polishing (CMP) followed. Multilayer wiring was then formed by 3 μm copper plating with conventional patterning. A stress-relieving polymer was used as the interlayer dielectric instead of CVD-SiO_2 to avoid silicon cracking due to thermal expansion mismatch with the Cu TSV. The chips were mounted on microbumps that were made by electroplating.

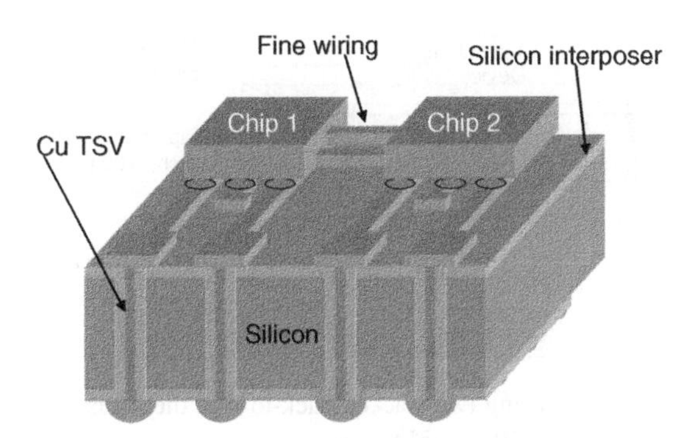

Figure 6.17 Two-chip side-by-side 2.5D TSV integration using silicon interposer. (Based on M. Sunohara *et al.*, (Shinko), ECTC, 27 May 2008 [32].)

Existing designs of 2D memory and logic chips can be reused in designs of 2.5D stacked chips, thereby providing higher-density chips without initially requiring redesign. This design reuse of 2D stacked memory and logic in 3D intellectual property (IP) blocks was discussed in February of 2012 by Dresden University of Technology [33]. They showed how to integrate established 2D IP blocks into 3D chips without altering the layout of the 2D IP block. The overhead of the integration proposed was shown to be small compared with the benefits of the 3D integration.

A memory and system architecture for 400 Gb/s networking using a 2.5D packaging solution was described by Cypress in February of 2014 [34]. The high-capacity SRAM was implemented as a separate die that was integrated with the ASIC/ASSP on a 2.5D interposer-based SIP. The interposer provided fine-pitch routing of a large number of signals between closely placed chips, which reduced the loading on the drivers. The drivers consumed only 1 mW/Gb/s of power. A JEDEC high-bandwidth memory interface was used. The printed circuit board (PCB) routing complexity was reduced, and the required 9600 MT/s for the 400 Gb/s line card was achieved. For similar products an ASIC/ASSP die could be integrated with multiple memory dice that could be SRAM or SRAM and SDRAM.

True 3D ICs using TSV technology involve two or more dice connected together vertically using TSV. This technology was described by Cadence in December of 2012 [35]. An example is shown in Figure 6.18 in which one chip containing TSV is attached to a SIP substrate face down using conventional flip-chip technology. A second face-down chip is attached to the first using microbumps. This is an example of a back-to-face configuration for the two stacked chips [35]. A summary of the advantages of 3D ICs with TSVs includes lower cost because chips at different levels of technology can be used, higher speed because interconnects can be shortened, smaller footprints, and reduced power usage through smaller I/O drivers.

Another example of true 3D TSV interposer technology used for 3D heterogeneous integration was discussed in May of 2011 by ITRI. A TSV passive interposer was used to support high-power chips, such as MPU and logic on its top side, and low-power chips, such as memory on its bottom side [36]. Special underfills were needed between the Cu-filled interposer and all of the chips. A special underfill was used between the TSV interposer and the high- and low-power flip chips. Ordinary underfills were needed between the interposer and the organic substrate. An illustration of the use of a TSV interposer with both high- and

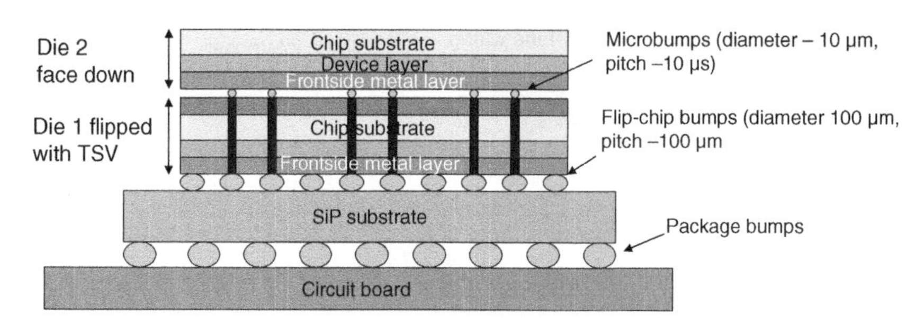

Figure 6.18 3D TSV technology using two stacked back-to-face die. (Based on white paper posted at http://www.cadence.com, December 2012 [35].)

low-power chips, including logic and memory, is shown in Figure 6.19 [36]. As well as discussing a TSV interposer with chips on both sides, ITRI discussed 3D IC integration using a TSV interposer. The Cu-filled interposer acted as a stress relief buffer, reducing stress acting on the Cu-low-κ pads on product chips.

Testing of a passive TSV interposer technology used in a 3D IC integration system was discussed in May of 2012 by Siliconware Precision [37]. The intent of the TSV interposer was to provide high-density and heterogeneous integration using conventional chips. The interposer was tested for thermal-mechanical stress and warpage induced by thermal expansion mismatch during the fabrication process. This mismatch can affect TSV cracking or C4 chip connection bump cracking after bonding to the organic substrate. The study involved a step-by-step process simulation using the finite element method to investigate stress and warpage behaviors resulting from processing. Both full array dummy bumps and thinner top chips were used to reduce microbump stress. In addition, a low coefficient of thermal expansion organic substrate core, high coefficient of thermal expansion mold compound, and thinner top chips helped to lower package warpage. These finite element models were meant to help provide a guideline to the designer of 3D IC integrated SIP with passive TSV interposer.

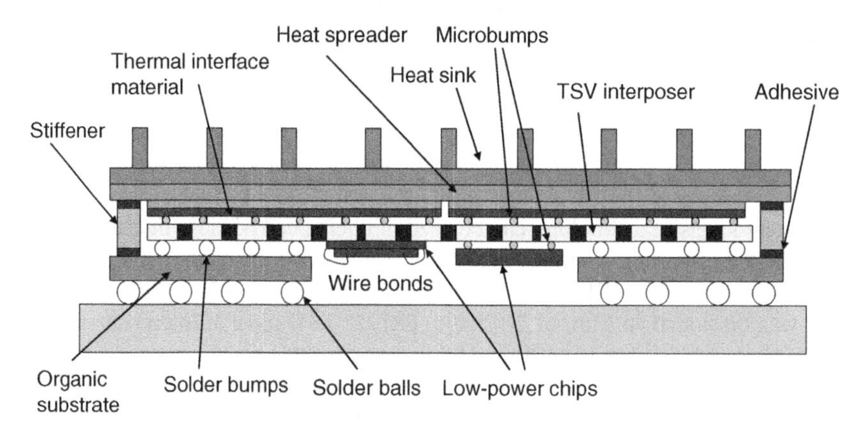

Figure 6.19 TSV interposer with high- and low-power chips. (Based on J.H. Lau, (ITRI), ASME Interpack, 21 May 2011 [36].)

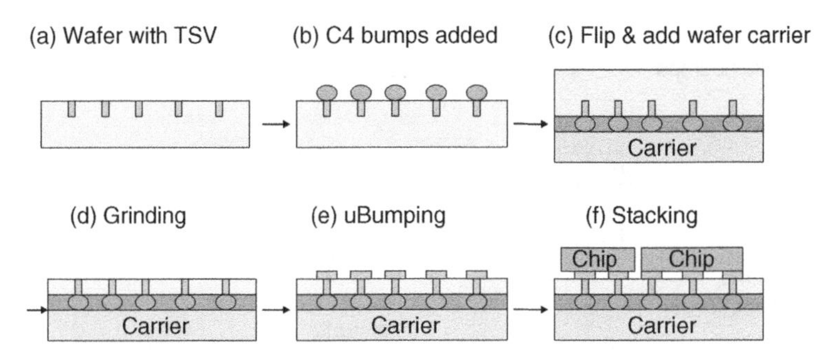

Figure 6.20 Cu TSV 3D chip-on-wafer-on-substrate technology process. (Based on J.Y.C. Sun, (TSMC), VLSI Technology Symposium, June 2013 [38].)

6.3.2 Process Fabrication Methods and Foundries for Early 2.5D and 3D Integration

A Cu TSV 3D chip stacking technology, called chip-on-wafer-on-substrate, was discussed by TSMC in June of 2013 [38]. An illustration of this technology is shown in Figure 6.20 [38]. In (a) the wafer (interposer) is shown with TSV and in (b) has C4 bumps added. It is then in (c) inverted on a carrier and in (d) the backside ground down to expose the TSV. The exposed TSV are then microbumped in (e), and in (f) the active chips are flipped onto them. These active chips are known good dice (KGD). Because the chips are placed side by side on a passive interposer with the TSV connections, this is a 2.5D technology.

Both homogeneous and heterogeneous 3D TSV chips have been integrated by TSMC. Heterogeneous has been defined as having three different types of chips on the TSV module, and homogeneous usually means a memory stack of identical chips integrated on the TSV module [38].

The GlobalFoundries Singapore wafer fab was expected to make 2.5D chips using silicon interposers, and it was reported in April of 2012 that GlobalFoundries was installing equipment to make vertical TSV copper interconnects in its Fab 8 in New York [39]. Production was expected in the second half of 2013 for 3D stacks using 20 nm and 28 nm process technology. It was working with several packaging companies including Amkor to develop process flows so that the packaging companies were not locked out of the TSV loop. A tool suite was being developed with system suppliers to create the TSV. The first 20 nm test wafers with TSV were expected in late 2012, with reliability data available early in 2013. The first commercial run of 20 nm and 28 nm wafers with TSV was expected to start in the second half of 2013, with full production ramp in 2014.

One challenge was expected to be development of common metrology tools and processes. The thick wafers with 5 μm deep TSVs were expected to be shipped to packaging companies that would then thin the wafers and do back-side metallization. Electronic design automation (EDA) tools were being developed to control placement of blocks for vertical connections in logic and memory designs. The 20 nm process would be based on high-κ metal gate technology.

A 200 mm prototyping fab for TSV 3D solutions was announced by CEA-LETI for its Open 3D technology partners in February of 2012 [40]. The Open 3D program included 3D design,

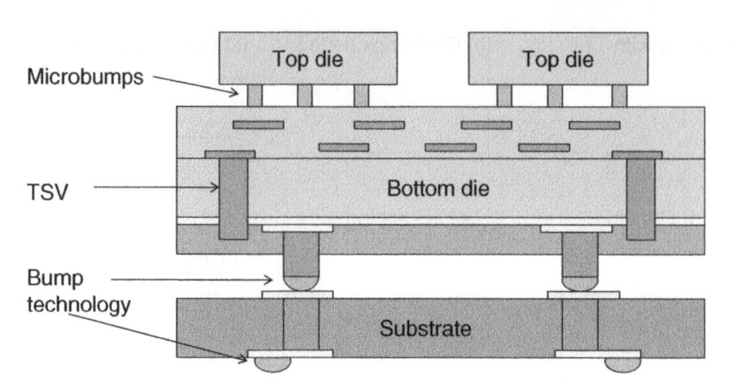

Figure 6.21 Illustration of some open 3D processes offered by CEA-LETI for prototyping. (Based on CEA-LETI press release, 1 February 2012 [40].)

layout, TSV implementation, components assembly, reliability tests, and final packaging. The Open 3D fab was located on the Minatec campus in Grenoble, France, and was expected to be operational in 2012 for 300 mm wafers. It was expected to be based on mature technologies to moderate costs. Modules available included TSV with a 1 : 3 aspect ratio, chip-to-wafer interconnects using microbump technology, chip-to-substrate bump technology interconnects, redistribution layers, underbump metallurgy and temporary bonding, and thinning and debonding processes. These processes are illustrated in Figure 6.21 [40]. Applications targeted included biomedical, aeronautics and space, consumer, and defense and security.

Applied Materials and A*STAR, in March of 2012, announced the opening of a 3D chip packaging lab in Singapore [38]. The lab had 14 000 ft^2 of cleanroom, which included a 300 mm production line and 3D packaging tools developed by A*STAR. The packaging lab was designed for other companies to experiment with 3D packaging and work with the Applied Materials equipment. The lab included the equipment for TSV chip stacking and for using a silicon transposer to connect two or more chips together where the bond pads don't match.

6.3.3 Integration with TSV Using a High-κ–Metal Gate CMOS Process

A 3D integration process using TSV and chip-on-wafer (CoW) techniques was described by TSMC in December of 2012 [41]. This 3D process compared polysilicon and high-κ–metal gate CMOS wafers that were thinned and stacked with little to no degradation. A schematic cross-section of a poly gate and a metal gate device near a TSV is shown in Figure 6.22.

The effect of TSV-induced mechanical stress on ΔI_{dsat} for the high-κ–metal gate device was smaller than for poly gate devices with the same channel length. It was also shown that ΔI_{dsat} for the high-κ–metal gate device was proportional to the diameter of the TSV squared and was independent of TSV orientation, device polarity, and distance from the TSV [41].

A 28 nm low-power CMOS full SoC platform with high-κ–metal-gate technology was discussed by STMicroelectronics in December of 2011 [42]. This platform included 3D integration for high-data-rate interfaces based on a wide I/O using TSV. The large memory size was achieved with an eDRAM macro. The CPU critical path speed was enhanced by a triple-gate oxide scheme using a high-κ–metal gate combined with a 20 fF/μm^2 MIM solution for decoupling capacitance. A radio frequency (RF) device suite was also developed to enable high-performance analog cells.

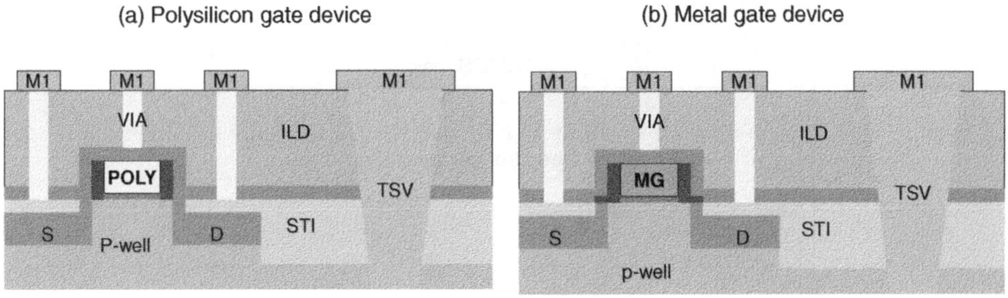

Figure 6.22 Schematic cross-section of a TSV near (a) a poly-gate device; and (b) a metal gate device. (Based on T. Lo *et al.*, (TSMC), IEDM, December 2012 [41].)

The wide I/O can be used with new interfaces that use a stacked die structure, achieving a bandwidth of 12. 8 Gb/s. By using a 512-bit bus at 200 MHz, interface with DDR3 can be achieved using TSV with a "via-middle" integration scheme. The TSV diameter is 6 μm, and the depth is 50 μm. The SRAM DC performance showed that the TSV process did not disturb the surrounding logic block because it was sensitive to transistor matching, local noise, and mechanical stress.

The eDRAM macro used a TiN–ZiO_2–TiN capacitor placed over the low-κ (COLK), which permitted a good-yielding, high-density cell with 0.08 mm^2/Mb. A decoupling capacitor available in the SoC technology was used for a low-cost technology. Access time of 400 MHz and leakage of 10 fA was achieved using process–design macrocell co-development. The technology has thin oxide transistors with multiple voltage thresholds and thick oxide transistors for multiple power supplies.

6.3.4 Processor with Deep Trench DRAM TSV Stacks and High-κ–Metal Gate

Processors with embedded deep trench DRAM can be used in TSV stacks. In December of 2012, IBM provided a close-up scanning electron microscope (SEM) image of a thinned lower die in 32 nm DRAM technology that showed the interconnect layers, TSV, and deep trench, illustrating the size incongruity of the TSV and other chip features. A sketch of this SEM image is shown in Figure 6.23 [43].

A high-performance eDRAM deep trench "P7" processor-type prototype made in a 3D stackable 32 nm high-κ–metal gate technology using copper TSV with C4 stacking technology was discussed by IBM in June of 2011 [44]. A post "through-via" processing functional test was used to show that latency <1.5 ns and 500 MHz operation were both preserved in the stacking operation. Features of the chips and stack are shown in Table 6.4 [44].

A schematic cross-section of the two DRAM chips is shown in Figure 6.24. The top chip shown in this figure is 128 Mb, and the bottom chip is 96 Mb [44].

A major challenge for 3D sequential integration of chip substrates is the control of molecular bonding. Well-controlled molecular bonding permits obtaining a very high-quality top active layer. In December of 2011, CEA-LETI, STMicroelectronics, and EPFL discussed 3D sequential integration of die substrates [45]. Solid-phase epitaxy was used to match the performance of the top FET processed at a low temperature of 600 °C with the bottom FET

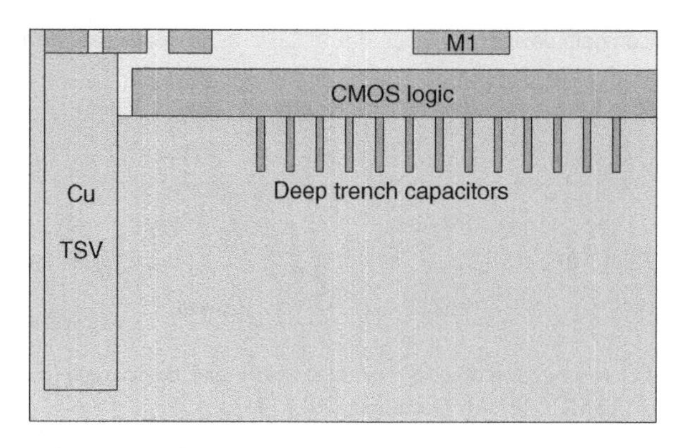

Figure 6.23 Sketch of SEM showing relative size of TSV and deep trench capacitors. (Based on S.S. Iyer, (IBM), IEDM, December 2012 [43].)

Table 6.4 Feature Table of 32 nm High-κ–Metal SOI eDRAM [44].

Chip	Cell Size	0.039 μm^2
	Retention	40 μs @ 105 °C
Macro	Macro size	0.136 mm^2/Mb
	Performance	tRAC = 1.5 ns
		tRC = 2.0 ns

devices. The development of a stable salicide was considered to permit retaining bottom performance after the top FET had been processed. Because 3D sequential integration offered fine-grain circuit partitioning at the transistor scale, it opened a new field of applications and design. It enabled increasing both density and performance without requiring aggressive scaling. Its main technologies were molecular bonding and a low-temperature top FET process that could permit designing 3D transistors, matching the targets of advanced technology nodes using low access resistance, salicide, scaled effective oxide thickness (EOT), optimized threshold voltage, and mobility boosters. Various SoC and FPGA applications could be served.

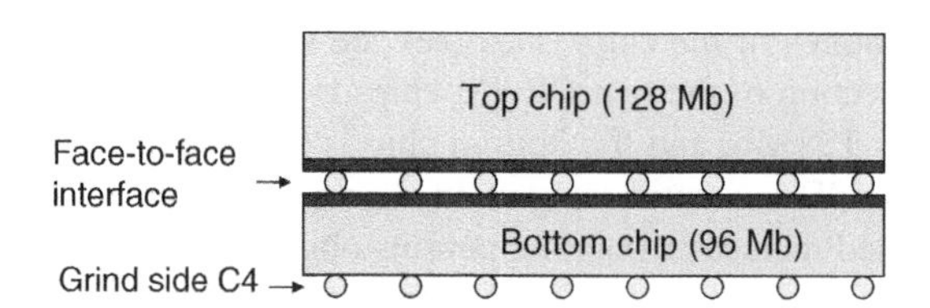

Figure 6.24 Schematic cross-section of two stacked DRAM chips. (Based on J. Golz *et al.*, (IBM), VLSI Circuits Symposium, June 2011 [44].)

6.4 Process and Fabrication Issues of TSV 3D Stacking Technology

There are various process and fabrication issues with 3D integration using vertical TSV to connect stacked memory and logic systems. These include the high aspect ratio of the TSV, which can affect uniform filling of the via, and copper pumping, which is an effect of thermal expansion mismatch between copper TSV and the surrounding dielectric and silicon. Copper pumping results in plastic deformation of the copper TSV, which leads to a pump-up effect during high-temperature post-TSV processing. This effect can lead to delamination and cracking of the dielectric and silicon near the TSV.

6.4.1 Using Copper TSV for 3D Stacking

Various challenges surrounding copper TSV for use in 3D dice integration were investigated by IBM in April of 2012 [46]. Issues discussed included differential thermal expansion mismatch between Cu and Si, high aspect ratios for TSV, and integration and reliability issues. A TSV structure that answered these issues was shown using CMOS with a high-κ–metal gate. Data from test structures showed no "copper pumping" or other negative effects on neighboring devices or interconnects. Functional 3D prototypes using stacked, embedded DRAMs demonstrated no impact from the TSV processing. The advantages of 3D integration claimed were improving system performance, reducing wiring delay, improving memory bandwidth, reducing power consumption, enabling integration of heterogeneous chips, and enabling smaller form factors. A copper TSV provided a low parasitic channel for power and for signals. Embedded DRAM chips in 32 nm high-κ metal gate with TSV were stacked and characterized to gain understanding of these issues.

In this process, the TSVs were made in the back end of line (BEOL) in a "via-last" process, which was chosen to ensure it was "node-agnostic." TSVs were formed using a deep silicon reactive ion etch (RIE) followed by conformal oxide deposition. After the deposition of liner and seed layers, a void-free, bottom-up, copper plating was used to form the TSV conductor. Planarization followed before continuing the standard process flow. After the BEOL flow was complete, the wafer was attached to a glass handler and thinned, and then a backside wiring layer was constructed over an oxide–nitride isolation layer. The TSVs were then exposed, and the wafers were diced and assembled either FTF or FTB. The process used was a 90 nm DRAM and 32 nm logic, both SOI and bulk. Die stacking was either FTF or FTB. Key attributes of the 3D TSV process are shown in Table 6.5 [46].

Several reliability issues were checked. Long etch times during TSV formation and the thinning steps could crack the silicon. The large aspect ratio could result in voids in the TSV.

Table 6.5 Attributes of the 3D TSV Process [46].

# Wiring Levels	5–12
BEOL Dielectrics	Dielectric constant (κ): 4.1 > 2.4
Wiring Levels Below TSV	3–9
TSV Form Factor	Annular, circular
TSV Diameter/Depth	6–20 μm/50–100 μm
Minimum TSV Pitch	50 μm
Pb-Free C4 Pitch	50–186 μm

Thermal expansion mismatches between the copper and silicon could cause delamination during annealing operations in the BEOL. Optimization of the various processes was needed to avoid voiding, silicon cracking, and delamination.

The test chip had 3000 TSV links. Thermal cycling was between −65 and +150 °C, and resistance was found unchanged. Wiring structures were made adjacent to and above the TSV to check for leakage between the structures due to cracks or extrusions of the interlayer dielectric. Leakage current was low until break-down occurred at 300 V. The effect on the TSV of stress engineering to improve mobility in sub–45 nm nodes was checked by placing a FET array near the TSV. The FET was found unaffected by the proximity or orientation of the channel region to the TSV. A process change was made to eliminate a proximity effect on threshold voltage. Dice with embedded deep trench DRAM were made using TSVs, and no influence on the chips due to the TSVs were observed in performance or retention.

6.4.2 Air Gaps for High-Performance TSV Interconnects for 3D ICs

An air gap can be integrated into the TSV sidewall liner to form high-performance interconnects for 3D stacking. TSV sidewall air gap formation in a 28 nm high-κ metal gate CMOS logic process was discussed by TSMC in June of 2013 [47]. Compared with conventional TSV structures, this air gap TSV structure showed reduced capacitance, reduced TSV-induced stress, and, as a result, reduced keep-out zone for CMOS devices. A 3D schematic of a TSV with integral sidewall air gap is shown in Figure 6.25 [47].

A simulation confirmed that the inserted air gap reduced the TSV proximity effect and preserved the CMOS device performance. The structural integrity of the TSV structure was demonstrated by its good electrical characteristics and its compatibility with the BEOL metallization, which followed the TSV process. The process sequence of the TSV with integral air gap consisted of TSV etch, TSM liner deposition, copper filling and CMP, air gap formation, air gap closure, and final CMP. This was followed by BEOL metallization.

Depending on the air gap dimensions, significant capacitance reduction up to 18% was measured. The integrity of the TSV with air gap was verified by a stable TSV-to-substrate electrical leakage current. The integral air gap also did not cause any topographical variation for subsequent processes, which confirmed that the integrity of the metallization on top of the TSV

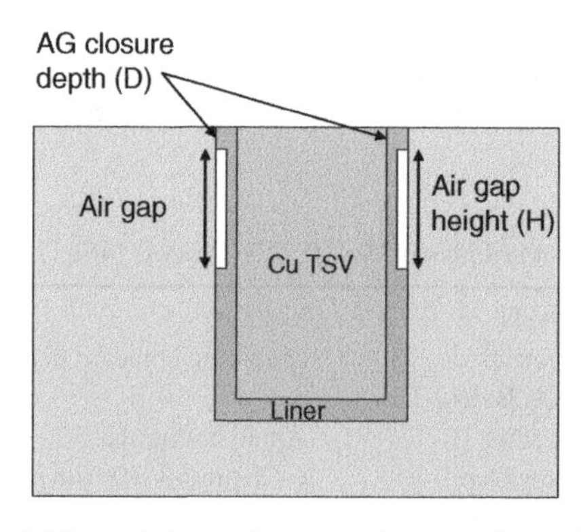

Figure 6.25 3D schematic of TSV with integral air gap, showing air gap height and air gap top closure depth. (Based on E.B. Liao *et al.*, (TSMC), VLSI Technology Symposium, June 2013 [47].)

with air gap was comparable to conventional TSVs. Parasitic capacitance and TSV-induced stress were significantly reduced by the inclusion of the air gap. The air gap structure and the processes associated with it were found to have no impact on the TSV leakage current. They also did not lead to any process incompatibility with the BEOL process.

6.5 Fabrication of TSVs

6.5.1 Using TSVs at Various Stages in the Process

As chips using lateral scaling move into sub–20 nm geometries, they become more difficult to fabricate and more expensive. The potential for using 3D integration to reduce the die footprint becomes increasingly attractive as an alternative solution because, in addition to a small footprint, it offers high performance, low power consumption, and multifunction integration. Three variations of the TSV process are commonly used. The "via-first" process is done during the front end of line (FEOL). In the "via-middle" process, the vias are formed after the FEOL and before the BEOL. These processes can have a very small TSV pitch but are complex to make. The "via-last" TSV process is done during or after the BEOL and is normally a simpler process, but it has a larger TSV pitch with a larger parasitic capacitance, which can range up to 300 fF/TSV [48]. Via-first and via-middle TSV processes support logic-to-logic stacking with dense TSV. The via-last TSV is more compatible with applications requiring a limited number of TSVs, less than 10 000, such as wide-I/O memory stacks.

The various TSV integration methods were described by UMC in April of 2012 [49]. These include via-first, via-last and via-middle integration. The via-first method of integration etches and fills the vias in the substrate before the CMOS FEOL and BEOL process. The via-first method is illustrated in Figure 6.26 [49]. In via-first technology, the vias are etched into the substrate, lined, and Cu-filled before the CMOS logic is formed. After the FEOL+BEOL CMOS process is completed, a carrier is attached and the backside of the wafer is thinned so the TSVs are exposed. The backside of the wafer is then bonded on the front side of the second wafer, "back to front." Companies that have explored this method include IBM, NEC, and Oki.

The via-middle method of via formation is done after the FEOL and before the BEOL. Spaces are left in the FEOL process for the via trenches, which are filled and planarized before the BEOL metallization. After the BEOL process, the wafer is bonded F2F and thinned. This method has been studied by UMC, TI, TSMC, IMEC, GlobalFoundries, Tezzaron, and others. The via-middle method is shown in Figure 6.27.

The via-last method is done after the BEOL+FEOL CMOS process but before bonding. The spaces left for the vias are etched, filled, and planarized. A carrier is attached to the front side of

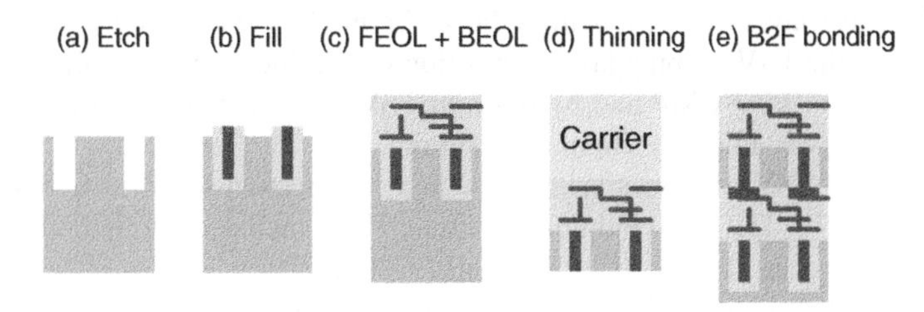

Figure 6.26 Via-first method for TSV before the CMOS integration. (Based on T.C. Tsai *et al.*, (UMC), 27th Annual Advanced Metallization Conference, April 2012 [49].)

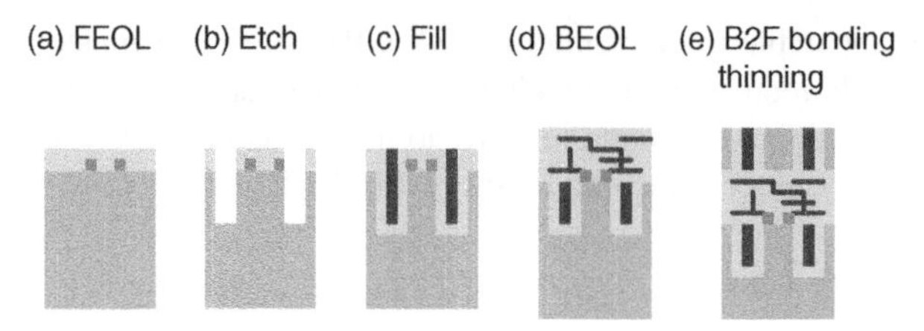

Figure 6.27 Via-middle method after FEOL and before BEOL. (Based on T.C. Tsai *et al.*, (UMC), 27th Annual Adv. Metallization Conference, April 2012 [49].)

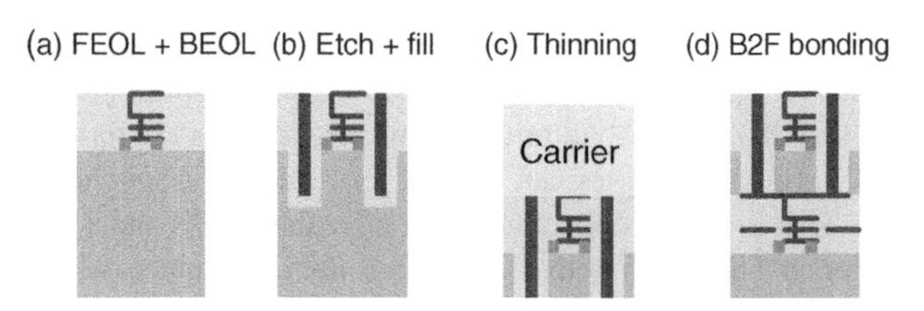

Figure 6.28 Via-last method after FEOL + BEOL. (Based on T.C. Tsai *et al.*, (UMC), 27th Annual Advanced Metallization Conference, April 2012 [49].)

the wafer, and the backside of the wafer is thinned to expose the TSV. The front of a second wafer is then bonded to the back of the thinned wafer, back to front. The exposed TSVs connect the circuitry on the two wafers. The via last method after FEOL+BEOL is shown in Figure 6.28.

This via-last approach was demonstrated by Infineon and several others. The via-last process can also be done after FEOL, BEOL, bonding the wafers F2F, and thinning. The via is then etched from the back and filled. This method has been demonstrated by Samsung, IBM, and several labs.

An example of via-middle TSV structures was discussed by UMC in April of 2012 [49]. These 28 nm technology, 70 μm deep, and 10 μm diameter via-middle TSV structures were formed using a CMP process. The CMP process included Cu polishing, Cu barrier/isolation layer polishing, and dielectric SiN stop layer polishing. To prevent formation of Cu extrusion and voids after the BEOL processes, pre-TSV CMP anneal and electrochemical deposition of Cu were evaluated to enlarge the TSV CMP process window. A less than 15 nm thickness range control for interlayer dielectric within the wafer and less than 10 nm Cu extrusion level without voids after the TSV capping layer deposition were achieved by annealing at 400 °C for 10 minutes before the TSV CMP and by uniform TSV electromechanical deposition profile electroplating in a lower-impurity chemical solution.

The three types of TSV processes using cross-layer connection schemes were described in June of 2013 by NTHU, ITRI, and Fukuoka Institute. These included via-first, via-middle, and via-last processes and are illustrated in Figure 6.29 [48].

The via-first and via-middle TSVs are formed before the BEOL. These processes can have very small TSV pitch, but they are complex to make. The via-last TSV process is done during or after the BEOL and is normally a simpler process but has a larger TSV pitch and a larger

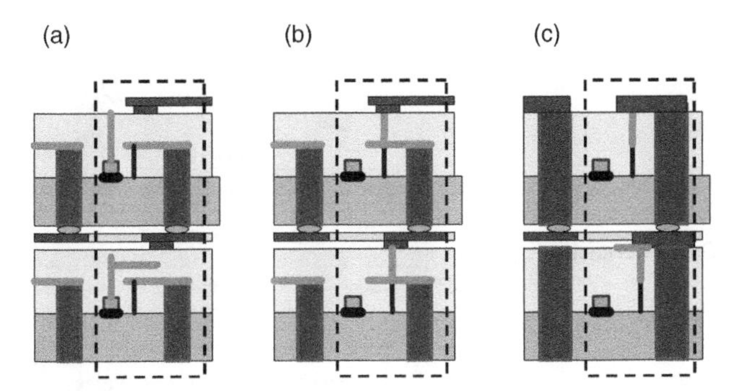

Figure 6.29 Three TSV technologies: (a) via first/middle with buffer; (b) via first middle (direct connect); and (c) via last. (Based on M.F. Chang *et al.* (2013) (NTHU, ITRI, Fukuoka Institut), *IEEE Journal of Solid-State Circuits*, 48(6), 1521 [48].)

parasitic capacitance up to 300 ff/TSV. Via-first and via-middle TSV processes support logic-to-logic stacking with dense TSV. The via-last TSV is more compatible with applications requiring a limited number of TSVs, less than 10 000, such as wide-I/O memory stacks [48].

6.5.2 Stacked Chips using Via-Middle Technology

The technical challenges of TSV process integration as used in high-density, fast, low-power, handheld computing applications was discussed by Samsung in October of 2011 [50]. These challenges include the impact on the BEOL, on the FEOL, and on the yield. These issues were potentially solved by using TSV made during the middle of the process, the via-middle technology.

A number of companies, including the large commercial foundries, have explored via-middle technology. A number of the technical methods and challenges have been described. Via-middle technology was one of the first to make it into commercial production. Tezzaron offers a two- and three-layer stack process using via-middle technology. Contact and vias are tungsten (W) with copper interconnects and copper bondpoints in the top metal [51]. A schematic cross-section illustrating the Tezzaron three-layer stack is shown in Figure 6.30.

3D TSV technology with via-middle TSV was described by Samsung for use with a 32GB DDR3 registered dual-inline-memory module in October of 2011 [50]. It indicated that bandwidth is currently doubling each year for mobile computing devices, and the use of a wide I/O with short connections using TSV in the middle of the chip can help achieve this bandwidth. The via-middle TSV technology was used in conjunction with microbumps and flip chip assembly. The via size for the TSV was 50 μm. The TSV can be made by deep RIE followed by isolation oxide deposition, a barrier seed, Cu electroplating, and CMP.

As a cost-effective alternative to deep RIE, a repeated isotropic etch process could be used. This process uses a mask with an isotropic etch, polymer deposition, and then isotropic etch, which is then repeated. A scalloped effect results, as shown in Figure 6.31 [50]. This effect can be minimized by using smaller etch steps. The TSV can be accessed from the backside of the wafer by thinning the finished wafer.

The proximity effects to active devices of isolated and arrayed via-middle Cu TSV were reported by TI in June of 2012 for eight-layer metal 28 nm CMOS [52]. TSV proximity greater

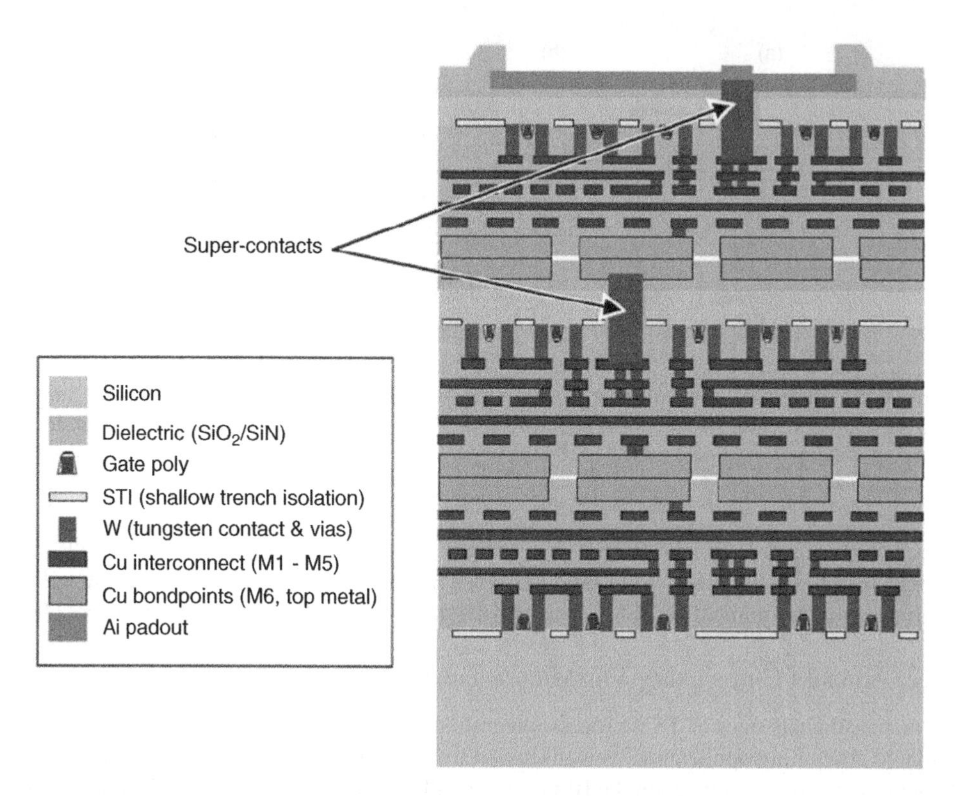

Figure 6.30 Schematic cross-section showing Tezzaron three-layer stack technology. (Based on white paper posted at http://www.tezzaron.com/technology/FaStack.htm, January 2012 [51], with permission of Tezzaron.)

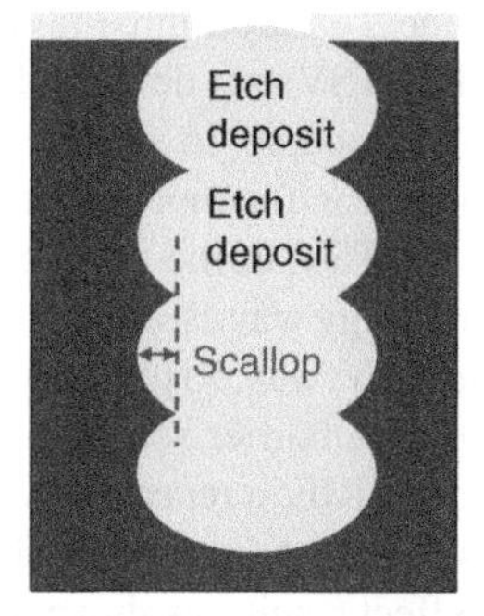

Figure 6.31 Cross-section showing scalloped effect of repeated polymer deposition followed by isotropic etch process with hardmask. (S. Cho, (Samsung), SEMATECH Symposium Korea, 27 October 2011 [50].)

than 4 μm was electrically measured at 27 and 105 °C. The largest shift in on-current (I_{on}) was 2.3%, which was less than that from other competing sources, such as dual stress liner boundaries, which are about 10%. It was shown that for proximity greater than 1.5 μm, the impact of the TSVs is negligible. The interaction with overlying interconnects is reduced by optimizing the post-TSV plating anneal and by introducing a TSV unit cell intended to minimize the impact of the TSV on the local environment.

For a typical TSV plating anneal, the zero stress state of the Cu TSV is expected at over 105 °C, so the impact of TSVs on surrounding silicon for a temperature range of 27 to 105 °C is tensile. Insertion of compressive stacked trench isolation between the TSV and device will act to buffer this impact. The impact on electrical properties of FETs between 4 and 16 μm from TSVs was found negligible in a 28 nm CMOS layout. For wide-IO memory–logic interface applications using a $40 \times 50\ \mu m^2$ JEDEC TSV array, the ESD and decoupling capacitors can be placed adjacent to the TSV so that CMOS logic circuitry does not require placement closer than 4 μm.

"Node-agnostic" Cu TSVs integrated with a high-κ–metal gate were discussed by IBM in December of 2011 [53]. Embedded DRAMs were also integrated into the functional 3D modules using a via-middle technology. Device and functional data showed no significant impact from the TSV processing or proximity. Devices used included 4 to 12 metal layers. It also included low-κ interlayer dielectric on SOI and on bulk silicon wafers. To form the TSV, blind vias less than 100 μm deep were etched with near-vertical sidewalls at a minimum pitch of 50 μm. A conformal oxide insulator had good coverage of the TSV bottom. Sputter deposition was used for a barrier and seed layer. Void-free, bottom-up copper plating was used followed by CMP. After processing the TSVs, additional BEOL processing was done. Thermal cycling for >500 cycles and thermal stress test at >275 °C for 1500 hours showed no degradation of the TSV or the BEOL structures. A thick Cu wiring level can result in bowing and stress in the interlayer dielectric and was tuned to minimize wafer bow. Wafers had additional processing including thinning. Modules were built both FTF and FTB.

A 28 nm low-power CMOS SoC platform using via-middle technology was discussed by STMicroelectronics Crolles in December of 2011 [54]. A high-κ–metal-gate architecture was used, which included 3D integration for high data rate interfaces using a wide-I/O based on TSV architecture. The required large memory capacity was implemented with an eDRAM macro. The CPU critical path speed was enhanced by using a triple-gate oxide scheme on high-κ metal gate combined with a $20\ fF/\mu m^2$ MIM for decoupling capacitance. An RF device suite was also developed to enable high-performance analog cells. The wide I/O can be used with the new TSV interfaces using a stacked die structure to achieve a bandwidth of 12.8 Gb/s. The 200 MHz 512-bit bus interface with DRAM can be achieved using via-middle TSV technology. The TSV diameter is 6 μm and depth is 50 μm. The SRAM DC performance showed that the TSV process did not disturb the surrounding logic block, which was sensitive to transistor matching, local noise, and mechanical stress. The eDRAM macro used a TiN–ZiO_2–TiN capacitor placed over the low-κ in a capacitor-over-low-κ process that permitted a good-yielding, high-density cell with $0.08\ mm^2/Mb$ density. The eDRAM was integrated in an SoC using a decoupling capacitor to provide a low-cost technology. Access time of 400 MHz and leakage of 10 fA were achieved by using process–design macrocell co-development. The technology offered thin oxide transistors with multiple voltage thresholds and thick oxide transistors for multiple power supply interfaces.

6.6 Energy Efficiency Considerations of 3D Stacked Memory–Logic Chip Systems

6.6.1 Overview of Energy Efficiency in 3D Stacked Memory–Logic Chip Systems

3D stacked DRAM systems can have high bandwidth to main memory, but the power density and temperatures on the chip can be a challenge that needs to be dealt with. Energy efficiency is also an important factor for 3D chips in mobile systems that operate with a battery. Architecture-level models for 3D stacked systems are needed for effective design of energy-efficient systems. Consideration needs to be given to the effect of heat from the CPU core, which affects the data retention of DRAMs stacked with the CPU.

6.6.2 Energy Efficiency for a 3D TSV Integrated DRAM–Controller System

Energy efficiency for a 3D integrated DRAM and SoC system using TSV was discussed by the Universities of Kaiserslautern and Bologna in April of 2013 [55]. TSV technology permits 3D integration of wide-I/O dice and heterogeneous stacking of different chips, which provides higher bandwidth and lower energy than planar approaches. The 3D DRAM architecture was studied using an optimized 2Gb 3D DRAM. Results showed 83% lower energy/bit than a 2Gb low-power DDR2 × 32-bit planar device.

The team proposed an energy-efficient 3D DRAM subsystem integrated with SoC that consisted of an SDR/DDR 3D DRAM controller and an attached 3D DRAM cube with fine-grained access and a flexible wide-I/O interface. The 3D DRAM subsystem cube used up to eight DRAM layers with a vertical channel architecture. A schematic of this architecture is shown in Figure 6.32 [55].

The energy efficiency was analyzed using a synthesizable model of the SDR/DDR 3D DRAM channel controller and also using functional models of the 3D stacked DRAM

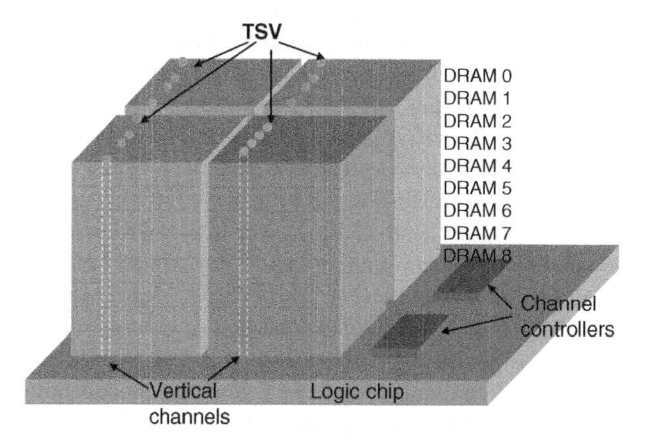

Figure 6.32 Schematic diagram of 3D cube architecture with eight DRAM layers. (Based on C. Weis *et al.*, (University of Kaiserslautern, University of Bologna), *IEEE Transactions on CAD of ICs and Systems*, 32(4), 597 [55].)

including a power-estimation engine. Several different DRAM families were studied, with densities ranging from 256Mb to 4Gb per channel. Energy-optimized accesses to the 3D DRAM were found to have up to 50% energy savings compared with standard accesses. In a mobile SoC subsystem running at 200 MHz with wide-I/O DRAMs (128-bit I/O), a 12.8 GB/s bandwidth was possible. When the full bandwidth was not needed to save energy, a flexible bandwidth and adaptable burst length could be used. This permitted efficient handling of a large range of access sizes.

Main contributions of the study were detailed architecture for 3D DRAMs, co-optimization of the memory controller and 3D DRAM, fine-grained access to the 3D DRAM, which led to power reduction, an energy consumption proportional to the size of a memory access, and implementation of a 3D DRAM controller that satisfied the requirements for a flexible interface. The inputs of the 3D DRAM generator model were process information, DRAM capacity, I/O width, interface, TSV pitch and diameter, number of power TSVs, and technology node. The outputs of the model were timings, maximum frequency, active and standby power, self-refresh power, and area per 3D layer [55].

Energy-efficiency considerations are also important for design optimization for mobile systems. It has been found that 3D integration of heterogeneous memory and logic layers has higher bandwidth at lower energy consumption than 2D integration. In March of 2012, the Technical University of Kaiserslautern also discussed energy efficiency in 3D heterogeneous DRAM stacked dice systems using TSV [56]. This study used a 3D stacked dice subsystem that included SDR/DDR 3D DRAM controller, 3D DRAM cube with fine-grained access, and flexible wide-I/O interface. A synthesizable model was made of the DRAM channel controller, while a functional model was made of the 3D stacked DRAM. The stacked DRAM model had an accurate power estimation engine that was used to study various DRAM families, including wide-I/O DDR/SDR, low-power DDR, and low-power DDR2. Densities ranged from 256Mb/channel to 4Gb/channel. The resulting 3D DRAM subsystem had an average 37% power savings compared to standard accesses in a 2D DRAM subsystem. A model was designed that co-optimized 3D memory and 3D controller architecture.

Energy efficiency can be improved by changing the architecture of 2D DRAM chips when put into 3D stacked DRAM systems using TSV connections. To leverage the big internal bandwidth of the TSVs, the DRAM chip must be rearchitected. The system concerns should be considered in order to build effective 3D DRAMs out of redesigned 2D DRAM chips. The CACTI-3DD model from HP Labs is an architecture-level integrated power, area, and timing modeling framework for designing 3D dice-stacked off-chip DRAM main memory. Described in March of 2012, the CACTI-3DD [57] includes TSV models and 3D integration models that enable analysis of a wide range of DRAM designs. Results showed that the rearchitected DRAM die achieved major improvements in power and timing compared to coarse-grained 3D dice-stacked DRAM.

An analysis of 3D systems with stacked DRAM focused on increasing energy efficiency was conducted in March of 2012 by Boston University [58]. A comprehensive evaluation framework was proposed to manage the interplay between performance, energy, and temperature in 3D multicore systems, which included a detailed analysis of the DRAM layers. A memory management policy targeted at applications with spatial variations in DRAM accesses was presented and used to perform mapping of memory access to DRAM banks with temperature as a variable. Evaluations using this tool on 16-core 3D systems running parallel applications showed up to 88.5% improvement in energy delay product compared to equivalent 2D systems.

An issue in energy efficiency is the design for thermal isolation between the controller and stacked TSV DRAMs. In May of 2012, Rambus discussed thermal isolation design for a double-sided flip chip package with disaggregated memory chips integrated with TSV on one side of an organic substrate and a memory controller on the other side. It was determined that co-design of all physical layers was required to optimize the integrated 3D package within the electrical and manufacturing constraints. Design of the power delivery network was described, with a prelayout design strategy being used that optimized the power delivery network across 11 power domains to meet impedance targets [59].

6.6.3 Adding an SRAM Row Cache to Stacked 3D DRAM to Minimize Energy

Adding an SRAM row cache to a stacked 3D DRAM system can improve performance and energy efficiency. In December of 2011, Intel explored designing a heterogeneous stack of an SRAM row cache and a DRAM [21]. A new floor plan and various architectural techniques were developed to exploit the benefits of 3D stacking technology.

Die stacking of SRAM row cache and 3D DRAM using TSVs was also discussed in August of 2011 by Intel and Georgia Institute of Technology [60]. The team used 3D integration to provide scaling and also permit heterogeneous die stacking within one DRAM package. A folded bank design was used for the DRAM, which permitted the integration of a small SRAM row cache on a logic layer while short TSVs integrated several DRAM dice vertically with high-bandwidth, low-latency, and low-power interconnects. A multiple-chip design permitted one unique logic die within the 3D DRAM stack, which was used for off-chip interface circuits and the SRAM row cache.

Simulations showed that by tightly integrating a small row cache with its corresponding DRAM array, the performance could be improved by 30% and the dynamic energy by 31% for memory-intensive applications. Open rows on DRAM consumed energy that could be reduced using an SRAM cache to maintain several active DRAM rows. The feasibility of a row cache was shown to be better with 3D IC technology. These DRAM layers and the interface layer were connected using a TSV bus located in the middle of a chip. A TSV bus was used per bank, and four banks were stacked vertically to share the TSV bus. An SRAM row cache was placed in the logic layer below the four banks. The baseline 3D DRAM design used in the study consisted of four DRAM dice and one logic interface die connected by TSVs. The baseline DRAM was made as four stacked DDR2 DRAM, each with eight banks with a dedicated interface layer. Figure 6.33 shows four stacked DRAM die with four banks and an interface chip [21].

The bus width can be reduced by using multiple banks of TSVs, but this consumes silicon area. Another solution is to use divided or folded banks. Folding the banks reduces the length of wires between the sense amps and the TSV bus and is scalable. As the number of DRAM layers increases, the banks can be folded into more layers to continue reducing wire length. The final proposed floor plan, shown in Figure 6.34 [21], was based on a folded bank architecture, where each bank of the SRAM row cache was aligned with its corresponding DRAM bank.

The SRAM bank was put next to the TSVs to minimize the energy consumed in transferring an entire row to an SRAM bank. Circuit- and architectural-level simulation was provided to

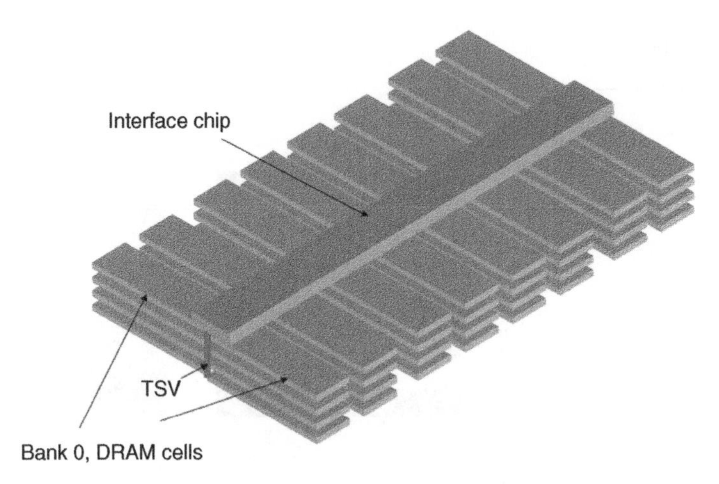

Figure 6.33 Baseline DRAM illustration showing a flipped four DRAM TSV stack with four banks per DRAM and an interface chip. (Based on D.H. Woo *et al.* (2011) (Intel), *IEEE Transactions on VLSI Systems*, 99(1) [21].)

evaluate the energy used in various cache operations. This was a scalable design. In the future, as the number of DRAM layers increase, one bank can be folded into more layers to reduce the wire length of each chip.

In order to minimize the energy consumption of the 3D stacked chip without violating the error rate limitation, it is necessary that the refresh period and ECC be controlled with an awareness of temperature. In May of 2011, KAIST discussed the effects of temperature, refresh

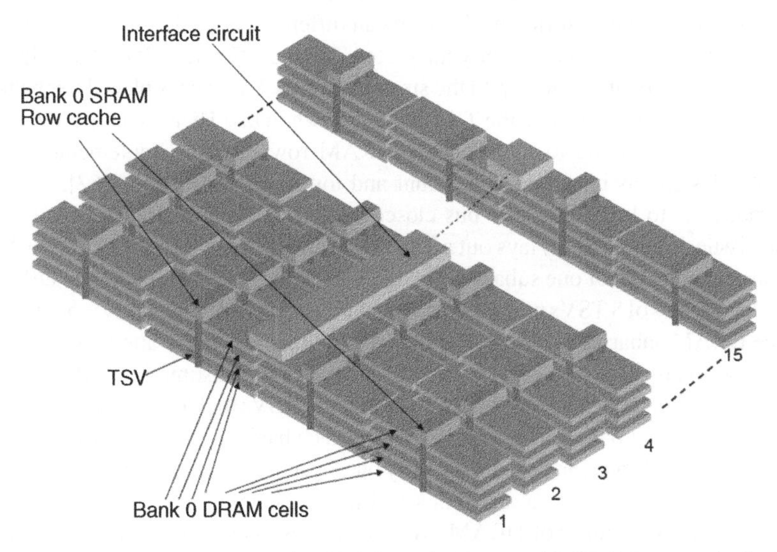

Figure 6.34 Final floor plan type showing folded DRAM banks with SRAM row cache banks. (D.H. Woo *et al.* (2011) (Intel), *IEEE Transactions on VLSI Systems*, 99(1) [21].)

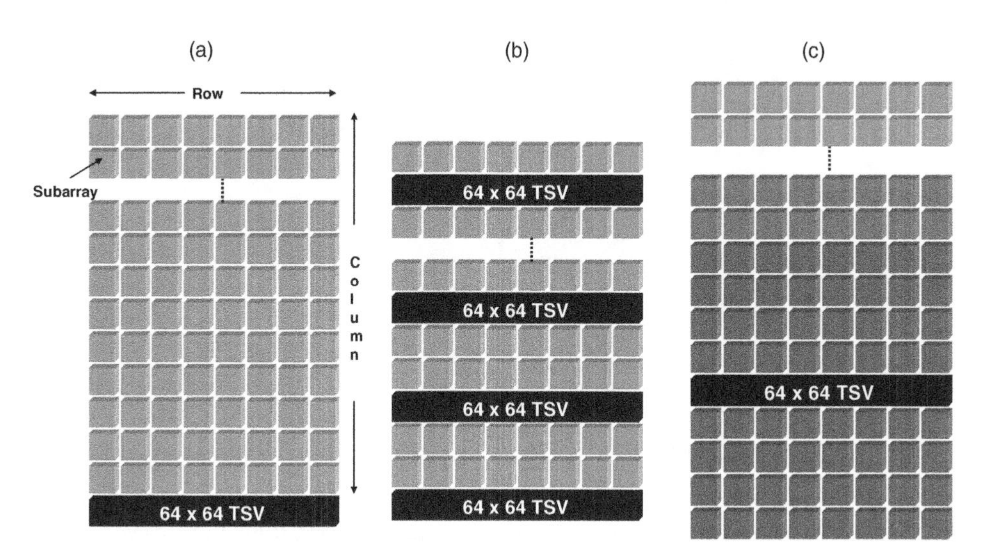

Figure 6.35 Different bank configurations for 3D TSV stacked DRAM. (Based on D.H. Woo *et al.* (2013) (Intel, Georgia IT), *IEEE Transactions on VLSI Systems*, 21(1), 1 [62].)

period, and ECC policy on the reliability and power consumption of 3D stacked eDRAM [61]. It was shown that an adaptive ECC policy with varying temperature achieved a reduction of energy consumption by up to 26% compared to a fixed ECC policy under given error rate constraints.

In January of 2013, Intel and Georgia Institute of Technology further discussed their SRAM row cache integrated using TSVs in a heterogeneous 3D DRAM [62]. Different methods of placing the TSV bus were examined, as shown in Figure 6.35. Different 3D designs of four half-banks are shown, where different shades mean different banks. Figure 6.35(a) is a simple wide TSV bus. Figure 6.35(b) is a tightly integrated TSV bus. Figure 6.35(c) has folded banks.

The data width is first increased up to the size of a row. A TSV bus is placed per bank so that four banks stacked vertically share the TSV bus as shown in (a). There is an SRAM row cache in the logic layer below the four banks. The SRAM row cache is implemented in a logic process. This design has increased wire count and low energy efficiency [62].

An alternative is to bring the TSV bus closer to the sense amplifiers to solve the energy-inefficiency issue. Figure 6.35(b) lays out a TSV bus per row of subarrays. Only eight TSVs can be placed along the width of one subarray. Bringing the entire 512 bits out in one DRAM cycle would require 64 rows of 8 TSVs per subarray. The TSV overhead for this array would be too high because the DRAM subarray is already very small, and the intent is to align a wide bus with a small subarray. In order to trade-off between dynamic energy consumption and area overhead, subarrays of one bank were made to share the same set of TSVs and to fold each bank vertically rather than stacking four banks vertically. By folding each bank, the length of wires between the sense amps and the TSV bus can be reduced. Only one layer of a bank actively uses the TSV bus at one time because each layer has a different set of rows belonging to the same bank.

In the future, as the number of DRAM layers increases, one bank can be folded into more layers, reducing the wire length of each die. The TSV bus can be placed in the middle of a bank to further reduce wire length. The new memory hierarchy is shown in Figure 6.36 [62].

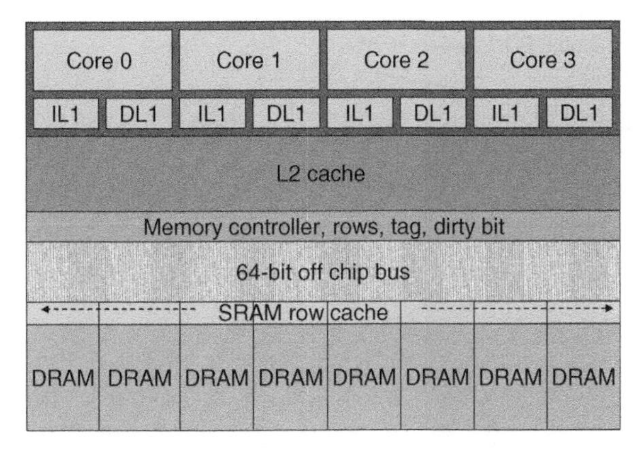

Figure 6.36 New memory hierarchy. (Based on D.H. Woo *et al.* (2013) (Intel, Georgia IT), *IEEE Transactions on VLSI Systems*, 21(1), 1 [62].)

In this hierarchy each DRAM chip has a very small SRAM row cache made in a CMOS logic process with TSV connecting the cache and the DRAM cells that provide an entire row of data across four memory bus cycles. The SRAM cache is also connected to a memory controller in the package through a conventional 64-bit off-chip bus. The effect of TSV pitch was examined from 2.5 to 10 μm pitch, and it was determined that neither latency nor energy was sensitive to the TSV pitch.

6.6.4 Power Delivery Networks in 3D ICs

Power delivery networks (PDNs) integrated on the chip can be used to reduce heat in a 3D IC that uses TSV. In December of 2012, Stanford, Monolithic 3D, and Rambus discussed a PDN to reduce heat in 3D IC using TSV [63]. Rather than using only TSVs to eliminate heat, with the silicon being a heat conduit to the TSVs, this study used both global and local PDN to remove trapped heat and dissipate it toward the heat spreader. An illustration of global and local PDN removing trapped heat in a stacked two-layer large-scale integration (LSI) chip is shown in Figure 6.37 [63].

Heat can escape laterally along silicon layers. However, silicon layers thinner than 1 μm have lower thermal conductivity than bulk silicon, so in advanced technology 3D ICs, additional means of heat dissipation must be used, such as interlayer vias (ILVs) and interlayer dielectrics (ILD). In addition, PDNs can improve heat removal through thermal conduction. It was shown that lower maximum chip temperature and reduced area used by ILVs was achieved using a PDN. An experimental device and simulations were used to quantify the heat dissipated using a PDN.

A verification study was done using two stacked SPARC T2 processor cores and an L2 cache bank stacked on top of a SPARC T2 processor core. The CACTI simulator was used to estimate power consumption for the L2 cache. When PDNs were used in addition to ILVs, the internal temperature of the stacks was significantly lower. The reduction in maximum chip temperature was more significant for applications with high-performance components on layer 2 of a two-layer 3D IC. PDNs were also shown effective for heat removal in monolithic 3D ICs for low-power applications using conventional air cooling.

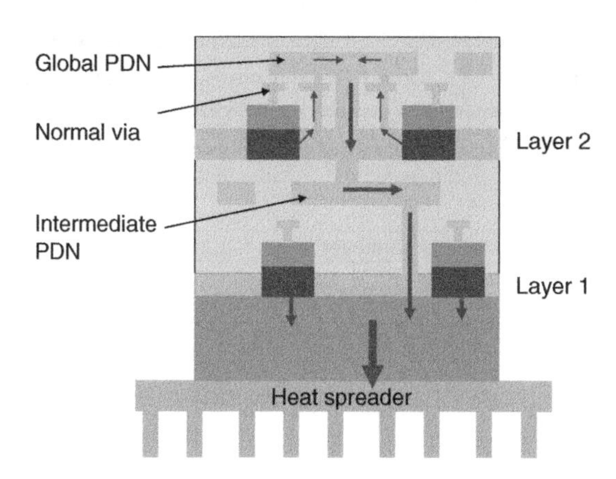

Figure 6.37 Schematic illustration of a power dissipation network in a two-layer 3D integrated circuit. (Based on H. Wei *et al.*, (Stanford University, Monolithic 3D, Rambus), IEDM, December 2012 [63].)

A study of partitioning and assembly for 3D ICs using logic and memory was discussed in December of 2012 by Rensselaer Polytech. The partition and assembly approach combined electromagnetic (EM) and analytical simulations to model and analyze several TSV-based 3D PDNs. These PDNs were made up of various stacked chips, interposer, and package substrate. To more closely approximate TSV-based 3D systems, frequency-dependent parasitic extraction using EM simulations and equivalent circuit models were combined to characterize the power performances. Various physical configurations were studied to form design ground rules for various 3D PDNs [64].

6.6.5 Using Near-Threshold Computing for Power Reduction in a 3D TSV System

Near-threshold computing has also been used to address the issues of power consumption in a stacked 3D TSV connected system. In near-threshold computing, cores are operated within about 200 mV of the threshold voltage to optimally balance power and performance. A 3930 DMIPS/W configurable 3D stacked system using near-threshold computing was described in February of 2012 by the University of Michigan [65]. The system used 64 ARM Cortex-M3 cores and SRAM memory. In this "Centip3De" system, cores were operated at 650 mV rather than 1.5 V. This improved measured energy efficiency by about 5.1 times. The lower power consumption of near-threshold computing makes it attractive for use in 3D design, because closely stacked chips have limited power-dissipation capability.

The SRAMs are operated at 870 mV and the logic at 670 mV in 130 nm technology. This higher voltage in SRAMs improves their speed. In the "Centip3De," four cores are connected to each cache, and each cache operates at four times the core frequency, communicating with the cores in round-robin fashion. This configuration automatically resolves coherence within the cluster, which reduces overhead.

The system uses two stacked dice with 64 ARM M3 near-threshold cores, making up 16 four-core clusters. Each cluster is connected to a four-way 1KB instruction cache and a four-way 8KB data cache. The caches communicate over a 3D TSV bus that connects them to DRAM controllers that back up the caches.

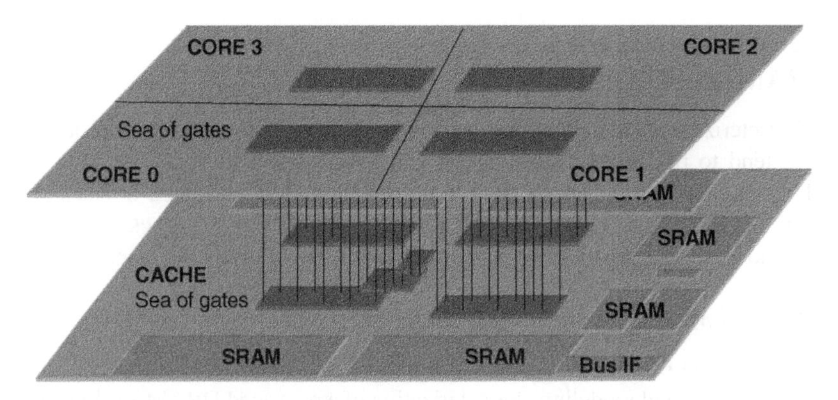

Figure 6.38 Schematic of cluster floor plan showing F2F connections with TSV lines between. (Based on D. Fick *et al.*, (University of Michigan), ISSCC, February 2012 [65].)

The cluster floor-plan is shown in Figure 6.38 [65]. It has F2F connections with connecting TSV lines between. Each cluster has 1591 F2F connections. The system bus operates at 160–320 MHz, which supplies up to 4.46 GB/s of memory bandwidth. A 40 nm ARM Cortex-A9 can achieve 8000 DMIPS/W. At peak system efficiency, the Centip3De achieves 3930 DMIPS/W in 130 nm technology.

The fabricated system has an unthinned cache layer and a thinned core layer with wire bonds connecting to TSVs on the backside, as shown in Figure 6.39 [65]. In the future, it is anticipated that the Centip3De will be expandable to four layers of cores/caches with two to three layers of stacked DRAM. Two pairs of the F2F bonded core/cache dice are thinned to 12 μm on the cache side and then bonded back-to-back to create a four-layer stack, which is thinned and diced. These dice are then bonded to a thinned two- to three-layer DRAM wafer. The DRAM stack uses the Tezzeron Octopus DRAM technology. The final system would have a maximum of 256 MB of shared DRAM connected through eight 128b DDR2 interfaces.

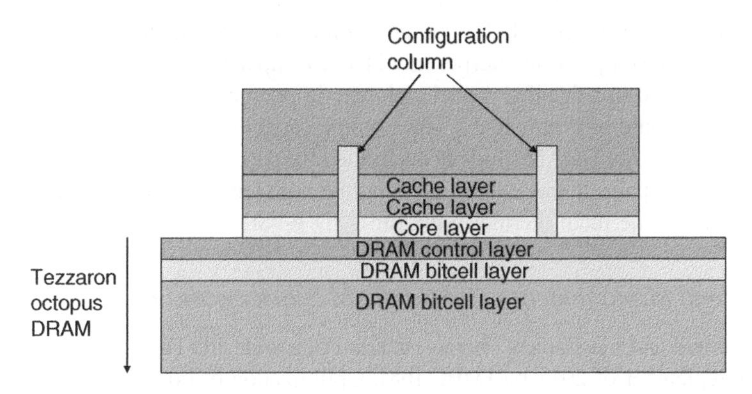

Figure 6.39 Future expanded Centip3De system with SRAM cache and stacked DRAM. (Based on D. Fick *et al.*, (University of Michigan), ISSCC, February 2012 [65].)

6.7 Thermal Characterization Analysis and Modeling of RAM–Logic Stacks

Thermal characterization and modeling is a major consideration in stacking memory and logic. Logic chips tend to run hotter than DRAM, and heat reduces the data retention time in a DRAM. Hot spots on the logic chip need to be modeled and considered in stacked DRAM architecture. Simulations of the thermal characteristics of the 3D stack can be helpful. Interposers and heat spreaders can be used to reduce thermal effects in stacked chips.

6.7.1 Thermal Management of Hot Spots in 3D Chips

Thermal management of hot spots is a primary issue with DRAM on thinned logic stacks. Thermal experimental and modeling characterization of a packaged DRAM-on-logic stack was discussed by IMEC in May of 2012 [66]. A DRAM die was stacked on a thinned logic die using copper–tin (CuSn) microbumps. The logic chip had integrated heaters and sensors that could cause hot spots. The thermal impact of logic hot spot dissipation on the temperature profile was characterized using experimental configurations that simulated both high-power and low-power systems. The use of the two configurations permitted calibration of a detailed finite element thermal model. The calibrated thermal models were then used to evaluate both the impact of the effective thermal conductivity of the microbump and underfill layer and the impact of the logic chip thickness on temperature distribution in the logic and DRAM dice. This evaluation was done for different cooling configurations of the dice stack.

6.7.2 Thermal Management in 3D Chips Using an Interposer with Embedded TSV

Thermal management is also necessary in 3D chips using an interposer with regularly embedded TSV. A thermal analysis of such a chip was described by ITRI in May of 2012 [67]. A simulation technique was used to analyze the thermal behaviors of a SIP for network systems based on a 3D IC structure. The configuration had a CPU chip on top and two DRAM chips on the bottom of the interposer. The interposer with chips was bonded on a bismaleimide–triazine (BT) substrate. The BT substrate was bonded on a PCB, and a metallic heat spreader was glued to the back of the CPU chip. For simulation purposes, equivalent models were used of the embedded TSV, bumps, bonds, and metallic traces. A slice model was used with four stacked chips on an interposer, with each chip having two heaters and TSVs. The results were compared to a detailed model and were in good agreement. It was found that the CPU temperature would be dominated by the cooling capability of a thermal module attached on the heat spreader. The DRAM chips beneath the interposer had a high temperature and an obstructed heat flow path. It was determined that the high temperature of the chips under the interposer would be the main thermal issue for a 3D IC SiP.

6.7.3 Thermal Management of TSV DRAM Stacks with Logic

A study of thermal and mechanical characterization of a wide-I/O DRAM-on-logic stack was discussed in September of 2011 by IMEC [68]. Optimization of both design and technology was necessary to produce working 3D DRAM-on-logic stacks. The study showed that the stack organization as well as TSV and microbump layout needed to be fine-tuned with the 3D

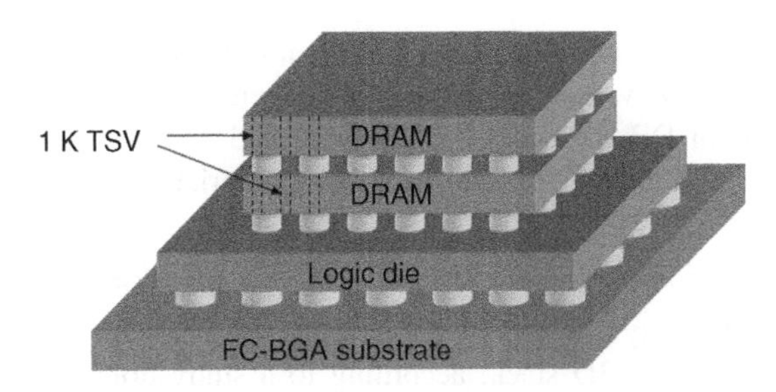

Figure 6.40 A typical mobile wide-I/O DRAM-on-logic stack. (Based on D. Milojevic *et al.*, (IMEC), CICC, 19 September 2011 [68].)

technology to meet mechanical and thermal challenges. A dedicated thermal and mechanical model was integrated into the design flow, and the data required from foundries for good results was defined. A mechanical sample was built using DRAM-on-logic, which was enhanced with sensors to collect information on temperature, stress, and performance of the electrical circuits in the 3D package. An example of a mobile wide-I/O DRAM-on-logic stack is shown in Figure 6.40 [68].

An experimental DRAM–logic sample showed a 40% change in device performance due to mechanical stress induced by stacking. In addition, a seven times higher hotspot temperature was measured in the DRAM–logic stack than in a 2D die. These results indicated that the design of DRAM-on-logic for mobile applications will require significant fine-tuning of various design and technology options in the course of building 3D stacks. A compact thermal and mechanical model was integrated into an early prototyping design flow to permit designers to validate key physical design assumptions before register-transfer level (RTL) design and tape-out.

Stacking multiple DRAM layers on processor cores in the same chip using TSV can dramatically improve system performance due to higher bandwidth and lower access latency. The higher core performance, however, increases power density, which can affect the reliability of the layers of the DRAM cube. In July of 2011, Boston University shared an evaluation of the performance and temperature tradeoffs of such a 3D DRAM cube and logic stack [69]. A comprehensive framework for exploring power, performance, and temperature characteristics of stacked DRAM layers on processor cores was developed. This framework permitted performance improvement along with power and thermal profiles to be quantified.

The performance in the application was improved by 72.6% on average compared with an equivalent 2D chip with off-chip memory. Power consumption per core increased by up to 32.7%. The increase in peak temperature permitted was limited to 1.5 °C because the lower-power DRAM layer shared the heat of the hotter cores. This resulted in reduced DRAM data retention at higher temperatures. In high-end systems, cost and space restriction were still necessary to compensate for the lack of efficient cooling in the chip system.

An analysis of 3D systems with stacked DRAM focused on increasing energy efficiency was further discussed in March of 2012 by Boston University [70]. The 3D stacked DRAM systems have high bandwidth to main memory, but the power density and temperatures on the chip need to be dealt with. A comprehensive evaluation framework was proposed to manage the interplay between performance, energy, and temperature in 3D multicore systems, which included a detailed analysis of the DRAM layers. Evaluations using this tool on 16-core 3D systems running parallel applications showed up to 88.5% improvement in energy delay product compared to equivalent 2D systems. A memory management policy targeted at applications with spatial variations in DRAM accesses was presented and used to perform mapping of memory access to DRAM banks with temperature as a variable.

6.7.4 Thermal Management of a 3D TSV SRAM on Logic Stack

Tier-to-tier thermal and supply crosstalk in the logic core also has an effect on performance and robustness of an SRAM in a 3D stack, according to a study done by Georgia Institute of Technology in June of 2013 [71]. This study integrated distributed-process variation-aware circuit analysis, resistor–capacitor (RC)-based thermal simulation, and distributed resistor–inductor–capacitor (RLC)-based PDN simulation. Results of the analysis showed that when the logic cores and SRAM were integrated in the 3D stack, the thermal and supply crosstalk degraded the SRAM performance and noise margin during read and write operations.

6.8 Testing of 3D Stacked TSV System Chips

Unless some form of on-chip test and repair is used, 3D system chips stacked with KGD can suffer significant yield loss. Macros for built-in self-test (BIST), built-in redundancy analysis (BIRA), and built-in self-repair (BISR) are commonly used for 3D stacks of memory and logic chips.

6.8.1 Using BIST to Reduce Testing for a Logic and DRAM System Stack

A 2.5D logic and DRAM GPS RF system chip with BIST was described by TSMC in June of 2013 [72]. A 65 nm RF receiver was integrated on a chip-on-wafer-on-substrate structure along with a 28 nm baseband processor and 40 nm DRAM. KGD were used for all chips, resulting in a known-good stack. The three chips were heterogeneously attached to a thin wafer using TSV, as shown in Figure 6.41 [72]. A BIST method was used to save screening time by over 90% compared to the conventional automated test equipment (ATE) full functional test.

6.8.2 Efficient BISR and Redundancy Allocation in 3D RAM–Logic Stacks

Yield is a key consideration for volume production of 3D stacked logic and RAM chips. A BISR method for performing repair of 3D RAMS using interdie redundancy was suggested by TSMC and National Central University of Taiwan in April of 2013. The interdie redundancy

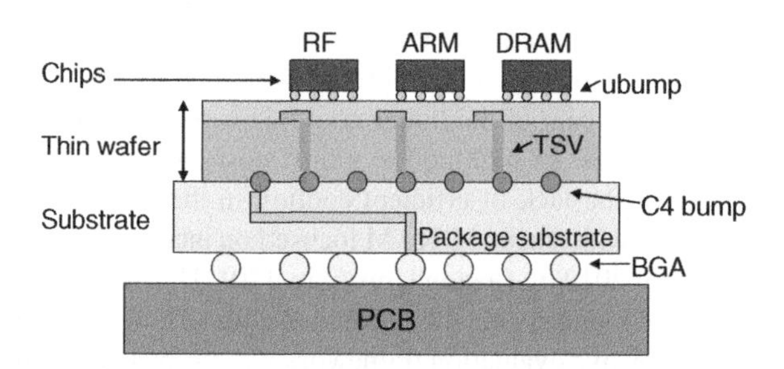

Figure 6.41 Chip-on-wafer-on-substrate 2.5D integration of RF receiver, AP, and DRAM. (Based on W.S. Liao *et al.*, (TSMC), VLSI Circuits Symposium, June 2013 [72].)

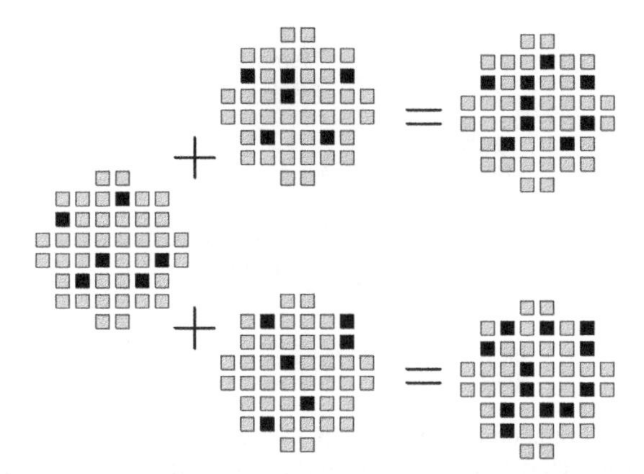

Figure 6.42 Illustration of different stacking strategies resulting in different yields. (Based on C.W. Chou *et al.*, (TSMC, National Central University), *IEEE Transactions on CAD of ICs and Systems*, 32(4), 572 [73].)

method was proposed to utilize unused repair circuits to improve yield [73]. In addition, three stacking flows for different bonding technologies with interdie redundancy was also proposed.

It was shown first that different yields for a 3D chip stack can depend on different stacking strategies. The potential for different yields for different stacking strategies depending on the location of defective die is illustrated in Figure 6.42 [73]. Using the same initial tested wafer, the upper combination resulted in nine stacked locations with defects, whereas the lower combination resulted in 12 stacked locations with defects.

For the postbond repair, unused redundancies of one die are used to repair the other die to increase the flexibility of redundancy. The BISR circuit is designed to support RAM repair in the prebond and postbond test phases. In the prebond phase, the BISR circuits on each die can allocate redundancy individually. In the postbond phase, the BISR circuits in two RAM dice can be integrated to allocate interdie redundancy to improve repair efficiency significantly. The BISR circuits in the two stacked dice work together to allocate interdie redundancy. Simulations showed that these yield-enhancement techniques effectively improved the yield of the 3D RAMs. A schematic illustration of the two-chip 3D RAM with TSV connections and BIST/BISR circuits is shown in Figure 6.43 [73]. The final interdie redundancy methods improved the yield of the 3D RAMs with little hardware overhead for the prebond test and repair. For a 1Mb RAM chip, for example, the area overhead of the BISR method was only 1.77%.

A BISR allocation method for repair of RAM circuits embedded in an SoC die was discussed in April of 2013 by the National Central University of Taiwan [74]. The SoC was just one of the chips in a 3D IC using TSV for interconnects. A memory BISR scheme was proposed to improve yield of the RAMS in the SoC, and a new BISR allocation method was proposed to allocate shared BISR circuits for RAMS in the SoC die such that the test and repair times and area of the BISR circuits were minimized.

To minimize the test and repair times of the RAMs in the prebond and postbond test phases, a test-scheduling engine was used to determine the prebond and postbond test sequences for the RAMs. These sequences took into consideration the corresponding test power constraints. A BISR minimization algorithm was also suggested to reduce the number of shared BISR circuits used for the RAMs, given constraints of prebond and postbond test sequences. The distance

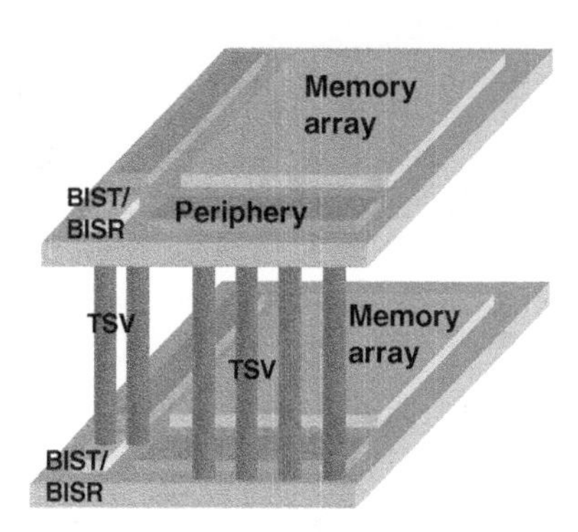

Figure 6.43 Schematic illustration of the two-chip TSV-connected 3D RAM with BIST and BISR. (Based on C.W. Chou *et al.*, (TSMC, National Central University), IEEE Transactions on CAD of ICs and Systems, 32(4), 572 [73].)

between the BISR circuit and the RAMs it served were taken into consideration. Simulation results showed that, compared with a dedicated BISR method where each RAM has its own BISR, there was a 35% area reduction with the shared BISR test method using the proposed allocation technique. A 1 mm distance constraint was imposed along with 500 mW and 600 mW prebond and postbond test power constraints.

Efficient test, redundancy analysis, and repair of 3D TSV-based RAMs is essential to minimize costs and maximize yields. A test and repair architecture for 3D ICs was proposed by the National Taiwan University of Science and Technology in April of 2013 [75]. This architecture consisted of stacked memory slave chips and a master processor chip. The chips were all KGD. KGD test and known-good-stack test were supported as well as final test and repair. Rather than including spare elements in each memory chip, a small redundant memory was included in the processor chip. The added redundancy could be used globally for repairing defective cells among other stacked memory chips. Each slave chip contained the BIST and BIRA modules for performing KGD and known-good-stack tests and redundancy analysis. The results of redundancy analysis were then used for die stacking, yield management, and for BISR after final test. A 1149.1-based test interface was added for each slave chip. Only four test pads were required for test and repair. Based on the results of the BIRA module, a simple matching algorithm was proposed to increase the stacking yield. Experimental results showed that the hardware overhead for an 8K 32-bit SRAM was only 2.6%, and the stacking yield was improved significantly.

The possibility of using the unused spare redundancy circuits in a memory integrated on a stacked 3D IC system to repair errors that occur after KGD test and assembly was discussed by ITRI in February of 2013 [76]. It was shown that the majority of repair spares are not used in initial chip repair. This off-chip repair method could help reduce the effects of the external inaccessibility of the memory die in such an assembly. A TSV would be used to access the redundancy control circuit. Unused spares are reactivated by overwriting their states, as if the corresponding fuses had been blown. Even if a row or column that has already been repaired is damaged again, it can be replaced with a new spare.

The results were simulated in a 65 nm process technology and showed that the maximum timing penalty of the off-chip repair was only 93 ps. The area overhead was estimated at 490 μm^2 per fuse set by using a 5 μm diameter TSV process. The yield improvement of a two-die stacked memory was shown to potentially be over 50%. The yield improvement was

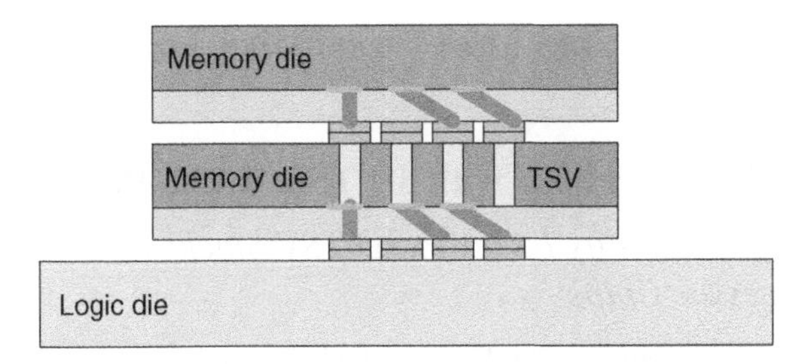

Figure 6.44 Memory and logic subsystem used for chip spares repair simulation. (Based on Y.F. Chou *et al.*, (ITRI), *IEEE Transactions on Circuits and Systems*, 60(9) [76].)

thought to more than compensate for the added cost of the repair. An example of the memory and logic die stack used in the simulation with the memory chips face down and bonded to the logic chip is shown in Figure 6.44 [76]. The relative sizes of the memory and logic chips can vary. The memory chips are connected using TSV.

6.8.3 Direct Testing of Early SDRAM Stacks

A 1.2 V 1Gb mobile SDRAM was described by Samsung in January of 2012 [8]. This chip stack used two 50 nm technology dice connected with 40 μm pitch TSV. In order to ensure TSV process stability and to test for mass production, test correlation techniques were developed to verify functions through microbumps and test pads. The microbump pad connectivity was tested using boundary scan. Existing test pad interfaces were equipped for wafer probe because the microbump pads were too small to be probed. A direct access mode using only 32 pins and four chip select (CS) pins could be used to test all four channels to support failure analysis. Direct access mode could also be used to test the DRAMS after package assembly, with additional pins connected directly to memory.

To improve frequency at low 1.2 V operating voltage, a read strobe (QS) function was added to provide more timing margin to controllers by optimizing input setup and hold time. The QS was SDR-aligned with the rising edge of the clock, which provided 50% power reduction compared to DDR. To further reduce power, the period of QS was two times that of the clock. QS was controlled by the extended mode register set. The SDR read strobe function is shown in the timing diagram in Figure 6.45. To reduce self-refresh current (IDD6), a dual period refresh

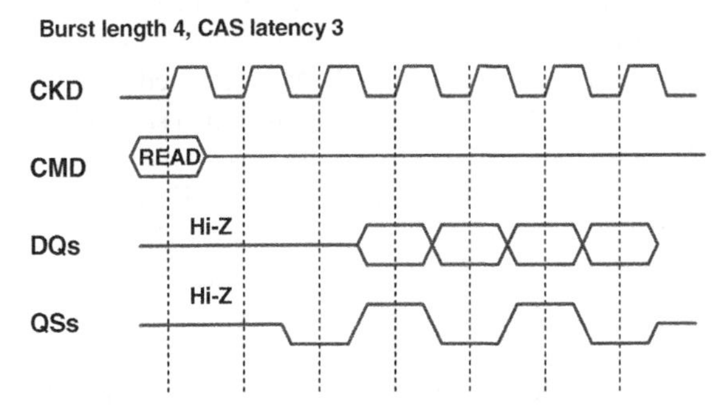

Figure 6.45 SDR timing diagram of read strobe function. (Based on J.S. Kim *et al.*, (Samsung), *IEEE Journal of Solid-State Circuits*, 47(1), 107 [8].)

method was used. The chip size adder, including metal fuses, registers, and logic, occupies 0.11% of the total chip area [8].

6.9 Reliability Considerations with 3D TSV RAM–Processor Chips

6.9.1 Overview of Reliability Issues in 3D TSV Stacked RAM–Processor Chips

Various reliability issues occur for 3D TSV stacked systems. Process variation in multiple chips requires careful consideration in 3D systems. Simultaneous switching noise in 3D needs to be modeled, and the placement of decoupling capacitors can be an issue. Placement and configuration of ECC needs to be considered. Issues relating to copper pumping of the TSV have already been considered in an earlier section.

6.9.2 Variation Issues in Stacked 3D TSV RAM–Processor Chips

Process variations in ICs that impact performance, leakage, and stability were discussed in June of 2012 by the University of Pittsburgh [25]. This is particularly an issue in high-density chips such as DRAMs. When DRAMs are stacked as last-level cache on multicore processors, variations in bank speed create nonuniform cache accesses in the 3D structure.

A model was developed for the process variation in a four-layer DRAM to characterize the latency and retention time variations among different banks to the core. Cache migration schemes were developed to use fast banks while limiting the cost of migration. They showed that there was a performance benefit in exploiting fast memory banks through migration. It was found that on average a variation-aware management can improve the performance over baseline by about 16.5%. It also put the system only 0.8% away in performance from an ideal memory with no process variation.

Variation became more of an issue in DRAMs with the advent of 3D stacked DRAM-processor chips, which tend to use very short-I/O TSV connections with low access latency. Process variation in 3D stacked DRAM design was considered by Utah State University in March of 2012 [77]. Techniques were presented for modeling the effect of random and spatial variation in large DRAM array structures. Sensitivity-based gate-level process variation models were combined with statistical timing analysis to estimate the impact of process variation on DRAM performance and leakage power. A block-based adaptive body biasing algorithm was used.

Data self-aligners have been proposed for use in 3D stacked DRAM and logic circuits to relieve the effects of variation. Both DRAMs and general logic circuits tend to have performance variation from chip to chip, with DRAMs estimated to have over 30% variation and logic circuits over 60% [78]. When both are combined in a stacked TSV system, variation can become an issue. An 800 Mb/s/pin delay-locked-loop (DLL)-based data self-aligner for use in the TSV interface for 3D DRAM stacks was proposed in February of 2012 by Hynix and Korea University [78]. When TSV are used to connect stacked dice, large access time (t_{AC}) variation can result in higher power consumption from short circuit currents due to data conflicts among the shared I/Os. An example of the short circuit current path is shown in the schematic circuit diagram in Figure 6.46 [78].

It was proposed that TSV DRAM intended for fast operation in mobile DRAM adopt a DLL-based data self-aligner. This would reduce the data conflict time among the stacked dice while

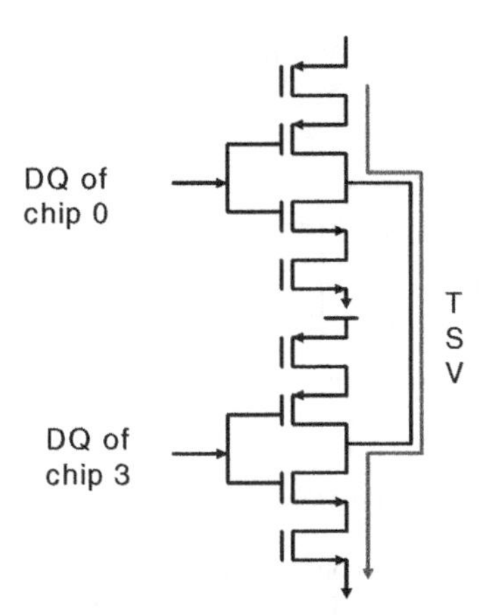

Figure 6.46 Example of I/O short circuit current path due to access time performance variation. (Based on H.W. Lee *et al.*, (Korea University, Hynix), ISSCC, February 2012 [78].)

consuming only 283.2 uW during read operation at 800 Mb/s/pin. A data self-aligner was built in 130 nm technology. The DLL power consumption in self-refresh mode was 4.98 uW when used with a leakage current reduction controller.

In March of 2013, SK Hynix and Korea University further discussed their DLL-based data self-aligner (DBDA) for TSV interface [79]. An issue with stacking die using TSV is that process mismatches cause data conflicts that decrease the data valid window and increase power consumption. The DBDA reduces data conflicts by automatically aligning data output timings without relying on control signals from the master die or requiring an extra signal in the stacked die. The DBDA reduces data conflict time due to process, voltage, and temperature (PVT) variations from 500 to 50 ps, which reduces short current from 3.62 to 0.41 mA.

The DBDA has two operation modes: synchronous self-align mode in which the data is aligned in the external clock domain and asynchronous self-align mode. Lock time for the DBDA is <20 cycles. A prototype DBDA implemented in 130 nm CMOS dissipated 247 uW for 800 Mb/s/pin. A leakage current controller was also proposed to reduce leakage current during power down mode by 90.5%.

6.9.3 Switching and Decoupling Noise in a 3D TSV-Based System

An estimate of the simultaneous switching noise on the 3D VDDQ PDN in a TSV-based GPU system was made by KAIST in December of 2011 [80]. The PDN impedance and pull-up impedance of the VDDQ PDN in the GPU system were estimated and analyzed. The systems included a GPU, four stacked DRAMs, a silicon interposer, and an organic package. The impedance estimation technique was based on a segmentation method. A balanced transmission line method was used for estimating the PDN impedance and the pull-up impedance of the 3D VDDQ PDN. The PDN impedance and pull-up impedance were also analyzed for variation in the number of power/ground TSVs.

Minimizing noise was a consideration in a miniaturized neural sensing microsystem using TSV developed by National Chiao Tung University along with China Medical University and the Advanced Semiconductor Engineering Group in February of 2013 [81]. This 180 nm

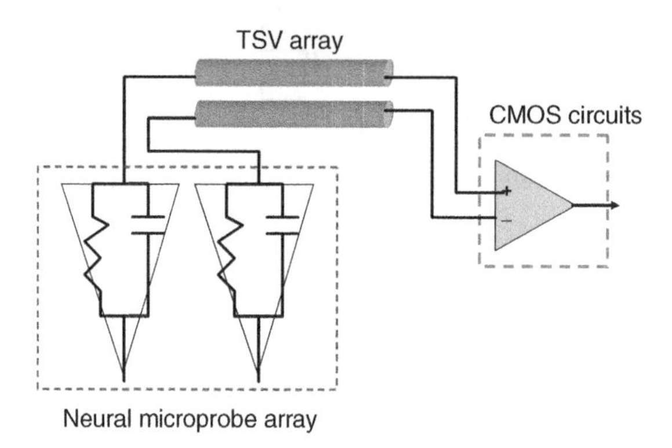

Figure 6.47 Illustration of low-noise TSV connections between CMOS circuits and neural array. (Based on C.W. Chang *et al.*, (National Chiao Tung University, China Medical University, Advanced Semiconductor Engineering Group), ISSCC, February 2013 [81].)

CMOS circuit included the sensor and logic circuitry for an implanted neural prosthesis intended for capturing brain signals.

The double-sided microsystem included sensor probes on one side of the chip and CMOS logic on the other side of the same substrate. TSVs formed low-impedance, low-noise connections with the shortest signal transmission distance between the microprobes and the CMOS logic, as illustrated in Figure 6.47 [81].

The 25 mm^2 chip had 480 microprobes divided into 4×4 sensing areas to form 16 channels. There were 16 TSV arrays that connected the microprobes to 16 read-out circuits on the other side of the chip. Provision was made for stacking other CMOS chips, including data storage capacity on the circuit side using further TSV 3D techniques. The TSV technique was a front-side, via-last, solid tube Cu TSV technology with 200 µm height and 30 µm diameter. A redistribution layer connected the TSV arrays to the circuit input pads on the CMOS side of the chip. After dicing, an encapsulation layer of parylene-C was added. The microprobe array was then formed on the other side. The platinum sensing probe tips were exposed and were the only part of the circuit that directly contacted brain tissue. A schematic cross-section of the chip is illustrated in Figure 6.48 [81].

The impedance of a single TSV is 5.5 mΩ and 34.2 fF, which is smaller than for longer transmission lines so that voltage drop and noise are reduced significantly. The low 3 dB

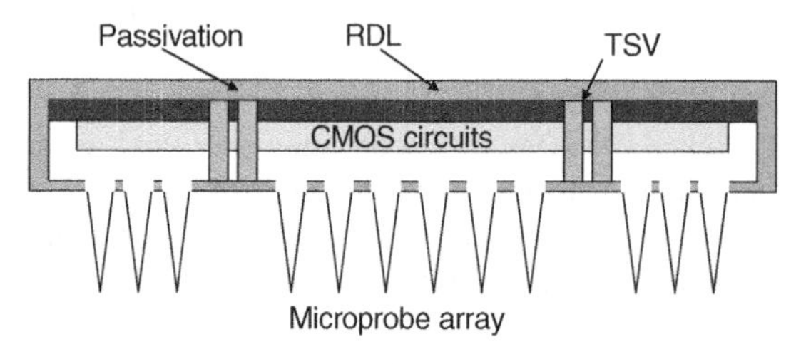

Figure 6.48 Schematic cross-section of low-noise neural probe chip with through-chip TSV. (Based on C.W. Chang *et al.*, (National Chiao Tung University, China Medical University, Advanced Semiconductor Engineering Group), ISSCC, February 2013 [81].)

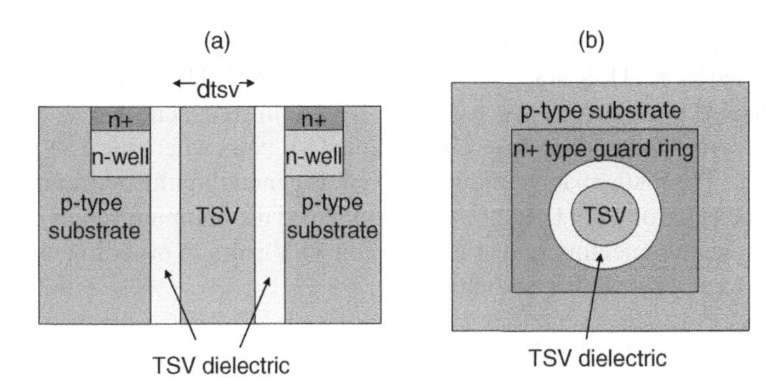

Figure 6.49 N+/− well guard ring to reduce coupling noise in TSV, showing (a) schematic cross-section; and (b) top-down view. (Based on K.D. Kim *et al.*, (Hynix, Seoul National University), VLSI Technology Symposium, June 2013 [82].)

frequency is 0.41 Hz to filter DC offset and flicker noise. Common mode noise from the human body and the power supply is suppressed. Total power of the 16-channel circuitry is 351.2 uW.

A guard ring can be used to reduce coupling noise in TSV. In June of 2013, SK Hynix and Seoul National University discussed using a new guard-ring technique as a shielding method to reduce TSV coupling noise [82]. The n+/− well guard ring was adjacent to the TSV dielectric surrounding the TSV. It used the inversion layer induced by a positive interface charge as a shield layer. This study determined the interface trap density, which is responsible for the interface charge between the TSV dielectric and the silicon substrate. The proposed guard-ring technique reduced coupling noise by a factor of about 3. A schematic cross-section (a) and a top-down view (b) of the n+/− well guard ring adjacent to the TSV dielectric is shown in Figure 6.49 [82].

Decoupling capacitor insertion is a common method used to handle power delivery integrity issues in high-performance ICs. In September of 2011, Rensselaer Polytech Institute discussed decoupling capacitor insertion strategy in 3D processor–DRAM integration [83]. Because the heat sink is against the processor die, the processor die must be on top of the DRAM dice, which are in turn stacked on the package substrate. A large number of TSVs are required to connect the processor chip and package for power and for I/O signals. These TSVs must pass through the DRAM chips and will both affect the DRAM design and cause increased power consumption overhead.

This study analyzed how to allocate these through-DRAM TSVs on the DRAM chip and studied their impact. The first TSV allocation strategy investigated was compatible with the normal DRAM architecture. Because there is a longer path between power/ground pads and the processor die, power delivery integrity may be an issue in high-performance systems using this strategy.

Another proposal was to use the 3D stacked DRAM chips to provide decoupling capacitors for the processor chip. This technique can leverage the capacitor fabrication ability of DRAMs and reduce the area penalty of decoupling capacitor insertion on the processor die. A simple uniform decoupling capacitor network design strategy was discussed. To demonstrate through-DRAM TSV allocation and the decoupling capacitor insertion strategy and tradeoffs, circuit SPICE simulations and computer system simulations were done to show the effectiveness of various design tradeoffs.

A new technique for reducing decoupling noise when placing DRAM dice over a processor chip was discussed by KTH, Kista, in December of 2011 [84]. The system modeled was two DRAM dice stacked over a single processor die. Decoupling capacitors were placed on each DRAM die and connected to the power distribution TSV pairs where the TSVs pass through the DRAM stack. The mathematical model, as well as generating the optimum values of the decoupling capacitance on each DRAM die, also modeled the optimum values of the effective resistance of the interconnecting power distribution TSV pairs in order to ensure the power integrity of the logic load during switching.

6.9.4 TSV-Induced Mechanical Stress in CMOS

The impact of thermomechanical stresses induced by copper TSV on fully depleted bulk FinFET devices in CMOS technology was modeled in December of 2012 by IMEC and KU Leuven [85]. Both n- and p-type FinFETs were affected by TSV proximity. The results were supported by the thermomechanical models for Cu TSV and by the four-point bending stress calibration. The concern was mechanical stresses in the silicon caused by different coefficients of thermal expansion of the copper TSV and the silicon substrate. These stresses may affect carrier mobility in the electronic devices, which in turn affects their performance.

The high-κ–metal gate FinFET technology studied had 40 nm fin height, 20 nm fin width, various fin lengths, and a TiN metal gate. The FinFET channel orientation was (110), resulting in sidewall crystal orientation of (110). A 1.2 GPa tensile strain SiN etch stop layer was used to improve the n-FinFET performance. The TSVs were implemented as via-middle type before the first Cu–damascene metal layer processing. TSV diameter was 5 μm, and depth was 50 μm—a 1 : 10 aspect ratio. Minimum TSV pitch was 10 μm. A 200 nm oxide liner electrically isolated the TSV, and a Ta–Cu barrier–seed layer was deposited by PVD. A bottom-up electroplating process was used to fill the TSVs without voids forming.

Due to the large difference in thermal expansion coefficients between Cu and Si, significant stress built up at the Cu TSV–Si interface during thermal processing steps. The stress in the silicon around the TSV had a cylindrical symmetry with a radial and a tangential component of stress, as shown in Figure 6.50 [85].

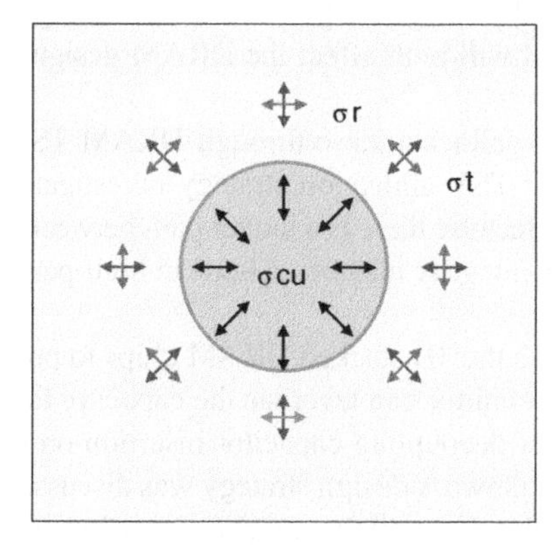

Figure 6.50 Stress in silicon around TSV, showing cylindrical symmetry. (Based on W. Guo *et al.*, (IMEC, KU Leuven), IEDM, December 2012 [85].)

The vertical stress component is negligible more than 1 μm away from the TSV edge. To first approximation, the stress is described by the Lame equation [85].

$$\sigma_r \approx -\sigma_\theta \approx \sigma_{Cu}\left(\frac{\phi}{r}\right)^2$$

Stress induced in the surrounding silicon is characterized by tensile stress in radial stress σ_r and compressive stress in the tangential direction σ_θ. Stress levels in radial and tangential directions are equal in amplitude, opposite in sign, proportional to the square of the via diameter φ, and inversely proportional to the square of the distance from the TSV to the device r.

The main impact of the TSV stress on the FinFET was on the drive current. Compared with planar devices, there was a reduced impact of the TSV-induced stress on the p-type FinFET devices. For the n-type FinFET, there was a higher impact than measured on n-type planar devices. The impact on the n-FinFET was similar to that of the p-FinFET but of the opposite sign. The impact on the planar device was up to 11.5%, while that on the FinFET was below 7.5%.

Minimizing local deformation induced around Cu TSV and CuSn–InAu microbumps in stacked 3D LSI is important to relieve stress. This stress is due to die thickness and subsurface structures formed after various stress-relief methods. NICHe, ASET, and Tohoku University investigated this issue in December of 2012 [86] and found more than a degree of local misorientation created in stacked LSI around the microbump region. This induces large tensile stress above the μ-bump region and small compressive stress in the bump space region. It also increases n-MOSFET mobility in the μ-bump and decreases the mobility in the bump-space region. Stress relief obtained by plasma etching or dry polishing was found to cause the device characteristics to deteriorate after stacking, while stress relief by CMP caused much less deterioration of device characteristics. For an LSI die/wafer thickness less than 50 μm, the Young modulus (E) and hardness (H) of the thinned die no longer acted like a bulk single-crystal silicon. This increased the reliability risks in a highly integrated 3D LSI.

To avoid proximity effects from Cu TSVs, arrays of TSVs are often isolated on the chip. In June of 2012, Texas Instruments discussed the impact of isolated/arrayed via-middle Cu TSVs in 8L metal 28 nm CMOS [87]. This was electrically measured for proximities >4 μm at 27 °C and 105 °C. The largest shift in I_{on}, at <2.3%, was less than that from other competing sources, such as dual stress liner boundaries, which was about 10%. It was shown that for proximity >1.5 μm, the impact of TSVs was negligible. The interaction with overlying interconnects was mitigated by optimizing the post-TSV plating anneal and by introducing a TSV unit cell intended to minimize the impact on local environment.

For a typical TSV plating anneal, the zero stress state of the Cu TSV is expected at greater than 105 °C, so the impact of TSVs on surrounding silicon over a temperature range of 27 to 105 °C is tensile. Insertion of compressive stacked trench isolation (STI) between the TSV and device will act to buffer this impact. Electrical properties of FETs between 4 and 16 μm of TSV are negligible compared to I_{on} shifts of 10% from DSL boundaries in a 28 nm CMOS layout. For Wide-IO memory–logic interface applications using a $40 \times 50\,\mu m^2$ JEDEC TSV array, ESD and decoupling capacitors can be placed adjacent to TSV so that CMOS logic circuitry does not require placement nearer than 4 μm.

A study using hard X-rays to determine the dependence of device reliability on lattice regularity in active silicon in high-density 3D systems that use TSVs and microbumps was discussed in December of 2013 by NICHe [88]. The data showed that the silicon lattice structure was severely deteriorated by the thermomechanical stress exerted by Cu TSV, CuSn μ-bumps, and the local deformation of very thin dies. It was found that the retention period for stacked memory chips degraded with a decrease in the chip thickness. The median retention time for a DRAM chip with thickness below 30 μm was found reduced by 50% from the retention period for a 100 μm thick DRAM chip. The DRAM retention time degradation was explained by the deterioration of the Young's modulus and the distorted silicon lattice structure in the very thin vertically stacked memory chip.

Reliability in 3D LSI associated with mechanical issues induced by Cu TSVs, μ-bumps, and crystal defects was discussed in December of 2013 by Tohoku University [89]. Other mechanical factors affecting reliability were the crystalline structure in thinned silicon wafers and metal contamination induced by Cu diffusion from TSVs and thinned backside surfaces. Mechanical stress induced by Cu TSVs and μ-bumps were found dependent on design rules and process parameters. DRAM retention characteristics were found severely degraded by silicon thinning, particularly below 30 μm thickness.

Guidelines for TSV integration of 10 nm node FinFET technology were discussed in December of 2013 by IMEC and Synopsys [90]. Key contributing factors to TSV, including "keep out zone" for FinFETS, were explored, and TCAD sub-band modeling of the impact of stress on mobility was verified by means of uniaxial wafer bending.

6.10 Reconfiguring Stacked TSV Memory Architectures for Improved Performance

6.10.1 Overview of Potential for Reconfigured Stacked Architectures

TSV stacked memory systems require a different architecture configuration than standalone memory and processor chips in order to make the best use of the many high-bandwidth connections. Reconfigured cache architectures can be used to optimize performance and power for different applications. An efficient network interface can be configured to exploit the potential bandwidth of a stacked memory on processor architecture. TSV can be multiplexed to increase the effective number of connections. A hybrid cache architecture partitioned among memories with difference characteristics can be used to improve the performance level for the application. ReRAM routing switches can be used for density improvement and power reduction. A dynamic configurable RAM can result in improved speed and power between chips.

6.10.2 3D TSV-based 3D SRAM for High-Performance Platforms

A 3D universal memory and logic platform for high-performance systems using TSV die-stacking technology was discussed in June of 2013 by NTHU, ITRI, and Fukuoka Institute [48].

This technology permits reuse of predesigned, pretested logic dies stacked with multiple memory layers in various configurations to form a universal memory capacity platform. A new concept was proposed to avoid the high parasitic load of TSV and cross-layer PVT variations that occur in large layers stacked with wide-I/O devices. This concept is a semi-master-slave (SMS) memory structure with self-timed differential TSV signal transfer, which improves the speed, power, and yield of the 3D memory devices and provides good scalability for the 3D stack.

The method provides a constant load logic SRAM interface across the stack, a high tolerance for variations in the cross-layer PVT, and at-speed prebonding KGD sorting. It uses a TSV load tracking method to achieve a small TSV voltage swing and to suppress power and speed overhead in cross-layer TSV signal communications resulting from large TSV parasitic loads. To verify the usefulness of the structures, a two-layer 32 kb 3D SRAM test chip with layer-scalable test modes was used that had a via-last TSV process with die-to-die bonding. The test chip confirmed the functionality and demonstrated superior scalability in a large stack with small speed overhead. Cross-layer connection schemes were used.

For SRAM and DRAM stacks, the current TSV sizes are too large to fit into the small SRAM/DRAM cell pitch; as a result, direct stacking and master–slave stacking have been used in these 3D memory designs. A direct stacking structure is not well suited to large stacks for high speed due to the parasitic load at the logic–memory interface being higher than that found in 2D chips. The master–slave structure has a fixed parasitic load at the logic–memory interface, but the yield is affected by variations in cross-layer processes and PVT.

This study involved a SMS structure and a self-timed differential TSV signal transfer method for a 3D memory tower. This structure had the following properties: constant-load logic–SRAM interface, suppressed TSV-induced power and speed overhead, high tolerance for cross-layer PVT variations, and at-speed prebonding KGD sorting, which improves speed and yield.

A two-layer 3D-SRAM with test modes was made using a via-last process with die-to-die bonding for verification of the feasibility of the method. Reusing the design, as shown in Figure 6.51, of a logic macro for both a small SRAM and a large-capacity SRAM was proposed [48].

For a direct stacked memory tower, 2D dice are used for stacking. Each memory layer has its own timer, decoder, controller, and I/O buffers. All layers share the same TSV I/O bus.

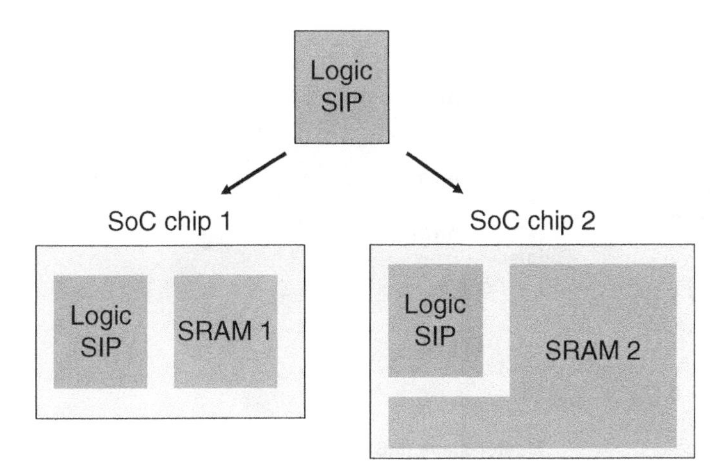

Figure 6.51 Illustration of reuse of design for logic and SRAM. (Based on M.F. Chang *et al.*, (NTHU, ITRI, Fukuoka Institute), *IEEE Journal of Solid-State Circuits*, June 2013 [48].)

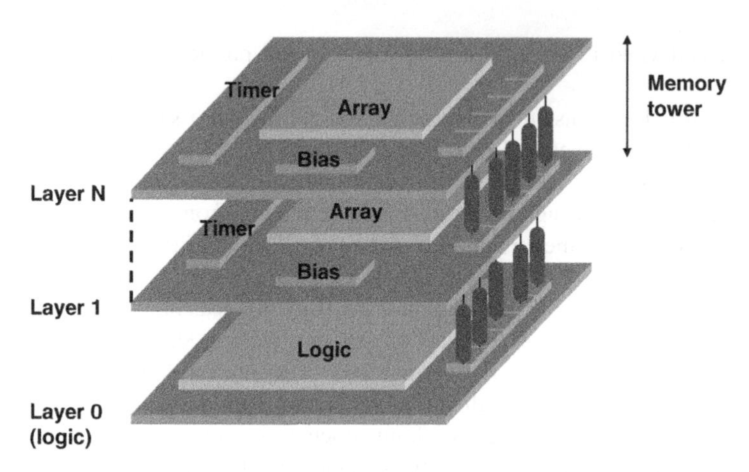

Figure 6.52 3D schematic of direct-stacked 3D memory tower in logic + memory system. (Based on M. F. Chang *et al.*, (NTHU, ITRI, Fukuoka Institute), *IEEE Journal of Solid-State Circuits*, June 2013 [48].)

An example of a direct stacked memory tower is shown in Figure 6.52 [48]. With direct stacking, the parasitic RC on a vertical signal path is the sum of all the TSV RCs. Cross-layer timing skews are large in advanced process nodes so that the logic–SRAM interface loading of the structure depends on the number of stacked layers.

The 3D master–slave memory tower structure requires circuit designs that are different from 2D designs. Each master and slave die has its own cell array, local write drives, local sense amps and decoders, I/O latches, and buffers. There is a global timing generator, and all layers share the same timing generated by the single master die, which saves area but leaves the structure subject to layer-to-layer PVT. A 3D master–memory slave stack is illustrated in Figure 6.53 [48]. KGD testing is more difficult with a master–slave design because each

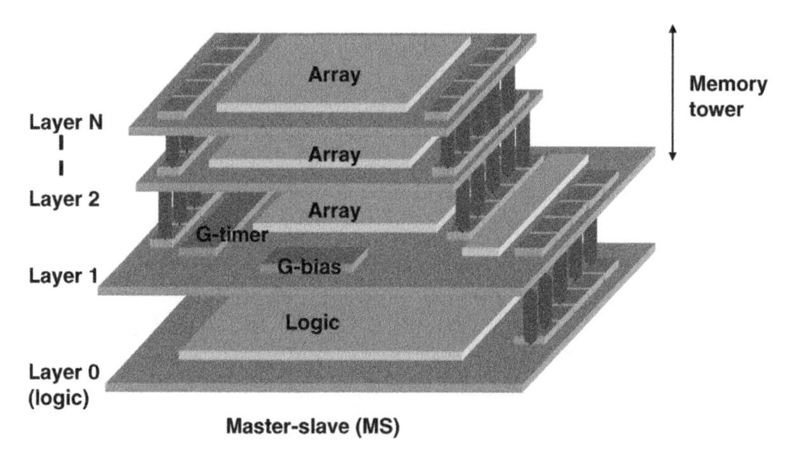

Figure 6.53 Illustration of 3D master–slave memory tower. (Based on M.F. Chang *et al.*, (NTHU, ITRI, Fukuoka Institute), *IEEE Journal of Solid-State Circuits*, June 2013 [48].)

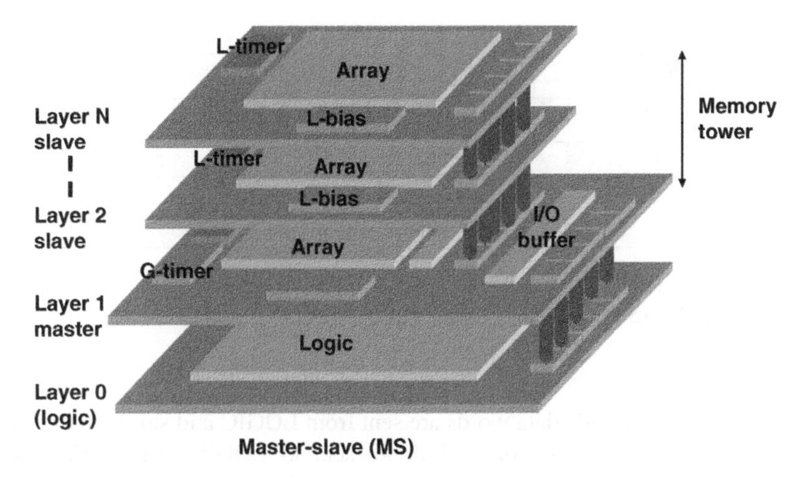

Figure 6.54 Illustration of a semi-master–slave logic and memory tower design. (Based on M.F. Chang *et al.*, (NTHU, ITRI, Fukuoka Institute), *IEEE Journal of Solid-State Circuits*, June 2013 [48].)

master–slave layer requires additional test logic. This study also proposed a SMS structure for a logic and memory 3D IC system, as shown in Figure 6.54 [48].

The SMS dice share address/command decoders and latches, I/O latches, and buffers to reduce the area. A buffered interface in the master layer provides a constant load logic–memory interface. Each slave layer has its own local time and bias generator along with a cell array and local sense amp. The global timer controls the pulse edges for signals timed to the clock signals. The local timer receives the positive edges and generates local signals and timings consistent with local PVT conditions. Local bias voltages and currents may differ between layers. This scheme improves speed, power, and yield of the 3D memory tower while permitting scalability of the memory stack. TSV-induced power and speed overhead is suppressed. There is tolerance of cross-layer PVT variations, and prebonded KGD sorting can be used.

6.10.3 Waveform Capture with 100 GB/s I/O, 4096 TSVs and an Active Si Interposer

In mobile applications where the requirements include very low power, high data bandwidth, stability, scalability, high capacity, and small footprint, new techniques for 3D stacking are required. In February of 2013, Kobe University and ASAET discussed a 3D test vehicle developed for TSV-based wide-I/O data communication in a three-tier 3D chip stack assembled in a 527-pin ball grid array (BGA) package on a system PCB [91]. The stack included an active silicon interposer. A schematic cross-section of the system is shown in Figure 6.55.

A digital control logic chip is placed at the base of the stack, an active silicon interposer in the middle, and a wide-I/O SRAM with 800KB capacity on top. All are in 90 nm CMOS technology. The process used was a via-last 50 µm pitch Cu TSV and chip-to-chip stacking technology. Vertical connections use a more than 7300, 20 µm diameter, 50 µm pitch TSV. Both bumps and microbumps are used. The 4096-bit wide-I/O TSV data bus is capable of 100 GB/s source-synchronous bidirectional data transfer at 200 MHz and 0.56 mW/Gb/s with a 1.2 V power supply. Core voltage is 1.2 V, and external I/O are at 3.3 V.

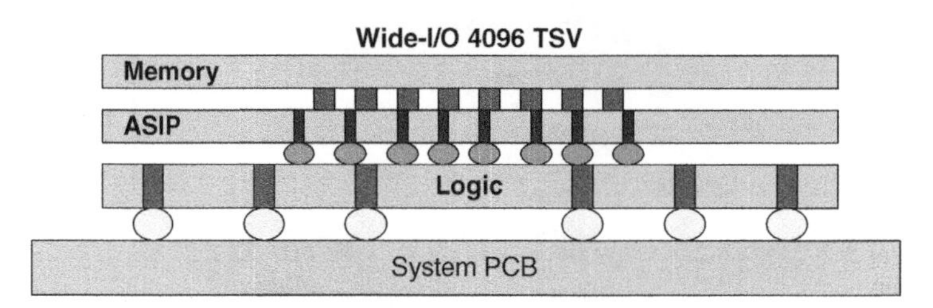

Figure 6.55 A schematic cross-section of the 3D TSV stacked memory system. (Based on S. Takaya *et al.*, (Kobe University, ASAET), ISSCC, February 2013 [91].)

In write mode, 4096-bit-wide data words are sent from LOGIC and stored in the 800KB of SRAM. In READ mode, words from MEM are sent to LOGIC. The BIST mechanism compares the received data bits with expected bits. The BIST operates at speed. The number of failed bits is stored in a fail register. I2C transactions and scan chains for test and debug operate through vertical communications channels in parallel with the data links. The wide-I/O bus is divided into eight parallel banks, with each bank having two TSV arrays of 64×7 bits and 64×6 bits with mini-I/O channels of 512 bits plus 16 bits for redundancy. Power and ground pins are placed every five columns. Each mini-I/O circuit consists of a pair of driver and receiver buffers and a bus keeper. Redundant bits and selectable driver strengths help adapt the bus to conditions in the chip stack.

It was confirmed that the wide-I/O performance was unchanged when using redundant bits for waveform capture, indicating that the input capacitance has a negligible effect on vertical signaling. The vertical power supply networks in the stack were shown to be completely integrated, and the eye diagrams showed good signaling quality in the data bus.

6.10.4 3D Stacked FPGA and ReRAM Configuration Memory

A 3D FPGA with stacked RRAM configuration memory was described in February of 2012 by Stanford University [92]. This RRAM configuration memory was compatible with CMOS and scalable. The transistors and interconnects were implemented in CMOS, while the configuration memory was stacked on top of the transistors and interconnects. A 1T2R cell was used, consisting of two programmable resistors and a shared select transistor. The resistor was made of a nitrogen-doped AlO_x-based resistance change film between two Al electrodes. The programmable resistor was compatible with the BEOL for CMOS and was scalable with the CMOS. The configuration memory cell layout area was 24 F^2. The two programmable resistors formed a resistive divider with a large margin for the desired logic level. The high resistance in the HR state limited the leakage current through the resistive divider.

6.10.5 Cache Architecture to Configure Stacked DRAM to Specific Applications

A configurable cache architecture using a small, fast SRAM and a large-capacity stacked DRAM was discussed by Kyushu University in August of 2011 [10]. The cache was intended to adapt the system to varying aspects of the application program behavior affected by the die stacking. Application programs of concern are those requiring large working data sets and high

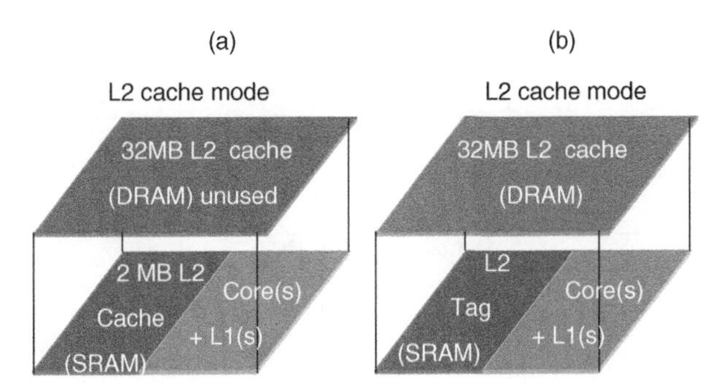

Figure 6.56 Illustration of hybrid cache concept for integrating SRAM and DRAM cache using (a) small but fast SRAM; and (b) slow but large DRAM. (Based on K. Inoue *et al.*, (Kyushu University), MWSCAS, 7 August 2011 [10].)

memory bandwidth, such as high-quality video processing and data mining. These applications require relatively large caches. The use of multicore processors makes the use of large cache more difficult. A potential solution is to use many on-chip caches. This approach is made promising by 3D stacking because a large last-level cache with high bandwidth becomes possible. However, access to DRAM is slow, and TSVs have large load capacitance, so activating a large number of TSVs consumes a significant amount of energy.

The architectural technique proposed is a SRAM–DRAM hybrid cache architecture using adaptive optimization in which the characteristics of target applications are considered in order to effectively use the stacked memory. Based on the demand for cache capacity, the cache attempts to select an appropriate operation mode. This hybrid cache concept is illustrated in Figure 6.56 [10].

Conventional cache architecture has either a large DRAM L2 cache, which reduces cache miss rates but makes access time longer, or a medium-size L2 SRAM cache, which has faster access time but higher cache miss rates. The proposed hybrid cache combines the two approaches into a cache that supports two modes. The first is an SRAM cache mode in which the SRAM implemented on the lower layer works as a conventional 2MB fast SRAM L2 cache and a stacked DRAM that is not used (due to a gated power supply). The second is a DRAM cache mode in which the stacked 32MB DRAM works as a data memory of the L2 cache, and the SRAM is used as a tag RAM.

An algorithm for deciding which cache mode to select to optimize a given application has been developed. If the application requires large cache capacity, the hybrid cache uses the stacked DRAM as data memory of the last level cache. If a large cache capacity is not required, the small SRAM is used. Future work will be required on techniques to optimally decide the operation mode for a given application.

6.10.6 Network Platform for Stacked Memory–Processor Architectures

Stacking memory layers on top of a multiprocessor layer can potentially reduce wiring delay and increase bandwidth. However, an efficient on-chip communication platform must be integrated into the logic layer to optimize this capability. In June of 2011, the University of Turku, Finland, discussed a 3D IC network interface intended to exploit the bandwidth

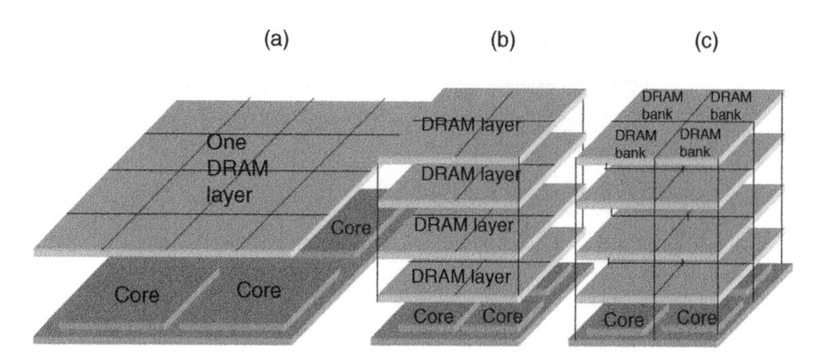

Figure 6.57 Three cases for DRAM stacking on multicore processor: (a) one DRAM layer stacked on one logic layer; (b) four DRAM layers stacked on one logic layer in 1/4 the footprint; and (c) one DRAM bank dedicated to each core. (Based on M. Daneshtalab *et al.*, (University of Turku), ReCoSoC, 20 June 2011 [93].)

potential of stacked memory-on-processor architectures [93]. An on-chip network platform for the logic layer was presented that used an efficient network interface to exploit the potential bandwidth of the stacked memory-on-processor architecture. It has been experimentally demonstrated that the platform with network interface increases performance. The three cases considered are illustrated in Figure 6.57 [93]. Figure 6.57(a) shows a conventional DRAM stacking with one DRAM layer stacked on top of the logic layer. This has an advantage if the memory size is small. In Figure 6.57(b), TSVs are used for a vertical bus through four DRAM layers to link them to the logic layer. The footprint is reduced, and the TSVs are fast and dense. Figure 6.57(c) shows stacked 3D DRAM with the main memory divided into four banks, with each bank aligned to one core slice. The four banks stacked above a core are expected to be used most frequently by that core. This option requires more area be reserved for peripherals on the interface layer, resulting in a more complex layout and lower area efficiency.

A modular communication platform is designed for the logic layer to scale bandwidth among the processors. In addition, the on-chip network is equipped with a streamlined network interface to provide efficient communication between the processor and memories. It is configured so requirements for local memory do not need to travel through the on-chip network.

6.10.7 Multiplexing Signals to Reduce Number of TSVs in IC Die Stacking

TSVs used in vertical die stacking of ICs tend to be limited in number due to manufacturing and reliability concerns. This limitation constrains partitioning at the system level, which affects routing area used by the TSV and can reduce the advantages of vertical stacking. A method to increase the effective number of interdie connections in the 3D ICs was proposed by the University of Massachusetts in July of 2011 [94]. This method involves multiplexing two signals originating on one die using the system clock. The signals are recovered by a combination of positive and negative edge-triggered flip-flops in the destination chip. An extension was also proposed where the signals need not originate on the same die. This multiplexing of TSVs permits doubling of interdie connections with little area or performance overhead.

6.10.8 3D Hybrid Cache with MRAM and SRAM Stacked on Processor Cores

A 3D hybrid cache architecture integration in which the shared last level of L2 cache is made up of an MRAM layer and an SRAM layer stacked on the processing cores was described in February of 2011 by Pennsylvania State University [95]. A control theory–centric approach was proposed to partition this shared hybrid L2 cache space dynamically among concurrently running applications so that application-level performance targets were met. At each time interval the two layers of the hybrid L2 cache were partitioned based on the cache demands made by the controllers of the applications to satisfy the specified performance targets. The feedback control methodology was evaluated using various workloads. The architecture was found able to satisfy the specified performance by partitioning the hybrid cache space among the co-running applications.

6.10.9 CMOS FPGA and Routing Switches Made with ReRAM Devices

A CMOS FPGA based on RRAM devices was discussed in September of 2010 by the University of Albany [96]. The array consisted of 1T1R ReRAM structures that can be made using a CMOS-compatible process. These devices can be used for FPGA block memories. In addition, ReRAM routing switches were developed to replace CMOS routing switches, which resulted in significant density increase and power reduction. Simulations show that 2D and 3D FPGAs provided two to three times more improvement in area with 20% lower power consumption than CMOS FPGAs.

6.10.10 Dynamic Configurable SRAM Stacked with Various Logic Chips

A dynamic reconfigurable 3D SRAM chip was discussed by NEC in January of 2010 [97]. This chip consisted of an on-chip SRAM from an SoC that was moved to a separate chip and stacked with a logic chip using 3D packaging technology. On the memory chip, RAM macros were arrayed and connected using a 2D mesh network of interconnects. A memory-specific network of crossbar interconnects reduced the area overhead by 63% and the latency by 43%. The signal lines between the two chips were directly connected by 10 μm pitch interchip electrodes. This resulted in fast, low-power transmission between the chips [97].

6.11 Stacking Memories Using Noncontact Connections with Inductive Coupling

6.11.1 Overview of Noncontact Inductive Coupling of Stacked Memory

An alternative to using TSVs for interconnection between stacked chips is to use an inductive-coupling interface between chips with no physical circuit connection. The evolution of this technology has occurred relatively recently, with much of the development occurring at Keio University. The "ThruChip Interface" (TCI), developed by Keio University, can be made in conventional CMOS technology and provides similar performance to TSVs without the need to build vias through the chips. The interface uses a "near-field" wireless connection of electromagnetic signals up to 50 GHz for distances shorter than 1 mm. These were discussed in June of 2012 by Keio University [98]. Signals attenuate in proportion to the cube of distance. There is no crosstalk, so simple transceivers can be integrated to reduce chip area. There is

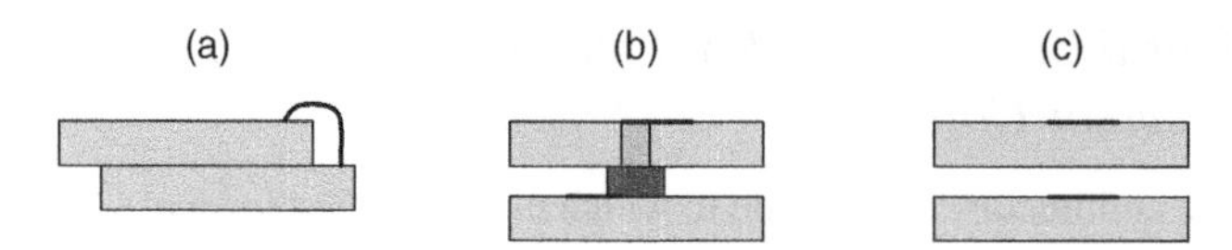

Figure 6.58 Chip connection by (a) wire bonding; (b) TSV; and (c) TCI inductive coupling. (Based on T. Kuroda, (Keio University),VLSI Technology Symposium, June 2012 [98].)

thermal tolerance, no process added to conventional CMOS, no need for ESD protection, no interposer required so chip packing can be tight, and no need for a level shifter because connections are AC. This trend is essentially a move from mechanical to electrical connections, as illustrated in Figure 6.58 [98].

The TCI uses inductive coupling with a magnetic field, which can penetrate a semiconductor chip, which an electric field cannot do. The TCI imposes no restriction on number of chips stacked and whether they are face up or face down. Any application that has been considered for TSV integration can also use TCI with comparable performance. The layout area is small, using only 3% of the area of a conventional CMOS I/O. Chip stacking height is low because there is no need to use bumps between the chips. Performance is comparable to that obtained using TSV.

Energy consumption is smaller than using a conventional DDR interface. The low power dissipation is a result of the ESD protection device being eliminated, the multidrop bus being formed without increasing the capacitive load, and transmission power being controlled according to the communication distance.

Signal degradation caused by eddy current and chip misalignment has been shown to be negligible. Crosstalk is also negligible if coils are placed at intervals larger than the size of the coil. The coils can be placed anywhere on the chip without interference with other circuits, and the effect on the environment due to electromagnetic susceptibility (EMS) or electromagnetic interference (EMI) has not been shown to be an issue. The measured bit-error-rate is less than 10^{-14}. Noncontact wafer-level testing can be used and reduces testing cost. Most applications shown to date use wire bonding for power delivery. Wire bonding is widely available and inexpensive, although wireless power delivery by inductive coupling is expected to open new applications.

The TCI interfaces use coils integrated in a conventional CMOS process, using normal wiring on IC chips that are then stacked. AC current changes in the transmitting coil on one chip are sensed as voltage in the receiver coil on another chip. Figure 6.59 shows a circuit schematic of the transmitter and receiver circuitry [98]. The inductive relationship between change of current in the transmitting coil t and voltage in the receiving coil r is given by

$$V_r = (K/L_t L_r)\mathrm{d}I_t/\mathrm{d}I$$

where the transmit current is I_t and the receive voltage is V_r.

6.11.2 *Early Concepts of Inductive-Coupling Connections of Stacked Memory Chips*

Early discussions of inductive-coupling connections of stacked memory chips occurred in February of 2009 among the University of Tokyo, Keio University, and Toshiba [99]. They

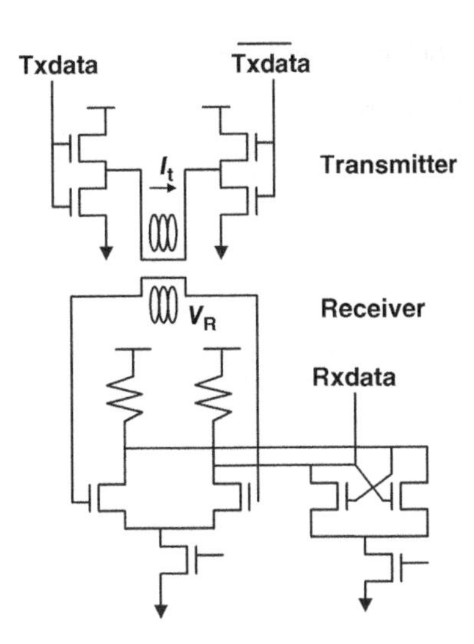

Figure 6.59 Transmitter and receiver circuitry for TCI interface. (Based on T. Kuroda, (Keio University), VLSI Technology Symposium, June 2012 [98].)

discussed a wireless communication technique that lets a controller chip stacked with a stack of 64 NAND flash chips communicate with the NAND flash at a data rate of 2 Gb/s. The system used relayed transmission with an inductor. The wireless interface let a capacitive ESD protection device be removed, resulting in a twofold power reduction and 40-fold reduction in I/O circuit layout area. Bond wires were used for the power supply and wireless interface for data access, using less than 200 wires as opposed to over 1500 wires in the conventional stack. For conventionally stacked chips, a spacer between layers would be required to provide headroom for the bonding wires. A schematic illustration of the proposed memory and controller stack with inductive communications is shown in Figure 6.60 [99].

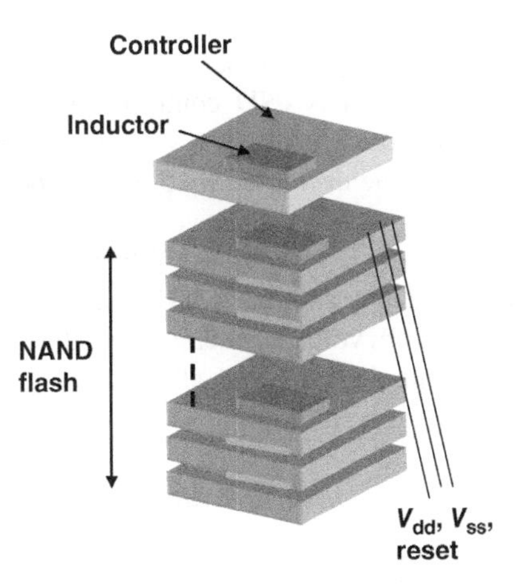

Figure 6.60 Proposed memory and controller stack with inductive communications. (Based on Y. Sugimori *et al.*, (Keio University, University of Tokyo), ISSCC, February 2009 [99].)

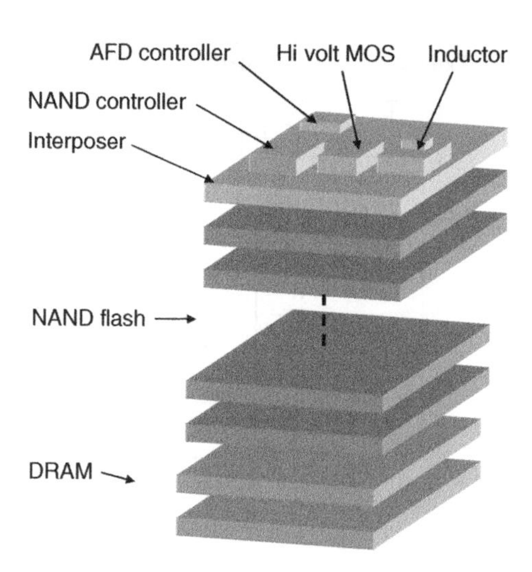

Figure 6.61 3D integrated SSD with NAND stack, DRAM, NAND controller, and high-voltage generator circuits. (Based on K. Ishida *et al.*, (University of Tokyo, Toshiba), ISSCC, February 2009 [100].)

The reduction in capacitance lets 64 NAND flash chips be used rather than the conventional eight chips. The inductors emitted a magnetic field both upward and downward, which enabled data delivery upward and downward for memory read and write. The bit error rate was found to be less than 10^{-12}.

Different stacking methods were considered to permit the wire bonds to be attached including staggered, half-turned, and terraced. Test chips were made in 180 nm CMOS and stacked using both the half-turned and staggered stacking techniques. Because each chip was 60 μm thick, communication distance was 120 μm. A square shield was used with a 400 μm side. A bit error rate of $<10^{-12}$ was measured. The average energy consumption per chip was 15 pJ/b.

Toshiba and the University of Tokyo in February of 2009 discussed a 3D integrated SSD consisting of NAND chips, DRAM, a NAND controller and inductor, and a low-power program voltage generator integrated with SiP. The stack is illustrated in Figure 6.61 [100].

The die size of the NAND could be decreased by 5 to 10% because no charge pump was needed.

A 56 nm 16Gb NAND flash memory chip was used. When a write command was input to the NAND, the ready/busy signal went low and the NAND went into a busy state. The program voltage was supplied from the program voltage generator, and the program pulse was applied to the memory cells. The verify-read detected that all memory cells were successfully programmed, and the ready/busy signal went high.

6.11.3 Evolution of Inductive-Coupling Connections of NAND Flash Stacks

For stacking 128 NAND flash chips and a controller chip in a single package for SSD applications, Keio University in February of 2010 discussed a terraced spiral chip stacking

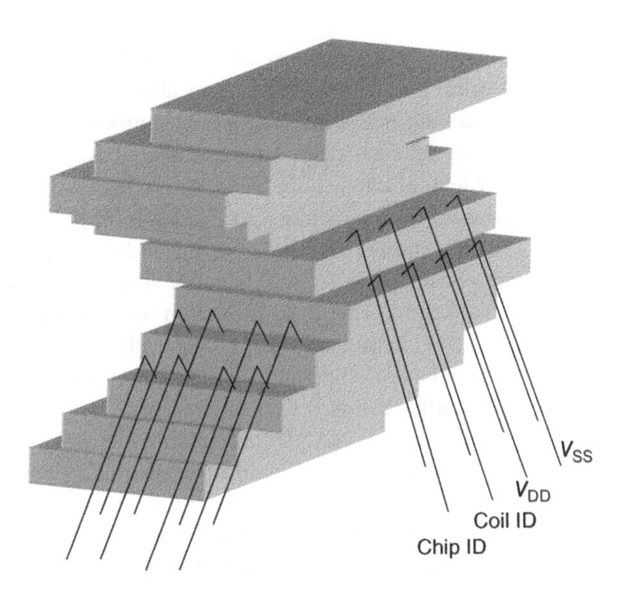

Figure 6.62 Schematic of spiral stair stacking scheme. (Based on M. Saito *et al.*, (Keio University), ISSCC, February 2010 [101].)

methodology [101]. The spiral stair stacking scheme required no spacer chips and had a total height of 3.9 mm between chips. It provided space for bonding the power, ground, and reset wires to each chip. The controller chip accessed a memory chip by relayed transmission using inductive-coupling transceivers. The previous method, without the spiral, required a spacer, resulting in a total height of 6.0 mm between chips. The average communication distance was also shortened, and transmission power was reduced to 60%. A large 1.1 mm coil is used so that the communication could be relayed at every eighth chip. This reduced the number of transceivers activated for chip access to 1/4 compared to stacks where relay occurs every second chip with a coil of 0.2 mm diameter. A schematic diagram of the spiral stair stacking scheme is shown in Figure 6.62 [101].

Energy consumption for a random access was reduced to 1.8 pJ/b/chip, which is 33% of the consumption of the conventional method. A large coil was placed over the memory core by using the third metal layer, which was not used other than for reinforcing power supply in the source lines. Capacitive and inductive interference between the chip access and the memory read/write was reduced by placing the square coil diagonal to the bit-line and word-lines.

This inductive-coupling programmable bus for NAND flash SSD memory access was also described in January of 2010 by Keio University and the University of Tokyo [102]. Compared to a conventional SSD, the inductive-coupling wireless interface reduced power consumption and I/O circuit layout area while achieving a data rate of 2 Gb/s in a 180 nm CMOS process. Using the wireless interface, one package contained 64 chips, reducing the package footprint to 1/8 its original size.

Because the controller chip was stacked on top of the memory chips, access required bidirectional capability. An inductive-coupling up/down repeater was used, which consisted of two pairs of inductive-coupling transmitter/receivers with two enable signals, Txen and Rxen.

By setting the enable signals correctly, the memory could be both written and read. A shield was used to reduce crosstalk from unintentional communications.

Half-turned and staggered stacking was used to provide headroom for bonding the power supply wires. The memory state of each chip was controlled based on a code sequence transmitted by the controller, which eliminated the need for a chip ID number. Packet-based communications protocol was used for the memory access. Multibit parallel data and control signals were multiplexed and transmitted serially over the inductive-coupling link. For a package with 64 chips, power consumption for the inductive link was 307 mW compared to 557 mW for the conventional packaging, energy consumption was 15 pJ/bit/chip compared to conventional 27 pJ/bit/chip, and I/O circuit area was 3645 μm^2/chip compared to the conventional 145 744 μm^2/chip. The number of chips and the data rate remained the same for the inductive-coupled link and conventional link.

A three-coil channel method for stacking NAND flash memory using an inductive-coupling programmable bus was discussed by Keio University in September of 2010 [103]. This method permitted random access for read and write. Transmit power was reduced by 47% compared to previous methods using shields. A new coil layout permitted the coils to interleave with logic interconnects, resulting in a 91% area reduction with only a 17% transmit power increase. Relayed data transmission was at 1.6 Gb/s.

6.11.4 TCI for Replacing Stacking with TSV Connections

A wireless CMOS transceiver using inductive coupling with the TCI interface was discussed by Keio University in December of 2010 as a method for replacing TSV electrical connections [104]. TCI was claimed to be less expensive and lower power than TSV while reducing layout area. Performance of the two was comparable. ESD protection was not needed because TCI has no contact. Data rate per coil was 11Gb/s/ch. Using 1000 channels a data rate of 8Tb/s was obtained. Energy consumption was 0.01 pJ/b. The measured bit error rate was $<10^{-14}$, which is comparable to that of wired communications.

A high-bandwidth inductive-coupling interface for use in stacking NAND flash memory was discussed in February of 2011 by Keio University [105]. This interface had bandwidth per unit area of 2.7 Gb/s/mm^2 and energy consumption per chip of 0.9 pJ/b/chip. This is 10 times the bandwidth and half the energy consumption of Keio's presentation [99] on this type of circuit in February 2009. A relayed transmission method that used one coil was proposed to reduce the number of coils in a data link from three to one. Coupled resonation was used for clock and data recovery, which resulted in eliminating a source-synchronous clock link. This reduced the number of coils needed to form a channel from six to one, resulting in improvement in data rate, layout area, and energy consumption [105].

Clock and data recovery (CDR) was used to eliminate the source-synchronous clock link, resulting in reduction of the required coils from two to one. The CDR method required a global reference clock distribution among all the stacked chips. Because the distribution network remained active for random memory access, power consumption had to be reduced. This was accomplished by using coupled resonation for the reference clock link with frequency and phase synchronized by inductive coupling between the oscillating coils with high Q factor, implying high coupling efficiency, so the power dissipation was reduced significantly.

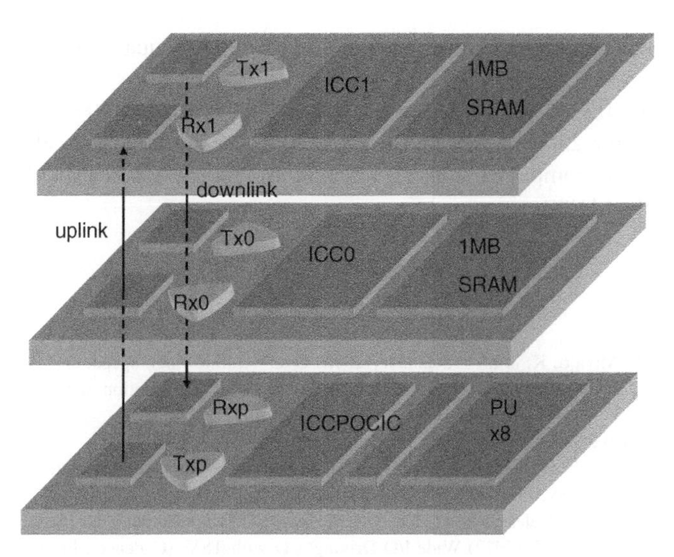

Figure 6.63 Block diagram of three-chip 3D inductive coupling system. (Based on M. Saen *et al.*, (Hitachi, Renesas, Keio University), *IEEE Journal of Solid-State Circuits*, 45(4) [106].)

6.11.5 Processor–SRAM 3D Integration Using Inductive Coupling

In April of 2010, Hitachi, Renesas, and Keio University discussed a 3D system integration of a processor chip and two memory chips using inductive coupling [106]. A 3D communication link using three chips and maintaining small area and low power consumption required both a short link distance and the prevention of signal degradation due to unused inductors. To this end, they developed a 3D wire-penetrated multilayer structure with a shorter link distance and an open skipped inductor method of suppressing signal degradation. They also proposed a new memory access control method. In addition, they showed that the three chips could be successfully AC coupled using inductive coupling. The power of the link was 1 pJ/b, and the area efficiency was 0.15 mm^2/Gb/s. These are the same as in two-chip integration. A block diagram of the 3D inductive-coupling system with three chips is shown in the diagram in Figure 6.63. This inductive-coupling link has a 600MHz clock, 16-bit data, and a control signal [106].

6.11.6 Optical Interface for Future 3D Stacked Chip Connections

Further out in time, an optical interface could be used to connect 3D stacked chips. In May of 2012, the University of Ferrara and Columbia University discussed a bandwidth-scalable optical layer for a 3D multicore processor [107]. Because the performance of future chip multiprocessors will only scale with the number of integrated cores if there is also an increase in memory access efficiency, the scalable optical layer complements efforts ongoing elsewhere on photonically integrated bandwidth-rich DRAM devices. This study presented network partitioning options and bandwidth scalability techniques with technology and layout in mind. The main contribution is in the characterization and quantification of interaction effects among the technology platform, layout constraints, and network-level quality metrics of a possible optical network.

In December of 2013, STMicroelectronics and Luxter discussed a 300 mm silicon photonics platform designed for 25 Gb/s and above applications at the three typical communication wavelengths that were compatible with 3D integration. Process features and device characteristics were described [108].

References

1. Sun, C., Fujii, H., Miyajo, K. *et al.* (22 January 2013) Over 10-times high-speed, energy efficient 3D TSV-integrated hybrid ReRAM/MLC NAND SSD by intelligent data fragmentation suppression, (University of Tokyo). ASP-DAC.
2. Courtland, R. (2012) 3-D Chips Grow Up. IEEE Spectrum (January), p. 123.
3. Semtech (2010) Semtech and IBM Join Forces to Develop High-Performance Integrated ADC/DSP Platform Using 3D TSV Technology. Press release, December 8.
4. Merritt, R. (2012) JEDEC Releases Next-Gen DRAM Spec. EE Times (September 25).
5. Greenberg, M. and Bansal, S. (2012) Wide I/O Driving 3-D with TSV, (Cadence). EE Times (March 9).
6. Micron. (2012) Consortium to Accelerate Dramatic Advances in Memory Technology Announces New Members. Press Release, June 27.
7. Iyer, S.S. *et al.* (June 2009) Process-design considerations for three dimensional memory integration, (IBM). VLSI Technology Symposium.
8. Kim, J.S. *et al.* (2012) A 1.2 V 12.8 GB/s 2 Gb mobile wide-I/O DRAM with 4 × 128 I/Os using TSV based stacking, (Samsung). *IEEE Journal of Solid-State Circuits*, **47**(1), 107.
9. Jeong, S.W., Cho, H.D., Hwang, J.W., and Kwon, H.K. (June 2013) Perspectives on mobile devices and their impact on semiconductor technologies, (Samsung), VLSI Circuits Symposium.
10. Inoue, K., Hashiguchi, S., Ueno, S. *et al.* (7 August 2011) 3D implemented SRAM/DRAM hybrid cache architecture for high-performance and low power consumption, (Kyushu University). MWSCAS.
11. Lin, C.H., Hsieh, W.T., Hsieh, H.C. *et al.* (9 October 2011) System-level design exploration for 3-D stacked memory architectures, (ITRI). CODES+ISSS, p. 389.
12. Black, B., Nelson, D.W., Webb, C., and Samra, N. (11 October 2004) 3D processing technology and its impact on iA32 microprocessors, (Intel), ICCD, p. 316.
13. Franzon, O. *et al.* (December 2013) Applications and design styles for 3DIC, (North Carolina State University, Synopsys). IEEE IEDM.
14. University of Rochester. (2008) 3-D Computer Processor: 'Rochester Cube' Points Way To More Powerful Chip Designs. Science News (September 17).
15. Kim, D.H. *et al.* (February 2012) 3D-MAPS: 3D massively parallel processor with stacked memory, (Georgia IT, KAIST, Amkor). IEEE ISSCC.
16. Jeddeloh, J. and Keeth, B. (June 2012) Hybrid memory cube new DRAM architecture increases density and performance, (Micron). VLSI Technology Symposium.
17. Pang, L.T., Restle, P.J., Wordeman, M.R. *et al.* (June 2012) A shorted global clock design for multi-GHz 3D stacked chips, (IBM). VLSI Circuits Symposium.
18. Wordeman, M., Silverman, J., Maier, G., and Scheuermann, M. (February 2012) A 3D system prototype of an eDRAM cache stacked over processor-like logic using through-siliconvias, (IBM). IEEE ISSCC.
19. Fang, K., Zhang, Z., and Zhu, Z. (10 October 2011) Memory architecture for integrating emerging memory technologies, (University of Illinois). PACT, p. 403.
20. Changra, N., Durodami, L., and Riko, R. (30 May 2012) 3D stacking: Where the rubber meets the road, (Qualcomm). ICICDT, p. 1.
21. Woo, D.H., Seong, N.H., and Lee, H.H.S. (2011) Pragmatic integration of an SRAM row cache in heterogeneous 3-D DRAM architecture using TSV, (Intel). *IEEE Transactions on VLSI Systems*, **21**(1), 1.
22. Zhang, T. *et al.* (7 August 2011) Leveraging on-chip DRAM stacking in an embedded 3D multi-core DSP system (Penn State U., NTHU, ITRI). MWSCAS.

23. Sekiguchi, T., Ono, K., Kotabe, A., and Yanagawa, Y. (2011) 1-Tbyte/s 1-Gbit DRAM architecture using 3-D interconnect for high-throughput computing, (Hitachi). *IEEE JSSC*, **46**(4), 828–837.
24. Burns, J.L. (June 2011) 3D integration from the viewpoint of high-end server system design, (IBM). VLSI Technology Symposium.
25. Zhao, B., Du, Y., Yang, J., and Zhang, Y. (2012) Process variation aware non-uniform cache management in 3D Die stacked multicore processor, (University of Pittsburgh). *IEEE Transactions on Computers*, **PP**(99), 1.
26. Dong, X., Wu, X., Xie, Y. *et al.* (2011) Stacking magnetic random access memory atop microprocessors: an architecture-level evaluation, (Penn State Unversity). *IET Computers & Digital Techniques*, **5**(3), 213.
27. Fujii, H. and Miyaji, K. (June 2012) x11 Performance increase, x 6.9 endurance enhancement, 93% energy reduction of 3D TSV-integrated hybrid ReRAM/MLC NAND SSD by data fragmentation suppression, (University of Tokyo). VLSI Circuits Symposium.
28. Takeuchi, K. (August 2011) Green high performance storage class memory & NAND flash memory hybrid SSD system, (University of Tokyo). ISLPED.
29. Kannan, S., Kim, B., Cho, S.B., and Ahn, B. (20 May 2012) Analysis of propagation delay in 3-D stacked DRAM, (University of Alabama, University of Ulsan, Yeungnam University). ISCAS.
30. Ching, H.H. *et al.* (22 April 2013) A case study: 3-D stacked memory system architecture exploration by ESL virtual platform, (ITRI). VLSI DAT.
31. Kim, D., Yoo, S., Lee, S. *et al.* (14 March 2011) A quantitative analysis of performance benefits of 3D die stacking on mobile and embedded SoC, (Postech). DATE, pp. 1–6.
32. Sunohara, M., Tokunaga, T., Kurihara, T., and Higashi, M. (27 May 2008) Silicon interposer with TSVs (through silicon vias) and fine multilayer wiring, (Shinko). ECTC, p. 847.
33. Knechtel, J. (2012) Assembling 2-D blocks into 3-D chips, (Dresden University of Technology). *IEEE Transactions on CAD of IC & Systems*, **31**(2), 228.
34. Masheshwari, D. (February 2014) Memory and system architecture for 400 Gb/s networking and beyond, (Cypress Semiconductor). IEEE ISSCC.
35. Cadence (December 2012) 3D ICs with TSVs—design challenges and requirements. White paper posted at http://www.cadence.com.
36. Lau, J.H. (21 May 2011) TSV interposer: the most cost-effective integrator for 3D IC integration, (ITRI). ASME Interpack.
37. Kao, N., Chen, E., Lee, D., and Ma, M. (29 May 2012) Development of through silicon via (TSV) interposer for memory module flip chip package, (Siliconware Precision Ind. Co.). ECTC, p. 1461.
38. Sun, J.Y.C. (June 2013) System scaling and collaborative open innovation, (TSMC). VLSI Technology Symposium.
39. Merritt, R. (2012) GlobalFoundries Installs Gear for 20-nm TSVs. EE Times (April 6).
40. CEA-Leti (2012) Leti Offers "Open3D" TSV Prototyping Program. Press release, February 1.
41. Lo, T. *et al.* (December 2012) Thinning, stacking, and TSV proximity effects for poly and high-κ/metal gate CMOS devices in an advanced 3D integration process, (TSMC). IEDM.
42. Arnaud, F. *et al.* (5 December 2011) Technology-circuit convergence for full-SOC platform in 28 nm and beyond, (STMicroelectronics Crolles). IEDM.
43. Iyer, S.S. (December 2012) The evolution of dense embedded memory in high performance logic technologies, (IBM). IEDM.
44. Golz, J. *et al.* (June 2011) 3D stackable 32 nm high-κ/metal gate SOI embedded DRAM prototype, (IBM). VLSI Circuits Symposium.
45. Batude, P. *et al.* (December 2011) Advances, challenges and opportunities in 3D CMOS sequential integration, (CEA-LETI, STMicro, EPFL). IEEE IEDM.
46. Kothandaraman, C. *et al.* (15 April 2012) Copper through silicon via (TSV) for 3D integration, (IBM). IRPS.
47. Liao, E.B. *et al.* (June 2013) An integrated air gap structure to achieve high-performance TSV interconnects for 28 nm 3D-IC integration, (TSMC). VLSI Technology Symposium.
48. Chang, M.F. *et al.* (2013) A high layer scalability TSV-based 3D-SRAM with semi-master-slave structure and self-times differential-TSV for high performance universal-memory-capacity platforms, (NTHU, ITRI, Fukuoka Institute). *IEEE Journal of Solid-State Circuits*, **48**(6), 1521.
49. Tsai, T.C. *et al.* (April 2012) CMP process development for the via-middle 3D TSV applications at 28 nm technology node, (UMC). 27th Annual Advanced Metallization Conference, vol. 92, p. 29.
50. Cho, S. (27 October 2011) Technical challenges in TSV integration to Si, (Samsung). SEMATECH Symposium Korea.

51. Tezzaron (January 2012) Tezzaron FaStack stacking technology. White paper posted at http://www.tezzaron .com.
52. West, J., Choi, Y.S., and Vartuli, C. (June 2012) Practical implications of via-middle Cu TSV-induced stress in a 28 nm CMOS technology for wide-IO logic-memory interconnect, (Texas Instruments). VLSI Technology Symposium.
53. Farooq, M.G. *et al.* (December 2011) 3D copper TSV integration, testing and reliability, (IBM). IEEE IEDM.
54. Arnaud, F. *et al.* (5 December 2011) Technology-circuit convergence for full-SOC platform in 28 nm and beyond, (STMicroelectronics Crolles). IEDM.
55. Weis, C., Loi, I., Benini, L., and Wehn, N. (2013) Exploration and optimization of 3-D integrated DRAM subsystems, (University of Kaiserslautern, University of Bologna). *IEEE Transactions on CAD of ICs and Systems*, **32**(4), 597.
56. Weis, C., Loi, L., Benini, L., and Wehn, N. (12 March 2012) An energy efficient DRAM subsystem for 3D integrated SoCs, (TU Kaiserslautern). DATE, p. 1138.
57. Chen, K., Li, S., Muralimanohar, N. *et al.* (12 March 2012) CACTI-3DD architecture-level modeling for 3D die-stacked DRAM main memory, (HP Labs). DATE.
58. Meng, J. and Coskun, A.K. (12 March 2012) Analysis and runtime management of 3D systems with stacked DRAM for boosting energy efficiency, (Boston University), DATE, p. 611.
59. Secker, D., Ji, M., Wilson, J. *et al.* (29 May 2012) Co-design and optimization of a 256-GB/s 3D IC package with a controller and stacked DRAM, (Rambus). ECTC, p. 857.
60. Woo, D.H., Seong, N.H., and Lee, H.H.S. (7 August 2011) Heterogeneous die stacking of SRAM row cache and 3-D DRAM: An empirical design evaluation, (Intel, Georgia IT). MWSCAS.
61. Yun, W., Kang, K., and Kyung, C.M. (15 May 2011) Thermal-aware energy minimization of 3D-stacked L3 cache with error rate limitation, (KAIST). ISCAS, pp. 1672–1675.
62. Woo, D.H., Seong, N.H., and Lee, H.H.S. (2013) Pragmatic integration of an SRAM row cache in heterogeneous 3-D DRAM architecture using TSV, (Intel, Georgia IT). *IEEE Transactions on VLSI Systems*, **21**(1), 1.
63. Wei, H., Wu, T.F., Sekar, D. *et al.* (December 2012) Cooling three-dimensional integrated circuits using power delivery networks, (Stanford University, Monolithic 3D, Rambus). IEDM.
64. Xu, Z. and Lu, J.Q. (December 2012) Hybrid modeling and analysis of different through-silicon-via (TSV)-based 3D power distribution networks, (Rensselaer Polytech). IEDM.
65. Fick, D. *et al.* (February 2012) Centip3De: A 3930DMIPS/W configurable near-threshold 3D stacked system with 64 ARM Cortex-M3 cores, (University of Michigan). IEEE ISSCC.
66. Oprins, H., Cherman, V., Vandevelde, B. *et al.* (29 May 2012) Numerical and experimental characterization of the thermal behavior of a packaged DRAM-on-logic stack, (IMEC). ECTC, p. 1081.
67. Chien, H.C. *et al.* (29 May 2012) Thermal evaluation and analyses of 3D IC integration SiP with TSVs for network system applications, (ITRI). ECTC, p. 1866.
68. Milojevic, D., Oprins, H., Ryckaert, J. *et al.* (19 September 2011) DRAM-on-logic stack—calibrated thermal and mechanical models integrated into pathfinding flow, (IMEC). CICC.
69. Meng, J., Rossell, D., and Coskun, A.K. (25 July 2011) Exploring performance, power, and temperature characteristics of 3D systems with on-chip DRAM, (Boston University). IGCC, p. 1.
70. Meng, J. and Coskun, A.K. (12 March 2012) Analysis and runtime management of 3D systems with stacked DRAM for boosting energy efficiency, (Boston University), DATE, p. 611.
71. Yueh, W., Chatterjee, S., Trivedi, A.R., and Mukhopadhyay, S. (2013) Performance and robustness of 3-D integrated SRAM considering tier-to-tier thermal and supply crosstalk, (Georgia Institute of Technology). *IEEE Transactions on Components Packaging and Manufacturing Technology*, **3**(6).
72. Liao, W.S. *et al.* (June 2013) 3D IC heterogeneous integration of GPS RF receiver, baseband and DRAM on CoWoS with system BIST solution, (TSMC), VLSI Circuits Symposium.
73. Chou, C.W., Huang, Y.J., and Li, J.F. (2013) A built-in self-repair scheme for 3-D RAMs with interdie redundancy, (TSMC, Nat. Central U.). *IEEE Transactions on CAD of ICs and Systems*, **32**(4), 572.
74. Hou, C.S. and Li, J.F. (April 2013) Allocation of RAM built-in self-repair circuits for SOC dies of 3D ICs, (National Central University), VLSI Test Symposium.
75. Lu, S.K., Lu, U.C., Pong, S.W., and Cheng, H.C. (April 2013) Efficient test and repair architectures for 3D TSV-based random access memories, (National Taiwan University of Science and Technology), VLSI-DAT.
76. Chou, Y.F., Kwai, D.M., Shieh, M.D., and Wu, C.W. (2013) Reactivation of spares for off-chip memory repair after die stacking in a 3-D IC with TSVs, (ITRI). *IEEE Transactions on Circuits and Systems*, **60**(9).

77. Desai, S., Roy, S., and Chakraborty, K. (19 March 2012) Process variation aware DRAM design using block based adaptive body biasing algorithm, (Utah State University). ISQED.
78. Lee, H.W. *et al.* (February 2012) A 283.2 uW 800 Mb/s/pin DLL-based data self-aligner for through-silicon via (TSV) interface, (Korea University, Hynix). IEEE ISSCC.
79. Lim, S.B., Lee, H.W., Song, H., and Kim, C. (2013) A 247 uW 800 Mb/s/pin DLL-based data self-aligner for through silicon via (TSV) interface, (SK Hynix, Korea University of Seoul). *IEEE Journal of Solid-State Circuits*, **48**(3), 711.
80. Kim, K., Pack, J.S., Kim, H. *et al.* (12 December 2011) Impedance of power distribution networks in TSV-based 3D-ICs, (KAIST). EDAPS, p. 1.
81. Chang, C.W. *et al.* (February 2013) Through-silicon-via-based double-side integrated microsystem for neural sensing applications, (National Chiao Tung University, China Medical University, Advanced Semiconductor Engineering Group). IEEE ISSCC.
82. Kim, K.D. *et al.* (June 2013) A New guard-ring technique to reduce coupling noise from through silicon via (TSV) utilizing inversion charge induced by interface charge, (SK Hynix, Seoul National University). VLSI Technology Symposium.
83. Wu, Q. and Zhang, T. (2011) Design techniques to facilitate processor power delivery in 3-D processor-DRAM integrated systems, (Rensselaer PI). *IEEE Transactions on VLSI Systems*, **19**(9), 1655.
84. Ahmad, W., Chen, Q., Zheng, L.R., and Tenhunen, H. (7 December 2011) Decoupling capacitance for the power integrity of 3D-DRAM-over-logic system, (KTH, Kista). IEEE Electronics Packaging Technology Conference, p. 590.
85. Guo, W. *et al.* (December 2012) Impact of through silicon via induced mechanical stress on fully depleted bulk FinFET technology, (IMEC, KU Leuven). IEDM.
86. Murugesan, M. *et al.* (December 2012) Minimizing the local deformation induced around Cu-TSVs and CuSn/InAu-microbumps in high-density 3D-LSIs, (NICHe, ASET, Tohoku University). IEE IEDM.
87. West, J., Choi, Y.S., and Vartuli, C. (June 2012) Practical Implications of via-middle Cu TSV-induced stress in a 28 nm CMOS technology for wide-IO logic-memory interconnect, (Texas Instruments). VLSI Technology Symposium.
88. Murugesan, M., Fukushima, T., Bea, J.C. *et al.* (December 2013) Revisiting the silicon-lattice in the high-density 3D LSIs in the perspective of device reliability, (NICHe). IEEE IEDM.
89. Lee, K.W., Murugesan, M., Beal, J. *et al.* (December 2013) Characterization and reliability of 3D LSI and SIP, (Tohoku University). IEEE IEDM.
90. Guo, W. *et al.* (December 2013) Copper through silicon via induced keep out zone for 10 nm node bulk FinFET CMOS technology, (IMEC, Synopsys). IEEE IEDM.
91. Takaya, S. *et al.* (February 2013) A 100 GB/s wide I/O with 4096b TSVs through an active silicon interposer with in-place waveform capturing, (Kobe University, ASAET). IEEE ISSCC.
92. Liauw, Y.Y., Zhang, Z., Kim, W. *et al.* (February 2012) Nonvolatile 3D-FPGA with monolithocally stacked RRAM-based configuration memory, (Stanford University). IEEE ISSCC.
93. Daneshtalab, M., Ebrahimi, M., Liljeberg, P. *et al.* (20 June 2011) High-performance on-chip network platform for memory-on-processor architectures, (University of Turku). ReCoSoC.
94. Buttrick, M. and Kundu, S. (4 July 2011) Mitigating partitioning, routing, and yield concerns in 3D ICs by multiplexing TSVs, (University of Massachusetts). ISVLSI.
95. Sharifi, A. and Kandemir, M. (9 February 2011) Automatic feedback control of shared hybrid caches in 3D chip multiprocessors, (Pennsylvania State University). PDP, p. 393.
96. Tanachutiwat, S., Liu, M., and Wang, W. (September 2010) FPGA based on integration of CMOS and RRAM, (University Of Albany). *IEEE Transactions on VLSI Systems*, **19**(11), 2023.
97. Saito, H. *et al.* (2010) A chip-stacked memory for on-chip SRAM-rich SoCs and processors, (NEC). *IEEE Journal of Solid-State Circuits*, **45**(1), 15.
98. Kuroda, T. (June 2012) Near-field wireless connection for 3D-system integration, (Keio Univniversity). VLSI Technology Symposium.
99. Sugimori, Y. *et al.* (February 2009) A 2 Gb/s 15pJ/b/chip inductive-coupling programmable bus for NAND flash memory stacking, (Keio University, University of Tokyo). IEEE ISSCC.
100. Ishida, K. *et al.* (February 2009) A 1.8 V 30 nJ adaptive program-voltage (20 V) generator for 3D-integrated NAND flash SSD, (University of Tokyo, Toshiba). IEEE ISSCC.
101. Saito, M., Miur, N., and Kuroda, T. (Feb. 2010) A 2 Gb.s 1.8pJ/b/chip inductive-coupling through chip bus for 128-die NAND-flash memory stacking, (Keio University). IEEE ISSCC.

102. Saito, M. *et al.* (2010) 2 Gb/s 15 pJ/b/chip inductive-coupling programmable bus for NAND flash memory stacking, (Keio University, University of Tokyo). *IEEE Journal of Solid-State Circuits*, **45**(1), 134.
103. Saito, M., Yoshida, Y., Miura, N. *et al.* (2010) 47% power reduction and 91% area reduction in inductive-coupling programmable bus for NAND flash memory stacking, (Keio University). *IEEE Transactions on Circuits and Systems*, **57**(9), 2269.
104. Kuroda, T. (December 2010) ThruChip interface (TCI) for 3D integration of low-power system, (Keio University). IEDM.
105. Miura, N., Take, Y., Saito, M. *et al.* (February 2011) A 2.7 Gb/s/mm^2 0.9pJ/b/chip 1-coil/channel ThruChip interface with coupled-resonator-based CDR for NAND flash memory stacking, (Keio University), IEEE ISSCC.
106. Saen, M. (2010) 3-D system integration of processor and multi-stacked SRAMs using inductive-coupling link, (Hitachi, Renesas, Keio University). *IEEE Journal of Solid-State Circuits*, **45**(4), 856.
107. Ramini, L., Bertozzi, D., and Carloni, L.P. (9 May 2012) Engineering a bandwidth-scalable optical layer for a 3D multi-core processor with awareness of layout constraints, (University of Ferrara, Columbia University). NOCS.
108. Boeuf, F. *et al.* (December 2013) A multi-wavelength 3D-compatible silicon photonics platform on 300 mm SOI wafers for 25Gv/s applications, (STMicroelectronics, Luxtera). IEEE IEDM.

Index

Vertical 3D Memory Technologies, First Edition. Betty Prince.

Printed and bound by CPI Group (UK) Ltd, Croydon, CR0 4YY
16/04/2025
14658395-0002